Lecture Notes in Networks and Systems

Volume 413

AF906089

The series "Lecture Notes in Networks and Systems" publishes the latest developments in Networks and Systems—quickly, informally and with high quality. Original research reported in proceedings and post-proceedings represents the core of LNNS.

Volumes published in LNNS embrace all aspects and subfields of, as well as new challenges in, Networks and Systems.

The series contains proceedings and edited volumes in systems and networks, spanning the areas of Cyber-Physical Systems, Autonomous Systems, Sensor Networks, Control Systems, Energy Systems, Automotive Systems, Biological Systems, Vehicular Networking and Connected Vehicles, Aerospace Systems, Automation, Manufacturing, Smart Grids, Nonlinear Systems, Power Systems, Robotics, Social Systems, Economic Systems and other. Of particular value to both the contributors and the readership are the short publication timeframe and the world-wide distribution and exposure which enable both a wide and rapid dissemination of research output.

The series covers the theory, applications, and perspectives on the state of the art and future developments relevant to systems and networks, decision making, control, complex processes and related areas, as embedded in the fields of interdisciplinary and applied sciences, engineering, computer science, physics, economics, social, and life sciences, as well as the paradigms and methodologies behind them.

Indexed by SCOPUS, INSPEC, WTI Frankfurt eG, zbMATH, SCImago.

All books published in the series are submitted for consideration in Web of Science.

For proposals from Asia please contact Aninda Bose (aninda.bose@springer.com).

More information about this series at https://link.springer.com/bookseries/15179

Brahim Lejdel · Eliseo Clementini ·
Louai Alarabi
Editors

Artificial Intelligence and Its Applications

Proceeding of the 2nd International Conference on Artificial Intelligence and Its Applications (2021)

Springer

Editors
Brahim Lejdel
Department of Computer Science
University of Echahid Hamma Lakhdar
El-Oued, Algeria

Eliseo Clementini
Department of Industrial and Information
Engineering and Economics
University of L'Aquila
L'Aquila, Italy

Louai Alarabi
Department of Computer Science
Umm Al-Qura University
Makkah, Saudi Arabia

ISSN 2367-3370 ISSN 2367-3389 (electronic)
Lecture Notes in Networks and Systems
ISBN 978-3-030-96310-1 ISBN 978-3-030-96311-8 (eBook)
https://doi.org/10.1007/978-3-030-96311-8

This Springer imprint is published by the registered company Springer Nature Switzerland AG
The registered company address is: Gewerbestrasse 11, 6330 Cham, Switzerland

Preface

The artificial intelligence will be more and more used through the world in short time, because it offers all opportunities to develop any country. In this time, the developed countries try to use the smart applications to offer different services to its citizens. So, the Smart applications are used in the entire world to solve the different problems of cities as electricity, water, pollution and others. This book aims to show the importance of deployed technologies and research niches in the context of the considerable development of information and communication technologies at both the domestic and urban levels.

When we take smart cities as a good application of artificial intelligence, the opportunities offer, to decision makers make it possible to create an efficiently and sustainably smart city using all the smart applications available to them in order to offer valuable services to citizens. The objective of using the smart application is to proposed service and/or a reduction of the costs for the citizens while adopting an artificial application to develop the country in all dimension. The general goal of a smart city is to improve the quality of life of all citizens in the city and in the countryside, in a sustainable way and respectful of the environment. Artificial intelligence is an indispensable answer in the supply of a smart city, from an environmental and technological point of view, using the smart and green applications, which will have not any pollution. Artificial intelligence technologies are more than likely to bring multiple benefits to the deployment and growth of smart and sustainable cities. The citizens can know in real time the invoice of electricity and adapt it according to their consumption. In addition, they can manage its smart home in any place in the world using Internet connection. The use of artificial intelligence today makes it possible to improve the management of cities by optimizing flows, in reduced time.

In this book, the authors want to clarify the different concepts and issues of artificial intelligence as smart cities, energy, control system and robotics. These issues are related to the development of information and communication technologies using the application of artificial intelligence, especially the Internet of things (IoT) which is becoming increasingly important to offer a city smarter and sustainable.

Contents

Machine Learning Based Indoor Localization Using Wi-Fi and Smartphone in a Shopping Malls

Kamel Maaloul[1], Nedioui Med Abdelhamid[2], and Brahim Lejdel[2(✉)]

[1] Informatique Departement LABTHOP Laboratory, Eloued University, El Oued, Algeria
[2] Computer Science Department, Eloued University, El Oued, Algeria
Brahim-lejdel@univ-eloued.dz

Abstract. The availability of sensors in smartphones has led to indoor positioning solutions. However, the accuracy of these techniques remains uneven as a straightforward solution to indoor positioning. Solutions based on Wi-Fi signal strength work in favor of the idea of controlling infrastructure costs. Our work attempts to explore other learning algorithms and make more robust trade-offs between accuracy and power. Our work also focuses on using classification-based learning algorithms to achieve higher accuracy. By using methods to select the appropriate model and using more complex on-device learning algorithms. Accurate indoor positioning, based on general sensors and user permission, allows for a great location based experience. Machine learning (ML) based methods are also used to improve the quality and efficiency of services. To verify the accuracy of the models, we reviewed the algorithms using several comparisons between a variety of machine learning approaches. We have verified the system's performance using measurements of a smartphone's Wi-Fi RSS (Really Simple Syndication) sensor. Evaluation results show that the gradient boosting method achieves the best internal feature localization accuracy of more than 95%.

Keywords: Indoor localization · Smartphone sensors · Gradient boosting · Machine learning · Wi-Fi

1 Introduction

Nowadays indoor positioning technology is considered advanced. Therefore, it can be used to locate visitors accurately, especially in shopping malls. Indoor positioning based on Wi-Fi with the help of access points and smart devices about a person's device or location is a well-developed and successful thing. The widespread use of smartphones has increased the demand for several important services including location-based internal localization service such as navigation. Although GPS is known as the best method for external positioning, internal

B. Lejdel et al. (Eds.): AIAP 2021, LNNS 413, pp. 1–10, 2022.
https://doi.org/10.1007/978-3-030-96311-8_1

localization still faces many challenges [1]. The most used methods are localization with the help of Wi-Fi, acoustics, Bluetooth, cellular network, visible light, GPS, etc.

GPS signals cannot penetrate large buildings and navigate indoor spaces such as shopping malls, due to walls, ceilings, windows and doors that greatly reduce GPS signals transmitted by radio waves [2]. Among the features in the internal positioning, supporting shopping centers and providing advanced customer service, such as increasing the spread of shoppers in all areas of the mall, and restructuring the shape of stores [3].

The development of artificial intelligence algorithms and the increase of data to improve the quality and efficiency of services provided to customers in shopping centers has made the use of methods based on machine learning (ML) [4]. Machine learning algorithms can be used to classify data into different categories or predict regressions with a continuous variable by learning from training data [5].

So we offer the use of specially supervised machine learning (ML) methods to process these large amounts of data. By training the classifier on the collected data. User location can be predicted on another level. We suggest applying several machine learning methods, to solve this task due to the huge amount of features available in indoor environments, such as Wi-Fi RSS values, magnetic field values and other sensor data.

The rest of the paper is organized as follows. In Sect. 2 we present some previous work relevant to internal positioning in general. Section 3 describes some of the machine learning models we used. Section 4 provides details of implementation and trial. Section 5 discusses the performance outcomes of the approach used. Section 6 concludes the paper.

2 Related Work

The indoor localization can be determined by determining the traffic of visitors. In this section, we mention some previous studies related to our topic.

H. Salamah et al. [6] proposed approaches were developed using Android-based smart phone with IEEE 802.11 WLANs. The evaluation of the proposed approaches was compared with SVM, DT, RT and K-NN classifiers. The results highlighted that the proposed approach reduced the computational strain by 70% with the help of RT classifier in the static environment and by 33% when K-NN was used for classification. It was noted that the location correctness was improved in event of using K-NN and RT classifiers.

In this paper [7] the author used a Recurrent Neural Network (RNN)-based approach to Wi-Fi fingerprinting and then fingerprinting for ILBs. So select the objects in different paths and make the relationship between the RSSI (Received Signal Strength Indicator) values received in the path. Filtering is also used for RSSI data input and output positions.

Zhao et al. [8] suggest a method for obtaining data using a smartphone accelerometer and magnetometer. The identification of visitors' movements using a machine learning classifier. Then, the data is matched against the fingerprint database by the closest Euclidean approximation.

In the paper of Jiang et al. [9], suggested an accurate method for forecasting shopping. It uses the XGBoost machine learning algorithm, to predict which stores customers are currently in. The Global Positioning System (GPS) information provided by the mobile terminals to the customers and the WiFi information throughout the shopping mall were used.

G. Jiang et al. [10] has developed a mass profiling system based on WiFi distance estimation using LightGBM. By applying multi-dimensional measurement technology to the distance matrix between marital persons within the same group in a multi-storey campus building and shopping mall.

3 Machine Learning-Based Indoor Localization

In this section, we briefly introduce some of the machine learning algorithms used in order to approximate concepts.

3.1 Algorithms

Some classification techniques were used to find the best and most accurate internal locations.

3.1.1 Naive Bayes (NB)

Naive Bayes (NB) classifiers are a family of simple probabilistic classifiers based on applying Bayes' theorem with strong (naive) independence assumptions between the features. Naïve Bayes classifiers are highly scalable, requiring a number of parameters linear in the number of variables (features/predictors) in a learning problem [11].

3.1.2 Support Vector Machine (SVM)

SVM is a supervised learning model with associated learning algorithms that analyze data used for classification and regression. Given a set of training examples, each is marked as belonging to one or the other category. An SVM training algorithm builds a model that assigns new data measurements to one category or the other, making it a non-probabilistic binary linear classifier [12]. The objective of the support vector machine algorithm is to find a hyperplane in an N-dimensional space (N-the number of features) that distinctly classifies the data points.

3.1.3 K-Nearest Neighbors (KNN)

KNN is a non-parametric method used for classification and regression. In both cases, the input consists of the k closest training examples in the feature space. KNN works by finding the distances between a query and all the examples in the data, selecting the specified number examples (K) closest to the query, then votes for the most frequent label (in the case of classification) or averages the labels (in the case of regression) [13].

3.1.4 Random Forest (RF)

The random forest method is used for classification and regression analysis. So that it is generated by random data extraction, and each result is combined to make a prediction model [14]. The random forest algorithm is an algorithm that is used as a classifier to calculate the result of a positioning. By designing the random forest using several decision trees.

3.1.5 Gradient Boosting (GB)

The gradient boosting method is used to strengthen a model that produces weak predictions, for example a decision tree. Gradual reinforcement trees are produced when the decision tree is the weak learner [15]. So that it designs the model in a phased manner like other enhancement methods, and generalizes it by allowing the optimization of the differentiable arbitrary loss function [16].

3.2 Features

In our work, the features are the measurements obtained, and they are the value of the RSS. We select the appropriate features so that new features can be created and that determines good prediction accuracy for machine learning. WiFi Received Signal Strength (RSS) in the considered environment is used to build radio maps using WiFi fingerprinting approach [17]. Wi-Fi RSS values provide the core data as they contribute the most to the performance of the ML methods. The smartphone scans the surrounding Wi-Fi access points, obtains and registers the RSS values of each access point. As illustrated in Fig. 1.

Wi-Fi RSS values depend on the distance between the smartphone and the Wi-Fi access points. Normally, the Wi-Fi RSS values in our datasets were between −20 dBm and −90 dBm. RSSI is also used in Wi-Fi networks in the CSMA/CA channel allocation algorithm between several terminals: Wi-Fi radio channels being half-duplex and shared, a transmitter must verify, before transmitting, that the radio channel is free by measuring the RSSI on this channel 1 [18]. It is also used to measure the received signal in order to precisely orient television antennas or satellite antennas. Wi-Fi RSSI based indoor localization is one of the standard approaches for indoor localization. It is able to utilize the RSSI measurements received from a large number of access points (APs) that are already built in construction [19].

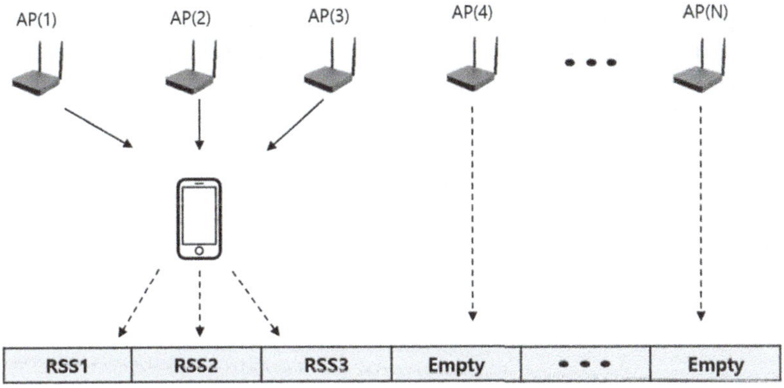

Fig. 1. The sparsity of a raw Wi-Fi RSSI set

4 Methodology

4.1 Architecture System

The architecture of the implemented system of the data flow and the different components as illustrated in Fig. 2. Sensor and Wi-Fi RSS values are measured by the smartphone and received. We then perform the data training process offline to pass the collected data to the Training Model, which applies different machine learning algorithms to build the models. The trained models are then optimized and transfered on the smartphone for online experiments [20].

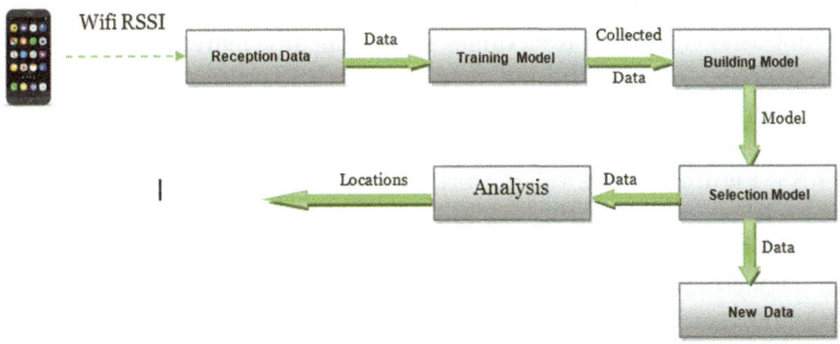

Fig. 2. The architecture of the implemented system and experiment flow diagram from data collection to classification

4.2 Datasets

The UJIIndoorLoc database covers three buildings of Universitat Jaume I with 4 or more floors and almost 110.000 m^2 (https://archive.ics.uci.edu/ml/datasets/ujiindoorloc). It can be used for classification, e.g. actual building and floor identification, or regression, e.g. actual longitude and latitude estimation. It was created in 2013 by means of more than 20 different users and 25 Android devices. The database consists of 19937 training/reference records (trainingData.csv file) and 1111 validation/test records (validationData.csv file). The 529 attributes contain the WiFi fingerprint, the coordinates where it was taken, and other useful information. Each WiFi fingerprint can be characterized by the detected Wireless Access Points (WAPs) and the corresponding Received Signal Strength Intensity (RSSI) [21].

5 Experiment and Results

The computer used for the experiment was as Intel i5-4460 with 8 GB RAM and GTX 1050 card. After training the ML models, we provided them with testing data set for prediction.

We discuss the accuracy of the models when using different classifiers and features. Then we compared their accuracy, it is not possible to identify a single measure that would provide a fair comparison in all possible applications. We focused on measures of prediction accuracy by means of percentages of results obtained for the correct identification of the internal location [22].

When common graphs are generalized to data sets of larger dimensions, we get even graphs. This is useful for exploring the correlations between multidimensional data. We have used a linear dimensionality reduction technique that can be used to extract information from a high dimensional space by projecting it into a low dimensional subspace which is Principal Component Analysis (PCA). You can take advantage of this to speed up machine learning algorithm training and testing time considering that the data has a lot of features, and machine learning algorithm learning is very slow (As shown in Fig. 3).

We first used the Wi-Fi RSS values in the machine learning algorithms for the training process. We also did not increase the RSS values so as not to get bad accuracy through signal interference. We directly tested 6 values Wi-Fi RSS. Next, we compare the classification accuracy, recall and precision when using Wi-Fi RSS. Figure 4 shows the performance evaluation of the selected classifiers obtained with different feature combinations. We trained five models based on NB, SVM, KNN, RF, and GB. The best performance is reached by the gradient boosting, which achieves more than 95% of instances correctly classified by Wi-Fi RSS. The accuracy is improved in all tested classifiers. Wi-Fi RSSI scaling varies by internal locations. But they remain close according to the proximity of the studied areas.

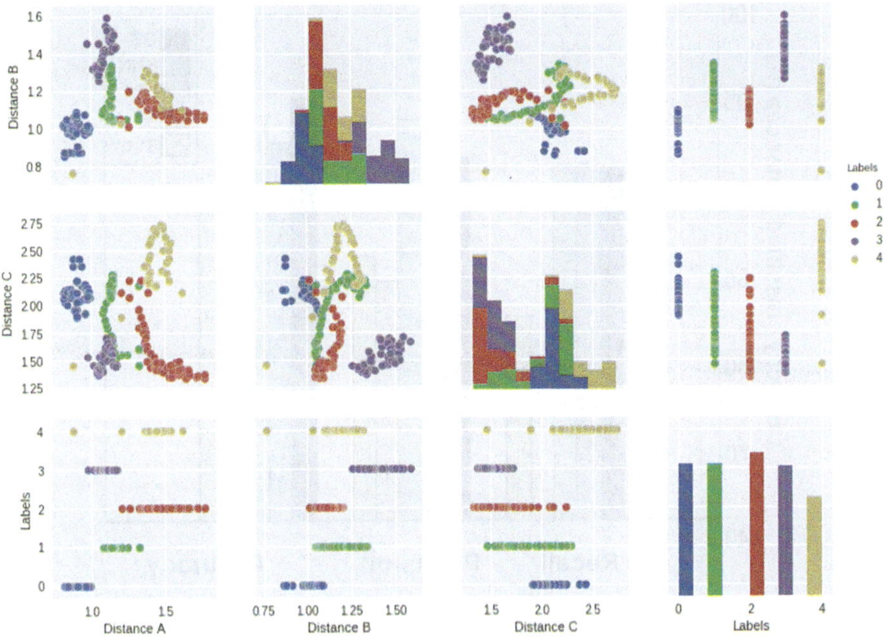

Fig. 3. Principal component analysis using scikit-learn.

This can lead to convergence of results or errors and problems in classification. From Fig. 3, we can see that GB and NB outperform the others in terms of accuracy. This is because GB is used for classification problems and reduces bias error. Gradient boosting does very well because it is a robust out of the box classifier (regressor) that can perform on a dataset on which minimal effort has been spent on cleaning and can learn complex non-linear decision boundaries via boosting. The main difference we found in the study with others is that gradient boosting requires more care in the setup. While it's entirely possible to apply RF and end up performing decently, with very little chance of overfitting, it's pretty much pointless to train xgboost without cross-validation. You will need to set the maximum depth of the trees. Another thing to consider is the feasibility of running those algorithms on such a large amount of data.

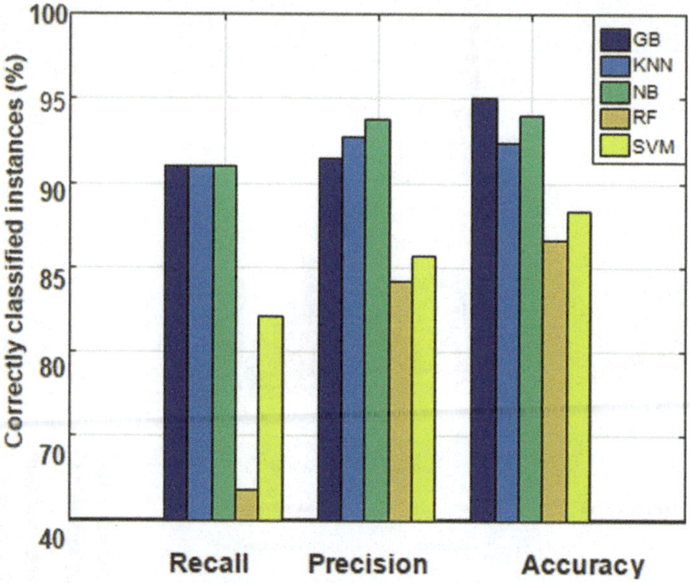

Fig. 4. Prediction performance when using different models

6 Conclusion

In this paper, machine learning approaches have been investigated for indoor location positioning. In this work we analyzed the performance of 5 predictors in internal feature localization using machine learning methods. We have verified the system's performance using measurements of a smartphone's Wi-Fi RSS sensor. Evaluation results show that the gradient boosting method achieves the best internal feature localization accuracy of more than 95%.

7 Future Work

In the future, we will improve the verification procedure for more ultra-accurate machine learning algorithms and combine this work with an internal tracking system to locate a target with high accuracy. Our plan also includes working on a method for deter-mining the access point, which can increase the accuracy of the indoor translation.

References

1. Han, K., Yu, S.M., Kim, S.-L.: Smartphone-based indoor localization using wi-fi fine timing measurement, pp. 1–5 (2019). https://doi.org/10.1109/IPIN.2019.8911751
2. Liu, W., Guo, W., Zhu, X.: Map-aided indoor positioning algorithm with complex deployed BLE beacons. ISPRS Int. J. Geo-Inf. **10**(8), 526 (2021)
3. Maheepala, M., Joordens, M.A., Kouzani, A.Z.: A low-power connected 3D indoor positioning device. IEEE Internet Things J. (2021). https://doi.org/10.1109/JIOT.2021.3118991
4. Ullah, Z., Al-Turjman, F., Mostarda, L., Gagliardi, R.: Applications of artificial intelligence and machine learning in smart cities. Comput. Commun. **154**, 313–323 (2020)
5. Christodoulou, E., Ma, J., Collins, G.S., Steyerberg, E.W., Verbakel, J.Y., Van Calster, B.: A systematic review shows no performance benefit of machine learning over logistic regression for clinical prediction models. J. Clin. Epidemiol. **110**, 12–22 (2019)
6. Pérez, M.D.C., et al.: Android application for indoor positioning of mobile devices using ultrasonic signals, pp. 1–7. IEEE (2016)
7. Hoang, M.T., Yuen, B., Dong, X., Lu, T., Westendorp, R., Reddy, K.: Recurrent neural networks for accurate RSSI indoor localization. IEEE Internet Things J. **6**(6), 10639–10651 (2019)
8. Zhao, M., Qin, D., Guo, R., Wang, X.: Indoor floor localization based on multi-intelligent sensors. ISPRS Int. J. Geo-Inf. **10**(1), 6 (2021)
9. Jiang, H., He, M., Xi, Y., Zeng, J.: Machine-learning-based user position prediction and behavior analysis for location services. Information **12**(5), 180 (2021)
10. Jiang, G., et al.: WiDE: WiFi distance based group profiling via machine learning. IEEE Trans. Mob. Comput. (2021). https://doi.org/10.1109/TMC.2021.3073848
11. García-Díaz, V., Espada, J.P., Crespo, R.G., G-Bustelo, B.C.P., Lovelle, J.M.C.: An approach to improve the accuracy of probabilistic classifiers for decision support systems in sentiment analysis. Appl. Soft Comput. **67**, 822–833 (2018)
12. Lee, S., Mohr, N.M., Street, W.N., Nadkarni, P.: Machine learning in relation to emergency medicine clinical and operational scenarios: an overview. Western J. Emerg. Med. **20**(2), 219 (2019)
13. de Souza, J.V., Gomes Jr., J., Souza Filho, F.M., Oliveira Julio, A.M., de Souza, J.F.: A systematic mapping on automatic classification of fake news in social media. Soc. Netw. Anal. Min. **10**(1), 1–21 (2020). https://doi.org/10.1007/s13278-020-00659-2
14. Rahman, M.S., Rahman, M.K., Kaykobad, M., Rahman, M.S.: isGPT: An optimized model to identify sub-golgi protein types using SVM and random forest based feature selection. Artif. Intell. Med. **84**, 90–100 (2018)
15. Zhang, Y., Haghani, A.: A gradient boosting method to improve travel time prediction. Transp. Res. Part C: Emerg. Technol. **58**, 308–324 (2015)
16. Fafalios, S., Charonyktakis, P., Tsamardinos, I.: Gradient boosting trees (2020)
17. Ye, X., Huang, S., Wang, Y., Chen, W., Li, D.: Unsupervised localization by learning transition model. Proc. ACM Interact. Mob. Wearable Ubiquit. Technol. **3**(2), 1–23 (2019)
18. Sung, C., Chae, S., Kang, D., Han, D.: Estimating AP location using crowdsourced wi-fi fingerprints with inaccurate location labels. In: Proceedings of the 2nd International Conference on Vision, Image and Signal Processing, pp. 1–6 (2018)

19. Labinghisa, B.A., Lee, D.M.: Neural network-based indoor localization system with enhanced virtual access points. J. Supercomput. **77**(1), 638–651 (2021)
20. Zhao, Z., Braun, T., Pan, Z., et al.: Conditional probability-based ensemble learning for indoor landmark localization. Comput. Commun. **145**, 319–325 (2019)
21. Montoliu, R., Sansano, E., Torres-Sospedra, J., Belmonte, O.: Indoorloc platform: a public repository for comparing and evaluating indoor positioning systems. In: 2017 International Conference on Indoor Positioning and Indoor Navigation (IPIN), pp. 1–8 (2017)
22. Khokhar, Z., Siddiqi, M.A.: Machine learning based indoor localization using wi-fi and smartphone. J. Independent Stud. Res. Comput. **18**(1) (2021)

A Comparative Study Between the Two Applications of the Neural Network and Space Vector PWM for Direct Torque Control of a DSIM Fed by Multi-level Inverters

O. F. Benaouda[1](✉), M. Mezaache[1], R. Abdelkader[1], and A. Bendiabdellah[2]

[1] Research Center in Industrial Technologies CRTI, P.O. Box 64, 16014 Cheraga, Algiers, Algeria
{o.benaouda,m.mezaache,r.abdelkader}@crti.dz

[2] Diagnostic Group, LDEE Laboratory, Faculty of Electrical Engineering, University of Sciences and Technology of Oran MB, El-Mnaouer, BP 1505, 31000 Oran, Algeria

Abstract. Nowadays, thanks to the development of control and power electronics, the dual stator induction machine DSIM has become among the most important multi-phase machines included in industrial application, this is due to its positive features among them is its high reliability and reduce both losses and rotor torque ripple.

This paper aims to apply both techniques of artificial intelligence represented by the neural network algorithm NNA and the Space Vector PWM SVM for direct torque control DTC of the DSIM to improve the machine performance and control algorithms DTNC and DTC-SVM.

Generalization capacity, the parallelism of operation, computational speed, and learning capacity all these features made it possible to exploit the neural network algorithm to control the machine. Fixed switching frequency obtained, dispensed with the vector selection table and the hysteresis controller, the three pros allowed the inclusion of SVM technique in DTC strategy.

Three-level NPC inverters are included to feed the DSIM. A several of the results obtained prove the two applied techniques (NNA, SVPWM) in improving the quality of both electromagnetic torque and flux and the dynamic responses of the DSIM.

Keywords: DSIM · Neural network algorithm NNA · Space vector PWM SVM · DTNC · DTCSVM · Three-level NPC inverter

1 Introduction

The continuous increase in industrial progress has led to an increase in the demand for the use of many multi-phase machines; this is due to its high reliability compared to the ordinary three-phase machine, in addition to the successful choice of reducing the weight of windings [1, 2], the power divided between phases enables the voltage of each phase to be reduced and reduces distortions of phases currents. Most of the multi-level

B. Lejdel et al. (Eds.): AIAP 2021, LNNS 413, pp. 11–21, 2022.
https://doi.org/10.1007/978-3-030-96311-8_2

machines have high torque-free of undulations [3], the most important feature is the continuity of these machines' operation in the event of a fault in one or more phases [4, 5].

The dual stator induction machine DSIM is more common among the family of multi-phases machines; the DSIM modeling with its control has been suggested by the authors [4, 6].

There are many DSIM control strategy, the most famous of which is the direct torque control DTC that was proposed in the 1980s [7], the DTC strategy has a set of features including rapid dynamic response, robust performance, non-use of a mechanical sensor, simple implementation, control the speed, flux, and torque of the motor at the same time. However, there are some negative features, including the sensitivity of the DTC to physical variables such as the stator resistance, in addition to the emergence of ripple at the level of each electromagnetic torque and stator flux. Researchers are still seeking to improve the performance of the DTC strategy for DSIM [8].

Due to advances in power electronics and control, several topologies of static converters have emerged [9]; these topologies provide a power source and adjustable speed with good performance [10].

The three-level NPC inverter is one of the most popular topologies in variable speed systems [11], the standard DTC strategy requires a hysteresis controller and switching table, these requirements make this strategy has constant switching frequency, This contributed to the suggestion of many studies urging inclusion the Space Vector PWM SVM technique within the DTC strategy to control the switching frequency to reduce the torque ripple, and improve the quality of stator flux with a wide speed range [12].

To overcome torque ripple caused by the switching table applied to the DTC strategy, a switching table based on several artificial intelligence techniques, such as neural network, has been proposed to obtain optimal switching table patterns [13].

This paper proposes a comparative study between the two applications of the Space Vector PWM SVM and the neural network included in the DTC strategy for the DSIM;

First, a DSIM modeling will be proposed. Second, the control strategies DTC-SVM and DTNC will be applied for DSIM. Third, a comparative study will be conducted between the two aforementioned applications.

2 Modeling of DSIM

The DSIM is a very complex system, thanks to the Park transformation, the mathematical model of the nine nonlinear differential equations has been simplified, this is to control the speed, and the torque and the stator flux at the same time.

Whereas, the following equations represent of the DSIM model in space α-β;

$$\begin{cases} \overline{v}_s = v_{\alpha s} + jv_{\beta s} \\ \overline{i}_s = i_{\alpha s} + ji_{\beta s} \\ \overline{i}_r = i_{\alpha r} + ji_{\beta r} \\ \overline{\varphi}_s = \varphi_{\alpha s} + j\varphi_{\beta s} \\ \overline{\varphi}_r = \varphi_{\alpha r} + j\varphi_{\beta r} \end{cases} \quad \begin{cases} \overline{V}_s = r_s\overline{i}_s + p\overline{\varphi}_s \\ 0 = r_r\overline{i}_r + p\overline{\varphi}_r - j\omega_r\overline{\varphi}_r \\ \overline{\varphi}_s = L_s\overline{i}_s + L_m\overline{i}_r \\ \overline{\varphi}_r = L_m\overline{i}_s + L_r\overline{i}_r \\ \Gamma_{em} = \frac{1}{2}P\left(\varphi_{\alpha s}i_{\beta s} - \varphi_{\beta s}i_{\alpha s}\right) \end{cases} \tag{1}$$

$$[T_6]^{-1} = \frac{1}{\sqrt{3}}\begin{bmatrix} \cos(0) & \cos\left(\frac{2\pi}{3}\right) & \cos\left(\frac{4\pi}{3}\right) & \cos(\gamma) & \cos\left(\gamma + \frac{2\pi}{3}\right) & \cos\left(\gamma + \frac{4\pi}{3}\right) \\ \sin(0) & \sin\left(\frac{2\pi}{3}\right) & \sin\left(\frac{4\pi}{3}\right) & \sin(\gamma) & \sin\left(\gamma + \frac{2\pi}{3}\right) & \sin\left(\gamma + \frac{4\pi}{3}\right) \\ \cos(0) & \cos\left(\frac{4\pi}{3}\right) & \cos\left(\frac{2\pi}{3}\right) & \cos(\pi - \gamma) & \cos\left(\frac{\pi}{3} - \gamma\right) & \cos\left(\frac{5\pi}{3} - \gamma\right) \\ \sin(0) & \sin\left(\frac{4\pi}{3}\right) & \sin\left(\frac{2\pi}{3}\right) & \sin(\pi - \gamma) & \sin\left(\frac{\pi}{3} - \gamma\right) & \sin\left(\frac{5\pi}{3} - \gamma\right) \\ 1 & 1 & 1 & 0 & 0 & 0 \\ 0 & 0 & 0 & 1 & 1 & 1 \end{bmatrix} \tag{2}$$

The angle between the stator flux and the rotor flux is γ, Fig. 1 shows the input and output of the DSIM system.

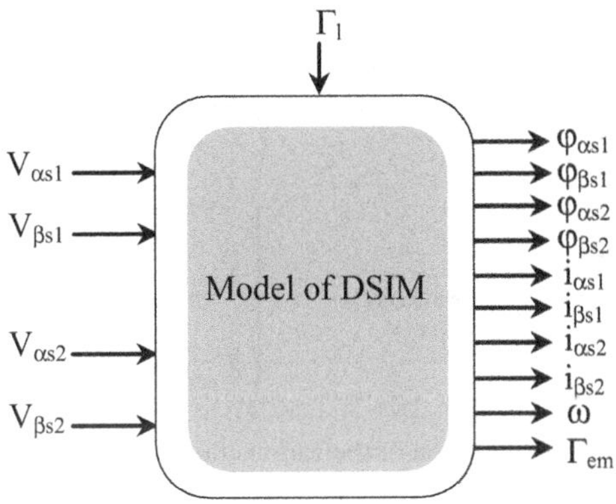

Fig. 1. Model of DSIM in the α-β coordinate plan

The state system is obtained as $\dot{X} = AX + BU$, with the matrices A and B given as:

$$A = \begin{bmatrix} 0 & 0 & 0 & 0 & -r_{s1} & 0 & 0 & 0 \\ 0 & 0 & 0 & 0 & 0 & -r_{s2} & 0 & 0 \\ 0 & 0 & 0 & 0 & 0 & 0 & -r_{s2} & 0 \\ 0 & 0 & 0 & 0 & 0 & 0 & 0 & -r_{s2} \\ r_r h_1 & -r_r h_2 & \omega L h_1 & -\omega L h_2 & -a & -b & -\omega & 0 \\ -r_r h_2 & r_r h_1' & -\omega L h_2 & \omega L h_1' & -c & -d & 0 & -\omega \\ -\omega L h_1 & \omega L h_2 & r_r h_1 & -r_r h_2 & \omega & 0 & -a & -b \\ \omega L h_2 & -\omega L h_1' & -r_r h_2 & r_r h_1' & 0 & \omega & -c & -d \end{bmatrix} B = \begin{bmatrix} 1 & 0 & 0 & 0 \\ 0 & 1 & 0 & 0 \\ 0 & 0 & 1 & 0 \\ 0 & 0 & 0 & 1 \\ Lh_1 & -Lh_2 & 0 & 0 \\ -Lh_2 & Lh_1' & 0 & 0 \\ 0 & 0 & Lh_1 & -Lh_2 \\ 0 & 0 & -Lh_2 & Lh_1' \end{bmatrix} \tag{3}$$

With:

$$\begin{cases} a = [h_3 h_1 - r_r l_m h_2],\ b = [r_r l_m h_2 - h_2 h_4],\ c = \left[r_r l_m h_1' - h_2 h_3\right],\ d = \left[h_4 h_1' - r_r l_m h_2\right] \\ h_1 = \frac{l_m(l_r + l_{s1}) + l_r + l_{s2}}{y},\ h_1' = \frac{l_m(l_r + l_{s1}) + l_r + l_{s1}}{y} \\ h_2 = \frac{l_m l_r}{y},\ h_3 = [r_r(l_{s1} + l_m) + r_{s1} \cdot (l_{s1} + l_m)]\, h_4 = [r_r(l_{s2} + l_m) + r_{s2} \cdot (l_r + l_m)] \\ y = [l_m \cdot (l_r + l_{s1}) + l_r l_{s1}] \cdot [l_m \cdot (l_r + l_{s2}) + l_r l_{s2}] - l_m^2 \cdot l_r^2 L = l_m + l_r \end{cases} \tag{4}$$

3 DTC Strategy Control

3.1 Control Strategy of Flux

The DSIM relies on two three-level inverters separated from each other to be feed, to control the two stator fluxes $\varphi_{s\alpha}$, $\varphi_{s\beta}$ at the same time, the angle between them must be 30°, as shown in Fig. 2, the half resulting flux φ_{sres} is equivalent to modules of the individual fluxes φ_{s1} and φ_{s2}. The estimated flux and the currents measurement contribute to the calculation of the electromagnetic torque.

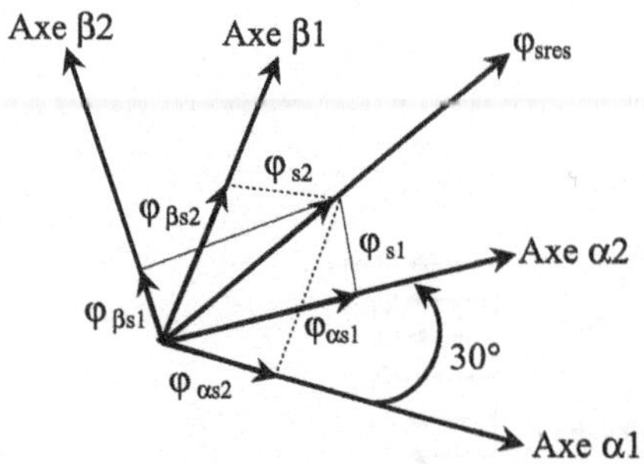

$$\begin{cases} |\varphi_{s1}| = \sqrt{\varphi_{\alpha s1}^2 + \varphi_{\beta s1}^2} \\ |\varphi_{s2}| = \sqrt{\varphi_{\alpha s2}^2 + \varphi_{\beta s2}^2} \\ \vec{\varphi}_s = \vec{\varphi}_{sres} = \vec{\varphi}_{s1} e^{-j15^0} + \vec{\varphi}_{s2} e^{+j15^0} \\ \Gamma_{em1} = P(\varphi_{\alpha s1} i_{\beta s1} - \varphi_{\beta s1} i_{\alpha s1}) \\ \Gamma_{em2} = P(\varphi_{\alpha s2} i_{\beta s2} - \varphi_{\beta s2} i_{\alpha s2}) \\ \Gamma_{em} = \Gamma_{em1} + \Gamma_{em2} \end{cases} \quad (5)$$

Fig. 2. Description of the various fluxes representation

4 Principal of DTC-SVM

The values of V_{ds1},V_{ds2}, V_{qs1},V_{qs2}, are calculated from the error values of φ_{s1}, φ_{s2}, $\omega_1$with Γ_{em1}, ω_1 with Γ_{em2} respectively, for the SVM Blocks, depend on the four voltages $V_{ds1,2}$,$V_{ds1,2}$, in Fig. 3, two three-level inverters are included in control strategy DTC-SVM.

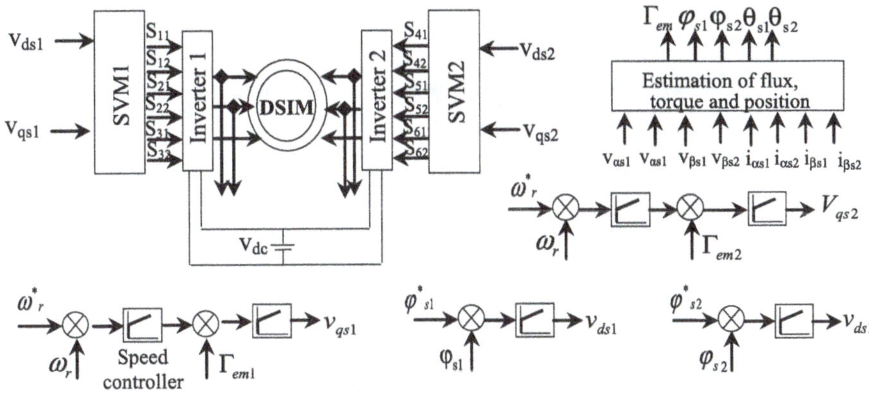

Fig. 3. Structure blocks of the DTC-SVM strategy for the DSIM.

5 DTC Based on Neural Network DTNC

5.1 Artificial Neural Network Principle

The neural network model is considered a non-linear function, this model is inspired by the neuron model proposed by McCulloch and Pitts, This model has a simple mathematical explanation derived from biological reality, it is a system that has a set of inputs and output, shown in Fig. 4

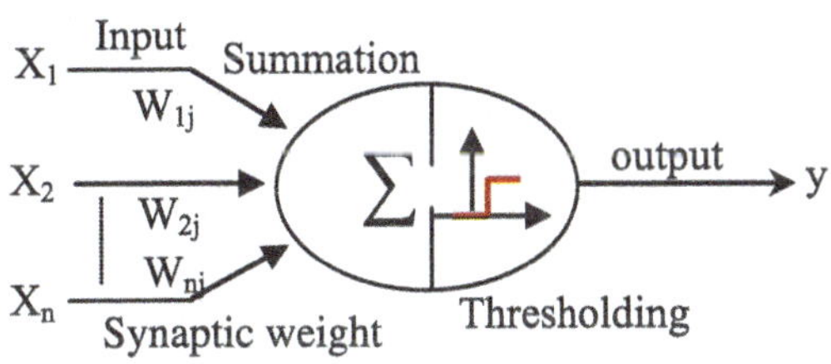

Fig. 4. Neural network multi-layer structure.

5.2 Principal of DTNC

To eliminate the faults caused by the voltage vectors within the switching table (ripple, harmonics,…etc.), a neural network algorithm based on reverse propagation learning is included in the DTNC strategy as shown in Fig. 5.

To describe the neural network used in the DTNC strategy, a matrix of inputs (torque error, flux error, sectors) and outputs (pulses) must be implemented, as shown below:

% Input matrices (E_Torque, E_Flux, Sector).

P = [311; 211; 111; 011; 301; 301; 101; 001; 312; 212; 112; 012; 302; 202; 102; 002; 313; 213; 113; 013; 303; 203; 103; 003; 314; 214; 114; 014; 304; 204; 104; 004; 315; 215; 115; 015; 305; 205; 105; 005; 316; 216; 116; 016; 306; 206; 106; 006; 317;

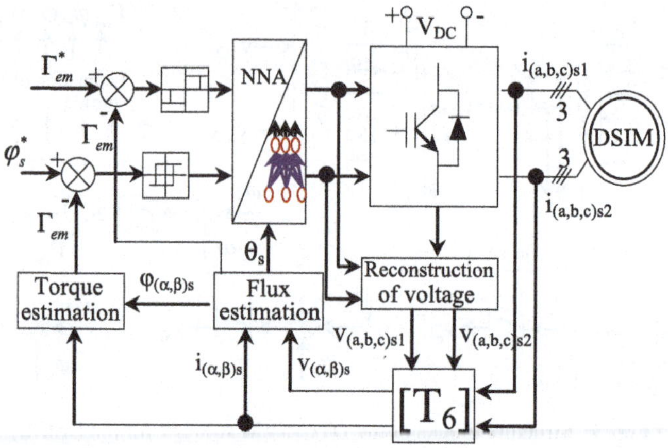

Fig. 5. Blocks structure of the DTNC strategy for the DSIM.

217; 117; 017; 307; 207; 107; 007; 318; 218; 118; 018; 308; 208; 108; 008; 319; 219; 119; 019; 309; 209; 109; 009; 3110; 2110; 1110; 0110; 3010; 2010; 1010; 0010; 3111; 2111; 1111; 0111; 3011; 2011; 1011; 0011; 3112; 2112; 1112; 0112; 3012; 2012; 1012; 0012]

% Output matrices (Switching case or pulses)

P = [110; 110; 100; 101; 010; 011; 111; 001; 010; 110; 100; 100; 011; 011; 101; 010; 010; 110; 100; 011; 001; 000; 101; 011; 010; 110; 110; 001; 001; 100; 100; 011; 011; 010; 110; 001; 101; 111; 100; 001; 011; 010; 101; 001; 100; 110; 001; 001; 011; 010; 101; 100; 000; 110; 101; 001; 011; 011; 100; 100; 110; 010; 101; 101; 001; 011; 100; 010; 111; 010; 100; 101; 001; 001; 110; 110; 011; 011; 100; 100; 101; 001; 110; 010; 000; 011; 110; 100; 101; 101; 010; 010; 011; 001]

b2 = min(a); b1 = max(a); RNA = newff([b2′ b1′], [24 3],{'logsig' 'logsig'}); %bdids.IW{1} = A′;

%bdids.LW{2, 1} = B′; %bdids.LW{3, 2} = C; RNA.trainParam.epochs = 207; [RNA, tr] = train(RNA, a′ ,d′); %gensim(RNA).

Where RNA: Artificiel Neural Network.

6 Simulation Results and Discussion of DTC-SVM and DTNC

Figures 6, 7, 8 and 9 illustrate the behavior of the structure of the direct torque control applied to the DSIM supplied by voltage inverters at three levels for switching by Space Vector PWM (SVM) with the corrector multi-level of torque and stator flux.

These results show a good performance of the torque which precisely follows its reference value, this precision depends on the variation of the load.

The flux path describes a circle, the use of torque three-level corrector allows good control of the variation of the torque. Torque fluctuations change between the two values 19.7 and 20.3 N.m, the sectoral representation is for a position of the flux is detected in space, broken down into twelve symmetrical sectors.

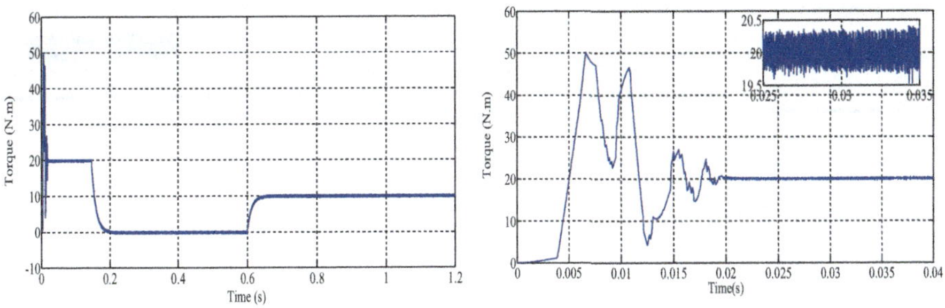

Fig. 6. Change and evolution of torque versus time of DTC-SVM strategy

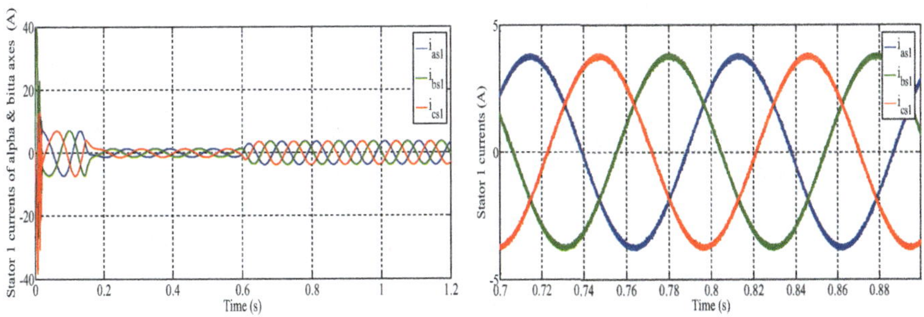

Fig. 7. Stator1 currents of α and β phases of DTC-SVM strategy.

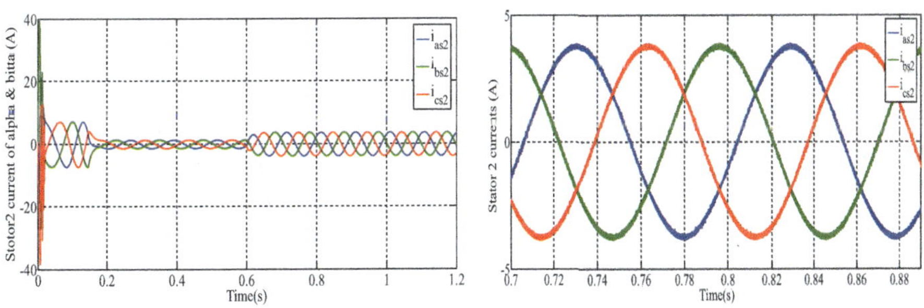

Fig. 8. Stator2 currents of α and β phases of DTC-SVM strategy.

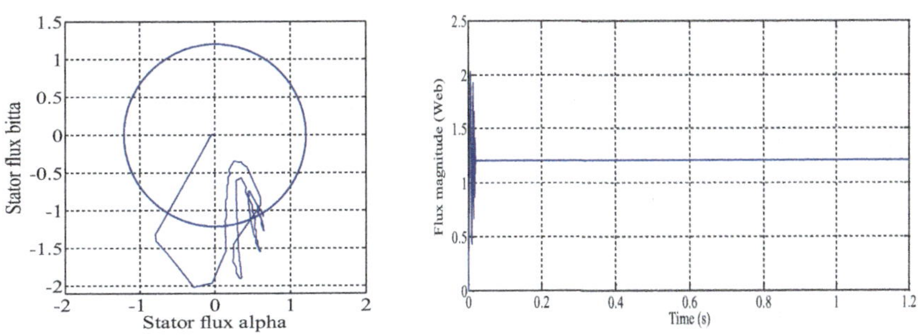

Fig. 9. Evolution of flux versus time for a reference $\varphi_{s0} = 1.2$ Web of DTC-SVM strategy.

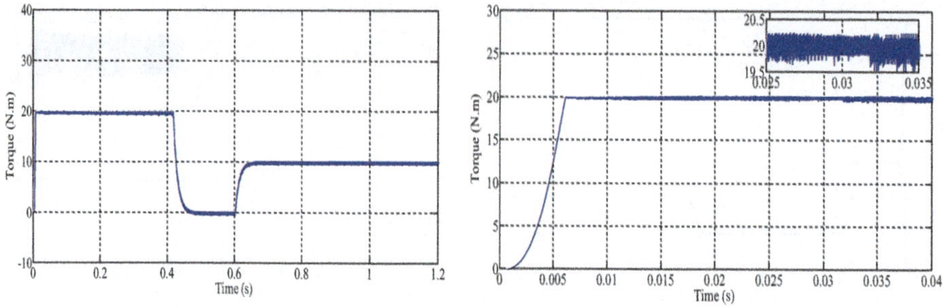

Fig. 10. Change and evolution of torque versus time of DTNC strategy

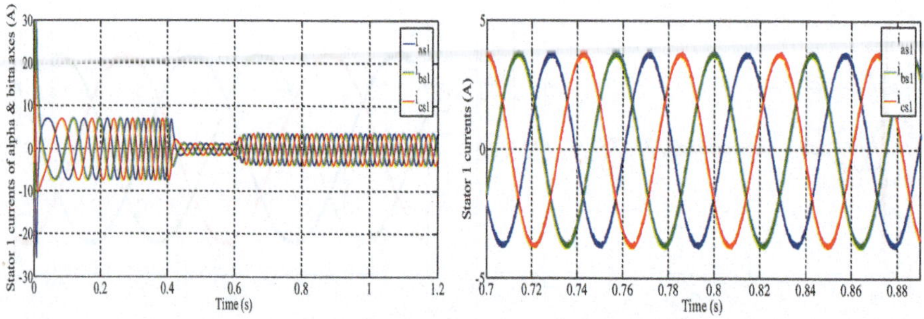

Fig. 11. Stator2 currents of α and β phases of DTNC strategy

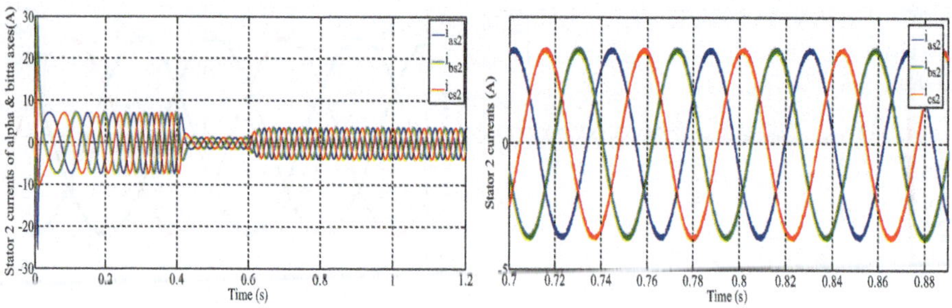

Fig. 12. Stator2 currents of α and β phases of DTNC strategy

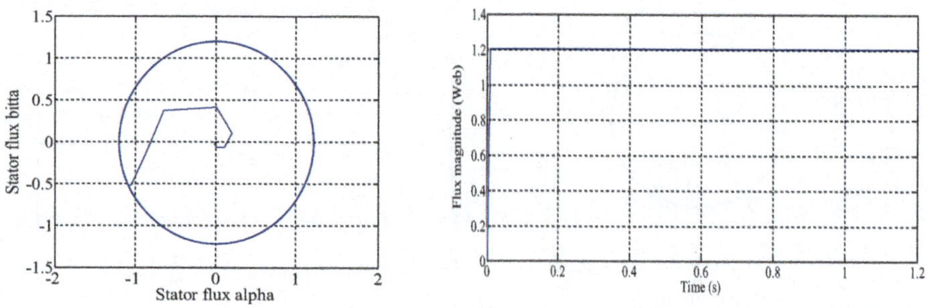

Fig. 13. Evolution of flux versus time for a reference $\varphi_{s0} = 1.2$ Web of DTNC strategy.

The simulation results of Fig. 10, 11, 12 and 13 show better performance than that obtained by DTC-SVM. It is interesting to notice in Fig. 6 a torque response dynamic with a very fast transient regime Fig. 10. The stator flux presents a very good response Fig. 13, where we notice that there is less overshoot compared to those of the DTC-SVM, see the magnifying effect of flux Fig. 9. Figure 13, shows a fast transient of the static flux modulus that to a perfectly circular shape without any steady-state ripple where the torque and flux follow their references with static errors which are virtually zero. As well as a significant attenuation of current ripples which appear sinusoidal Fig. 12, the DTNC strategy achieves a constant switching frequency.

From these results, it can be seen that the performance of the system, controlled by a neural controller, is unsatisfactory, despite the online adaptation of the neural network.

This phenomenon is because there is not a general rule for choosing the parameters of the neural network (the learning rate, the number of neurons in the hidden layer), as well as the weighting values in the cost function. It is usually difficult to determine this choice just based on trial and error.

7 Study Comparative Between the DTC-SVM and DTNC Strategies

The following table illustrates a comparative study between the two control strategies applied in this paper (Table 1).

Table 1. Adnavtages and disadvantages of DTC-SVM and DTNC strategies [14]

Strategy	Advantages	Disadvantages
DTC-SVM	- Flux and torque are well controlled - The sinusoidal stator currents - Constant switching frequency around 3 kHz	- The flux is slowly established - The algorithm is more complicated
DTNC	- The Torque is well controlled - The flux and the torque perfectly follow its references - The sinusoidal stator currents	- Variable switching frequency around 3 kHz - Problem of choice of apprentice

8 Conclusion

This paper aims to compare the dynamic performance of direct torque control strategies by the Space Vector PWM (DTC-SVM) and neural network algorithm (DTNC), also, to know the effect of the two proposed strategies on the dynamic performance of the dual stator induction machine (DSIM) fed by three-level NPC inverters.

The obtained results prove that the electromagnetic torque of DTNC has less fluctuation compared to the torque obtained at DTC-SVM.

Both stator currents of the proposed strategies have a sinusoidal form, and the DTNC strategy stator currents are of better quality than DTC-SVM stator currents.

The stator flux trajectory mesh of alpha and bitta axes has a thin circular trajectory in two proposed strategies.

On the negative side, it can be said that the DTC-SVM strategy algorithm is more complicated, there is a problem of choice of apprentice in DTNC strategy.

From the above, it can be said that the application of artificial intelligence technique by the neural network algorithm has led to improving the dynamic performance of the direct torque control strategy DTNC for the DSIM compared to the space vector PWM application performance.

Acknowledgment. This work was supported by Research Center in Industrial Technologies CRTI, P.O.Box 64, Cheraga 16014, Algiers, Algeria crti.dz.

References

1. Duran, M.J., Gonzalez-Prieto, I., Rios-Garcia, N., Barrero, F.: A simple fast and robust open-phase fault detection technique for six-phase induction motor drives. IEEE Trans. Power Electron. **33**(1), 547–557 (2018)
2. Levi, E.: Multiphase electric machines for variable-speed applications. IEEE Trans. Industr. Electron. **55**(5), 1893–1909 (2008)
3. Basak, S., Chakraborty, C.: Dual stator winding induction machine: problems, progress, and future scope. IEEE Trans. Industr. Electron. **62**(7), 4641–4652 (2015)
4. Kianinezhad, R., Nahid, B., Baghi, L., Betin, F., Capolino, G.A.: Modeling and control of six-phase symmetrical induction machine under fault condition due to open phases. IEEE Trans. Ind. Appl. **55**(5), 1966–1977 (2008)
5. Talaeizadeh, V., Kianinezhad, R., Seyfossadat, S.G., Shayanfar, H.A.: Direct torque control of six-phase induction motors using three-phase matrix converter. Energy Convers. Manag. **51**, 2482–2491 (2010)
6. Zhao, Y., Lipo, T.A.: Space vector PWM control of dual three-phase induction machine using vector space decomposition. IEEE Trans Ind Appl **31**(5), 1100–1109 (1995)
7. Benaouda, O.F., Bendiabdellah, A., Cherif, B.D.E.: Contribution to reconfigured multi-level inverter fed double stator induction machine DTC-SVM control. Int. Rev. Modell. Simul. **9**(5), 1–12 (2016)
8. Bojoi, R., Farina, F., Griva, G., Profumo, F., Tenconi, A.: Direct torque control for dual three-phase induction motor drives. IEEE Trans. Ind. Appl. **41**(6), 1627–1636 (2005)
9. Benaouda, O.F., Babess, B., Bouchakour, M., Kahla, S., Bendiabdellah, A.: Arc welding current Control using thyristor based three-phase rectifiers applied to gas metal arc welding connected to grid network. J. Eur. Syst. Autom. **54**(2), 335–344 (2021)
10. Benaouda, O.F., Bendiabdellah, A., Kahla, S.: Contribution to reconfiguration of fault-tolerant inverter applied to the wind park connected to the electrical network. Int. Rev. Modell. Simul. **9**(5) (2016). pp. 143–148, Bucarest (2019)
11. Ben Abdelghani, H., Ben Abdelghani, A.B., Belkhodja, I.S.: Three-level fault-tolerant DTC control for induction machine drives. In: 9th IEEE Annual System and Devices. Special Conference, pp. 1–6, March 2012

12. Krim, S., Gdaim, S., Mtibaa, A., Mimouni, M.F.: Real time implementation of high performance's direct torque control of induction motor on FPGA. Int. Rev. Electr. Eng. (IREE) **9**(5), 919–929 (2014)
13. Casadei, D., Profumo, F., Serra, G., Tani, A.: FOC and DTC: two viable schemes for induction motors torque control. IEEE Trans. Power Electron. **17**(5), 779–787 (2002)
14. Toufouti, R.: Contribution à la Commande Direct du Couple de la Machine Asynchrone, (Contribution to the Direct Torque Control of the Induction Machine). PhD Thesis, Mantoury University Constantine, Algeria (2008)

Interval Versus Histogram of Symbolic Representation Based One-Class Classifier for Offline Handwritten Signature Verification

Mohamed Anis Djoudjai(✉) and Youcef Chibani

Laboratoire d'Ingénierie des Systèmes Intelligents et Communicants, Faculty of Electrical Engineering, University of Sciences and Technology Houari Boumédiene, 32, El Alia, Bab Ezzouar, 16111 Algiers, Algeria
{ma.djoudjai,ychibani}@usthb.dz

Abstract. This paper proposes a comparison study of using Interval and Histogram of Symbolic Representation (ISR and HSR) based One-Class classifiers, namely OC-ISR and OC-HSR, respectively, applied to the offline signature verification. Usually, symbolic verification models are built straightforward from the feature space. The proposed work explores an alternative approach based on the use of feature-dissimilarities generated from Curvelet Transform (CT) for building the OC-ISR and the OC-HSR classifier. For the OC-ISR classifier, a new weighted membership function is proposed for computing the similarity values between a dissimilarity query vector and a targeted ISR model. The experimental evaluation performed on the well-known public datasets GPDS, CEDAR, and MCYT, reveals the proposed OC-ISR's superiority over the OC-HSR classifier. Moreover, the proposed verification model based on the OC-ISR classifier outperforms the last similar work reported in the literature on the GPDS-160 dataset by 0.99%, 0.8%, and 0.35% of Average Error Rate (AER) for 5, 8, and 12 reference signatures, respectively.

Keywords: SDA · Histogram · Interval · One-class classification · Dissimilarity · Signature verification

1 Introduction

Automating biometric recognition systems using offline handwritten signatures can offer two distinct applications which are signature identification and signature verification. The former aims to attribute an identity to a query signature belonging to a writer enrolled in a database. While the latter aims to verify the authenticity of a query signature allegedly belonging to a writer, whether it is a genuine or a forgery. Nevertheless, signature verification is a more challenging problem for researchers according to the state-of-the-art performances achieved during the last two decades, and therefore represents the focus of the present paper. Generally, an Offline Handwritten Signature Verification System (OHSVS) is composed of three main modules which are preprocessing, feature generation, and classification. Since the main contribution of the present paper concerns

B. Lejdel et al. (Eds.): AIAP 2021, LNNS 413, pp. 22–31, 2022.
https://doi.org/10.1007/978-3-030-96311-8_3

the classification module, the present paper focuses only on attempting to develop this module. Hence, the classification methods proposed in the literature for OHSVSs can be divided into two categories: Multi-Class Classifiers (MCCs) and One-Class Classifiers (OCCs). OCCs represents an alternative of MCCs when negative examples are not available during the training step. Hence, the OCC concept is desirable for signature verification cases, since only genuine signatures (positive class) contained in a bank database, for example, are available for training the OHSVS.

The classifiers proposed in the literature for OHSVSs are built following one of the two approaches: Writer Dependent (WD) and Writer Independent (WI). The WD consists of building a model for each writer using its genuine signatures. On the other hand, the WI approach is based on building one single model for all writers involved in the database. This later uses the dissimilarity concept where the pattern recognition problem becomes a bi-class problem namely target and reject class [1]. Several MCCs have been explored for building OHSVSs such as Hidden Markov Models (HMMs), Support Vector Machines (SVMs), Neural Networks, and Deep Learning or an ensemble of combined classifiers [1, 2]. On the other hand, few authors have explored the use of OCC such as the OC-SVM [2].

Recently, a new OCC based on Symbolic Data Analysis ($OC - SDA$ classifier) method has been introduced for OHSV. Generally, the symbolic models are constructed either *via* intervals (*ISR*) or histograms (*HSR*) [3] using exclusively straightforward features such as Curvelet Features and Local Binary Patterns (LBP) features [4]. In this investigation work, a comparative evaluation is proposed using the $OC - SDA$ classifier through its two models namely the $OC - ISR$ and $OC - HSR$ model, for offline signature verification. The symbolic verification models proposed in this work are constructed on the feature-dissimilarity space. Dissimilarities are generated from the Curvelet Transform (CT) feature space [5]. Moreover, a new membership function is proposed for computing the similarity values between a dissimilarity of the feature vector and the model. The proposed system is based on WI parameters where the same configuration parameters are set for all writers evolved into the database.

The remainder of this paper is organized as follows. Section 2 presents a brief review of Symbolic Data Representation (SDR) and its extension for classification. Next, a detailed description of the proposed system is presented in Sect. 3. To evaluate the performance of the proposed system, various experiments performed on three offline signature datasets: GPDS-300, CEDAR, and MCYT datasets, are presented in Sect. 4. Finally, a conclusion and perspective work is provided in the last section.

2 Brief Review of Symbolic Data Representation

Usually, the classes are represented by a simple sample mean (or median or the like). However, such a representation of classes doesn't provide a real description of *intra-class* variability. Thus, an alternative approach for representing the aggregation observation (i.e. samples) of the same class can be performed through the SDR concept either *via* intervals (*ISR*) or histograms (*HSR*). In the beginning, this concept has been proposed for analyzing (*SDA*) and clustering complex data [3]. Later, the *SDA* applicability has been extended for reducing large datasets and for one-class classification problems [4] namely the $OC - SDA$ classifier.

The base idea of symbolic representation models consists of representing symbolically each feature component by either a set of intervals (*ISR*) or histograms (*HSR*). Hence, the symbolic model according to the *ISR* concept can be described as follows:

$$ISR = \{If_1, If_2, \ldots, If_P\} \quad (1)$$

where If_k represents the feature interval associated to the kth feature component such that $k = \{1, 2, \ldots, P\}$, and P represents the size of the feature vectors. For generating the inferior and superior bounds of the feature intervals (If_k), different statistical metrics can be used such as *mean* and *standard deviation* [4].

On the other hand, a writer can be described symbolically using the *HSR* concept as follows:

$$HSR = \{(If_1^t, \pi_1^t); (If_2^t, \pi_2^t); \ldots; (If_P^t, \pi_P^t)\} \quad (2)$$

where If_k^t represents the tth feature subinterval of the kth feature component such that $t = \{1, 2, \ldots, N_{bins}\}$, and N_{bins} is the number of subintervals tuned experimentally. While π_k^t is the frequency probability attributed to the tth bin of the histogram HSR_k, associated to the kth feature interval (If_k), such that:

$$\pi_k^t = \frac{N_{fk}}{N} \quad (3)$$

where N is the number of reference signatures, and N_{fk} is the number of features found within the tth subinterval belonging to If_k.

3 Proposed System

The proposed verification scheme is presented in Fig. 1, and the details of each step are described in next sections.

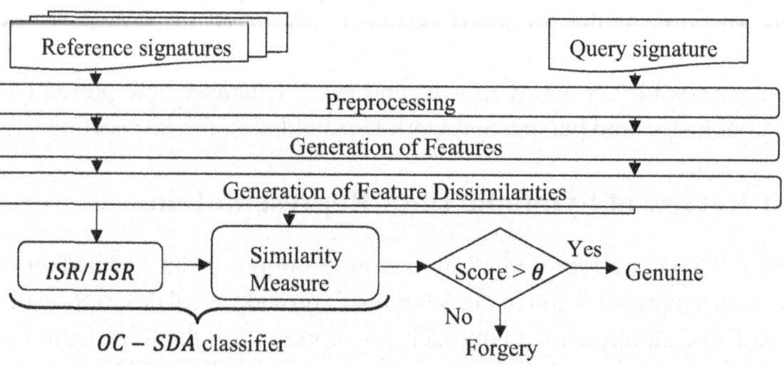

Fig. 1. Scheme of the proposed offline verification system.

3.1 Preprocessing

For this step, an efficient binarization method is specifically performed on the signature image using Local Iterative Method (LIM), followed by a simple signature extraction. LIM is performed through an iterative process for finding the binarization threshold in a sliding window using the mean and the standard deviation [6].

3.2 Generation of Features

For generating features, the Curvelet Transform (CT) is considered in this paper for its efficiency in extracting edges and other singularities along curves. Contrary to the wavelet transform, CT has a high degree of directional specificity elements contained into the curvelet pyramid [7]. Aiming to capture more effectively the local information, the signature image is subdivided into an equi-space grid images before applying the CT. Figure 2 depicts an example of generating a grid of 3 × 3.

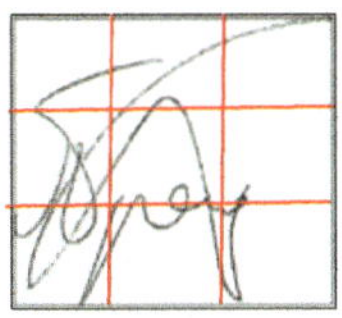

Fig. 2. Example of an equi-space grid image with 3 × 3.

Hence, a wrapping CT is performed on each grid at the scale j and orientation k, which allows generating curvelet coefficients namely $C_{j,k}$. Next, the energy E is calculated at each scale j and orientation k such as:

$$E(j,k) = \sum\nolimits_{t_1}\sum\nolimits_{t_1}\left|C_{j,k}(t_1, t_2)\right| \tag{4}$$

Finally, the feature vector is constructed by concatenating all energy components issued from all grid images.

3.3 Generation of Feature-Dissimilarities (GFD)

Usually, the building of symbolic verification models is based on using straightforward features. In this investigation work, an alternative approach is proposed using the feature-dissimilarities. It consists of performing an absolute difference between each pair of N reference feature vectors of size P namely F_i, and F_l, such that i (and l) = $\{1, 2, \ldots, N\}$ without redundancy i.e. $i \neq l$, such as $D_u = |F_i - F_l|$, where $u = \{1, \ldots, U\}$ and U is the total number of *intra-class* feature-dissimilarity vectors issued from N reference signatures. Each vector D_u of size P can be described as $D_u = \{d_{1u}, d_{2u}, \ldots, d_{Pu}\}$, where each feature-dissimilarity component d_{ku} is generated with the respect of its position such as: $d_{ku} = |f_{ki} - f_{kl}|$, where $k = \{1, 2, \ldots, P\}$. Finally, a matrix namely Matrix of Feature Dissimilarities (*MFD*) of size $P \times U$ is built for each writer containing all feature-dissimilarity components taking the following form:

$$\Theta = \{d_{ku}; k = 1, \ldots, P; u = 1, \ldots, U\} \tag{5}$$

MFD is then handled for creating the writer's model as described in the next section.

3.4 Building the *OC − SDA* Classifier

Basically, two steps are required for building the *OC − SDA* classifier: Creating the symbolic model and computing the similarity values which is described in the next section (verification process). In this work, two types of Symbolic Representation models (SRM) are considered namely the *ISR* and *HSR* models.

Creating the *ISR* Model. The first step consists of creating an interval of feature dissimilarities (instead of features) namely ID_k for each kth feature-dissimilarity component. More precisely, the inferior and superior bounds of ID_k are calculated for each kth column of the matrix Θ using simply the *minimum* and the *maximum* metric, such as:

$$ID_k = [d_k^-, d_k^+] \tag{6}$$

Then, each ID_k is symbolically represented by an adaptive weighted distribution function namely ϑ_k, inspired from the real distribution of training feature-dissimilarities. Its mathematic formula is described as follows:

$$\vartheta_k = \begin{cases} 1 & if \;\; d_k^- \leq d_{uk} \leq \mu_k \\ e^{-\lambda . d_{uk}} & if \;\; \mu_k < d_{uk} \leq d_k^+ \\ 0 & otherwise \end{cases} \tag{7}$$

where λ is a unique control parameter tuned experimentally during the design step, and μ_k is the mean value computed for each ID_k. Hence, the writing style of a writer according to the proposed *ISR* model is then defined as follows:

$$ISR = \{\vartheta_1, \vartheta_2, \ldots, \vartheta_P\} \tag{8}$$

Creating the *HSR* Model. Genuinely, the same interval of feature-dissimilarities ID_k provided in Eq. (6) is considered for this symbolic model, and modulated by symbolic histograms, as described in Sect. 2. Hence, the writing style of a writer is defined by a set P of symbolic feature-dissimilarity histograms namely HSR_k, such that $k = 1, \ldots, P$.

3.5 Verification Process

To verify the authenticity of a query signature represented by a query feature vector of size P namely F, the first step is to perform a straightforward absolute difference between F and the N reference feature vectors F_i such that $i = \{1, 2, \ldots, N\}$, which belongs to the claimed writer. This step allows providing N query vectors of feature-dissimilarities having P components namely $D_{qi} = \left\{d_{q1i}, d_{q2i}, d_{q3i}, \ldots, d_{qPi}\right\}$. Next, a specific similarity measure between all D_{qi} and the generated symbolic model of the claimed writer is performed such as:

$$Sim(D_{qi}, ISR) = \frac{1}{P}\left(\sum_{k=1}^{P} \vartheta_k\right) \tag{9}$$

or:

$$Sim(D_{qi}, HSR) = \frac{1}{P}\left(\sum\nolimits_{k=1}^{P} \sum\nolimits_{t=1}^{N_{bins}} \pi_k^t\right) \tag{10}$$

Consequently, N output scores ranged between 0 and 1 are then generated namely $S_q = \{s_{q1}, s_{q2}, s_{q3}, \ldots, s_{qN}\}$. In the sequel, a selection rule based on the *maximum* metric is performed for selecting only one representative output score namely s_{qmax}. Finally, the selected score (s_{qs}) is compared to a threshold θ for accepting or rejecting (i.e. genuine or forgery) the query signature (Sig_q) according to the following rule:

$$Sig_q \in \begin{cases} accepted \ \ if \ \ s_{qmax} > \theta \\ rejected \ \ \ \ otherwise \end{cases} \tag{11}$$

The threshold θ is tuned during the design step using the reference signatures for each writer.

4 Experimental Results

4.1 Dataset Description and Evaluation Criteria

Three offline handwritten signature datasets are used for evaluating the proposed system: GPDS, CEDAR, and MCYT. The GPDS signature dataset [8] contains 300 writers, each one has 24 Genuine Signatures and 30 Forgery Signatures designated as GS and FS, respectively. The CEDAR signature dataset [9], contains 55 writers where each one has 24 GS and 24 FS. While the MCYT dataset [10] which represents genuinely a part of a bimodal database is composed of 75 writers where each one has 15 GS and 15 FS. For the evaluation step, four well-known metrics are used which are "False Rejection Rate" (FRR), "False Acceptance Rate" (FAR), "Average Error Rate" (AER), and the Equal Error Rate (EER).

4.2 Experimental Setup

For designing the proposed system, a small set of M writers is selected randomly from the GPDS-300 dataset ($M = 30$). Next, among the 24 GS of each writer, only 5 GS ($N = 5$) are selected as reference signatures and used for building the symbolic model. The 19 remaining GS are used for finding the optimum configuration parameters by minimizing the AER. On the other hand, three signature datasets namely GPDS, CEDAR, and MCYT datasets, are considered during the testing phase for evaluating the proposed system.

After the preprocessing step, two parameters should be found during the design step namely N_x and N_y, which represents the number of image partitions performed per line and per column, respectively. Hence, the best configuration found during the design step is $N_x = 3$ and $N_y = 3$. For the classifier parameters, the optimal number of N_{bins} is required when using the $OC - HSR$ classifier. Thus, Table 1 shows the evolution of the AER and EER versus N_{bins} values using five reference signatures ($N = 5$). For better convenience, the obtained AER using the Global Threshold (GT) and Local Threshold (LT) are designated as AER_{GT} and AER_{LT}, respectively.

Table 1. Training results achieved by the $OC - HSR$ classifier for various number of bins.

N_{bins}	**3**	4	5	6	7
AER_{GT} (%)	**13.51**	16.29	20.70	22.98	24.14
AER_{LT} (%)	**12.13**	15.8	19.05	21.6	23.05
EER (%)	**11.49**	14.79	16.84	18.8	21.19

As can be seen, performances decrease gradually as long as N_{bins} increase. Hence, the best performances are obtained for $N_{bins} = 3$. Actually, the use of feature-dissimilarities justifies this result, since the range of feature-dissimilarity values of each component is small. Consequently, there is no need to subdivide again its value greater than 3. In the other hand, the optimal value of λ parameter is required for building the proposed $OC - ISR$ classifier. For better observing the effect of λ values, λ is taken within the interval [0.0001, 5]. Hence, Table 2 shows the evolution of the AER and EER versus λ values using five reference signatures ($N = 5$). For better convenience, only representative results are reported in the table.

Table 2. Training results achieved by the $OC - ISR$ classifier for various value of λ.

λ	0.0001	0.001	0.005	0.01	0.05	0.1	**0.5**	1	2	3	5
AER_{GT} (%)	11.21	11.55	11.43	10.38	10.12	9.85	**8.69**	9.54	10.04	10.28	10.64
AER_{LT} (%)	10.02	10.61	10.52	9.85	9.41	8.97	**7.57**	8.43	9.12	9.61	9.90
EER (%)	8.54	8.82	8.71	7.87	6.61	6.02	**5.42**	5.90	6.51	7.05	8.01

As clearly seen, the optimal value of λ is 0.5 corresponding to the best training verification performance offering 8.69%, 7.57% and 5.42% for AER_{GT}, AER_{LT} and EER, respectively. For better understanding the effect of λ in the verification process, Fig. 3 illustrates the real distribution of training feature-dissimilarities superimposed with the proposed weighted distribution function (ϑ_k) for different value of λ.

As can be clearly seen, when λ parameter takes small values, the width of ϑ_k shape is large (i.e. red color). While, when λ parameter takes high values, the width of ϑ_k shape is narrow (i.e. blue color). In contrast, almost the same shape of the real dissimilarity distribution is obtained for $\lambda = 0.5$ which corresponds exactly to the optimal λ value reported in Table 2.

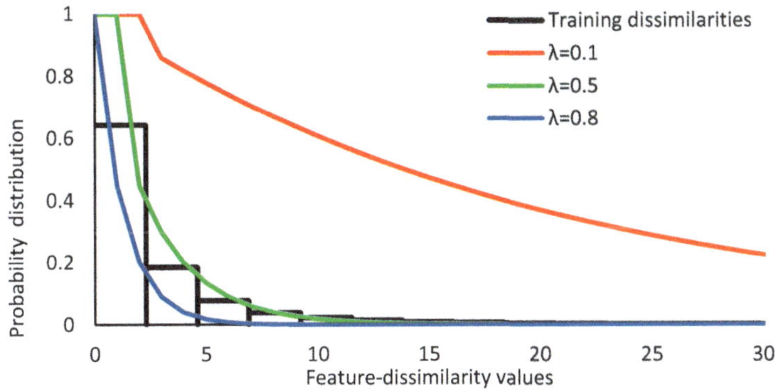

Fig. 3. The probability distribution of training feature dissimilarities superimposed with the proposed weighted membership function ϑ_k for three values of λ (low, optimal and high).

4.3 Experimental Evaluation

In this section, the results achieved on the GPDS-300 and GPDS-160 dataset are presented. Adding to that, a blind test is performed on the two signature datasets namely CEDAR and MCYT datasets using five reference signatures ($N = 5$) (Table 3).

Table 3. Verification performances achieved for both $OC - HSR$ and the $OC - ISR$ classifiers.

Dataset	GPDS-300		GPDS-160		CEDAR		MCYT	
Method	*HSR*	***ISR***	*HSR*	***ISR***	*HSR*	***ISR***	*HSR*	***ISR***
AER_{GT} (%)	23.16	**19.14**	22.45	**18.09**	19.23	**15.25**	14.28	**9.44**
AER_{LT} (%)	21.04	**17.06**	19.57	**16.62**	17.53	**12.87**	12.41	**7.97**
EER (%)	19.15	**15.34**	17.69	**12.48**	14.71	**10.84**	9.38	**5.54**

It is clearly shown that the $OC - ISR$ classifier allows obtaining the best performances on all used datasets. Indeed, 15.34% 12.48% 10.84% and 5.54% of EER are obtained for GPDS-300, GPDS-160 and CEDAR and MCYT dataset, respectively. Besides, the signature verification scheme using the local decision threshold allows getting as expected a better performance. The obtained performance on blind datasets especially on MCYT dataset, demonstrates the robustness and the flexibility of the proposed system even when few reference signatures are available.

4.4 Comparative Analysis

For better evaluating the performance of the proposed verification system based on the $OC - ISR$ classifier against the last similar work [4], Table 4 depicts the comparison of cross-validation results achieved on the GPDS-160 dataset using 5, 8, and 12 reference signatures.

Table 4. Comparative analysis from the last similar work on the GPDS-160 dataset.

Work	Descriptor	Classifier	Weighted function	N	AER_{LT} (%)	σ (%)
Alaei et al. [4]	LBP features	$OC - ISR$	Trapezium	5	17.61	0.9
	–	–	–	8	13.85	1.69
	–	–	–	12	11.47	1.99
Proposed	CT dissimilarities	$OC - ISR$	Exponential	5	**16.62**	**0.85**
	–	–	–	8	**13.05**	**0.93**
	–	–	–	12	**11.12**	**1.57**

As highlighted in Table 4, the best performances are achieved by the proposed system on the GPDS-160 dataset for different reference signatures. Indeed, an improvement of 0.99%, 0.8% and 0.35% in AER_{LT} is reported for 5, 8 and 12 reference signatures, respectively. Moreover, the stability of the proposed system is better, according to the standard deviation values. Furthermore, the proposed system requires adjusting the only OC-SDA classifier parameter which is set for all writers. In contrast, Alaei et al. [4] adjust the classifier parameter for each writer which requires more computations. Hence, these results show the effectiveness of the proposed suitable exponential weighted distribution function used for building the $OC - ISR$ classifier against the trapezium weighted function proposed in [4]. Adding to that, the use of dissimilarities seems more suitable for designing symbolic verification models than straightforward features. Indeed, it allows better defining the *intra-class* variability *via* only a few reference signatures.

5 Conclusion

This paper aimed to investigate the use of the $OC - SDA$ for handwritten signature verification. Usually, the symbolic verification models are built straightforward from features. For better capturing the *intra-class* variability, the dissimilarities generated from the curvelet transform are proposed for building the $OC - SDA$ classifier. Hence, two types of the $OC - SDA$ classifier are proposed in this work which are the $OC - ISR$ and the $OC - HSR$ classifiers. For the $OC - ISR$ classifier, a new weighted function based on decreasing exponential distribution is proposed, which is genuinely inspired from the real distribution of training dissimilarities. The experimental evaluation conducted on the three datasets namely GPDS, CEDAR and MCYT dataset, have shown an encouraging improvement offered by the proposed $OC - ISR$ over the $OC - HSR$ classifier. In addition, the proposed verification model based on the $OC - ISR$ classifier outperforms the symbolic verification model proposed in the last similar work. For future work, an interesting work is to use the deep learning for generating features to improve the verification process when using the $OC - ISR$ classifier.

Acknowledgement. This work was supported by the Direction Générale de la Recherche Scientifique et du Développement Technologique (DGRSDT) grant, attached to the Ministère de l'Enseignement Supérieur et de la Recherche Scientifique, Algeria.

References

1. Bertolini, D., Oliveira, L.S., Justino, E., Sabourin, R.: Reducing forgeries in writer-independent off-line signature verification through ensemble of classifiers. Pattern Recogn. **43**, 387–396 (2010). https://doi.org/10.1016/j.patcog.2009.05.009
2. Guerbai, Y., Chibani, Y., Hadjadji, B.: The effective use of the one-class SVM classifier for handwritten signature verification based on writer-independent parameters. Pattern Recogn. **48**(1), 103–113 (2015)
3. Billard, L., Diday, E.: Symbolic data analysis: definitions and examples. Technical report (2003). http://www.stat.uga.edu/faculty/LYNNE/Lynne.html
4. Alaei, A., Pal, S., Pal, U., Blumenstein, M.: An efficient signature verification method based on an interval symbolic representation and a fuzzy similarity measure. IEEE Trans. Inf. Forensics Secur. **12**(10), 2360–2372 (2017). https://doi.org/10.1109/TIFS.2017.2707332
5. Hadjadji, B., Chibani, Y., Nemmour, H.: An efficient open system for offline handwritten signature identification based on curvelet transform and one-class principal component analysis. Neurocomputing **265**, 66–77 (2017). https://doi.org/10.1016/j.neucom.2017.01.108
6. Djoudjai, M.A., Chibani, Y., Abbas, N.: Offline signature identification using the histogram of symbolic representation. In: The 5th International Conference on Electrical Engineering-Boumerdes (ICEE-B), Boumerdes, pp 1–6 (2017). https://doi.org/10.1109/ICEE-B.2017.8192092
7. Candès, E., Donoho, D.: Curvelets - a surprisingly effective non-adaptive representation for objects with edges. Curves and Surface Fitting, pp. 105–120. Vanderbilt University Press, Saint-Malo, Nashville (1999)
8. Vargas, J., Ferrer, M., Travieso, C., Alonso, J.: Off-line handwritten signature GPDS-960 corpus. In: Ninth International Conference on Document Analysis and Recognition (ICDAR), Curitiba, Brazil, pp. 764–768 (2007). https://doi.org/10.1109/ICDAR.2007.4377018
9. Kalera, M.K., Srihari, S., Xu, A.: Offline signature verification and identification using distance statistics. Int. J. Pattern Recogn. Artif. Intell. **18**(07), 1339–1360 (2004). https://doi.org/10.1142/S0218001404003630
10. Ortega-Garcia, J., et al.: MCYT baseline corpus: a bimodal biometric database. IEEE Proc. Vis. Image Signal Process. **150**(6), 395–401 (2003). https://doi.org/10.1049/ip-vis:20031078

Residual Neural Network for Predicting Super-Enhancers on Genome Scale

Sara Sabba[1(✉)], Meroua Smara[2], Mehdi Benhacine[2], and Amina Hameurlaine[3]

[1] Faculty of New Technologies of Information and Communication, Laboratory of Data Science and Artificial Intelligence (LISIA), Abdelhamid Mahri Constantine-2 University, 25016 Constantine, Algeria
sara.sabba@univ-constantine2.dz

[2] Department of Software Technologies and Information Systems, Faculty of New Technologies of Information and Communication, Abdelhamid Mahri Constantine-2 University, 25016 Constantine, Algeria
{meroua.smara,mehdi.benhacine}@univ-constantine2.dz

[3] Laboratory of Automatic and Robotic, Frères Mentouri Constantine-1 University, 25000 Constantine, Algeria
Amina.hameurlaine@umc.edu.dz

Abstract. Residual neural network (ResNet) is a Deep Learning model introduced by He et al. [13] in 2015 to enhance traditional Convolutional neural networks for computer vision problems. It uses skip connections over some layer blocks to avoid vanishing gradient problem. Currently, many researches are focused to test and prove the efficiency of the ResNet on different domains such as genomics.

In this paper, we propose a new ResNet model for predicting super-enhancers on genome scale. In fact, the prediction of super-enhancers (SEs) has prominent roles in biological and pathological processes; especially that related to the detection and progression of tumors. The obtained results are very promising and they proved the performance of our proposal compared to the CNN results.

Keywords: Deep learning · Residual neural network · Convolutional neural network · Bioinformatics · Transcriptional dysregulation · Super-enhancers · Oncogene · Cancer

1 Introduction

Transcription factors are proteins that bind DNA regulatory elements of genes called enhancers. They play critical roles in the control of cell type-specific gene expression programs [8,16,30]. Super-enhancers (SEs) are clusters of enhancers. They are formed by binding of high levels of enhancer-associated chromatin features that drive high level expression of genes encoding key regulators of cell identity [17,27].

B. Lejdel et al. (Eds.): AIAP 2021, LNNS 413, pp. 32–42, 2022.
https://doi.org/10.1007/978-3-030-96311-8_4

The identification of SEs is based on the differences in their ability to bind markers of promoter transcriptional activity [29], including cofactors such as mediators (MED1, MED12) and cohesions (Nipbl, Smc1), histone modification markers (H3K27ac, H3K4me1, H3K4me3, H3K9me3), chromatin regulators (Brg1, Brd4, Chd7), chromatin molecules (p300, CBP)and many additional transcription factors (Nr5a2, Prdm14, Tcfcp2l1, Smad3, Stat3 and Tcf3) [25,26].

Recently, many studies [23,28,29] proved that gene transcriptional dysregulation is one of the core tenets of cancer development that involves in noncoding regulatory elements, such as TFs, promoters, enhancers, SEs, and RNA polymerase II (Pol II). In particular, SEs play core roles in promoting oncogenic transcription to accelerate cancer development [4,29]. Recent research showed that cancer cells acquire super-enhancers at oncogene and cancerous phenotype relies on these abnormal transcription propelled by SEs [14,24]. Accordingly, it is important to understand super-enhancers and their components since they control much disease-associated sequence variation occurs in these regulatory elements [12,16,20] large amounts of data in order to better understand biological processes.

In fact, the massive evolution of biological data implies the need and necessity to develop new techniques and tools to classify and benefit from them. In this context, Deep Learning is actually an extremely active research area in Machine Learning and bioinformatics [7,21]. It algorithms proved their efficiency in many critical life situations. They allow predicting many diseases, treatments and biological phenomena from the analysis and interpretation of various types of data [1,9,10,22].

Many bioinformatics frameworks based on Machine Learning were developed in the literature to solve genomics problems. [32] proposed DeepEnhancer framework for predicting enhancers using convolutional neural networks (CNN). [33] developed a Deep Learning-based algorithmic framework, called DeepSEA to predict the noncoding-variant effects de novo from sequence. [3] used also deep convolutional neural networks to develop DeepBind approach for predicting the sequence specificities of DNA- and RNA-binding proteins. Likewise, SpliceFinder [31] and Splice2Deep [2] were designed to predict splice sites of human genomic using CNN model. The both works are trained and validated on some genomic sequences such us Homo sapiens, Oryza sativa japonica, Mus-musculus, Drosophila melanogaster, and Daniorerio. In fact, there are so many critical frameworks worthy of our interest that we cannot cite them all.

In this paper, we propose a new solution for predicting super-enhancers on genome scale, based on supervised Deep Learning technique. Our proposal, called ResSEN, predicts super-enhancers using residual neural networks (ResNet) model. The work aims to improve the results obtained by DEEPSEN method. In fact, there are three reasons behind this motivation: (i) first, ResNet was created to optimize the performance of CNN for avoiding vanishing gradient problem, (ii) second, to the best of our knowledge, there is no approach that uses ResNet for predicting super-enhancers has been proposed in the literature [13,15,30], and (iii) third, the obtained results proved the performance of our proposal compared to the DEEPSEN results.

2 Relatd Works

In the literature, there are few bioinformatics works based on Machine Learning proposed to predict super-enhancers of the genomes. [18] implemented and compared six different Machine Learning models to identify key features of SEs and to investigate their relative contribution in the prediction. The six models include: Random Forest, Support Vector Machine, k-Nearest Neighbor, Adaptive Boosting, Naive Bayes, and Decision Tree. To validate their idea, they used 10-fold stratified cross-validation, independent datasets in four human cell-types and a set of publicly available data. [5] proposed a new computational method called DEEPSEN for predicting super-enhancers based on convolutional neural network. The proposed method is trained and tested on 36 SEs features, where 32 ones are used by [18] and 4 others are selected from ChIP-seq and DNase-seq datasets.

3 ResSEN: Residual Neural Network for Predicting Oncogenic Super-Enhancers

3.1 Datasets

The public database used to train and test our approach is used in previous works of [18] and [5]. In fact, there are 36 features (see Table 1) incorporate publicly available ChIP-seq and DNase-seq datasets of mouse embryonic stem cells (mESC) taken from Gene Expression Omnibus (GEO).

Table 1. Features of datasets used by [5] and our approach.

Super-enhancers data type	Features
Histone modifications	H3K27ac, H3K4me1, H3K4me3, H3K9me3
DNA hypersensitive site	DNaseI
RNA polymeraseII	Pol II
Transcriptional co-activating proteins	p300, CBP
P-TFEb subunit	Cdk9
Sub-units of Mediator complex	Med12, Cdk8
Chromatin regulators	Brg1, Brd4 and Chd7
Cohesin	Smc1, Nipbl
Subunits of Lsd1-NuRD complex	Lsd1, Mi2b
Histone deacetylase	H-DAC2, HDAC
Transcription factors	Oct4, Sox2, Nanog, Esrrb, Klf4, Tcfcp2l1, Prdm14, Nr5a2, Smad3, Stat3, Tcf3
Sequence signatures	AT content, GC content, phastCons, phastConsP, repeat fraction

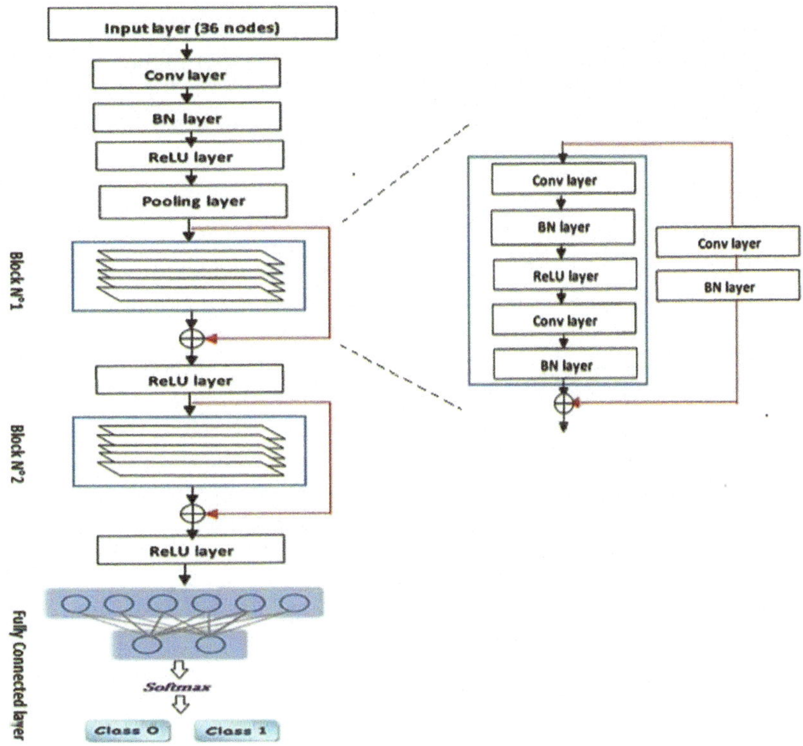

Fig. 1. ResSEN model

The datasets contain 11100 samples. Among them, 1119 are positive and 9981 are negative. To train, test and compare our ResSEN approach, we divided those samples into training datasets and test datasets. Where 90% (i.e. 9990) are used for training and 10% (i.e. 1110) are used for performance testing (see Table 2).

Table 2. Division of samples.

Datasets	Samples size	Positive samples	Negative samples
Training datasets	9990	1006	8984
Test datasets	1110	113	997

3.2 ResSEN Model

ResSEN architecture is composed of an input layer, a convolution layer, a pooling layer, two residual blocks and a fully connected layer.

3.2.1 Input Data

Thirty six (36) characteristics are used to predict the super-enhancers (see Table 1). So, there are 36 nodes in the input layer. The values of these nodes are normalized and standardized before they are transmitted to the next network layers.

3.2.2 Convolutional Layers

ResSEN is composed of five convolutional layers: i) a convolutional layer before the first residual block, ii) two convolution layers for each residual block ($2 \times 2 = 4$), and.

In the first convolutional layer we applied 64 filters of size 1×7, followed by Max-pooling with pool-size 1×3 and stride 1. The first residual block has two convolutional layers, we applied 128 filters of size 1×3 in the first one, and 256 filters of the same size 1×3 in the second one. The second residual block has also two convolutional layers. In the first layer, we applied 256 filters of size 1×3, while in the second layer we applied 512 filters of the same size 1×3.

Figure 2 illustrates the filters parameters' of the five convolution layers.

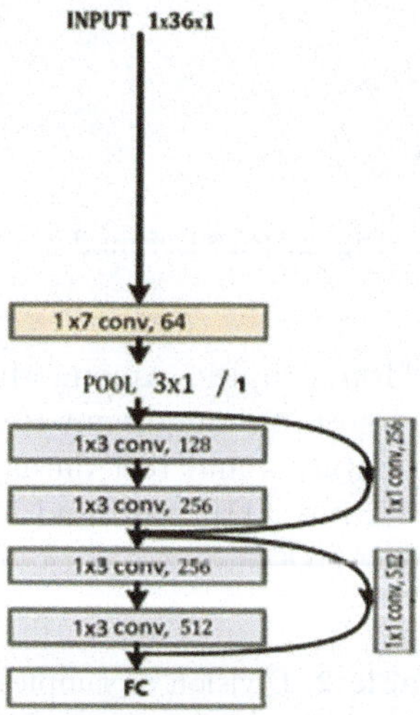

Fig. 2. Convolutional layers parameters of ResSEN

3.2.3 Batch Normalization Layer

Each convolutional layer is followed by a Batch Normalization layer (see Fig. 1). Batch Normalization (BN) is a technique that was introduced to improve the speed, performance, and stability of deep neural networks [11]. It is used to automatically normalize the input layer. Each input xi is normalized using the mean and the variance of the input sample.

3.2.4 Activation Layer

In ResNet, a non-linear activation function is generally used after each BN layer, to ensure the non-linearity of the model [11]. In ResSEN, we used ReLU (rectified linear unit) activation function:

$$ReLU(x) = max(0, x) \tag{1}$$

In fact, ReLU function has become the default activation function for ResNet allowing model to train easier and faster and perform better [8].

3.2.5 Add Identity

For each residual block, RcsSEN uses convolution block strategy to add the block's input to the block's output. This type of design requires that the block's output and its input have the same shape (size), so they can be added together. The output of the first block will be the input of the second and the output of the second block will be the input of the fully connected layer. To transform the block's input into the desired shape, we introduced 256 convolutions (256 filters) of size 1 × 3 for the first residual block and 512 convolutions (512 filters) of size 1 × 3 for the second residual block (see Fig. 2).

3.2.6 Fully Connected Layer

The fully connected (FC) layer of the ResSEN is structured as follow:

- The number of input neurons is 17408;
- The activation function is ReLU;
- The number of output layer is 2 neurons;
- The function used to calculate the probability of the output classes is: Soft max.

$$Softmax(x_j) = max\frac{e^{x_i}}{\sum_j e^{x_i}}, j \in \{1, 2, ...k\} \tag{2}$$

Where, k is the number of classes. Moreover, to obtain the predicted class A, we applied the *argmax* function to the *Softmax* function output:

$$A = argmax(Softmax(x_j)) \tag{3}$$

- So, if $A = 1$: the predicted class is positive, that's means, the presence of the super-enhancers in the genome;
- if $A = 0$: the predicted class is negative, that's means, the absence of super-amplifiers in the genome.

3.2.7 ResSEN Training

ResSEN training is based on supervised learning, which consists of calculating the optimal weights (weights) using the input matrix D (the data samples) and the output matrix A (the desired output or the class label) corresponding to D. D is a matrix of size N × 36 and A is a binary matrix of size N × 1, where N is the number of samples which is set to 11100. $A[i] = 1$ if the corresponding sample represents the super-enhancer class, otherwise, $A[i] = 0$.

During the training phase, ResSEN uses the cross entropy loss function to measure the difference between the calculated output and the desired output, and Backpropagation model and Adam method [19] to update network weight's.

4 Experiment Results and Comparison

In the context of binary classification, the evaluation of model's performance is based on some performance measures that are computed from the confusion matrix. Thus, to evaluate and compare the ResSEN performance, we calculated four measures: accuracy, recall, precision and F1-score (Table 3).

Table 3. Confusion matrix.

		Actuel class	
		−	+
Predicted class	+	**True Negatives**	**True Positives**
	−	**False Negatives**	**False Positives**

$$Accuracy = \frac{TP + TN}{TP + TN + FP + FN},$$

$$Recall = \frac{TP}{TP + FN}, Precision = \frac{TP}{TP + FN}$$

$$F1Score = 2 * Precision * Recall / Precision + Recall$$

The best results obtained by testing the best model of ResSEN on the test samples are presented in Table 4.

Table 4. ResSEN results.

Precision	Recall	F1-Score	Accuracy
94,79%	94,94%	94,85%	94,95%

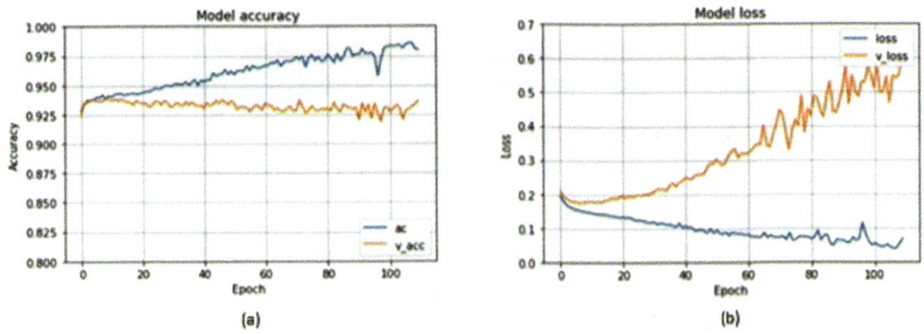

Fig. 3. (a) DeepSEN accuracy curve, (b) DeepSEN loss curve.

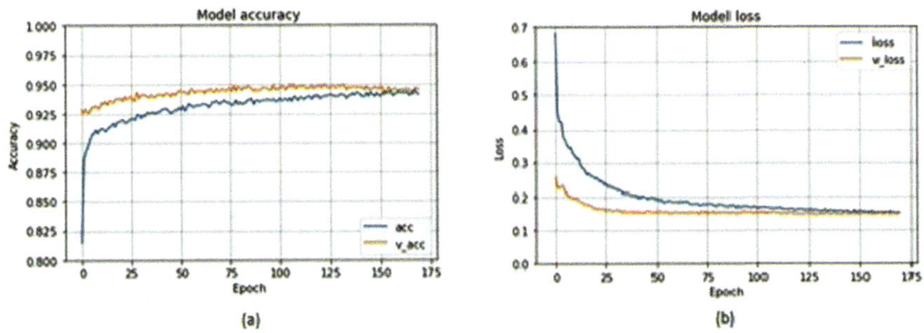

Fig. 4. (a) ResSEN accuracy curve, (b) ResSEN loss curve.

5 Comparison and Discussion

In order to ensure a fair comparison with our ResSEN algorithm, we re-executed the DeepSEN [6] using 90% of samples for training and 10% of samples for testing. The obtained results (see Fig. 5), show that the best model of DeepSEN achieves an accuracy of 93,64% and a precision of 90%. However, in both cases (validation with all the data sets or with 10% of samples) we noticed the presence of the overfitting problem. The latter is clearly modeled in the accuracy and loss curves that we generated after re-executed DeepSEN (see Fig. 3). Knowing that, the blue and the orange curves represent the development of accuracy/loss in the training phase and in the testing phase respectively.

Figure 4 shows the accuracy and loss curves of ResSEN model. In this case, there is no overfitting problem. We notice a harmonization between the curves generated in training and test phases.

Finally, the comparison graph (see Fig. 5) shows that our proposed model outperforms that of DeepSEN.

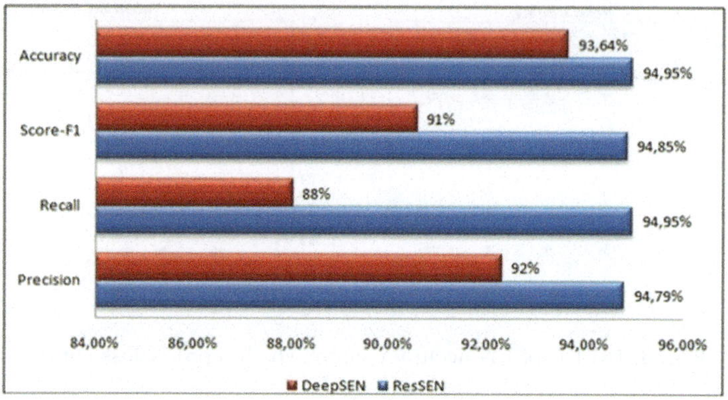

Fig. 5. Performance comparison graph of ResSEN and DeepSEN.

6 Conclusion

In this paper, we presented our prediction approach for detecting super-enhancers in human genomes, called ResSEN. This proposal is based on the ResNet Deep Learning technique aiming at improving the results of existing approaches.

The ResSEN was evaluated using 36 features of mESC datasets taken from Gene Expression Ominibus (GEO). The obtained results were compared with that obtained by DeepSEN approach which is based on CNN architecture. Comparison shows that our proposed ResSEN outperforms DeepSEN, and proves the effectiveness of ResNet architecture as a classifier of genomic problems.

References

1. Alazab, M., et al.: COVID-19 prediction and detection using deep learning. Int. J. Comput. Inf. Syst. Ind. Manag. Appl. **12**, 168–181 (2020)
2. Albaradei, S., et al.: Splice2Deep: an ensemble of deep convolutional neural networks for improved splice site prediction in genomic DNA. Gene X **5** (2020)
3. Alipanahi, B., Delong, A., Weirauch, M., Frey, B.J.: Predicting the sequence specificities of DNA- and RNA-binding proteins by deep learning. Nat. Biotechnol. **33**, 831–838 (2015)
4. Bradner, J.E., Hnisz, D., Young, R.A.: Transcriptional addiction in cancer. Cell **168**, 629–643 (2017)
5. Bu, H., Hao, J., Gan, Y., et al.: DEEPSEN: a convolutional neural network based method for super-enhancer prediction. BMC Bioinform. **20**, 1–9 (2019)
6. Bu, H., Hao, J., Gan, Y., et al.: DEEPSEN code (2019). https://github.com/1991Troy/DEEPSEN
7. Cao, Y., Geddes, T., Yang, J., Yang, P.: Ensemble deep learning in bioinformatics. Nat. Mach. Intell. **2**, 1–9 (2020)

8. Chen, S., Jia, Q., Tan, Y., Li, Y., Tang, F.: Oncogenic super-enhancer formation in tumorigenesis and its molecular mechanisms. Exp. Mol. Med. **52**, 713–723 (2020)
9. Ching, T., et al.: Opportunities and obstacles for deep learning in biology and medicine. J. Roy. Soc. Interface **15**(141), 20170387 (2018)
10. Esteva, A., et al.: A guide to deep learning in healthcare. Nat. Med. **25**(1), 24–29 (2019)
11. Furusho, Y., Ikeda, K.: ResNet and batch-normalization improve data separability. Proc. Mach. Learn. Res. **101**, 94–108 (2019)
12. Grossman, S.R., et al.: Identifying recent adaptations in large-scale genomic data. Cell **152**, 703–713 (2013)
13. He, K., Zhang, X., Ren, S., Sun, J.: Deep residual learning for image recognition. In: IEEE Conference on Computer Vision and Pattern Recognition (CVPR), Las Vegas, NV, pp. 770–778 (2016)
14. He, Y., Long, W., Liu, Q.: Targeting Super-Enhancers as a Therapeutic Strategy for Cancer Treatment. Front. Pharmacol. **10**, 361 (2019)
15. Alzantot, M., Wang, Z., Srivastava, M.: Deep residual neural networks for audio spoofing detection. arXiv:190700501v1 (2019)
16. Hnisz, D., et al.: Super-enhancers in the control of cell identity and disease. Cell **155**(4), 934–947 (2013)
17. Huang, J., et al.: Dissecting super-enhancer hierarchy based on chromatin interactions. Nat. Commun. **9**(943) (2018)
18. Khan, A., Zhang, X.: Integrative modeling reveals key chromatin and sequence signatures predicting super-enhancers. Sci. Rep. **9**, 1–15 (2019)
19. Kingma, D., Ba, J.: Adam: a method for stochastic optimization. In: International Conference on Learning Representations (2014)
20. Lee, T.I., Young, R.A.: Transcriptional regulation and its misregulation in disease. Cell **152**, 1237–1251 (2013)
21. Li, Y., Huang, C., Ding, L., Li, Z., Pan, Y., Gao, X.: Deep learning in bioinformatics: introduction, application, and perspective in the big data era. Methods **166**, 4–21 (2019)
22. Litjens, G., et al.: A survey on deep learning in medical image analysis. Med. Image Anal. **42**, 60–88 (2017)
23. Lu, J., et al.: MICAL2 mediates p53 ubiquitin degradation through oxidating p53 methionine 40 and 160 and promotes colorectal cancer malignance. Theranostics **8**(19), 5289–5306 (2018)
24. Mansour, M.R., et al.: Oncogene regulation. An oncogenic super-enhancer formed through somatic mutation of a noncoding intergenic element. Science (New York, N.Y.) **346**(6215), 1373–1377 (2014)
25. Ng, H.H., Surani, M.A.: The transcriptional and signalling networks of pluripotency. Nat. Cell Biol. **13**, 490–496 (2011)
26. Orkin, S.H., Hochedlinger, K.: Chromatin connections to pluripotency and cellular reprogramming. Cell **145**, 835–850 (2011)
27. Qu, J., et al.: Functions and clinical significance of super-enhancers in bone-related diseases. Front. Cell Dev. Biol. **8**, 534 (2020)
28. Sengupta, S., George, R.E.: Super-enhancer-driven transcriptional dependencies in cancer. Trends Cancer **3**, 269–281 (2017)
29. Tang, F., Yang, Z., Tan, Y., Li, Y.: Super-enhancer function and its application in cancer targeted therapy. NPJ Precis. Oncol. **4**(2), 1–7 (2020)
30. Tang, R., Lin, J.:. Deep residual learning for small-footprint keyword spotting. In: IEEE International Conference on Acoustics, Speech and Signal Processing (ICASSP), Calgary, AB, pp. 5484–5488 (2018)

31. Wang, R., Wang, Z., Wang, J., Li, S.: SpliceFinder: ab initio prediction of splice sites using convolutional neural network. BMC Bioinform. **20**, 1–13 (2019)
32. Xu, M., Ning, C., Ting, C., Rui, J.: DeepEnhancer: predicting enhancers by convolutional neural networks. In: IEEE International Conference on Bioinformatics and Biomedicine (BIBM), Shenzhen, pp. 637–644 (2016)
33. Zhou, J., Troyanskaya, O.: Predicting effects of noncoding variants with deep learning-based sequence model. Nat. Methods **12**, 931–934 (2015)

Machine Learning Algorithms for Big Data Mining Processing: A Review

Laouni Djafri[1,2(✉)] and Yacine Gafour[1,2]

[1] Department of Computer Science, Ibn Khaldoun University, Tiaret, Algeria
[2] EEDIS Laboratory, Djillali Liabes University, Sidi Bel Abbes, Algeria

Abstract. Big data mining is an excellent source of information and knowledge from systems to end users. However, managing such amounts of data or knowledge requires automation, which leads to serious consideration of the use of machine learning algorithms. Machine learning helps us make decisions if there is no right way to solve a problem identified in previous knowledge bases, and that is, too, one of the most widely used analysis and modeling tools for this purpose. In this work, we present an in-depth study that helps us to choose the best machine learning algorithms in order to process big data and extract knowledge from it, so that, this treatment can be very flexible, either in a simple system with sequential computing, or in a distributed system with parallel computing. To achieve this, we will, first and foremost, test the accuracy of the results provided by the classifiers; here we mean the strength and flexibility of a classifier when it comes to dealing with big data mining. Second, we will also test the execution speed for each classifier in complex cases; that is, when the classifier will not be sufficient to solve a particular problem in the context of big data mining, especially if all cases are dealt with quickly and efficiently. The results obtained in this paper demonstrated the superiority of certain classifiers over others in certain cases, and demonstrated their failure in other cases, the reason being due to the nature of the dataset, in particular the number of instances, the number of attributes , and the number of classes.

Keywords: Artificial intelligence · Big data mining · Supervised classification · Binary classification · Multi-class classification

1 Introduction

Machine learning and data mining are not the same, but cousins. Machine learning is a branch of artificial intelligence that provides systems that can learn from data. Machine learning is often used to classify data or make predictions, based on known properties in the data learned from historical data that's used for training [1]. Data mining is sorting through data to identify patterns and establish relationships. Generally, data mining (sometimes called knowledge discovery) is the process of analyzing data from different perspectives and summarizing it into

B. Lejdel et al. (Eds.): AIAP 2021, LNNS 413, pp. 43–55, 2022.
https://doi.org/10.1007/978-3-030-96311-8_5

useful information [2]. Data mining is the analysis of data for relationships that have not previously been discovered. It is an interdisciplinary subfield of computer science, the computational process of discovering patterns in large data sets ("Big Data") involving methods at the intersection of artificial intelligence, machine learning, statistics, and database systems. So, data mining works to provide insights and discovery of unknown properties in the data [3]. Machine learning can be carried out through either supervised learning or unsupervised learning methods [4]. The unsupervised learning uses algorithms that operate on unlabeled data, namely, the data input where the desired output is unknown. The goal is to discover structure in the data but not to generalize a mapping between inputs to outputs. The supervised learning (It is the subject of our concern) use labeled data for training. Labeled data are datasets where the input and outputs are known. The supervised learning method works to generalize a relationship or mapping between inputs to outputs [4]. There is an overlap between the two. Often data mining uses machine learning methods and vice versa, where machine learning can use data mining techniques [5].

The paper is structured according to the plan described as follows. Section 2, in this section, we present relevant recent work that addresses problems of big data mining classification, so that we will largely cover the machine learning methods used to deal with this problem, In particular the supervised algorithms. Then, the summary of the papers reviewed, is discussed in Sect. 3. Thereafter, experimental results are discussed in Sect. 4. The experimental section describes the two datasets (binary and multi class classification). Finally, a conclusion and future works is presented in Sect. 5.

2 Literature Survey

2.1 Big Data Analytics and Machine Learning

Today, in our world, whoever has more information has more power; this information is extracted from a large amount of data. Big data is generated terribly daily, unlike what we were living in at the end of the last century as the amount of data produced at that time is very small as the data is only generated when certain types of events occur, and you can live weeks and months without producing a single piece of data. But, today we can never do that because data is everywhere; it is produced by individuals, groups, companies, and even things that depend on the Internet, etc. Data analysis is extremely important, especially for companies for public and private companies of all types and services [6]. Companies use this analytics to make informed decisions about self-strategies, including recruitment, marketing, and branding. In general, these analyzes can be used to predict unknowns or what we call extrapolation, what makes the big data concept even more important is actually the concept of artificial intelligence [7]. We especially mention machine learning; thanks to its advantages such as being fast, automaticity, having no acquisition costs and saving on labor, it increases companies to be superior in competition.

2.1.1 Big Data Analytics

Since the advent of the Internet to this day, we have seen explosive growth in the volume, velocity and variety of data created daily [8]; this amount of data is generated by a variety of methods such as click stream data, financial transaction data, log files generated by web or mobile applications, sensor data from Internet of Things (IoT), in-game player activity and telemetry from connected devices, and many other methods [9,10]. This data is commonly referred to as "Big Data" because of its volume, the velocity with which it arrives and the variety of forms it takes. In 2001, Gartner proposed a three-dimensional or 3 Vs (Volume, Variety and Velocity) view of the challenges and opportunities associated with data growth [11]. In 2012, Gartner updated this report as follows: Big data is high volume, high speed, and/or wide variety of information resources that require new forms of processing to improve decision making [12]. Often times, these Vs are supplemented by a fourth V, is Veracity: How accurate is the data? [13,14]. We can extend this model to the Big Data dimensions over ten Vs: volume, variety, velocity, veracity, value, variability, validity, volatility, viability and viscosity [10,15–21]. Accordingly, the increasing digitization of our activities, the ever-increasing ability to store digital data, and the accumulation of information of all kinds, is generating a new sector of activity aimed at analyzing these large amounts of data.

Analytics is a broad term that encompasses the processes, technologies, frameworks and algorithms to extract meaningful insights from data. Raw data in itself does not have a meaning until it is contextualized and processed into useful information. Analytics is this process of extracting and creating information from raw data by filtering, processing, categorizing, condensing and contextualizing the data. This information obtained is then organized and structured to infer knowledge about the system and/or its users, its environment, and its operations and progress towards its objectives, this is known as big data mining [22,23]. Its main purpose is to extract and retrieve desired information or patterns from a large amount of data [24]. It is usually performed on a large amount of structured or unstructured data using a combination of techniques that make it possible to explore these large amounts of data, automatically or semi-automatically [25,26].

2.1.2 Machine Learning

Big Data Mining is a great source of information and knowledge from systems to other end users. However, managing such a large amount of data or knowledge requires automation, which leads to serious thinking about the use of machine learning techniques. Machine learning consists of many powerful algorithms for learning patterns, knowledge acquisition, and predicts future events. Specifically, these algorithms work by searching a group of possible predictive models to capture the best relationship between descriptive features and target functions in the dataset. Based on this, the machine learning algorithm makes the selection during the training process. The clear criterion for driving this choice is the search for data-compatible models [12,27]. We can then use this model to make predictions for new cases (instances) [28]. Therefore, Machine learning, which is one of the sub domains of artificial intelligence, aims to automatically extract and exploit the information present in the dataset, that is, equipping machines with human intelligence, so that they are able to make predictions based on a huge amount of data, which is an almost impossible task for a human being [29]. For example, machine learning plays a key role in better understanding and coping with the COVID-19 crisis, in which machine learning algorithms allow computers to mimic human intelligence and ingest large volumes of data to quickly identify models and information; these models are used to predict new observed values. After that, smart decisions can be taken to help us out of the crisis [30,31].

Machine learning algorithms are broadly classified into three categories: supervised, unsupervised and reinforcement learning [4]. In our work, we have relied on supervised algorithms in order to build predictive models; so that, it connects past and current datasets with the help of labeled data to predict future events [32]. We can simply say that supervised learning refers to known labels (predicted classes are known beforehand) as a set of samples to predict future events [33,34]. It is divided into three phases: the learning phase, the validation phase and the test phase. Supervised learning is also divided into two broad categories [35]: classification and regression. Classification algorithms are suitable for the system that produces discrete responses [36]. In other words, responses are categorical variables, whereas regression algorithms are algorithms that develop a model that relies on equations or mathematical operations based on the values taken from input attributes to produce a continuous value representing the output [35]. This means that, the input of these algorithms can take continuous and discrete values depending on the algorithm, whereas the output is a continuous value [36]. Supervised learning algorithms in the context of big data are more complex. Nowadays, there is a growing interest in social, economic, health, safety, and other issues that need to be solved using big data analysis and machine learning algorithms. These two concepts are starting to gain attention in many scientific researches. For example, but not limited to, in the business world, most decisions would be much easier if we can anticipate the likelihood, or propensity of customers to take different actions using machine learning algorithms. Successful applications of propensity modeling include predicting the likelihood of customers moving from one mobile operator to another, responding to particular marketing efforts, or purchasing different products [37]. Also, organizations can use the machine learning algorithms to better control and manage the situation in the event of risks [38]. In the healthcare world, these algorithms can help professionals make better diagnoses by tapping into large collections of historical examples on a scale beyond anything an individual might see in their career. For example, predicting optimal doses based on past dose data and associated outcomes [39]. In a similar study conducted by D. Nguyen et al. [40] in order to find out the optimal distribution of prostate cancer radiotherapy a patient will receive. Currently, if we are talking about fighting epidemic diseases and how to prevent them, we are talking more specifically about the Corona virus pandemic. In early 2020, coinciding with the emergence of this pandemic in China, December 2019 [41], and to this day, the machine learning algorithms is used terribly in most, if not we say, in all scientific research related to fighting this virus [30,31,33,34]. Therefore, big data mining and machine learning are two promising technologies used by many healthcare providers use to help medical experts in order to solve real problems. But, most of the works in this regard have centered on predictions for prevention and saving lives. Predictions are mainly based on supervised algorithms. For example, A. Ardakani et al. [41] they adopted the Deep Convolutional Neural Network method to build predictive models. Also, T. Ozturk et al. [42] they adopted Convolutional Neural Network method for prediction. A similar work done by L. Sun et al. [43] in which they used the SVM method. Another work presented by J. Wu et al. [44] in the same context, no less important than the other works, which is based on the random forests algorithm. Also, in the context of data mining, there is a comparative study for a better precision in the prediction of cardiovascular diseases carried out by R. Sharma and S. N. Singh [45], where several classifiers have been used including Naive Bayes, C-PLS, KNN and decision tree.

3 Summary of the Papers Reviewed

In this part, we have selected the most recent works that compare classifiers. Thus, we can choose the best and optimal classifier. But, before starting the summary of the papers reviewed and the experiences, we ask the following question: Why did we choose the six classifiers in our work?

We have selected the six classifiers because they are the most widely used and are highly suitable for big data analytics. To confirm our choices, we quote the following:

SVM (Support Vector Machine) over the past decade, SVM has been gradually integrated into the Big Data field. It solves big data classification problems. In particular, it can help multi-domain applications in a big data environment [46]. **ANN** (Artificial neural networks) constitute a realistic criterion in the Big Data field, thus knowledge of this field is of paramount importance for those who wish to extract significant information from the big data available to date [47]. **KNN** (K-Nearest Neighbors) is widely used in big data analytics, especially if it is developed more and more in order to give satisfactory classification results [48, 49]. **RF** (Random Forests) seem insensitive to over-fitting, this method generally does not require a lot of parameter optimization efforts. Random forests therefore avoid one of the main pitfalls of Big Data approaches in machine learning [50]. **LR** (Logistic Regression) gives better result for analyzing the big data [51]. **BN** (Bayesian Network) or (Naïve Bayes) can also be used in the Big Data field, it is very useful for generating synthetic data when the actual data is insufficient [52].

We have done a thorough research by investigating some of the related work to find out which of the classifiers works best to give an excellent classification result compared to other classifiers. For reference only, these classifiers used the same datasets and default hyper-parameters in each independent work. Table 1 summarizes the results obtained as follows:

From Table 1, we note that in most cases the RF classifier gives a good classification result than SVM and LR [55,56,59–61] , and in some cases the SVM classifier gives a good classification result than RF [54,58]. In other cases, the LR classifier gives us good results than RF and SVM [57,58]. We also see that in most cases these classifiers (RF, SVM, LR) give good classification results than KNN and NB [56,59].

Through all the works previously presented in Table 1, we note that when an ANN classifier exists among these classifiers it gives good classification results compared to other classifiers, despite the diversity of the nature and size of the dataset [53,54,58,62].

From here, we conclude that the ANN classifier is the best and first classifier for dealing with big data mining, followed by the RF classifier in the second place, followed to a lesser extent by SVM and LR, then NB in the last place.

4 Practical Proofs and Discussion

From Table 1, we note that some classifiers sometimes outperform others, and fail at other times.

In order to objectively discuss the work presented in Table 1 and to see the stability or volatility (fluctuation) of the classification results, we conducted two experiments to test these classifiers in the context of big data mining.

We developed our software with the tools and APIs under the Linux Ubuntu operating system 20.04.1, Pycharm CE 2020.2, Java development kit 11.0.8, Apache Spark 3.0.1, Python 3.8, PyQt5 desinger 5.

- **The first experiment:** In this experiment, we use the KDD Cup 2012 dataset[1], knowing that this dataset has two classes (binary classification). KDD Cup 2012 saved in the LIBSVM format, the size of this dataset is detailed in the following Table:
- **The second experiment:** In this experiment, we use the Mnist8m dataset, knowing that this dataset has ten classes (multi-class classification). To see more information on this database, visit this site[2] (Tables 2 and 3).

The results (classifier performance metrics) obtained are shown below:

From Table 4 which represents the big data mining binary classification, where we obtained a precision for binary classification equal to 100% using the ANN classifier, it is followed by LR in second place with an precision equal to 90.99%, then comes in third place the SVM classifier with an precision equal to 84.88%, and finally comes in fourth place the RF classifier with an precision equal to 83.45%.

---

[1] https://www.csie.ntu.edu.tw/~cjlin/libsvmtools/datasets/binary.html.

[2] https://www.csie.ntu.edu.tw/~cjlin/libsvmtools/datasets/multiclass.html.

Table 1. A survey to choose the best classifier.

References	Year	Classifiers used	Performance metrics	The best classifier
[53]	2020	ANN, SVM	**Accuracy** **ANN = 0.982** SVM = 0.941	ANN with a difference of (0.041) compared to the next classifier (SVM)
[54]	2020	ANN, SVM, RF	**Accuracy** **ANN = 0.8218** **SVM = 0.8229** RF = 0.7839	ANN and SVM with a very slight increase compared to ANN (difference = 0.0011)
[55]	2020	SVM, RF, NB, KNN	**Accuracy** SVM = 0.582609 **RF = 0.721739** NB = 0.553623 KNN = 0.628986	RF with a difference of (0.092753) compared to the next classifier (KNN)
[56]	2020	RF, LR, KNN	**Accuracy** **RF = 0.9321** LR = 0.9096 KNN = 0.9096	RF with a difference of (0.0225) compared to the next classifier (LR and KNN)
[57]	2020	LR, RF	**Accuracy** **LR = 0.80** RF = 0.79	LR with a difference of (0.01) compared to the next classifier (RF)
[58]	2020	LR, NB, ANN, RF, SVM	**Precision** LR = 0.530 NB = 0.515 **ANN = 0.700** RF = 0.450 SVM = 0.451	ANN is much better than other classifiers, with a difference of (0.17) compared to the next classifier (LR)
[59]	2020	KNN, LR, RF, SVM	**Accuracy** KNN = 0.6462 LR = 0.8232 **RF = 0.8461** SVM = 0.8050	RF with a difference of (0.0229) compared to the next classifier (LR)
[60]	2020	RF, KNN, SVM	**Accuracy** **RF = 0.9984** KNN = 0.9983 SVM = 0.9976	RF with a difference of (0.0001) compared to the next classifier (KNN)
[61]	2021	RF, KNN, LR, NB	**Accuracy** **RF = 0.8677** KNN = 0.6629 LR = 0.6192 NB = 0.6056	RF with a difference of (0.2048) compared to the next classifier (KNN)
[62]	2020	KNN, RF, SVM, NB, LR, ANN	**Accuracy** KNN = 0.8916 RF = 0.9460 SVM = 0.8418 NB = 0.8233 LR = 0.8408 **ANN = 0.9580**	ANN With a difference of (0.012) compared to the next classifier (RF)

In addition, from Table 5 which represents the big data mining multi-class classification, where we obtained a precision for multi-class classification equal to 100% using the SVM classifier. On the other hand, we got the two metrics ROC

Table 2. Characteristics of the KDD Cup 2012 dataset.

Data-set	Instance-training	Instances-validation	Features	Classes
KDD 2012 (2 Go)	119705032 (1.60 Go)	29934073 (458 Mo)	54686452	2

Table 3. Characteristics of the Mnist8m dataset.

Data-set	Instances training	Features	Classes
Mnist8m (2.75 Go)	8100000	784	10

Table 4. Big data mining binary classification.

Metrics (%)/Classifier	LR	RF	SVM	ANN
Precision	0.9099952241540803	0.8345043151705323	0.8488354118601478	1.0
Accuracy	0.9167988783529498	0.8121272365805169	0.8319844609906119	1.0
Recall	0.9107581787524190	0.8121272365805168	0.8319844609906117	1.0
F-measure	0.9133063016519345	0.8160949483720379	0.8366175902829752	1.0
RMSE	0.1088647507021327	0.3935667105928753	0.4098969858505704	0.0
MAE	0.2049216107001812	0.4099316501604129	0.1680155390093881	0.0
MSE	0.1702721106852340	0.9420281768726457	0.1775217779419465	0.0
ROC	0.6164675070485738	0.5060675170482035	0.5710370150412737	1.0
AUC	0.5777808410445922	0.3160580181014294	0.4462580781142544	1.0
Time (s)	94	9	6	5

Table 5. Big Data Mining multi-class classification

Metrics (%)/Classifier	LR	RF	SVM	ANN
Precision	0.88557010181677	0.77996042809628	1.0	0.91642815472205
Accuracy	0.88581392104896	0.73163912885459	0.94646349909959	0.91624279416751
Recall	0.88581392104896	0.73163912885459	0.94646349909959	0.91624279416751
F-measure	0.88553253087226	0.74477511561745	0.97249550226594	0.91620695452747
RMSE	0.13935667105928	0.21644947572987	0.23137956024766	0.21564348435074
MAE	0.40993165016041	0.48797841706505	0.05353650090040	0.31129196337741
MSE	0.19420281768726	0.24675405508314	0.05353650090047	0.14777890810444
ROC	0.59993165016041	0.51074125016131	0.0	0.73230124012973
AUC	0.48794202817687	0.37974402511608	0.0	0.68707544271661
Time (s)	370	32	52	71

and AUC equal to 0. This indicates an over-fitting problem. So, we conclude that SVM does not perform well in big data mining classification (multi-class classification). Thus, the ANN classifier comes in first place with an precision equal to 91.64% with a processing time equal to 71 s, then it is followed by LR with an precision equal to 88.55%, but the processing time long enough (370 s). After that, the RF classifier comes in third place with an precision equal to 77.99% and a somewhat acceptable processing time (32 s).

From the results of the two Tables 4 and 5 above, we noticed that there are machine learning classifiers which are affected by the structure of big datasets (binary or multi-class), such as SVM which has been shown to perform very well in binary classification, and largely failed in multi-class classification. In addition, there are some classifiers that work comfortably and give good results in both cases with rather low execution time, such as ANN. Besides, there are also other classifiers which give satisfactory results in binary and multi-class classification, but with slow execution time like LR.

5 Conclusion and Future Scope

In conclusion, our study shows that the majority of machine learning algorithms are influenced in terms of performance metrics and execution time. The ranking of these algorithms changes in terms of efficiency and effectiveness depending on the nature of the datasets used, particularly in the context of big data analytics. But the ANN classifier was less affected than other classifiers, as it retained its first place despite the diversity of data sets. Based on this, we consider the ANN classifier to be the best for dealing with big data mining. As future work, we can change the hyper-parameters and increase the number of hidden layers and/or increase the number of neurons per hidden layer to improve classification accuracy more and more, which we simply call the deep learning.

References

1. Bailly, S., Meyfroidt, G., Timsit, J.-F.: What's new in ICU in 2050: big data and machine learning. Intensive Care Med. **44**(9), 1524–1527 (2017). https://doi.org/10.1007/s00134-017-5034-3
2. Jayasri, N.P., Aruna, R.: Big data analytics in health care by data mining and classification techniques. ICT Express (2021). https://doi.org/10.1016/j.icte.2021.07.001
3. Smith, P.F., Zheng, Y.: Applications of multivariate statistical and data mining analyses to the search for biomarkers of sensorineural hearing loss, tinnitus, and vestibular dysfunction. Front. Neurol. **12**, 205 (2021). https://doi.org/10.3389/fneur.2021.627294. ISSN 1664-2295
4. Dasgupta, A., Nath, A.: Classification of machine learning algorithms. Int. J. Innov. Res. Adv. Eng. **3**(3), 6–11 (2016)
5. Dogan, A., Birant, D.: Machine learning and data mining in manufacturing. Expert Syst. Appl. **166**, 114060 (2020). https://doi.org/10.1016/j.eswa.2020.114060
6. Kushwaha, A.K., Kar, A.K., Dwivedi, Y.K.: Applications of big data in emerging management disciplines: a literature review using text mining. Int. J. Inf. Manag. Data Insights **1**(2), 100017 (2021). https://doi.org/10.1016/j.jjimei.2021.100017
7. Chui, K.T., Lytras, M.D., Visvizi, A., Sarirete, A.: An overview of artificial intelligence and big data analytics for smart healthcare: requirements, applications, and challenges, pp. 243–254. Academic Press (2021). https://doi.org/10.1016/B978-0-12-822060-3.00015-2
8. Sathyaraj, R., Ramanathan, L., Lavanya, K., Balasubramanian, V., Saira Banu, J.: Chicken swarm foraging algorithm for big data classification using the deep belief network classifier. Data Technol. Appl. (2020). https://doi.org/10.1108/DTA-08-2019-0146

9. O'Donovan, P., Leahy, K., Bruton, K., O'Sullivan, T. J.: Big data in manufacturing: a systematic mapping study. J. Big Data **20**(2) (2015). https://doi.org/10.1186/s40537-015-0028-x
10. Hariri, R.H., Fredericks, E.M., Bowers, K.M.: Uncertainty in big data analytics: survey, opportunities, and challenges. J. Big Data **6**(1), 1–16 (2019). https://doi.org/10.1186/s40537-019-0206-3
11. Chen, M., Liu, Y.: Big data: a survey, mobile networks and application. **19**(2), 171–209 (2014)
12. Erl, T., Khattak, W., Buhler, P.: Big Data Fundamentals: Concepts, Drivers and Techniques. Prentice Hall Press, Hoboken (2016)
13. Chan, J.O.: An architecture for big data analytics. Commun. IIMA **13**(2), 1–13 (2013)
14. Deutsch, R., Corrigan, D., Zikopoulos, P., Giles, J.: Harness the Power of Big Data: The IBM Big Data Platform. McGraw-Hill, New York (2013)
15. Khan, N., Shah, H., Badsha, G., Abbasi, A.A., Alsaqer, M., Salehian, S.: 10 Vs, issues and challenges of big data. In: International Conference on Big Data and Education ICBDE 2018, pp. 203–210 (2018)
16. Kayyali, D., Knott, S.V.: The big-data revolution in us health care: accelerating value and innovation. Mc Kinsey Company **2**(8), 1–13 (2013)
17. Katal, A., Wazid, M., Goudar, R.: Big data: issues, challenges, tools and good practices. In: Sixth International Conference on Contemporary Computing (IC3), pp. 404–409. IEEE (2013)
18. Ferguson, M.: Enterprise information protection-the impact of big data. IBM (2013)
19. Patgiri, R., Ahmed, A.: Big data: the v's of the game changer paradigm. In: IEEE 18th International Conference on High Performance Computing and Communications; IEEE 14th International Conference on Smart City; IEEE 2nd International Conference on Data Science and Systems (2016). https://doi.org/10.1109/HPCC-SmartCity-DSS.2016.8
20. IBM, The top five ways to get started with big data (2014)
21. Elgendy, N., Elragal, A.: Big data analytics: a literature review paper. In: Perner, P. (ed.) Advances in Data Mining. Applications and Theoretical Aspects, ICDM 8557 (2014)
22. Cen, T., Chu, Q., He, R.: Big data mining for investor sentiment. J. Phys. Conf. Scr. **1187**(5) (2019)
23. Che, D., Safran, M., Peng, Z.: From big data to big data mining: challenges, issues, and opportunities. In: Hong, B., Meng, X., Chen, L., Winiwarter, W., Song, W. (eds.) DASFAA 2013. LNCS, vol. 7827, pp. 1–15. Springer, Heidelberg (2013). https://doi.org/10.1007/978-3-642-40270-8_1
24. Oussous, A., Benjelloun, F.-Z., Lahcen, A., Belfkih, S.: Big data technologies: a survey. J. King Saud Univ. - Comput. Inf. Sci. (2017). http://dx.doi.org/10.1016/j.jksuci.2017.06.001
25. Xindong, W., Xingquan, Z., Gong-Qing, W., Wei, D.: Data mining with big data. IEEE Trans. Knowl. Data Eng. **26**(1), 97–107 (2014). https://doi.org/10.1109/TKDE.2013.109
26. Xingquan, Z., Ian, D.: Knowledge Discovery and Data Mining: Challenges and Realities. Hershey, New York (2007). ISBN 978-1-59904-252
27. Bailly, S., Meyfroidt, G., Timsit, J.: What's new in ICU in 2050: big data and machine learning. Intensive Care Med **44**, 1524–1527 (2018). https://doi.org/10.1007/s00134-017-5034-3

28. Klaine, P.V., Imran, M.A., Onireti, O., Souza, R.D.: A survey of machine learning techniques applied to self-organizing cellular networks. IEEE Commun. Surv. Tutor. **19**(4), 2392–2431 (2017). https://doi.org/10.1109/COMST.2017.2727878
29. Khan, B., Olanrewaju, R.F., Altaf, H.: Critical insight for MapReduce optimization in Hadoop. Int. J. Comput. Sci. Control Eng. **2**(1), 1–7 (2014)
30. An, C., Lim, H., Kim, D.: Machine learning prediction for mortality of patients diagnosed with COVID-19: a nationwide Korean cohort study. Sci. Rep. **10**, 1–11 (2020). https://doi.org/10.1038/s41598-020-75767-2
31. Goodman-Meza, D., Rudas, A., Chiang, J., Adamson, P., Ebinger, J., Sun, N.: A machine learning algorithm to increase COVID-19 inpatient diagnostic capacity. PLoS One **15**(9), e0239474 (2020). https://doi.org/10.1371/journal.pone.0239474
32. Mathkunti, N.M., Rangaswamy, S.: Machine learning techniques to identify dementia. SN Comput. Sci. **1**(3), 1–6 (2020). https://doi.org/10.1007/s42979-020-0099-4
33. Muhammad, L.J., Algehyne, E.A., Usman, S.S., Ahmad, A., Chakraborty, C., Mohammed, I.A.: Supervised machine learning models for prediction of COVID-19 infection using epidemiology dataset. SN Comput. Sci. **2**(1), 1–13 (2020). https://doi.org/10.1007/s42979-020-00394-7
34. Li, Y., Hai-Tao, Z., Jorge, G.: A machine learning-based model for survival prediction in patients with severe COVID-19 infection. medRxiv (2020). https://doi.org/10.1101/2020.02.27.20028027
35. James, G., Witten, D., Hastie, T., Tibshirani, R.: Statistical learning. In: An Introduction to Statistical Learning. Springer Texts in Statistics, vol. 103, 15–57. Springer, New York (2013)
36. Siirtola, P., Roning, J.: Comparison of regression and classification models for user independent and personal stress detection. Sensors **20**, 4402 (2020)
37. Coulet, A., Chawki, M., Jay, N., Shah, N., Wack, M., Dumontier, M.: Predicting the need for a reduced drug dose, at first prescription. Sci. Rep. **8**(1), 1–11 (2018). https://doi.org/10.1038/s41598-018-33980-0
38. Nguyen, D., et al.: A feasibility study for predicting optimal radiation therapy dose distributions of prostate cancer patients from patient anatomy using deep learning. Sci. Rep. **9**(1), 1–10 (2019). https://doi.org/10.1038/s41598-018-37741-x
39. Lalmuanawma, S., Hussain, J., Chhakchhuak, L.: Applications of machine learning and artificial intelligence for Covid-19 (SARS-CoV-2) pandemic: a review. Chaos Solit. Fractals **130**(1), 110059 (2020). https://doi.org/10.1016/j.chaos.2020.110059
40. Pham, Q., Nguyen, D.C., Huynh-The, T., Hwang, W., Pathirana, P.N.: Artificial intelligence (AI) and big data for coronavirus (COVID-19) pandemic: a survey on the state-of-the-arts. IEEE Access **8**, 130820–130839 (2020). https://doi.org/10.1109/ACCESS.2020.3009328
41. Ardakani, A.A., Kanafi, A., Acharya, U.R., Khadem, N., Mohammadi, A.: Application of deep learning technique to manage COVID-19 in routine clinical practice using CT images: results of 10 convolutional neural networks. Comput. Biol. Med. **121**, 103795 (2020). https://doi.org/10.1016/j.compbiomed.2020.103795
42. Ozturk, T., Talo, M., Yildirim, E.A., Baloglu, U.B., Yildirim, O., Rajendra Acharya, U.: Automated detection of COVID-19 cases using deep neural networks with x-ray images. Comput. Biol. Med. (2020). https://doi.org/10.1016/j.compbiomed.2020.103792
43. Sun, L., et al.: Combination of four clinical indicators predicts the severe/critical symptom of patients infected COVID-19. J. Clin. Virol. (2020). https://doi.org/10.1016/j.jcv.2020.104431

44. Wu, J., et al.: Rapid and accurate identification of COVID-19 infection through machine learning based on clinical available blood test results. medRxiv (2020). https://doi.org/10.1101/2020.04.02.20051136
45. Sharma, R., Singh, S.N.: Data mining classification techniques - comparison for better accuracy in prediction of cardiovascular disease. Int. J. Data Anal. Tech. Strategies **11**(4), 356–373 (2019)
46. Sadrfaridpour, E., Razzaghi, T., Safro, I.: Engineering fast multilevel support vector machines. Mach. Learn. **108**(11), 1879–1917 (2019). https://doi.org/10.1007/s10994-019-05800-7
47. Chiroma, H., et al.: Progress on artificial neural networks for big data analytics: a survey. IEEE Access **7**, 70535–70551 (2019). https://doi.org/10.1109/access.2018.2880694
48. Deng, Z., Zhu, X., Cheng, D., Zong, M., Zhang, S.: Efficient kNN classification algorithm for big data. Neurocomputing **195**, 143–148 (2016). https://doi.org/10.1016/j.neucom.2015.08.112
49. Xing, W., Bei, Y.: Medical health big data classification based on kNN classification algorithm. IEEE Access **8**, 28808–28819 (2020). https://doi.org/10.1109/ACCESS.2019.2955754
50. Djafri, L., Amar-Bensaber, D., Adjoudj, R.: Big data analytics for prediction: parallel processing of the big learning base with the possibility of improving the final result of the prediction. Inf. Discov. Deliv. **46**(3), 147–160 (2018). https://doi.org/10.1108/IDD-02-2018-0002
51. Dhamodharavadhani, S., Rathipriya, R.: Enhanced-logistic-regression-(ELR)-model-for-big-data. IGI Global (2019). https://doi.org/10.4018/978-1-7998-0106-1.ch008
52. Scutari, M., Vitolo, C., Tucker, A.: Learning Bayesian networks from big data with greedy search: computational complexity and efficient implementation. Stat. Comput. **29**(5), 1095–1108 (2019). https://doi.org/10.1007/s11222-019-09857-1
53. Fengying, M., Zhang, J., Liang, W., Xue, J.: Automated classification of atrial fibrillation using artificial neural network for wearable devices. Math. Probl. Eng. (2020). Article ID 9159158. https://doi.org/10.1155/2020/9159158
54. Miao, J., Zhu, W.: Precision-recall curve (PRC) classification trees. arXiv:201107640v1 [stat.ML] (2020)
55. Naseem, R., et al.: Performance assessment of classification algorithms on early detection of liver syndrome. J. Healthc. Eng. (2020). Article ID 6680002. https://doi.org/10.1155/2020/6680002
56. Eedi, H., Kolla, M.: Machine learning approaches for healthcare data analysis. J. Crit. Rev. **7**(4), 806–811 (2020). ISSN 2394-5125
57. Rustam, F., Mehmood, A., Ahmad, M., Ullah, S., Khan, D.M., Sang Choi, G.: Classification of shopify app user reviews using novel multi text features. IEEE Access **8**, 30234–30244 (2020). https://doi.org/10.1109/ACCESS.2020.2972632
58. Lamurias, A., Jesus, S., Neveu, V., Salek, R.M., Couto, F.M.: Information retrieval using machine learning for biomarker curation in the exposome-explorer. bioRxiv (2020). https://doi.org/10.1101/2020.12.20.423685
59. Zhang, X., Saleh, H., Younis, E.M.G., Sahal, R., Ali, A.A.: Predicting coronavirus pandemic in real-time using machine learning and big data streaming system. Complexity, Article ID 6688912 (2020). https://doi.org/10.1155/2020/6688912
60. Ghori, K.M., Imran, M., Nawaz, A., Abbasi, R.A., Ullah, A., Szathmary, L.: Performance analysis of machine learning classifiers for non-technical loss detection. J. Ambient Intell. Human. Comput. (2020). https://doi.org/10.1007/s12652-019-01649-9

61. Hanafy, M., Ming, R.: Machine learning approaches for auto insurance big data. Risks **9**, 42 (2021). https://doi.org/10.3390/risks9020042
62. Muhammad, Y., Tahir, M., Hayat, M., Chong, K.: Early and accurate detection and diagnosis of heart disease using intelligent computational Model. Sci. Rep. **10**, 19747 (2020). https://doi.org/10.1038/s41598-020-76635-9

Digital Text Authentication Using Deep Learning: Proposition for the Digital Quranic Text

Zineb Touati-Hamad(✉), Mohamed Ridda Laouar, and Issam Bendib

Laboratory of Mathematics, Informatics and Systems (LAMIS), University Larbi Tebessi, Tebessa, Algeria

{zineb.touatihamad,ridda.laouar,issam.bendib}@univ-tebessa.dz

Abstract. Nowadays, the detection of digital texts manipulation is a hot topic in the field of natural language processing and artificial intelligence. This type of text spreads quickly and inexpensively, which can cause great concern due to its negative impact on social life. The text authentication process has gained a great deal of interest. However, the authentication of Arabic texts is still under development. The Quranic text is one of the Arabic texts sensitive to change and the most vulnerable to falsification at all. In order to prevent misuse of this type of texts, in this research a deep learning approach based on LSTM network and the pretrained Word Embeddings has been developed for authentication one of the manipulations types of the Arabic Quranic texts. By building a model that enables Internet users to automatically validate the arrangement of the Quran content, the experimental results showed that the proposed approach can effectively improve the accuracy of text classification and achieve a significant time difference compared to previous works.

Keywords: Deep learning · LSTM · Natural language processing · Artificial intelligence · Authentication · Integrity · Quranic text · Arrangement

1 Introduction

Tamper text has been a classic problem since the advent of the Internet. With the rapid development of digital websites, publishing texts, including news, information, messages and citations have become a double-edged sword. False texts spread without censorship, and negatively affect the credibility of the sites and their content. Detecting text manipulation is a difficult and complex task [1]. Recent advances in artificial intelligence, including machine learning and deep learning techniques, facilitate the process of word processing and classifying it automatically and in real time by relying on training neural networks based on a group of data. Deep learning-based paradigms have surpassed traditional machine learning-based methods for various text classification tasks [2]. In this paper, we present a study of various applications of deep learning in the detection of the most common forms of text manipulation and identify the way these methods work and

B. Lejdel et al. (Eds.): AIAP 2021, LNNS 413, pp. 56–64, 2022.
https://doi.org/10.1007/978-3-030-96311-8_6

the possibility of applying them to texts of a sensitive nature such as the Holy Quranic text.

The rest of the paper is organized as follows: The second section presents the repeated works. The third section introduces the background. The fourth section provides the proposed methodology, and we discuss the experiments in the fifth section. Finally, we conclude this paper with its conclusion and future work.

2 Related Works

Recently, there is a lot of research that seeks to delve into the field of natural language processing. One of the most important areas of research is the process of detection and authentication certain texts, especially if they are of a negative nature, such as spam, fake news and paraphrasing. Recent studies [3, 4] have shown that it is effective to use deep neural networks in NLP research on text classification, due to its accurate results in texts understanding and analysis.

In this section, we present the four most common applications of deep neural networks for identifying manipulation of different text types.

2.1 Spam Detection

Technological development contributed to the development of electronic attack methods. Spam is the most common electronic crime that seeks to deceive victims in various ways. Deep learning approaches have been used in many researches such as [5] to solve this problem through the process of filtering messages/mail based on their content or source.

2.2 Fake News Detection

The main purpose of this type of data is to mislead opinion and spread deception. Deep learning approaches such as [6–9] seek to reduce the problem of spreading these lies, especially with the widespread adoption of communication sites as digital media by Internet users.

2.3 Paraphrase Detection

Paraphrasing can be discovered in each of the tasks of plagiarism, translation, summarization, and more. Identifying sentences that are similar in meaning or form is a very important matter that preserves the author's authentication and limits the methods of distortion. Traditional reformatting detection systems require a lot of extra effort, such as cleaning texts. These challenges were recently overcome with the deep neural network approach in the work of [10, 11], where the results showed impressive performance without the need for language experience.

2.4 Language Checking

Whether it is proofreading or spell checking, its aim is to discover various linguistic errors, as well as correct them. Based on the strengths of deep learning and its flexibility in dealing with errors such as repetition or non-idiomatic wording [12], it has been adopted in proofreading and spelling applications for many languages including Arabic [13], Malayalam [14], Hindi [15], Indonesian [16] and others.

The authentication of the English text has achieved good efficiency and validity and has reached high classification accuracy. It became more difficult with the methods of documenting the Arabic text. The Quranic text is considered one of the finest Arabic texts and is distinguished by the diacritical symbols, sequence of contents, and extreme sensitivity to change.

Deep learning algorithms can be applied in dealing with this text in the field of detecting distortion, in a more precise sense, authentication of verses and surahs. Some deep learning algorithms specialize only in dealing with sequences of texts. However, its applications with Quranic text sequences are still in their infancy. Therefore, in this paper we present an approach based on deep learning to authenticate the order of the content of the Holy Quran.

3 Background

Deep learning is a science based on neural networks that aims to process data with a high level of abstraction [17]. Deep learning algorithms have significantly improved in text recognition, analysis, classification, and so on. In general, data first goes through the preprocessing stage in order to filter it, followed by the data representation stage to be ready for the input stage for a specific type of deep learning algorithm for training.

3.1 Text Preprocessing

At this stage, the useful information is extracted from the text data. text units are first processed by dividing texts into words in the tokenization process. The results of this process allow the application of several other procedures to words such as deleting symbols, marks, punctuation and stop words. Some applications also require word abbreviation through the stemming process [18] or delimit word fragments through the lemmatization process [19]. The order of using these processes may change as needed, and they are sometimes dispensed with in some applications.

3.2 Text Representation

The goal of this stage is to encode the data so that it can be readable by the algorithms, each encoding is actually extractive features of the corresponding data.

There are many methods of data representation [20], including the traditional methods such as the Bag of word (BOW) and the term frequency-inverse document frequency (TF-IDF), which rely on encoding the word based on the number of times it appears in the text. More advanced methods work to preserve semantic relationships between words are known as Word Embeddings, including the word2vec model [21] that works on the representation of Words with close context in one space. A modified version of this model that works on paragraph level instead of words is known as Doc2vec [22]. Glove (Global Vectors) [23] is also an embedding model that fully exploits global statistical information on word frequency. Each of the previous models fails to provide a vector representation of words that has never appeared in the dictionary before. FastText [24] model was developed to incorporate this feature. The previous models produced a fixed vector for the word without taking into account the context in which the word was used. This deficiency was addressed with the ELMo (ELMomethod) [25] model.

3.3 Deep Learning Algorithms

In this section, we introduce the most popular deep learning algorithms used in natural language processing. Each algorithm has advantages and weaknesses. The type of algorithm is chosen according to the nature of the application, and it is also possible to combine more than one algorithm in one application.

Convolutional Neural Networks (CNN). CNN is a well-known network in deep learning, which is commonly used in image processing. However, the 1D-CNN version showed strong performance in word processing tasks. This network consists of several layers, including: the embedding layer that serves to represent the data in one-dimensional arrays, followed by the convolution layer, where a one-way filter is applied to extract features map, then a Pooling layer to preserve the important features [26].

Repetitive Neural Networks (RNN). RNN is a neural network architecture that specializes in word processing and sequential data. In this case, the neural network looks at the previous node information to assign them more weights, for better semantic analysis of the structures in the dataset [27].

Long-Term Memory (LSTM). LSTM is a special type of RNN that addresses the vanishing gradient problem and maintains long-term dependencies. LSTM uses gates to carefully organize the amount of information that will be allowed in each node state. In addition, the bidirectional network "Bi-LSTM" allows for back and forward information about the sequence at each time step to be obtained [27].

4 Proposed Methodology

Our goal is to build a system capable of authenticate the verses of the Holy Quran. Due to the many forms of distortion of the digital Quran texts, cases of distortion at the level of word order represent one of the most famous methods of distortion frequently used in the field of magic and sorcery. Moreover, it is often difficult to discover errors of order even for memorizers of the Holy Quran. Therefore, we are interested in this study to determine which verses are respected for arranging the words of the Quran as revealed in the Mushaf Al-Quran.

We begin our discussion by introducing the dataset. Next, we discuss the process of data representation. Finally, we present the proposed classification model.

4.1 Dataset

In this work we use the dataset built in [28] as a CSV file. The data set consists of 78,248 sentences for each category with a length of five words for each sentence and a tag indicating one of the balanced categories. These categories are restricted to three possibilities:

- **Ordered Quranic sentences:** This group represents the correct category of the Quran that collects the correct sequence of words.
- **Unordered Quranic sentences:** This group represents one of the forms of manipulation at the level of arrangement, which can be caused by chance / mockery through a random arrangement of Arabic words that gives a mixture of the Quran.
- **Inversed Quranic sentences:** Represent another kind of manipulation. It may be due to devices that do not support Arabic language, in this case the texts are displayed from left to right. Or it may be the result of setting in reverse 'Tankis' the Quran words.

4.2 Data Representation

In the beginning, we use the one-hot vector method to represent all the words of the Quran (14,870 words) with one hot vector bearing the number 1 which indicates the presence of the word, or the number 0 which indicates the absence of the word in the sentence. Next, we inject all five vectors sequentially as an input to the embedding layer to learn the weights and reduce the order from 14870 to 300 long numerical vectors.

4.3 Classification Model

There are a variety of text classification models. Each model has strengths and weaknesses that vary according to the type of text used. Regarding the Quranic text, the drawback of traditional methods is manual feature extraction, and also gives poor classification accuracy with sequential strings. The one-dimensional convolutional neural network (CNN-1D) can extract features automatically but in return, it loses context dependencies. In contrast, the LSTM network excels in maintaining the context of sequence words but the training is time-consuming. To take advantage of the power of these models, we propose to build a hybrid CNN-LSTM model for the Quranic verses class. We first represent the words as a vector using the organized layer and then apply a CNN to extract the features and then send them to LSTM to maintain the verses dependencies, and finally, we use the softmax activation function for classification.

The proposed hybrid deep learning model is implemented using the sequential model of the Python deep learning library Keras. The serial model consists of several layers. The embedding layer is the first layer, which produces an output of length 300. The output of the embedding layer is fed to a one-dimensional CNN (Conv1D) layer to extract local features using 32 filters of size 5 and using the corrected linear unit (ReLU) default activation function. The large feature vectors generated by the CNN are then aggregated by inserting them into the MaxPooling1D layer with a window size of 2. The aggregated feature maps are then fed into a LSTM layer which outputs the long-term dependent features of the input feature maps, while retaining memory. The output dimension is set to 64. Finally, the trained feature vectors are labeled using a dense layer that shrinks the output space dimension to 3, which corresponds to the label of classification (i.e., ordered, unordered or inversed). This layer implements the Softmax activation function.

5 Experiments

The work has been done on a Lenovo i7 8th Generation, RAM 8 Go, Hard disk: SSD 512 Go, touring on Windows 10.

5.1 Evaluation Measures

To measure the quality of the classification system, we use a confusion matrix to calculate Accuracy, Precision, Recall and F1-Score based on predictions of true positives (TP), false positives (FP), true negatives (TN), and false negatives (FN).

- **Accuracy:** is the number of Quranic texts that were correctly predicted divided by the total number of predicted Quranic texts.

$$Accuracy = \frac{TP + TN}{TP + TN + FP + FN} \tag{1}$$

- **Recall:** It is the percentage of actual positives (ordered/unordered/reversed) that are correctly classified.

$$Recall = \frac{TP}{TP + FN} \tag{2}$$

- **Precision:** is the percentage of positive predictions that are truly positives.

$$Precision = \frac{TP}{TP + FP} \tag{3}$$

- **F1-Score:** is the harmonic mean of precision and recall

$$F1\text{-}Score = \frac{2 * Precision * Recall}{Precision + Recall} \tag{4}$$

5.2 Experimental Results

After training the proposed model for 50 epochs and using a batch size of 64. We tested the model on 20% of the total dataset containing equal values for each category, we obtained a test accuracy of 99.9808%.

In Fig. 1 and Table 1, we show the results found for the proposed CNN-LSTM model in this work and the proposed LSTM model in the work of [28]. Starting with accuracy, the CNN-LSTM hybrid model was the best with a value of 99.9808%. For Precision,

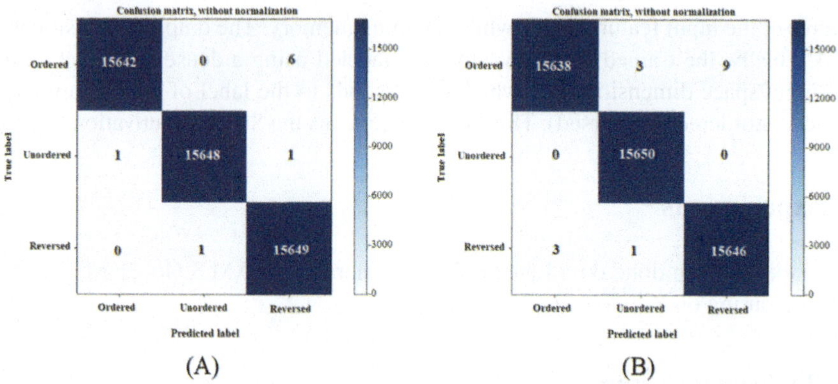

Fig. 1. (A): Confusion matrix of hybrid CNN-LSTM model, (B) Confusion matrix of LSTM model

Table 1. Comparison results

Model	Accuracy	Precision	Recall	F1 score
LSTM [28]	99.9800%	99.9633%	99.9633%	99.9600%
CNN-LSTM	**99.9808%**	**99.9666%**	**99.9633%**	**99.9600%**

CNN-LSTM gave the best score with a value of 99.9666%. Regarding the Recall, CNN-LSTM and LSTM share the best result with a value equal to 99.9633%. Likewise, for F1-Score, CNN-LSTM and LSTM share the best score with a value equal to 99.9600%.

Finally, we conclude that the hybrid model based on CNN and LSTM that we pr posed in this paper outperformed the LSTM model by giving the best results, and also recording less training time. These results confirm that the hybrid model represents a very promising solution for creating effective systems for classifying Quranic verses and detecting their manipulation.

6 Conclusion and Future Works

This paper aims to enhance the applications of deep learning in the field of the Arabic language. By taking the digitized Quranic text as a case study, we propose to address the problem of authenticating the integrity of the Quranic content arrangement. In order to take advantage of the advantages of deep learning models, we suggested incorporating the text classification method based on the CNN-LSTM hybrid model through an empirical comparative study.

Compared with LSTM, the hybrid model has the advantage of better extracting text context dependencies, improving accuracy, improving text classification performance, and reducing training time.

Based on this study, some future research can be suggested:

- Optimizing the datasets to test the extent of CNN-LSTM in enhancing the accuracy of Quranic text classification.
- At the same time, explore other deep learning techniques to classify the Quranic text.

Acknowledgment. This work was supported by the Algerian General Direction of Scientific Research and Technological Development (DGRSDT) and the Laboratory of Mathematics and Informatics System (LAMIS) of Tebessa university.

References

1. Ahmed, H., Traore, I., Saad, S.: Detecting opinion spams and fake news using text classification. Secur. Priv. **1**, e9 (2018)
2. Minaee, S., Kalchbrenner, N., Cambria, E., Nikzad, N., Chenaghlu, M., Gao, J.: Deep learning–based text classification: a comprehensive review. ACM Comput. Surv. CSUR **54**, 1–40 (2021)
3. Devlin, J., Chang, M.-W., Lee, K., Toutanova, K.: BERT: pre-training of deep bidirectional transformers for language understanding. ArXiv preprint arXiv:1810.04805 (2018)
4. Liu, X., He, P., Chen, W., Gao, J.: Multi-task deep neural networks for natural language understanding. ArXiv preprint arXiv:1901.11504 (2019)
5. Srinivasan, S., Ravi, V., Alazab, M., Ketha, S., Al-Zoubi, A., Kotti Padannayil, S.: Spam emails detection based on distributed word embedding with deep learning. In: Maleh, Y., Shojafar, M., Alazab, M., Baddi, Y. (eds.) Machine Intelligence and Big Data Analytics for Cybersecurity Applications. SCI, vol. 919, pp. 161–189. Springer, Cham (2021). https://doi.org/10.1007/978-3-030-57024-8_7
6. Nasir, J.A., Khan, O.S., Varlamis, I.: Fake news detection: a hybrid CNN-RNN based deep learning approach. Int. J. Inf. Manag. Data Insights. **1**, 100007 (2021)
7. Chokshi, A., Mathew, R.: Deep learning and natural language processing for fake news detection: a survey. Available SSRN 3769884 (2021)
8. Kaliyar, R.K., Goswami, A., Narang, P.: DeepFakE: improving fake news detection using tensor decomposition-based deep neural network. J. Supercomput. **77**(2), 1015–1037 (2020). https://doi.org/10.1007/s11227-020-03294-y
9. Kong, S.H., Tan, L.M., Gan, K.H., Samsudin, N.H.: Fake news detection using deep learning. In: 2020 IEEE 10th Symposium on Computer Applications & Industrial Electronics (ISCAIE), pp. 102–107. IEEE (2020)
10. Shahmohammadi, H., Dezfoulian, M., Mansoorizadeh, M.: Paraphrase detection using LSTM networks and handcrafted features. Multimed. Tools Appl. **80**(4), 6479–6492 (2020). https://doi.org/10.1007/s11042-020-09996-y
11. Agarwal, B., Ramampiaro, H., Langseth, H., Ruocco, M.: A deep network model for paraphrase detection in short text messages. Inf. Process. Manag. **54**, 922–937 (2018)
12. Xie, Z., Avati, A., Arivazhagan, N., Jurafsky, D., Ng, A.Y.: Neural language correction with character-based attention. ArXiv preprint arXiv:1603.09727 (2016)
13. Alkhatib, M., Monem, A.A., Shaalan, K.: Deep learning for Arabic error detection and correction. ACM Trans. Asian Low-Resour. Lang. Inf. Process. TALLIP **19**, 1–13 (2020)
14. Sooraj, S., Manjusha, K., Anand Kumar, M., Soman, K.P.: Deep learning based spell checker for Malayalam language. J. Intell. Fuzzy Syst. **34**, 1427–1434 (2018)
15. Singh, S., Singh, S.: HINDIA: a deep-learning-based model for spell-checking of Hindi language. Neural Comput. Appl. **33**(8), 3825–3840 (2020). https://doi.org/10.1007/s00521-020-05207-9

16. Zaky, D., Romadhony, A.: An LSTM-based Spell Checker for Indonesian Text. In: 2019 International Conference of Advanced Informatics: Concepts, Theory and Applications (ICAICTA), pp. 1–6. IEEE (2019)
17. LeCun, Y., Bengio, Y., Hinton, G.: Deep learning. Nature **521**, 436–444 (2015)
18. Jivani, A.G.: A comparative study of stemming algorithms. Int. J. Comp. Tech. Appl. **2**, 1930–1938 (2011)
19. Plisson, J., Lavrac, N., Mladenic, D.: A rule based approach to word lemmatization. In: Proceedings of IS, pp. 83–86 (2004)
20. Touati-Hamad, Z., Laouar, M.R., Bendib, I.: Quran content representation in NLP. In: Proceedings of the 10th International Conference on Information Systems and Technologies, pp. 1–6 (2020)
21. Mikolov, T., Chen, K., Corrado, G., Dean, J.: Efficient estimation of word representations in vector space. ArXiv preprint arXiv:13013781 (2013)
22. Le, Q., Mikolov, T.: Distributed representations of sentences and documents. In: International Conference on Machine Learning, pp. 1188–1196. PMLR (2014)
23. Pennington, J., Socher, R., Manning, C.D.: GloVe: global vectors for word representation. In: Proceedings of the 2014 Conference on Empirical Methods in Natural Language Processing (EMNLP), pp. 1532–1543 (2014)
24. Joulin, A., Grave, E., Bojanowski, P., Mikolov, T.: Bag of tricks for efficient text classification. ArXiv preprint arXiv:1607.01759 (2016)
25. Peters, M.E., et al.: Deep contextualized word representations. ArXiv preprint arXiv:1802.05365 (2018)
26. Jacovi, A., Shalom, O.S., Goldberg, Y.: Understanding convolutional neural networks for text classification. ArXiv preprint arXiv:1809.08037 (2018)
27. Jozefowicz, R., Zaremba, W., Sutskever, I.: An empirical exploration of recurrent network architectures. In: International Conference on Machine Learning, pp. 2342–2350. PMLR (2015)
28. Touati-Hamad, Z., Laouar, M.R., Bendib, I.: Authentication of Quran verses sequences using deep learning. In: 2021 International Conference on Recent Advances in Mathematics and Informatics (ICRAMI), pp. 1–4. IEEE (2021)

Prediction of Cancer Clinical Endpoints Using Deep Learning and RPPA Data

Imene Zenbout[1,2,3(✉)], Abdelkrim Bouramoul[1], and Souham Meshoul[4]

[1] Fundamental Informatics and its Applications Department, Misc Laboratory, Faculty NTIC, University Abdelhamid Mehri - Constantine 2, Constantine, Algeria
imene.zenbout@constantine2.dz, abdelkrim.bouramoul@univ-constantine2.dz
[2] National Center for Biotechnology Research, Constantine, Algeria
[3] Research Center for Scientific and Technical Information, Algiers, Algeria
[4] IT Department, CCIS-RC, Princess Nourah Bint Abdulrahman University, Riyadh, Kingdom of Saudi Arabia
sbmeshoul@pnu.edu.sa

Abstract. Whithin the advances in highthrouput technologies, handling vast and various cancer omic data requires more accurate and felexible models to either achieve a precise clinical decision or to discover new and relevant diagnostic, prognostic and theurapeutic genes. Reverse protein phase array (RPPA) data are considered to be more stable than gene expression data and contain less noisy inputs. The analysis of these type of data may help in accurately classify cancer types or predict more precise clinical and pathological endpoints. In this paper we construct a computational framework that combines deep learning and biological knowledge, to predict clinical cancer related outcomes and extract a set of discriminative features in cancer classification. The framework have been experimented on different cancer data sets and the results shows promising performance of deep learning and RPPA data in classifying predicting cancer pathological stage, progression free interval, and overall survival.

Keywords: Cancer classification · Reverse protein phase · Deep learning · Protein-protein interaction network

1 Introduction

The remarkable advances in high throughput technologies also came with a tremendous advances in measuring protein expression using the Reverse Phase Protein Array, or called RPPA. RPPA is a sensitive and powerful functional proteomic approach that captures cancer related molecular mechanism and aid in the development of therapies [6]. It allows the monitoring of hundreds to thousands of samples simultaneously, in order to the quantitative measuring of protein expression [12]. Previous research explored genomic and transcriptomic impact on cancer diagnosis yet these research mainly aid on a research level and on the

B. Lejdel et al. (Eds.): AIAP 2021, LNNS 413, pp. 65–74, 2022.
https://doi.org/10.1007/978-3-030-96311-8_7

monitoring of patients yet the clinical impact was weak and limited in comparison with the needs of targeted therapy [9], also the transcriptomic measurement of next generation sequencing came with the curse of small patient sample and huge genomic expression, which complicated the analysis and exploration of these data [3]. In contrast proteomic measuring is more efficient in capturing biological process [9] as well as handling RPPA data is easier because of the low dimensionaliy in comparison with transcriptomic data. The research on the impact of RPPA on cancer was mainly explored by medical community like the work of Mari et al. [9], where the authors presented a detailed research on the impact of RPPA on precision oncology. Also Mari et al. [8], explained the signal pathway profiling using RPPA and its clinical application. In cancer classification RPPA have been used to classify breast cancer in the paper of Negm O et al. [11]. Zhang et al. [8] used RPPA data set to classify ten most known cancer type, where the authors selected the most relevant 23 proteins in cancer type classification. Deep learning and machine learning have been used in the context of cancer classification mainly on transcriptomic dataset such us the use of adaptive neural fuzzy inference network [1,5], and autoencoders [4] for multiomic data analysis. Our work falls in the range of the first papers that explores the impact of RPPA data on targeting diagnostic endpoints in association with biological protein-protein interaction network (PPI), using a deep learning model. Where we used autoencoders for there relevancy in cancer research, mainly in omic data integration, gene expression analysis, and cancer type classification [4,7,13], As well as in cancer stage prediction [14]. The used autoencoders were trained by a set of proteins extracted from the PPI and, in order to map the RPPA features space into a reduced representation, that further used in training classifiers in predicting cancer clinical and pathological endpoints. The architecture was experimented on the PanCancer Atlas data on cancer pathological stage, progression free interval (PFI), and Overall Survival (OS).

The rest of the paper is structured as follows: Sect. 2 explains the architecture and details its steps. The experimental results were conducted and explained in Sect. 3. Finally we concluded our paper in Sect. 4 with overall overview and perspectives.

2 Proposed Architecture

Our predictive model consists of four phases (Fig. 1), the first phase is the data collection and preparation then we applied a proteins' filtration, where we select only proteins that appears in the PPI network from the string database. The third phase is a feature learning phase, in which we train a deep autoencoder to map the expression of the filtered set of proteins into a reduced new features space, After training the DAE, we pass to the last phase in order to train a classifier based on the learned feature space along with the corresponding endpoint.

2.1 Data Collection

From The Cancer Genome Atlas, we have collected the TCGA-RPPA pancancer data set along with the patients clinical outcomes. The pancancer project represent a set of cancer samples collected from different patients all around the US

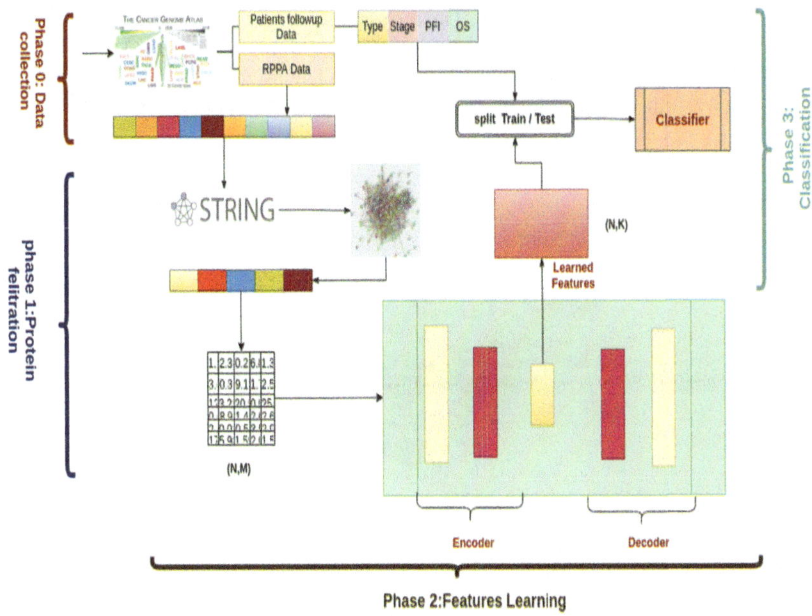

Fig. 1. Illustration of the proposed architecture

hospitals for more than 30 cancer type including some rare types [10]. Reverse phase protein array was used to measure the RPPA data set, were the experiments was applied on 7790 patient's sample and, and 199 protein. As for the patients followup data set it contains the clinical, pathological and all the follow up information of 11160 patient.

From the followup data set we defined three endpoints as classification targets namely:

- pathological type, which contain four stages and each stage is divided into sub-stages, in our case we adressed all substages as the global stage i.e. all sub-stages of stage 1 are considered as stage 1 cancer patient.
  Progression free interval (PFI) and overall survival (OS), where both were addressed as binary endpoints (0/1)

For each endpoint we extracted the set of patients barcode that have an available endpoint value, then, we concatenated the list of patients' barcode with the RPPA matrix in order to construct three expression matrix $S(N\_1 \times M)$, for stage prediction, $P(N\_2 \times M)$ for PFI prediction and $O(N\_3 \times M)$ for OS prediction. Where M is the list of expressed proteins which is initially 199, and N_* is the number of patients with available target value. Table 1 exhibits the datasets in numbers.

Table 1. Data description - 1

Data	Protein expression				Follow up data			
	Initial Data	Stage	PFI	OS	Initial Data	Stage	PFI	OS
Patients	7790	4954	7769	7677	11160	4954	7769	7677
Proteins	199				/			

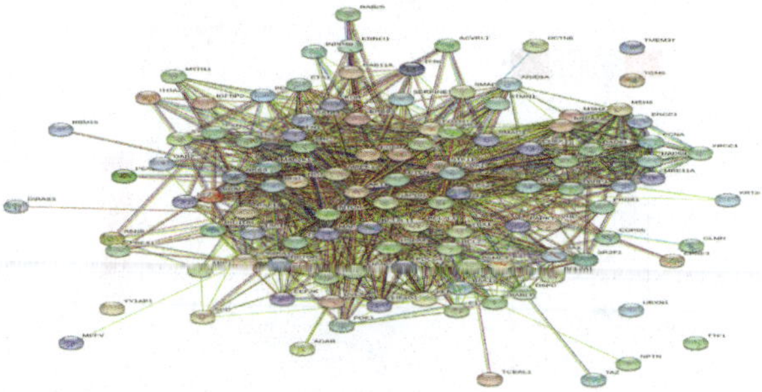

Fig. 2. Initial protein-protein network

2.2 PPI and Features Filtration

From the RPPA data sets we collected the set of proteins and constructed the protein-protein network (Fig. 2) using String tools, We downloaded the mapping matrix and the interaction-score matrix. Then, we merged the two matrices by the protein identifier and extracted the list $IM(K)$ of the interacted proteins i.e. proteins that posses an interaction with other proteins. This phase reduced the set of proteins from 199 to 97 interacted protein. After defining the list of most interacted proteins we used the matrices S, P, O, to drop the proteins expression that does not appear in IM, which led to re-scale the matrices dimension from (N_*, M) to (N_*, K). After the Construction of the matrices using in training our models, we moved to a preprocessing phase, where we replaced the missing protein expressions by using a KNN imputation and then we normalized the data set using log transformation. After we split the data sets 80% for training and 20% for testing. Table 2 shows the data statistics.

2.3 Deep Features Learning

The tackled features learning problem can be formulated as follows: given a matrix $P(N \times M)$, where vector p_{ij} of P represents the value of an expressed protein j for a sample i. Since deep features learning falls in the category of unsupervised learning, at this phase the samples' classes were ignored. Taking in count the matrix P dimensions, the deep autoencoder takes $P(X \times M)$ as an input and transform it successively through the encoder E (Eq. 1) layers into a reduced features representation $P1$ of range K, where $K < M$ at the bottleneck layer.

Table 2. Data description - 2

	Stage		PFI		OS	
Train	3963		6143		6141	
Test	991		1536		1536	
Endpoints	**Stage 1**	1409				
	Stage 2	1638	**Class 0**	4994	**Class 0**	5295
	Stage 3	1333	**Class 1**	2685	**Class 1**	2382
	Stage 4	574				

$$Encoder(P, \Phi_E) = P_1; P \in \mathbf{R}^{N\times M}, P_1 \in \mathbf{R}^{K\times M} \tag{1}$$

Φ_E is the encoder parameters responsible of transforming P into $P1$, the latter represents the linear transformation that contains all the necessary information of P. Following the autoencoder architecture The input P is later reconstructed into P' using the decoder$D$ output (Eq. 2)

$$Decoder(P_1, \Phi_D) = P'; P' \in \mathbf{R}^{N\times M}, P' \approx P. \tag{2}$$

Although, the decoder output is a primordial part in autouencoder, yet in our case it is not the output P' that matters but the learned features $P1$ space by the encoder. So, in order to assist the consistency of the $P1$ and the whole auto encoder performance in mapping P to a reduced features space, we evaluated the performance of $P1$ in reconstructing P into P' through the reconstruction loss error. In which we calculate the degree of similarity between the input P and the decoder D output P'. In our case we used the $mean\_absolute\_error(mae)$ (Eq. 3)

$$mae = 1/n \sum_{i=1}^{n} |y_i - y_i'| \tag{3}$$

As a result the smaller the loss is the more the AE architecture is capable to generate consistent reduced features space $P1$. So, the objective of training this autoencoder is to minimize the loss (Eq. 4) between P and P';

$$Loss(P, \Phi_D(\Phi_E(P))) = 1/n \sum_{i=1}^{n} |P - P'| \tag{4}$$

$$P' = \Phi_D(\Phi_E(P).$$

2.4 Classification

After training the autoencoder to map the input P into an output P' with a low loss score, we used the trained model to map our input data P into the reduced space data P'. Then using cross validation we split the data set into training and testing set (80%, 20% respectively). After we built a support vector

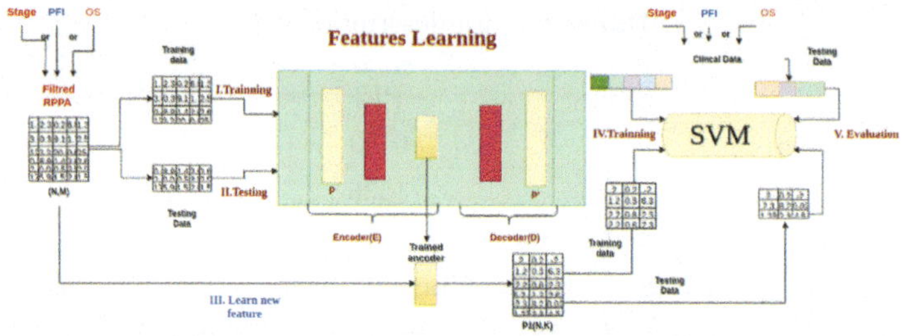

Fig. 3. Features learning and classification architecture

machine classifier and used the train and test data set to train and evaluate the svm perfomance as well as the performance of the features learned from the previous phases (PPI filtration and dimensionality reduction).

3 Results and Discussion

In order to evaluate the performance of the cancer related end points prediction we built three different instances of the proposed architecture *AE+SVM*, one for predicting stage of cancer, one for PFI score and the other for the OS score (Fig. 3). The experiments were conducted on a hp laptop with Intel Core i7-7500U CPU @ 2.70 GHz $\times$ 4, with ubuntu 18.04 operating system. We used the Keras [2] package to implement the autoencoder architecture.

3.1 Implement and Train the Features Learning Model

We built three features learning model each for a specific end point the encoder E is constructed of an input layer (96 nodes), two hidden layers (40, 30 nodes) and the bottleneck (20 nodes) layer that represent the output of the encoder D that will be in charge of transforming the data into a reduced features representation $P1$, the decoder D takes as input $P1$ and is built symmetrically to the encoder with two hidden layers and an output layer responsible of reconstructing the input P into P'. We used $softplus$ activation function to setup the layers weights and we trained the models using $adamax$ optimizer and batch training. After a set of training we set the architecture's parameters as illustrated in Table 3.

We trained the three instances in the follow scenarios:

- **Stage:** We trained the autoencoder on two rounds the first round we trained the model for 400 epochs using a batch size of 120 with a learning rate (lr) of 5^{-5}. As shown in Fig. 4, we notice that the model is training without overfitting. Then we reset the optimizer' lr to 1^{-4} and we trained the model again for 100 epochs, which dropped the loss value to 0.5.

Table 3. Architecture Parameters setting

	Stage	PFI	OS
Architecture	(96,40,30,20,40,96)		
Activation function	Softplus		
Optimizer	adamax		
Batch size	120		
Learning rate	**round 1:** 5^{-3} **round 2:** 1^{-4}	**round 1:** 5^{-4} **round 2:** 1^{-4}	**round 1:** 5^{-3} **round 2:** 1^{-4}
Epochs	**round 1:** 400 **round 2:** 100		
Loss	**Train:** 0.5864 **Test:** 0.5992	**Train:** 0.636 **Test:** 0.6371	**Train:** 0.5580 **Test:** 0.5741

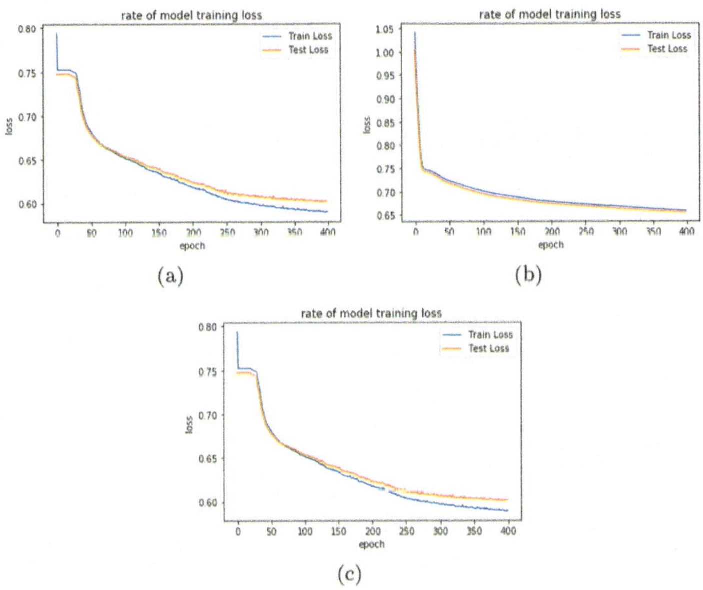

Fig. 4. Training performance based on reconstruction loss, (a): Stage, (b): PFI, (c): OS

– **PFI/OS:** In the same way we trained the autoencoder in the first round for 400 on batch size equals to 120 with a lr equals to 5^{-4} for PFI and 5^{-3} for OS. The loss training values shows that the model is training without an unnoticeable overfitting. After we reset the optimizer' lr to 1^{-4}, and we scored a loss score of 0.63,0.57 for PFI and OS respectively.

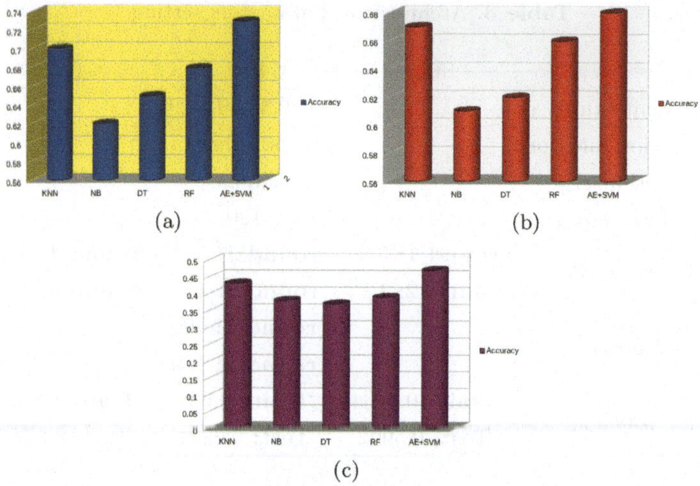

Fig. 5. Classification performance based on Accuracy (a): OS, (b): PFI, (c): Stage

3.2 Classification Results

After training the models for each endpoint we construct the new reduced data set and train/test split the data set in order to train and evaluate an SVM classifier to associate each sample to its corresponding endpoint. To ass the performance of the predictive model and the consistency of the learned features we performed a comparison with some classic classification models namely, K-nearest-neighbors (KNN), Decision Trees (DT), Random Forest (RF), and Gaussian Naive Bayes (NB). The Comparison was established using the following metrics: Accuracy, Precision, Recall, f1_score. The histogram plot on Fig. 5, shows that our proposal outperforms all the other models where we was able to score approximately 0.5%, 0.68%, 0.74% accuracy rate for Stage, PFI,and OS. Whereas, all the other models scored noticeably lower results.

In addiction to accuracy we collected the macro average classification report of the other metrics to capture, where our model's ups and falls. The overall performance of our proposed model in front of the comparison models on predicting pathological stage (Table 4) was positive along all metrics, citing that KNN had a fair closeness to our results. As for PFI and OS (Table 5, Table 6) our model was able to outperform the other model only in precision yet in PFI NB was able to score the best results with Recall measure and KNN with f1_score. While in OS, KNN and NB scored the best recall and KNN scored the best f1_score.

As general discussion we adress three points, the first one is the performance of data set in training the autoencoder, where we notice the absence of overfitting between training and testing data, with the note we have applied neither regularization, nor penalty dropouts on layers' input. We assume that this phenomena is due to the protein filtration based on the biological knowledge (PPI),

Table 4. Performance in predicting pathological stage for RPPA data

	Precision	*Recall*	*f1_score*
KNN	0.42	0.4	0.39
NB	0.38	0.36	0.38
DT	0.35	0.35	0.35
RF	0.21	0.32	0.24
AESVM	**0.5**	**0.41**	**0.39**

Table 5. Performance in predicting PFI score for RPPA data

	Precision	*Recall*	*f1_score*
KNN	0.62	0.62	**0.62**
NB	0.61	**0.63**	0.6
DT	0.58	0.58	0.58
RF	0.33	0.50	0.40
AESVM	**0.66**	0.54	0.5

Table 6. Performance in predicting OS score for RPPA data

	Precision	*Recall*	*f1_score*
KNN	0.65	**0.63**	**0.64**
NB	0.62	**0.63**	0.61
DT	0.6	0.6	0.6
RF	0.34	0.50	0.40
AESVM	**0.72**	0.6	0.6

that was responsible of eliminating nose input and dropping outliers that may leads to a misleading learning.

The second and third points resumes in the low prediction rate and the weak performance of the proposed model in PFI and OS on Recall and f1_score metrics, where we address this falls to the lack of data for stage prediction, where there is not enough data set for each stage, which leads to poor learning. As well as for the high unbalance between the two classes, which also leads to poor learning and weak discrimination between the samples of each classes, especially for cancers that have high correlation.

4 Conclusion

The most crucial phase when dealing with cancer related biological data, whether its is transcriptomic or proteomic is the selection of a representative, not noisy feature space. in this paper we tried to curate our RPPA data following two steps. The first was to filter the dataset based on biological background then to extract a small features set using unsupervised deep learning in order to make the classifier learns from data that have a high discriminative ratio and my play the role of in silico molecular signatures. Despite the curse of the unbalanced data sets in terms of endpoits classes, we were able to notice a interesting performances of our proposal that may further helps us on improving those results by data collection or using other biological background such as signaling pathways.

References

1. Bilalović, O., Avdagić, Z.: Robust breast cancer classification based on GA optimized ANN and ANFIS-voting structures. In: 2018 41st International Convention on Information and Communication Technology, Electronics and Microelectronics (MIPRO), pp. 0279–0284 (2018). https://doi.org/10.23919/MIPRO.2018.8400053
2. Chollet, F.: Keras (2015). https://github.com/fchollet/keras
3. Fakoor, R., Ladhak, F., Nazi, A., Huber, M.: Using deep learning to enhance cancer diagnosis and classification. In: Proceedings of the International Conference on Machine Learning, vol. 28. ACM, New York (2013)
4. Franco, E.F., et al.: Performance comparison of deep learning autoencoders for cancer subtype detection using multi-omics data. Cancers **13**(9), 2013 (2021)
5. Haznedar, B., Arslan, M.T., Kalinli, A.: Optimizing ANFIS using simulated annealing algorithm for classification of microarray gene expression cancer data. Med. Biol. Eng. Comput. **59**(3), 497–509 (2021)
6. Li, J., et al.: Explore, visualize, and analyze functional cancer proteomic data using the cancer proteome atlas. Cancer Res. **21**(77), 51–54 (2017)
7. Macías-García, L., Luna-Romera, J.M., García-Gutiérrez, J., Martínez-Ballesteros, M., Riquelme-Santos, J.C., González-Cámpora, R.: A study of the suitability of autoencoders for preprocessing data in breast cancer experimentation. J. Biomed. Inform. **72**, 33–44 (2017)
8. Mari, M., Tesshi, Y.: Signaling pathway profiling using reverse-phase protein array and its clinical applications. Expert Rev. Proteom. **14**(7), 607–615 (2017)
9. Masuda, M., Yamada, T.: Utility of reverse-phase protein array for refining precision oncology. In: Yamada, T., Nishizuka, S.S., Mills, G.B., Liotta, L.A. (eds.) Reverse Phase Protein Arrays. AEMB, vol. 1188, pp. 239–249. Springer, Singapore (2019). https://doi.org/10.1007/978-981-32-9755-5_13
10. Nawy T.A.: Pan-cancer atlas. Nat. Methods **15**(407), 291–304 (2018)
11. Negm, O., et al.: Clinical utility of reverse phase protein array for molecular classification of breast cancer. Breast Cancer Res. Treat. **155**(1), 25–35 (2016)
12. Spurrier, B., Ramalingam, S., Nishizuka, S.: Reverse-phase protein lysate microarrays for cell signaling analysis (2008)
13. Way, G.P., Greene, C.S.: Extracting a biologically relevant latent space from cancer transcriptomes with variational autoencoders. In: Pacific Symposium on Biocomputing 2018: Proceedings of the Pacific Symposium, pp. 80–91. World Scientific (2018)
14. Zenbout, I., Bouramoul, A., Meshoul, S.: Targeted unsupervised features learning for gene expression data analysis to predict cancer stage. In: Proceedings of the Tenth International Conference on Computational Systems-Biology and Bioinformatics, pp. 1–7 (2019)

Clustering Educational Items from Response Data Using Penalized Pearson Coefficient and Deep Autoencoders

Khadidja Harbouche[1], Nassima Smaani[2], and Imene Zenbout[3](✉)

[1] LRSD Laboratory, University Ferhat Abbes Setif 1, El Bez, Setif, Algeria
khadidja.harbouche@univ-setif.dz

[2] University Ferhat Abbes Setif 1, El Bez, Setif, Algeria
nassima.smaani@univ-setif.dz

[3] Fundamental Informatics and its Applications Department, MISC Laboratory, Faculty NTIC, University Abdelhamid Mehri - Constantine 2, Constantine, Algeria
imene.zenbout@constantine2.dz

Abstract. Educational data mining techniques are very useful to analyze learner performance in purpose to optimize the approach of item-to-skill mapping. Therefore computing a degree of similarity between items using different measures based on the performance of the learner toward items, enhance the clustering of different items into knowledge components. This paper proposes a computational framework to group the elements of the corresponding knowledge component. The first phase of the framework represents a variation of Pearson coefficient to measure item similarity by applying a penalty score that is calculated from the number of hints taken by the learner during solving two items. The second phase applies a dimensionality reduction using deep auto encoders to improve the clustering accuracy. The experimental results show that clustering based on the penalized Pearson coefficient and the deep dimensionality reduction (PPC+DDR) outperforms basic clustering based on different similarity methods, with approximately +0.2 in Mean silhouette coefficient.

Keywords: Educational data mining · Learner model · Machine learning · Deep learning · Item-to-skill mapping · Clustering

1 Introduction

Course curricula are usually organized in a meaningful sequence that evolves from relatively simple lessons to more complex ones. Incorporating prerequisite skill structures into education systems helps identify the order in which concepts should be presented to learners to maximize their success. Many skills have a strong causal relationship in which one skill must be presented and mastered by the learner before another (hierarchy of skills according to prerequisites). To sequence learning in an intelligent tutoring system (ITS), we refer to prerequisite

B. Lejdel et al. (Eds.): AIAP 2021, LNNS 413, pp. 75–85, 2022.
https://doi.org/10.1007/978-3-030-96311-8_8

structure as the relationships among skills that place strict constraints on the order in which skills can be acquired. Recent interest in computer assisted education promises large amounts of data from students solving items: questions, problems, parts of questions ... That data are used to create student models. These models represent an estimate of skill proficiency at a given point in time [8]. Student models are often used to personalize instruction in tutoring systems or to predict future student performance. Prior work has investigated how to discover prerequisites among items without considering their mapping into skills [3,6]. Item-to-skill mappings (also called Q-matrices) are desirable because they allow more interpretable diagnostic information. They are standard representation used to specify the relationships between individual test items and target skills. There are two approaches to item-to-skills mapping: Model-based approach, and Similarity-based ones that are based on the assumption that learners will tend to perform similarly on items that require the same skill, so we need to identify similarity between pairs of items. Our work falls into the second approach (Similarity-based). In this paper, we present a PPC-DDR architecture based on: Firstly, a proposed measure, we call penalized Pearson coefficient PPC, to calculate similarity score between two items; and secondly, a deep auto encoder to reduce the dimensionality of the Item to Item similarity matrix (Deep Dimensionality Reduction DDR). A set of experimental results were conducted to evaluate the proposed approach on the corresponding data set.

The rest of the paper is structerd as follow: Sect. 2, review some related works in educational datamining, Then we descriped, and explained our proposal in Sect. 3. We conducted a set of experiments and comparison in Sect. 4 to evaluate our proposal. Finally, we concluded our work in Sect. 5 with a general overview and perspectives.

2 Related Works to Educational Items Clustering

As mentioned above, item-item similarity is a crucial phase to perform a relevant educational item clustering. Therefore, many researches have been introduced in the literature, where:

Jirí Rihák and Radek Pelánek applied an automatic computation of item similarities based on learners' performance data using different measures such as pearson yule, jaccad with different setting and different levels. The results showed that The Pearson correlation coefficient is a good default choice [12]. Dharaneeshwaran et al. Calculated the User-item Similarity using Pearson's and Cosine Correlation by using three different techniques for computing similarities for obtaining a recommendation for them [7]. Pelánek et al. proposed a systematic approach to study similarity measures, they made an evaluation of similarity measures for several introductory programming environments [2]. Nazaretsky et al. proposed a new item similarity measure named 'Kappa learning' which compute the similarity between items based on the response data giving by the learner, they took in consideration improving the sequence of the knowledge skills [11]. Bjørn et al. developed a framework using an artificial neural network model to show the main differences between various types of similarity measures [10].

3 Proposed Architecture

Regrouping items into KC based only on similarity measuring through learner performance correct/incorrect answers, may leads to assigning items to the wrong cluster. Taking the assumption where for two items the answer of the learners were correct yet after asking for a lot of hints. Here if we only based on the correctness ratio we may assign the two items under one cluster yet looking to high hint asking frequency, we can't ignore that the learner was not able to answer the items without taking a considerable set of hints, which open the question about the similarity between the two items, if it is really high as it was computed or not. Illustrating that assumption, lets take the two items $i1$ and $i2$ where $i1 : (7 - 5)$ and $i2 : (5 - 7)$, the two items can be yield in the subtraction knowledge but to achieve the answer of the second item the learner need to have a prerequisites in negative number skill. So, he will ask for aid in order to answer the second item correctly, which reduce the similarity between i1 and i2, here we propose to take in consideration the number of hints asked by the learner to answer the two items since we don't have a certain order on knowledge retrieval, So, the learner may answer $i1$ than $i2$ or the reverse in order to build a certain skill. Therefor we applied a penalty score on the coefficient similarity between $i1$ and $i2$.

3.1 Penalized Pearson

Lets denote $L_{u,i}$ the learner-item matrix, where u represents the learner and i the item. From this matrix we compute the contingency matrix (Table 1) of each item i to item j as well as counting the following combination:

- $Nb_{correct}$: is the number of the passage of the items i and j with the outcome = correct given by the learner.
- $Nb_{incorrect}$: is the number of the passage of the items i and j with the outcome = incorrect given by the learner.

Table 1. item-item contingency matrix

		$item_i$	
		Incorrect	Correct
$item_j$	Incorrect	a	b
	Correct	c	d

- Nb_{hint}: number of hint asked by the learner during solving the items i and j
- N: total passage on the two items $\sum(Nb_{correct}, Nb_{incorrect}, Nb_{hint})$. Next we define a coefficient λ_{ij} that is calculated following Eq. 1. λ value is between 0 and 1.

$$\lambda = \frac{Nb}{N} \tag{1}$$

The final step is to calculate the item similarity score between i and j. As we mentioned above we proposed a Pearson variation that is based on a penalty on the score between the two items. Taking The Pearson formula as it is defined in (Eq. 2), after we apply a penalty on it based on λ.

$$Ps = \frac{(a \times d) - (c \times b)}{\sqrt{(a+b) \times (a+c) \times (b+d) \times (c+d)}} \tag{2}$$

When calculating the number of hints the more the learner asks for hints the bigger λ is i.e.λ is closer to one, so multiplying the Pearson score directly to λ will apply a higher penalty (λ closer to zero) on items with a few or no hints. To over come this we multiply the similarity score by $(1 - \lambda)$ to reverse the penalty score. Therefore we obtain the formula in Eq. 3

$$Ps = \frac{(a \times d) - (c \times b)}{\sqrt{(a+b) \times (a+c) \times (b+d) \times (c+d)}} \times (1 - \lambda) \tag{3}$$

The Algorithm 1, explains the steps that we followed in order to calculate the item-item similarity matrix based on the proposed PPC.

3.2 Dimensionality Reduction

After constructing the item similarity matrix, it may hold lot of noise due to its high dimensionality, that leads to poor results in clustering items, therefor ye apply a dimensionality reduction using deep auto encoder to learn a new features' representation. Though before proceed to the dimensionality reduction phase we passes through a data preprocessing as follows:

Eliminating Irrelevant Items: We eliminated items that have a number of missing values that is equal to or more than 40% missing values.

Imputation of Missing Values: Missing values is a well-known problem in data science that need to be handled because they reduce the quality for any of our performance metrics. We Imputed our data for completing missing values using k-Nearest Neighbors imputer, where Each sample's missing values are imputed using the mean value from k-neighbors found in the dataset. Then we used Z score to normalize it (Fig. 1).

Algorithm 1: Penalized Pearson Coefficient to construct item-item similarity

Data: learner-item, a, b, c, d, N, Nb_{hint}, λ
Result: item-item
for *each* $item_i$ *in leaner-item* **do**
    for *each* $item_j$ *in learner-item* **do**
        Initialize to zero (a, b, c, d, N, Nb_{hint});
        for *each learner in learner-item* **do**
            Get $Infoitem_{l,i}$;
            Get $Infoitem_{l,j}$;
            Update(a, c, b, d);
            Update(Nb_{hint});
            Update(N);
        end
        if $Nb_{hint}=0$ **then**
            $\lambda=0$
        end
        else
            Update lambda (Eq. 1)
        end
        Calculate PPC (Eq. 3);
        Assign $item - item_{i,j}$
    end
end

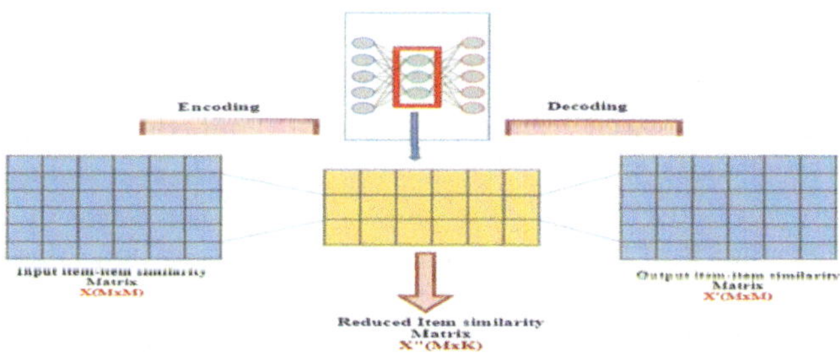

Fig. 1. Dimensionality reduction auto encoder

3.2.1 Deep Auto Encoder

Our architecture is based on Unsupervised learning, where we used deep autoencoders to reduce the dimensionality [1] of item-item matrix for improving the clustering phase by eliminating any noises data and to learn better descriminative reduced features space.

Let's denote the item-item matrix I of range $M \times N$, each i represents an item that to be further clustered and each j is the item feature used for training a clustering model. The proposed auto encoder is a symmetric deep model where the encoder takes the input X and encoded it into a latent space of range K

Table 2. Data set statistics

Number of students/learners	36
Number of unique steps	3,735
Total number of steps	24,890
Total number of transactions	71,553
Total student hours	334.24

where $K < M$. The decoder serves to reconstruct the input X as the output X', where $X \approx X'$. The consistency of the autoencoder was evaluated using the mean-square-error (mse) (Eq. 4), that calculates the difference between X and X'.

$$mse = 1/n \sum_{i=1}^{n} (y_i - y_i')^2 \tag{4}$$

4 Experimental Results

In order to evaluate the correctness of our proposal we conducted an experimental scenario on an educational dataset and draw a set of comparison to evaluate the performance of the proposed topology. The deep learning architecture was implemented using keras package [4].

4.1 Data Collection and Preparation

From the DataShop, (https://pslcdatashop.web.cmu.edu/) educational data repository we collected the. *'FrenchLanguage2'*, which is a dataset (Table 2) corresponds to an e-learning French course. One of the steps in the problem solving pipeline that is of huge importance is data preparation. Missing this step would cause the results of the analysis to be misleading. In this step we have 3 phases:

1. Choice of features that we use during our work: In our work, we focus on learner performance, which it is related to the item, and the outcome provided by the learner: (correct/incorrect) or taking a hint. The description of features is described in Table 3.

2. Elimination of items passed by just one student: Before, the construction of the learner-item matrix we have to keep only those items that have been taken at least by one learner (Eq. 5):

$$number_item = M_{item,learner}. \qquad |M_{item_learner} > 0 \tag{5}$$

3. Construction of the Learner-Item matrix: The learner-item matrix contains the counts of how many times a student passed through the same item and the outcome of each transaction whether its correct, incorrect, or asking hint.

Table 3. Final dataset description

Feature	Description	Total number
Anon student id	ID of the student	36
Problem name	Items	749
Outcome	The response giving by the student (correct/incorrect) Or taking a Hint	–

Table 4. Parameters of deep auto encoder

Parameter	Value
Activation function	tanh
Epochs	400
batch-size	100
Optimizer	sgd
loss	mse (0.5)
encoder	(200, 100, 50, 30)
decoder	(50, 100, 200)

4.2 Dimestionality Reduction Autoencoder Training

We construct a deep symmetric autoencoder with an input layer of 737 nodes. Three middle layers with 200, 100, and 50 nodes respectively, besides a bottleneck layer with 30 nodes that represents the latent space used in clustering. An decoder with three middle layer with 50, 100, 200 nodes respectively and an output layer composed of 737 nodes. The model was trained using a stochastic gradient descent optimizer [14] and mini-batch training. To train our model and select the best parameters settings we used hold-out cross validation and split our data set into 80% training and 20% testing, where we trained the autoencoder on the training set and test it's performance on unseen dataset through thousands of executions and we found that our model converge to it's performance (Fig. 2) using the parameters illustrated in Table 4.

Although, the objective of using an autoencoder here was not the reconstruction of the input X, yet evaluating the similarity between X and X' still the best factor to observe and decide whether the learned latent space was able to map the input towards the output with a little amount of loss. Here our trained model achieved an error score of 0.5, without a noticeable overfitting, which is considered a very promising score that allows us to use the learned latent space matrix I' of range $M \times K$.

4.3 Clustering

The last step of our experimentation is the clustering using the kmeans model as well as the implementation of WCSS to define the optimal number of clusters.

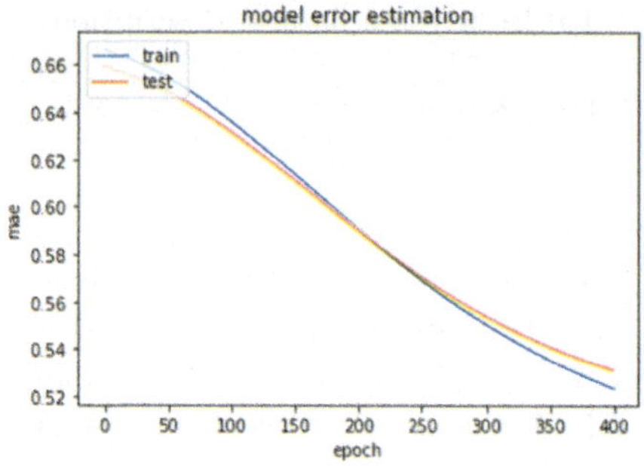

Fig. 2. Training loss of deep auto encoder

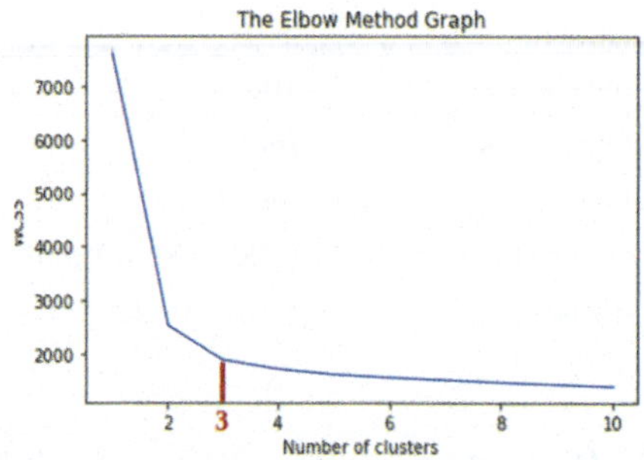

Fig. 3. WCSS elbow graph to define clusters

As shown in Fig. 3 that plots the elbow graph of the WCSS algorithm among 15 cluster, the edge falls between four and two clusters so the best option of cluster number is three.

4.4 Results and Discussion

We performed a comparison of our proposal with state of the art methods by evaluating the following three metrics: Mean Silhouette Coefficient (MSC) [13], Calinski-Harabasz Index (CHI) [9], and Davies-Bouldin score (BDS) [5]. After the choice of the metrics we selected three state of the arts similarity measurements methods, namely Pearson, Yule and Kappa. Table 5 exhibit the overall performance of our penalised dimensionality reduction approach compared to the other methods across all metrics, where we can visualize the huge superiority between the four models. The PPC-DDR was able to score a MCS of 0.42 compared to 0.20 by Pearson, and 0.1, 0.14 using Yule and Kappa respectively. In terms of CHs and DBS, PPc-DDR performance was also higher than the performance of the other models. It is really surprising the fact that the performance' differences between our proposal and the other models, and may open a perplex about whether this is real or imaginary. To address this question, we assumed that this scoring gap can be due to not applying dimensionality reduction after

Table 5. Performance of PPC-DDR and other similarity models

	MCS	CHs	DBS
PPC-DDR	**0,42**	**1031.04**	**1.17**
Pearson	0,20	146.22	1.36
Yule	0.21	191.3	1.61
Kappa	0,14	160.19	2.01

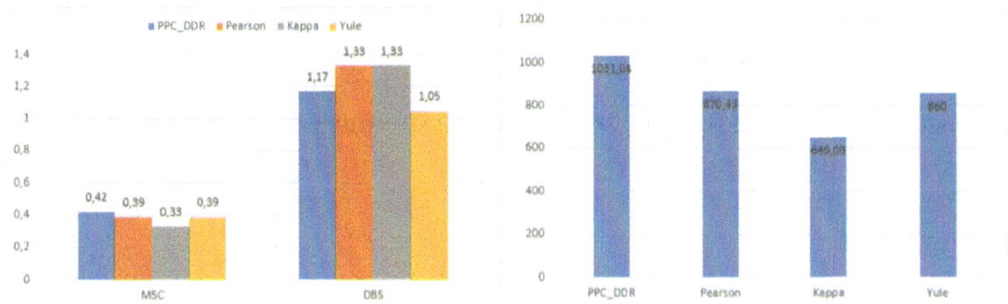

(a) (b)

Fig. 4. Performance of the methods with dimensionality reduction (a): MSC and DBS metrics, (b): CHs metric

using pearson, yule and kappa before clustering. To check this assumption, we draw another test for two reason, the first to check the difference of performance between the models with and without dimensionality reduction. Besides to test if our model can still perform better in same conditions as the other algorithms. To perform this test used the before mentioned similarity measuring methods in addition to dimensionality reduction on the outputs of those methods. The results shown in Fig. 4, visualize the performance of our PPC-DDR and the other methods plus dimensionality reduction, where we can notice the huge improvement in pearson, yule and kappa performance, when applying dimensionality reduction compared to the previous results as well as, it is clear that our model still able to compete with the three methods, in which we notice that the proposed approach outperformed the other methods in terms of MSC, and CHs. Yet we also notice that yule plus deep dimensionality reduction was able to achieve better score in terms of DBS. This second comparison allowed us to understand the deep impact of dimensionality reduction on the clustering phase. As well as we assume that superiority of the yule algorithm maybe due to the tested data set, therefore using other dataset may help to visualize the performance of our proposal.

5 Conclusion

In this work we have explored the integration of data mining in learning systems, that aims to optimize the learning methods and to handle the large pool of items (questions, problems) and their diversity. Collecting data about learners' performance can be used to get insight into item properties. therefore, analysing item similarities, can be used as input to cluster items or to visualize their correlations ratio. In the context Our proposed Penalized Pearson coefficient to measure item-similarity and deep features learning, achieved very promising performance in item to skill mapping. Considering the characteristics and limitations of the used dataset in terms of item to learner order of presentation and number of test we believe that working on the order in which concepts should be presented to learners to optimize their success may help to improve the performances of the proposal and helps in investigating more avenues that enhance the curriculum methodology.

References

1. Dong, S., Wang, P., Abbas, K.: A survey on deep learning and its applications. Comput. Sci. Rev. **40**, 100379 (2021)
2. Pelánek, R, et al.: Measuring item similarity in introductory programming. In: Proceedings of the Fifth Annual ACM Conference on Learning at Scale, June 2018
3. Vuong, A., Nixon, T., Towle, B.: A method for finding prerequisites within a curriculum. In: Educational Data Mining 2011. Jiawei Han and Micheline Kamber (2010)
4. Chollet, F.: Keras (2015). https://github.com/fchollet/keras
5. Davies, D.L., Bouldin, D.W.: A cluster separation measure. IEEE Tran. Pattern Anal. Mach. Intell. **PAMI–1**(2), 224–227 (1979)
6. Desmarais, M.C., Meshkinfam, P., Gagnon, M.: Learned student models with item to item knowledge structures. User Model. User-Adap. Inter. **16**(5), 403–434 (2006)
7. Dharaneeshwaran, Nithya, S., Srinivasan, A., Senthilkumar, M.: Calculating the user-item similarity using Pearson's and cosine correlation. In: 2017 International Conference on Trends in Electronics and Informatics (ICEI), pp. 1000–1004 (2017)
8. Kass, R.: Student modeling in intelligent tutoring systems-implications for user modeling. In: Kobsa, A., Wahlster, W. (eds.) User Models Dialog Systems. Symbolic Computation (Artificial Intelligence), pp. 386–410. Springer, Heidelberg (1989). https://doi.org/10.1007/978-3-642-83230-7_14
9. Kozak, M.: "A dendrite method for cluster analysis" by Calinski and Harabasz: a classical work that is far too often incorrectly cited. Commun. Stat.-Theory Methods **41**, 2279–2280 (2011)
10. Mathisen, B.M., Aamodt, A., Bach, K., Langseth, H.: Learning similarity measures from data. Progr. Artif. Intell. **9**(2), 129–143 (2019). https://doi.org/10.1007/s13748-019-00201-2
11. Nazaretsky, T., Hershkovitz, S., Alexandron, G.: Kappa learning: a new item-similarity method for clustering educational items from response data. In: Proceedings of the 12th International Conference on Educational Data Mining. International Educational Data Mining Society (2019)

12. Rihák, J., Pelánek, R.: Measuring similarity of educational items using data on learners' performance. In: 10th International Conference on Educational Data Mining, pp. 16–23. International Educational Data Mining Society, Wuhan (2017)
13. Wang, F., Franco-Penya, H.-H., Kelleher, J.D., Pugh, J., Ross, R.: An analysis of the application of simplified silhouette to the evaluation of k-means clustering validity. In: Perner, P. (ed.) MLDM 2017. LNCS (LNAI), vol. 10358, pp. 291–305. Springer, Cham (2017). https://doi.org/10.1007/978-3-319-62416-7_21
14. Yang, J., Yang, G.: Modified convolutional neural network based on dropout and the stochastic gradient descent optimizer. Algorithms **11**(3), 28 (2018)

Rational Function Model Optimization Based On Swarm Intelligence Metaheuristic Algorithms

Oussama Mezouar[1](✉), Fatiha Meskine[1], and Issam Boukerch[2]

[1] Communication Networks, Architecture and Multimedia Laboratory, University of Djillali Liabes, Sidi Bel Abbes, Algeria
{oussama.mezouar,fatiha.meskine}@univ-sba.dz

[2] Space Technics Center, Algerian Space Agency, Arzew, Oran, Algeria

Abstract. The Rational Function Model (RFM) is progressively being familiar in the mapping and photogrammetric research domain and is been widely used as an approximate to rigorous Models. According to its ability to preserve the complete accuracy of various types of physical sensors and its independence of sensors and platforms, it can be calculated with any coordination system. Nevertheless, the RFM coefficients are also known as rational polynomial coefficients (RPCs) depending on a large number of ground control points which make the model susceptible to over parameterization errors and time-consuming. In addition, the RPCs have no physical meaning, and as a result, selecting the best combination of RPCs is rather difficult. The swarm intelligent based meta-heuristic algorithms optimization seems to be an effective approach to overcome this problem. This paper focuses on the application of recent swarm intelligence based meta-heuristic algorithms for RFM optimization. The most popular algorithms belonging to this category are ant colony algorithm, genetic algorithms, and particle swarm optimization. Furthermore in this research, we propose a parallel hybrid meta-heuristic optimization algorithm that combines the genetic algorithm and particle swarm optimization concepts to overcome the swarm intelligent limitations for RFM optimization. The different algorithms are applied on two data sets provided from the Algerian satellite (ALSAT2). The results demonstrated that the proposed method is more accurate compared to the suggested tested swarm based meta-heuristic methods.

Keywords: Rational function model · Ant colony algorithms · Genetic algorithms · Particle swarm optimization · Hybrid algorithms

1 Introduction

One of the most important sources of geographic information systems (GIS) is high-resolution satellite imagery that has been used in many applications such as remote sensing and topographic maps. But raw images usually contain some significant geometrical distortions which cannot be used directly in GIS without ortho-rectification.

B. Lejdel et al. (Eds.): AIAP 2021, LNNS 413, pp. 86–99, 2022.
https://doi.org/10.1007/978-3-030-96311-8_9

Ortho-rectification is a standard process for correcting the geometric distortions and relief displacement errors introduced during image acquisition [1, 2]. The high accuracy potential of ortho-rectification depends on the relationship between images and object spaces [3]. To this end, a wide variety of mathematical models have been developed which generally fall into two main models: physical models (rigorous or sensor dependent) and non-physical (empirical or sensor independent) ones [4]. The most frequently used non-parametric model is based on 3D rational polynomials and which are known in literature as the Rational Function Model (RFM). Unlike rigorous models, RFM does not need any physical understanding of the sensor or satellite. RFM has received increasingly more acceptance because of its simple form, convenience for using, low requirement for specialized knowledge and not depending on the imaging parameters of specific satellites.

There are two methods for solving RFM, namely the dependent-terrain method and the independent-terrain one. In the case of the independent terrain, RFM is solved by using the physical sensor model which requires the availability of some information on the sensor (attitude and orbital parameters). Otherwise, in the dependent-terrain approach, RFM necessitates a large number of accurate and well-distributed ground control points (GCPs), which is a time-consuming and costly process. In addition, the RFM coefficients, also known as the rational polynomial coefficients (RPC), have no physical significance which makes it difficult to find their best combination [5]. To overcome these problems, the binary form of meta-heuristic algorithms may be helpful in optimizing and determining the optimum RPCs.

Nowadays, computational intelligence and meta heuristic algorithms have become increasingly popular in computer science, artificial intelligence, machine learning, image processing and data mining. Most algorithms in computational intelligence and optimization are based on swarm intelligence [6] that explores the behavior of natural entities (consisting of many individuals) in order to build artificial systems for solving problems of practical relevance. Therefore, among population-based we distinguish swarm intelligence-based metaheuristics such as Ant Colony Optimization (ACO), Genetic Algorithms (GA), and Particle Swarm Optimization (PSO) which are the most popular optimization algorithms that often have an excellent ability for self-organization, self-learning, or self-memory [7]. A few studies have been conducted on the use of artificial intelligence algorithms to solve RFMs optimization such as GA [8–10] and PSO [11, 12].

In this paper, we examine these three meta-heuristic algorithms ACO, GA and PSO applied for RFM optimization in which a comparative study is performed. Moreover, we have proposed a parallel hybrid algorithm that combines the GA and PSO concepts in order to overcome their limitations. The rest of the paper is organized as follows: the next section presents the theoretical concept of RFM, Sect. 3 presents a review of different meta-heuristic algorithms utilized for RFM optimization, Sect. 4 gives the different experimental tests with a discussion, and finally we finish with a conclusion.

2 Rational Function Model (RFM)

RFM is a mathematical model that uses a ratio of two polynomial functions to define the relationship between a point in-ground space (X, Y, Z) and its corresponding in image space (r, c) or vice versa as indicated in the following equation [13]:

$$r = \frac{P_1(X, Y, Z)}{P_2(X, Y, Z)}, c = \frac{P_3(X, Y, Z)}{P_4(X, Y, Z)} \tag{1}$$

The 3D polynomial function P_i (i = 1, 2, 3, 4) is defined as:

$$P_n = \sum_{i=0}^{m1} \sum_{j=0}^{m2} \sum_{k=0}^{m3} a_{ijk} X^i Y^j Z^k \tag{2}$$

where:

$$n = 1, 2, 3, 4$$

$$0 \le m1 \le 3$$

$$0 \le m2 \le 3$$

$$0 \le m3 \le 3$$

$$m1 + m2 + m3 \le 3$$

a_{ijk} indicates the RFM coefficients that are referred to as Rational Polynomial Coefficients (RPCs) or Rational Function Coefficients (RFCs) [1, 13]. To determine the RPCs values, we use a set of *ground control points* (GCPs) for which the (r, c) and (X, Y, Z) coordinates are known and must be taken into account, the first coefficients in the denominators (P_2 and P_4) are supposed to be 1. In result, there are 39 RPCs in each equation: 20 in the numerator and 19 (with the constant 1) in the denominator. So a minimum of 39 GCPs are needed to solve the 78 coefficients. It is necessary to consider errors in the reference points, which include not only GCPs but also check points in estimating the accuracy of the results [11].

The linearized RFM type can be used to solve unknown RPCs [13], as seen below:

$$P_1(X, Y, Z) - rP_2(X, Y, Z) = 0 \tag{3}$$

$$P_3(X, Y, Z) - cP_4(X, Y, Z) = 0 \tag{4}$$

Using n as the number of GCPs, the above equations can then be written as follows:

$$y = Ax + e \tag{5}$$

where:

A: design matrix
y: observations vector
e: residuals vector
x: vector of RPCs.

RPCs can be determined using the least-squares (LS) method, as shown below:

$$x = (A^T A)^{-1} A^T y \tag{6}$$

3 Swarm Intelligence Based Meta-heuristic Algorithms Applied for RFM Optimization Review

The swarm intelligence meta-heuristic search algorithms adopted in this paper as well as their improved schemes for RFM optimization have been demonstrated in this section.

3.1 Binary Ant Colony Optimization (BACO)

The ant colony optimization algorithm was inspired by the walk of ants on the search for food [14]. The binary version of ant colony algorithm (BACO) is applied for RFM optimization. In this section BACO is described, the algorithm starts by creating the environment in which our ant-like agents will evolve [15, 16]. Figure 1 illustrates this environment, which is composed of two lines (sequences of 0s and 1s). The ants start the search process from the root, as is described in the following figure.

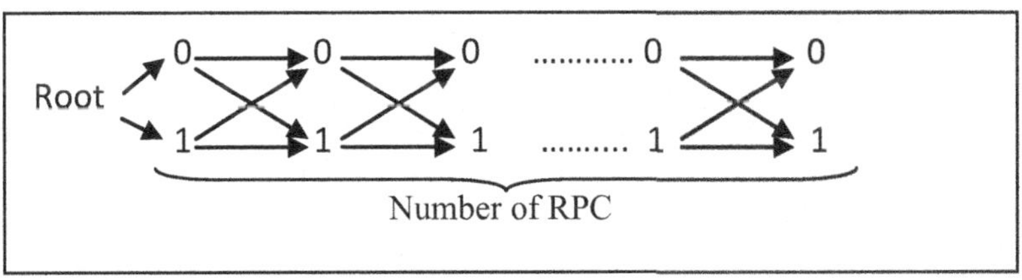

Fig. 1. Ant colony search space

At the tth iteration, N ants produce solutions in the form of binary sequences by coming in the field from the left and exiting from the other side, as a result, each ant formed, while crossing the field, a solution (RFM structure), the length of the field is equal to the dimension of the RFM structure (number of RPCs). The artificial ant (k) at each node chooses either 0 or 1 as the next node to walk according to the pheromone vector $T(t)^k = \tau_{1x}^k(t) + \tau_{ix}^k(t) + \cdots + \tau_{nx}^k(t)$, where τ_{ix}^k is the pheromone value for x which is a binary number (0, 1) from the sequence i = 1, …, n of the kth ant. The pheromones are updated regularly during the search, and initially set to value (τ_{init}) at the start of the search. An ant chooses which way to go when visiting a node (0 or 1) based on the transformation probabilities *p* [16]:

$$x_{0,1}(t) = \frac{(\tau_{0,1}(t))^{\alpha}}{\sum_{l\epsilon N_i^k}(\tau_{0,1}(t))^{\alpha}} \tag{7}$$

$$p_{0,1}^k(t) = \frac{(x_{0,1}(t))}{\sum_{l\epsilon N_i^k}(x_{0,1}(t))} \tag{8}$$

where parameter α controls the relative weight of the pheromone trail in the probability computation. Then the ant chooses the next step based on the probability *p*, repeating the procedure until it amounts to the last bit. The process continues until N ants have finished their walk across the field, and as a consequence, N solutions are produced.

The binary string discovered during the N ants walk through the field is considered as solutions to the problem (RFM optimization) and assessed by the fitness function. After that, the quantity of pheromone at all connections is evaporated using (Eq. 9), where the initial evaporation rate is set to ρ value.

$$\tau_{ix}(t+1) = (1-\rho)\tau_{ix}(t) \tag{9}$$

Finally, the pheromone reinforces at the connection with value proportional to each solution as in (Eq. 10) which is bounded between pheromone minimum τ_{min} and pheromone maximum τ_{max}.

$$\tau_{0,1}^k(t) = \tau_{0,1}^k(t) + \frac{1}{(1+e^{(\text{fitness value})^k})^2} \tag{10}$$

3.2 Genetic Algorithms

The Genetic algorithms (GAs) are heuristic search optimization algorithms developed firstly by J. Holland [17] based on the concept of natural selection and genetics inspired by Darwin's theory. They represent an intelligent exploitation of a random search that combines both exploration and exploitation in an optimal way. Due to these interesting properties, GAs present a robust technique that can deal successfully with a wide range of problem areas including those applied in remote sensing and especially for RFM optimization [8, 9].

GAs are an iterative process that evolves under a population of individuals or chromosomes generated randomly which represents a potential solution of the problem. In computing, chromosomes can be represented by strings. The most commonly used alphabet of the strings is binary, but other alphabets are also used, e.g, integer or real valued numbers, depending on which is the most suitable for a given problem. Each individual is a candidate solution to the optimization problem which is assigned a fitness value based on the objective function of the problem. In the selection step, the best fits of the individuals are chosen and are more likely to survive in the next generation. Therefore, the mechanism of reproduction involves crossover and mutation operations in order to generate a new population of individuals. The purpose of those operations is to modify the chosen solutions and select the most appropriate offspring to pass on the succeeding generation until no better fitted solutions are possible. The process is iterated from one generation to another until a defined termination criterion has been reached as the number of generations or a satisfactory fitness level.

3.3 Particle Swarm Optimization

Particle swarm optimization (PSO) is one of the most common meta-heuristic optimization algorithms based on social intelligence and cooperative behavior shown by a swarm of birds or fish while searching for food. The first version of the particle swarm algorithm was developed by James Kennedy and Russell Eberhart in 1995, while working in continuous search space [18]. The application of the standard PSO is not complicated, the algorithm is characterized by two fundamental parameters: the position (location) and velocity of each particle, where the velocity of particle i is bounded between velocity minimum and velocity maximum [Vmin, Vmax], and calculated at iteration t (starting with 0) using the following equation [19, 20]:

$$\mathrm{v_i(t+1) = w(t)v_i(t) + c_1r_1(p_i(t) - x_i(t)) + c_2r_2\big(p_g(t) - x_i(t)\big)} \tag{11}$$

where *w(t)* is time-varying inertia weight for RFM optimization and it is a decreasing function of iterations in this research, as shown in Eq. 12, which is described and used in [20].

$$\mathrm{w(t)} = w_{min} + (w_{max} + w_{min}).\frac{t_{max} - t}{t} \tag{12}$$

$\mathrm{t_{max}}$ is the maximum number of iterations, w_{min} and w_{max} are two constant experimental parameters, c_1 and c_2 signify the acceleration factors. Generally, c_1 equals c_2, r_1 and r_2 are two random numbers within a range of [0, 1]. P_g denotes the best particle of the swarm giving the best objective function value (best solution) and the best previous position of the ith particle is represented as P_i *and* x is the present position (solution) of particle i. When the velocity is determined, the position of the particle i is updated from:

$$\mathrm{x_i(t+1) = x_i(t) + v_i(t+1)} \tag{13}$$

PSO was used in many fields of research and applications, but some problems of optimization are solved in the discrete space and not in the continuous search space. It is for this reason that Kennedy suggested a binary (discrete) version of the particle swarm optimization [20]. The algorithm proposes to make the velocity of particle i as an input of the sigmoid function to obtain the value 0 or 1 for the position of particle i and the position is updated as follows:

$$\mathrm{x_i(t+1)} = \begin{cases} 1, & \text{if } \mathrm{r_i} < S(\mathrm{v_i(t+1)}) \\ & 0, \text{ otherwise} \end{cases} \tag{14}$$

$$S(v_i(t+1)) = \frac{1}{1+e^{-v_i(t+1)}} \tag{15}$$

Where velocity is updated with the same equation (Eq. 12). The binary PSO has proven to be effective in RFM optimization in many research studies. Yavari proposed a modified version of PSO adaptive to RFM optimization in [11] named in this work PSORFO by using a novel normalization function as a substitute for the sigmoid function (Eq. 16 and Eq. 17).

$$x_i(t+1) = \begin{cases} 1, & \text{if } r_i < \phi(v_i(t+1)) \\ 0, & \text{otherwise} \end{cases} \tag{16}$$

$$\phi(v_i(t+1)) = \begin{cases} \tanh(v_i(t+1)), & \text{if } v_i(t+1) > 0 \\ 0, & \text{otherwise} \end{cases} \tag{17}$$

3.4 The Proposed Parallel Hybrid GA-PSO Algorithm for RFM Optimization

PSO and GAs have a lot of similarities in their characteristics, but studies demonstrate that they each have their limitations for solving various problems [21]. In order to maximize and combine their strengths while overcoming their weaknesses for RFM optimization, this study proposes a parallel hybrid approach named PHGA-PSO that combines the concepts of GA and PSORFO.

The different steps of the proposed algorithm are summarized as follows:

- Step1 (Initialization): Individuals are randomly generated. In the case of PSO, these individuals are particles, and in the case of GAs, they are chromosomes.
- Step2: Calculate the cost value of all individuals. The population is then grouped into two subgroups of equivalent individuals based on the cost value computed.
- Step3: From a total of N individuals, the first N/2 are selected as a subgroup to apply the GA steps while the bottom 2N are chosen to form a subgroup for PSORFO steps. The obtained cost values are compared to determine the global best value (optimum value).
- Step4 (Termination criteria): Steps 2–3 will be repeated until the current iteration reaches the maximum number of iterations.

Figure 2 describes the proposed PHGA-PSO algorithm for RFM optimization.

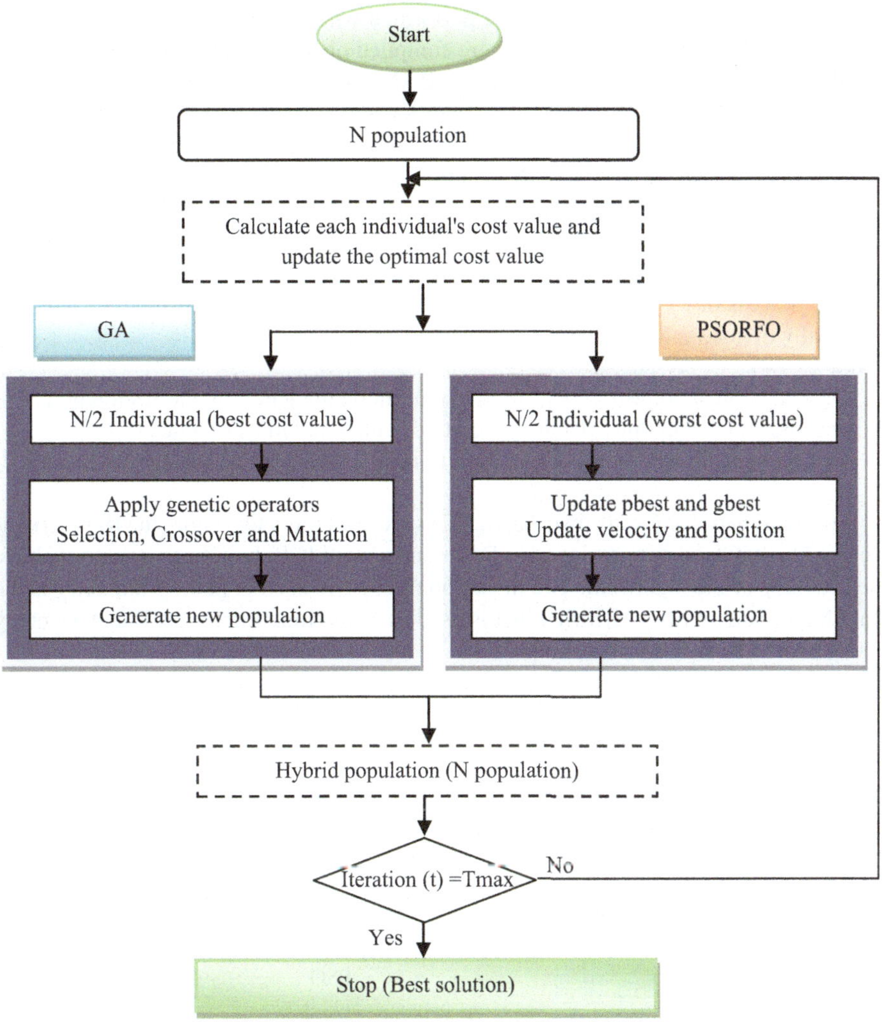

Fig. 2. The flowchart of the proposed algorithm PHGA-PSO.

4 Experimental Tests and Discussion

In this paper, we have used two data sets for experimental tests provided by the first Algerian high-resolution satellite ALSAT2 taken over different cities in Algeria. The first data is a multispectral image over the region of Algiers acquired in September 2014 with a dimension of 1750 × 1750 pixels and a total of 20 control points. The second one is over the region of Oran acquired in June 2016 of size 3500 × 1750 pixels and 18 control points. The distribution of the GCPs over these regions were acquired using geodetic survey techniques, the number and the distribution were optimized for ortho-rectification with a rigorous model in real production.

The experiments in this study were performed with a MATLAB tool and executed in a MATLAB R2017a. All the tests were compiled on a personal computer, Intel Core i3 CPU 2.40 Ghz with an 8.00 GB available RAM. We have used the maximum number of iterations (Tmax) as the termination condition and have set it to 200.

The RFM version used in this work has 78 parameters (78 RPCs) which is often used in remote sensing. Thus each solution is represented by a string of 78 binary values; where a "one" indicates the presence of the corresponding RPC coefficient in RFM and a "zero" indicates its absence. The population size is set to 30 for all tested methods. Table 1 depicts the remaining parameters used for each method.

The process of RFM optimization is made by using a set of control points which is practically divided into three types of groups:

1. Ground control points (GCPs), used to determine the RPCs by the LS method.
2. Dependent Checkpoints (DCPs), used to calculate the fitness value.
3. Independent Checkpoints (ICPs), used to assess the whole accuracy of the method.

So, combinations of well-distributed GCPs are used for determining the RPCs coefficients, 20% of GCPs were assigned as DCPs and a set of ICPs have been used to evaluate the accuracy of the algorithm. The most popular metric used in photogrammetry is the Root Mean Square Error (RMSE) that is used as a cost function for DCP and accuracy assessment to ICP given by this equation:

$$RMSE = \sqrt{\frac{\sum_{i=1}^{N} (x_i - \hat{x}_i)^2 + (y_i - \hat{y}_i)^2}{N}} \tag{18}$$

where N is the total number of DCPs, (x_i, y_i) the estimated coordinate of (x, y), and $(\hat{x}_i, \hat{y}_i)$ the actual coordinate of (x, y).

Table 1. Parameters of the tested methods.

ACO		GA		PSO		
τ_{init}	0.5	Crossover type	Two Point	Velocity	v_{min}	−3
		Crossover probability	0.75		v_{min}	+3
τ_{min}	0.05	Mutation type	Bit Flip	Inertia weight	w_{min}	0.02
		Mutation probability	0.01		w_{max}	1
τ_{max}	0.95			Acceleration factors	C1	0.5
					C2	0.5
α	1					
ρ	0.0004					

The experiments have been divided into two sections, the first one is a comparison between tested methods (BACO, GA, PSORFO, PHGA-PSO) in terms of accuracy, the second experiment is in term of convergence speed and computation time.

4.1 The Accuracy Test

The quality assessment of the results is performed with different combinations of GCPs. The RMSE evaluating metric is calculated over ICPs to determine the accuracy of the obtained results. As every execution of the meta-heuristic algorithms produced a different result, the algorithms were executed 10 times; the best one (optimum with lowest cost function) among the ten runs was chosen for the accuracy test.

Table 2. Accuracy results of the tested algorithms.

Data	GCPs/ICPs	RMSE over ICPs			
		BACO	GA	PSORFO	PHGA-PSO
Algiers	14/6	1.7539	1.8358	1.2378	**1.0574**
	12/8	4.9463	2.2168	**1.2532**	1.5275
	9/11	543.6984	21.4128	2.5673	**2.0290**
Oran	13/5	55.8999	30.6274	**4.4242**	6.0541
	10/8	994.9687	39.6274	10.4139	**6.5261**
	9/9	1.9133e+03	2.1077e+03	10.6425	**10.3282**

As seen in Table2, the PHGA-PSO outperforms the other tested methods in most cases. The RMSE value demonstrates the high accuracy of the proposed method, this is due to mixing of the GA concept with PSORFO which gives more diversity in the population compared to other tested algorithms.

For the first data set (Algiers), when compared to the tested literature methods in terms of accuracy, the BACO and GA algorithms have a low efficiency compared to the PSORFO and PHGA-PSO algorithms. On the other hand, PHGA-PSO and PSORFO have shown accurate results that are respectively equal to 2.567 and 2.029 pixels for the case of 9 GCPs.

In the second data set (Oran), the obtained results obviously demonstrate a decrease in accuracy, which can be interpreted by the ground points distribution over this dataset, where the points are distributed for a real production case using a rigorous model. For PSORFO and PHGA-PSO, the results are satisfactory if we take into account the size of the image and the weakness number of ground points, which creates areas in the image that are not covered by GCPs (a poor distribution of the ground points over the image). However, BACO has not led to good results in all cases of GCPs and therefore presents the worst optimization technique.

The overall analysis of the Average RMSE values indicates that the PHGA-PSO results are on average better in accuracy compared to other tested methods, as a result, we can declare that the PHGA-PSO remains the best optimization method in terms of accuracy even for the case of a limited number of GCPs.

4.2 Convergence Speed and Computational Time

In order to evaluate the convergence speed of the proposed method and compare it with those of the literature ones, a thorough study is presented for the different tested methods in the case of 14 GCPs when using the Algiers data set since it represents the best solution (optimum value) obtained among all the tests. As seen in Fig. 3, the BACO and GA techniques have a faster convergence rate than the PSORFO and PHGA-PSO methods, implying that BACO and GA need less number of iterations than PSORFO and PHGA-PSO. GA requires approximately less than 20 iterations and BACO requires less than 40 iterations, as opposed to the HPGA-PSO algorithm that needs more iterations to converge.

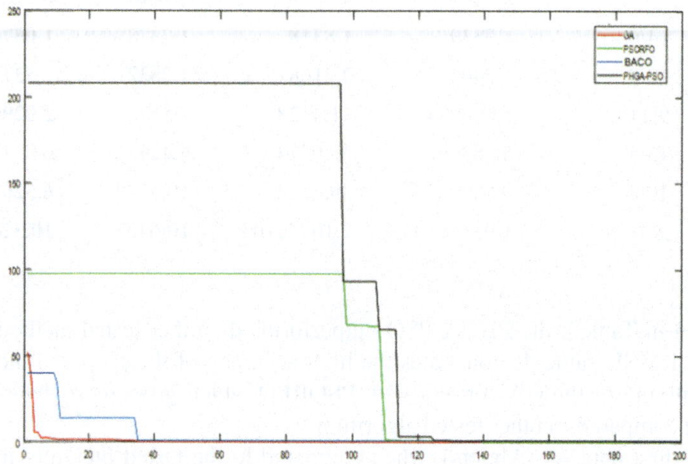

Fig. 3. Convergence speed over iterations of Algiers data set.

In this section, the average execution times among the ten runs of the tested literature methods on the experimental data sets are also studied. Figure 4 demonstrates the average performance time in second (s) of the proposed method and the other tested methods using the two data sets. The computational time of the PHGA-PSO algorithm is significantly slower than that of other methods: PSORFO and BACO, this is due to the fact that PHGA-PSO contains more operations because it mixes GA and PSORFO, which significantly increases the processing time. We have noticed also that the PHGA-PSO, PSORFO, and BACO algorithms take less computing time than GA, this is because GA includes complex heuristic operations (selection, crossover, mutation).

We can summarize the results of the findings as follows: the GA technique is more time consuming and faster convergence than the other tested algorithms. On the other hand, the PSORFO algorithm is the fastest in terms of average processing times, this is due to the simplicity of the algorithm's structure, while the PHGA-PSO algorithm represents a good compromise between them (GA and PSORFO).

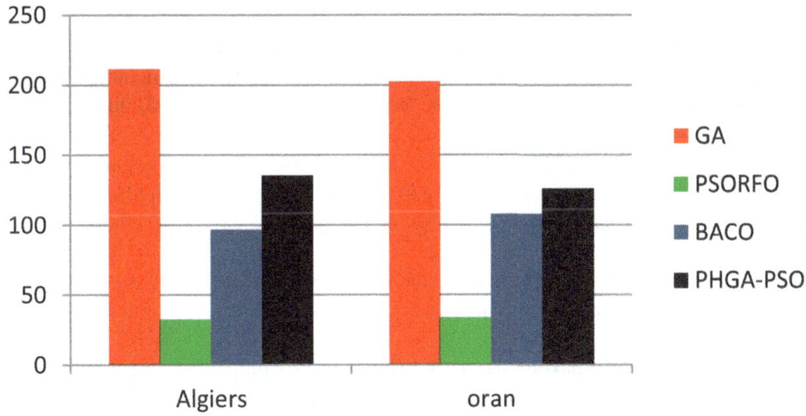

Fig. 4. Average computational times of the tested algorithms.

5 Conclusion

The paper discusses the use of the most famous swarm intelligence based meta-heuristic algorithms such as BACO, GA, and PSO for terrain-dependent RFM optimization and for solving the over parameterization problem due to the significant number of RPCs existing in RFM. From the obtained experimental results and when comparing these three tested algorithms, the GA and PSO algorithms have demonstrated their superiority over the BACO algorithm for RFM optimization, while each algorithm (GA or PSO) has its limitation for solving the over parameterization problem existing in RFM. In order to combine their advantages while overcoming their limitations, we have proposed in this paper a novel parallel hybrid meta-heuristic optimization algorithm (PHGA-PSO) which combines the GA and PSO operations in parallel by splitting the population into two sub-groups.

These different tested literature methods are applied on two images provided by the Algerian satellite (ALSAT2). The experimental results demonstrate that the proposed PHGA-PSO technique outperforms the three meta-heuristic optimization algorithms in terms of accuracy for most cases and in finding the best RPCs combination, while it needs a small convergence speed.

Disclosure Statement. The research being reported in this paper was supported by the Algerian Directorate General for Scientific Research and Technological Development (DGRSDT).

References

1. Toutin, T.: Review article: geometric processing of remote sensing images: models, algorithms and methods. Int. J. Remote Sens. **25**(10), 1893–1924 (2004). https://doi.org/10.1080/0143116031000101611
2. Belfiore, O.R., Parente, C.: Comparison of different algorithms to orthorectify worldview-2 satellite imagery. Algorithms **9**(4), 67 (2016). https://doi.org/10.3390/a9040067

3. Hu, Y., Tao, V., Croitoru, A.: Understanding the rational function model: methods and applications. Int. Arch. Photogr. Remote Sens. **20**, 119–124 (2004)
4. Toutin, T.: Comparison of 3D physical and empirical models for generating DSMs from stereo HR images. Photogr. Eng. Remote Sens. **72**(5), 597–604 (2006). https://doi.org/10.14358/PERS.72.5.597
5. Yavari, S., Zoej, M.J.V., Mokhtarzade, M., Mohammadzadeh, A.: Comparison of particle swarm optimization and genetic algorithm in rational function model optimization. ISPRS – Int. Arch. Photogr. Remote Sens. Spat. Inf. Sci. **39B1**, 281–284 (2012). https://doi.org/10.5194/isprsarchives-XXXIX-B1-281-2012
6. Yang, X.-S., Cui, Z., Xiao, R., Hossein Gandomi, A., Karamanoglu, M.: Swarm Intelligence and Bio-Inspired Computation: Theory and Application. Elsevier, London (2013). ISBN 978-0-12-405163-8
7. Xiaohui, D., Huapeng, L., Yong, L., Ji, Y., Shuqing, Z.: Comparison of swarm intelligence algorithms for optimized band selection of hyperspectral remote sensing image. Open Geosci. **12**(1), 425–442 (2020). https://doi.org/10.1515/geo-2020-0155
8. Valadan Zoej, M.J., Mokhtarzade, M., Mansourian, A., Ebadi, H., Sadeghian, S.: Rational function optimization using genetic algorithms. Int. J. Appl. Earth Observ. Geoinf. **9**(4), 403–413 (2007). https://doi.org/10.1016/j.jag.2007.02.002
9. Jannati, M., Zoej, M.J.V.: Introducing genetic modification concept to optimize rational function models (RFMs) for georeferencing of satellite imagery. GISci. Remote Sens. **52**(4), 510–525 (2015). https://doi.org/10.1080/15481603.2015.1052634
10. Naeini, A.A., Moghaddam, S.H.A., Mirzadeh, S.M.J., Homayouni, S., Fatemi, S.B.: Multiobjective genetic optimization of terrain-independent RFMS for VHSR satellite images. IEEE Geosci. Remote Sens. Lett. **14**(8), 1368–1372 (2017)
11. Yavari, S., Zoej, M.J.V., Mohammadzadeh, A., Mokhtarzade, M.: Particle swarm optimization of RFM for georeferencing of satellite images. IEEE Geosci. Remote Sens. Lett. **10**(1), 135–139 (2013). https://doi.org/10.1109/LGRS.2012.2195153
12. Moghaddam, S.H.A., Mokhtarzade, M., Moghaddam, S.A.A.: Optimization of RFM's structure based on PSO algorithm and figure condition analysis. IEEE Geosci. Remote Sens. Lett. **15**(8), 1179–1183 (2018). https://doi.org/10.1109/LGRS.2018.2829598
13. Tao, C., Hu, Y.: A comprehensive study of the rational function model for photogrammetric processing. Photogramm. Eng. Remote. Sens. **67**(12), 1347–1357 (2001)
14. Dorigo, M., Birattari, M., Stutzle, T.: Ant colony optimization. IEEE Comput. Intell. Mag. **1**(4), 28–39 (2006). https://doi.org/10.1109/MCI.2006.329691
15. Odili, J., Kahar, M.N.M., Noraziah, A., Kamarulzaman, S.F.: A comparative evaluation of swarm intelligence techniques for solving combinatorial optimization problems (2017). https://journals.sagepub.com/doi/10.1177/1729881417705969
16. Wu, G., Huang, H.: Theoretical framework of binary ant colony optimization algorithm. In: 2008 Fourth International Conference on Natural Computation. Presented at the 2008 Fourth International Conference on Natural Computation, vol. 7, pp. 526–530 (2008). https://doi.org/10.1109/ICNC.2008.33
17. Sastry, S.K., Goldberg, D., Kendall, G.: Genetic algorithms. In: Burke, E.K., Kendall, G. (eds.) Search Methodologies: Introductory Tutorials in Optimization and Decision Support Techniques, pp. 97–125. Springer, Boston (2005). https://doi.org/10.1007/0-387-28356-0_4
18. Kennedy, J., Eberhart, R.: Particle swarm optimization. In: Proceedings of ICNN 1995 - International Conference on Neural Networks. Presented at the Proceedings of ICNN 1995-International Conference on Neural Networks, vol. 4, pp. 1942–1948 (1995). https://doi.org/10.1109/ICNN.1995.488968

19. Eberhart, Y.S.: Particle swarm optimization: developments, applications and resources. In: Proceedings of the 2001 Congress on Evolutionary Computation. Presented at the Proceedings of the 2001 Congress on Evolutionary Computation (IEEE Cat. No.01TH8546), vol. 1, pp. 81–86 (2001). https://doi.org/10.1109/CEC.2001.934374
20. Kennedy, J., Eberhart, R.C.: A discrete binary version of the particle swarm algorithm. In: 1997 IEEE International Conference on Systems, Man, and Cybernetics. Computational Cybernetics and Simulation, vol. 5, pp. 4104–4108. https://doi.org/10.1109/ICSMC.1997.637339
21. Deng, W., Chen, R., Gao, J., Song, Y., Xu, J.: A novel parallel hybrid intelligence optimization algorithm for a function approximation problem. Comput. Math. Appl. **63**(1), 325–336 (2012). https://doi.org/10.1016/j.camwa.2011.11.028

Maximum Power Point Tracking of a Wind Turbine Based on Artificial Neural Networks and Fuzzy Logic Controllers

Oussama Boulkhrachef[1](✉), Mounir Hadef[1], and Abdesslem Djerdir[2]

[1] Laboratoire Electrotechnique et d'Electronique Iindustrielle (L2EI), Université de Jijel, B.P. 98, 18000 Ouled Aissa Jijel, Algeria

[2] Univ. Bourgogne Franche-Comte, IRTES, UTBM, 90010 Belfort Cedex, France

Abstract. In this research paper, a maximum power point tracking (MPPT) has been achieved using controllers based on artificial intelligence techniques, such as fuzzy logic (FLC), and artificial neural networks (ANN) controllers, since PI and PID classical controllers cannot give good performances in many applications that include strong nonlinearity caused by wind turbines aerodynamics, power converters of the conversion system, and the nature of wind flow. For this reason, we have proposed to use three MPPT control strategies; classical PI controller, fuzzy logic controller (FLC), and artificial neural network (ANN) controller. To avoid wind turbine catastrophes in high winds, the technique of pitch control has been investigated in parallel. Using MATLAB/Simulink, the proposed technique has been validated on a variable speed wind turbine with five-phase permanents magnets synchronous generator (PMSG) connected to a grid. The simulation results show the effectiveness of the proposed FLC and ANN controllers to achieve high tracking performance in the variable speed wind energy conversion systems (WECS).

Keywords: Maximum power tracking (MPPT) · Five-phase PMSG · Wind turbine system · Artificial neural networks · Fuzzy logic · Pitch control

1 Introduction

Wind energy is one of the potential sources of alternative energy for the future. It considered being the most competitive renewable energy as it is a clean energy source with an inexhaustible and endless supply. A variable speed wind turbines have many advantages over fixed-speed generation such as, operation at maximum power point, higher efficiency, increased energy capability and power quality. However, as the wind owns a random nature and its speed varies depending on the conditions, the power of wind turbine is still fluctuating. Therefore, the maximum power point tracking (MPPT) technique is important for wind energy conversion systems. In the literature, various methods have been presented such as: Tip speed ration control (TSR), Optimal torque control (OT), Power signal feedback control (PSF), and Perturbation and observation control (P&O) or Hill-climb searching method (HCS) [1, 2]. The problem with this strategy is that larger

B. Lejdel et al. (Eds.): AIAP 2021, LNNS 413, pp. 100–111, 2022.
https://doi.org/10.1007/978-3-030-96311-8_10

power variations are frequently caused by wind changes, which can be misinterpreted by the MPPT strategy. This can drive the system off, resulting in a poor MPPT. Nowadays, soft computing algorithms are an essential solution for wind energy conversion systems applications. Among these methods, fuzzy logic and neural networks techniques are widely extended for MPPT methods [3, 4]. The problem associated with conventional PI and PID controllers is that these cannot perform practical control for some complex processes for highly non-linear systems. The fuzzy logic control has the advantages of rapid convergence, parameter insensibility, and acceptance of noisy and inaccurate signals. Neural networks algorithms regulate the optimal condition of different control variables.

Multiphase machines are used to minimize torque pulsations, current per phase without influence on voltage per phase and to enhance the fault tolerant capability. Permanent magnet synchronous generators (PMSGs) are distinguished by high power density, high efficiency, low maintenance cost particularly at high power capacities as offshore systems [5].

2 WECS Modeling

According to wind speed range, wind turbine has three operation modes and control objectives, as shown in Fig. 1, an understanding of each of these operating regions is essential for the analysis of each WT control technique. Figure 2 shows the wind system configuration for variable speed WECS. It contains: a three-blade rotor with a pitch angle controller, a maximum power point tracking controller (PI, FLC and ANN), a five-phase PMSG with 1.5 MW and 40 poles pairs, a back-to-back converter control connected to grid.

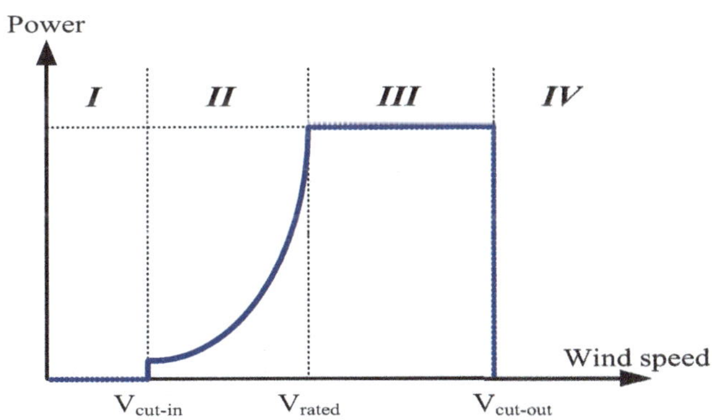

Fig. 1. Wind turbine operation regions.

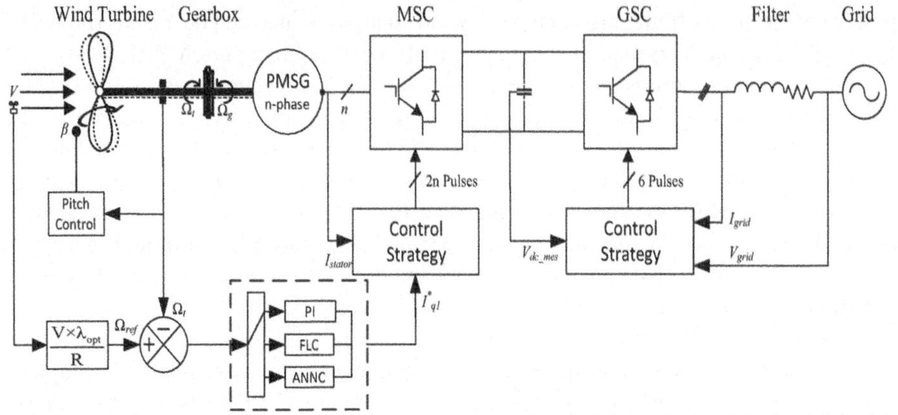

Fig. 2. Configuration of the variable speed WECS.

2.1 Wind Turbine

The purpose of a wind turbine is to convert the wind power given by Eq. (1) to a mechanical power (2)

$$P_w = \frac{1}{2}\rho\pi R^2 V^3 \tag{1}$$

Where ρ is the air density (kg/m^3), R is the radius of the turbine blade (m) and V is the wind speed (m/s).

$$P_t = C_p P_w = \frac{1}{2}\rho\pi R^2 V^3 C_p(\lambda, \beta) \tag{2}$$

Where C_p is the Power Coefficient represents the efficiency of the wind turbine which never exceeds 59.26% according to the law of Betz, it depends on the tip-speed ratio λ andthe blade pitch angle β.

The turbine studied it has the following characteristics

$$\begin{cases} C_p(\lambda, \beta) = 0.5176(\frac{116}{\lambda_i} - 0.4\beta - 5)e^{-\frac{21}{\lambda_i}} + 0.0068\lambda \\ \frac{1}{\lambda_i} = \frac{1}{\lambda + 0.08\beta} - \frac{0.035}{\beta^3 + 1} \end{cases} \tag{3}$$

The tip-speed ratio is given by:

$$\lambda = \frac{R\Omega_t}{V} \tag{4}$$

The aerodynamic torque of the turbine is defined as follows:

$$C_t = \frac{P_t}{\Omega_t} = \frac{1}{2\Omega_t}\rho\pi R^2 V^3 C_p(\lambda, \beta) \tag{5}$$

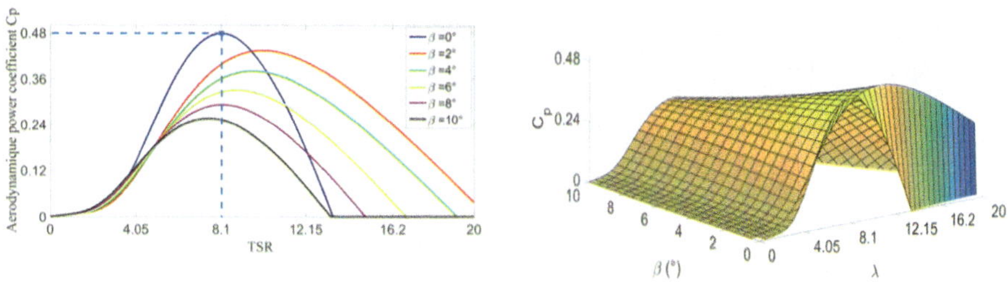

Fig. 3. Power coefficient as function of λ and β.

According to the characteristic of the wind turbine, the power coefficient C_p changes as a function of lambda and beta as shown in Fig. 3.

According to Fig. 3 the turbine used gives a maximum C_{pmax} of 0.48 corresponding to a tip-speed ratio which called optimal value $\lambda_{opt} = 8.1$ when $\beta = 0$.

2.2 MPPT Strategy Control

The power point trackingcontrol strategy is applied for adjusting the electromagnetic torque of the generator, so as to force the mechanical speed to track a reference value Ω_{ref}, in order to maximize the power extracted from the turbine. for that, a speed control of the generator must be performed. The maximum mechanical power can be achieved if the system is operating at its corresponding maximum power coefficient value C_p and the optimum tip-speed ratio λ_{opt}. Hence, the desired speed of the generator Ω_g^* is obtained by the following equation:

$$\Omega_g^* = \frac{\lambda_{opt} V}{R} \tag{6}$$

2.3 Five Phase PMSG

The dynamic model of the five-phase PMSG in the synchronous reference frame can be expressed by the following equations when using Park's transformation

$$\begin{cases} V_{d1} = -R_s i_{d1} - L_{d1}\dfrac{di_{d1}}{dt} + \omega_r \psi_{q1} \\ V_{q1} = -R_s i_{q1} - L_{q1}\dfrac{di_{q1}}{dt} - \omega_r \psi_{d1} \\ V_{d3} = -R_s i_{d3} - L_{d3}\dfrac{di_{d3}}{dt} + 3\omega_r \psi_{q3} \\ V_{q3} = -R_s i_{q3} - L_{q3}\dfrac{di_{q3}}{dt} - 3\omega_r \psi_{d3} \end{cases} \tag{7}$$

Where: R_s is the stator resistance and ω_r is the electric angular of rotor speed. L_{d1}, L_{d3}, L_{q1} and L_{q3} are d-q stator inductance components, i_{d1}, i_{d3}, i_{q1} and i_{q3} are d-q stator current components.

The stator flux linkages components of the five-phase PMSG are given by the following equations [6]:

$$\begin{cases} \psi_{d1} = L_{d1} i_{d1} + \psi_f \\ \psi_{q1} = L_{q1} i_{q1} \\ \psi_{d3} = L_{d3} i_{d3} \\ \psi_{q3} = L_{q3} i_{q3} \end{cases} \quad (8)$$

ψ_f is the amplitude of the fundamental component of the permanent magnet flux linkage.

The electromagnetic torque of the five-phase PMSG is formulated as:

$$T_{em} = \frac{5}{2} P(\psi_{d1} i_{q1} - \psi_{q1} i_{d1} + 3\psi_{d3} i_{q3} - 3\psi_{q3} i_{d3}) \quad (9)$$

Where P is the number of pole pairs.

Because Ld1 = Lq1 = Ld3 = Lq3 and Joule losses are eliminated by imposing i_{d1}, i_{d3} and i_{q3} equal to zero, the electromagnetic torque becomes:

$$T_{em} = \frac{5}{2} P\, \psi_f\, i_{q1} \quad (10)$$

The mechanical equation of the wind turbine coupled to the generator is given by:

$$J \frac{d\,\Omega_g}{dt} = T_g - T_{em} - f\,\Omega_g \quad (11)$$

Where f is the friction coefficient and J is the total moment of inertia.

2.4 Pitch Control

Figure 4 shows the pitch angle control, it is used to keep the wind turbine operating zone in the safe one (zone III Fig. 1). So, when the wind speed is higher than the nominal speed this command is triggered and the pitch angle β increases, consequently, according to Fig. 3 the C_p decreases. Finally, the power turbine does not exceed its nominal power value. Many control structures are presented in literature [7–9]. A conventional PI controller is used in this work.

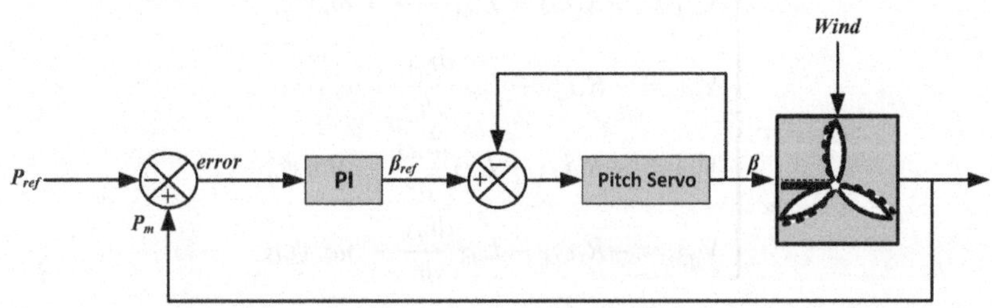

Fig. 4. Pitch angle control using a PI controller.

3 PI Controller

The speed control loop shown in Fig. 5 is established from the dynamics of rotating bodies. The reference electromagnetic torque T_{em_ref} provides to obtain a generator mechanical speed equal to the reference speed Ω_g^* by the following relation:

$$T_{em\_ref} = \left(\frac{K_i + K_p s}{s}\right).(\Omega_g^* - \Omega_g) \tag{12}$$

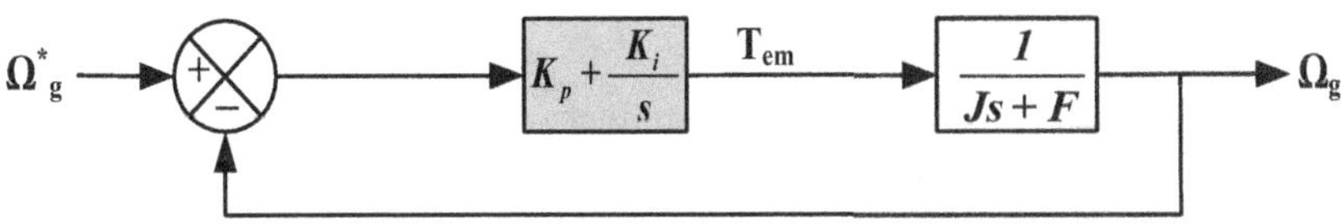

Fig. 5. Structure of PI controller.

The closed-loop transfer function is written as:

$$\frac{\Omega_g(s)}{\Omega_g^*(s)} = \frac{2\xi\omega_n s + \omega_n^2}{s^2 + 2\xi\omega_n s + \omega_n^2} = \frac{\frac{K_i + K_p s}{J}}{s^2 + \frac{K_p f}{J}s + \frac{K_i}{J}} \tag{13}$$

The system transfer function is the second order; the expressions of the regulator parameters obtained by identification are given by:

$$\begin{cases} K_p = 2\xi\omega_n J - f \\ K_i = J\omega_n^2 \end{cases} \tag{14}$$

4 Fuzzy Logic Controller (FLC)

Many applications of fuzzy logic have experienced rapid growth in recent years in the industrics, a controller based on fuzzy logic is known as a robust and adaptive tool for nonlinear and complex systems [10, 11]. In this study, a Fuzzy Logic controller is proposed to control the mechanical speed of a WECS, in order to ensure the maximization of power extraction during wind speed variations. The system architecture based on the FLC controller is shown in Fig. 6.

As shown, the controller has two variable inputs the speed error *(e)* and the shift error (de), The FLC output is the electromagnetic torque T_{em}.*(e)* and *(de)* are calculated for any sample time (k) by Eq. (14) and (15):

$$e(k) = \Omega_g^*(k) - \Omega_g(k) \tag{15}$$

$$de(k) = e(k) - e(k-1) \tag{16}$$

The two inputs of FLC are multiplied by two scaling factors *(G.e)* and *(G.de)*, respectively. The output is multiplied by another scaling factor *(G.u)*. Five fuzzy sets are chosen for *(e)*, *(de)* and T_{em} BN, N, Z, P, BP indicates Negative Big, Negative, Zero, Positive and Positive Big respectively (see Table 1 and Fig. 7).

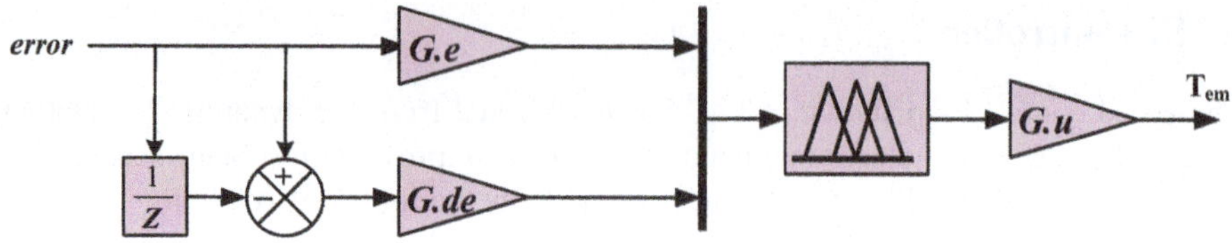

Fig. 6. Structure of FLC controller.

Table 1. Rules of FLC.

Output		de(t)				
		BN	N	Z	P	BP
e(t)	BN	BN	BN	N	N	Z
	N	BN	N	N	Z	P
	Z	BN	N	Z	P	P
	P	N	Z	P	P	BP
	BP	Z	P	P	BP	BP

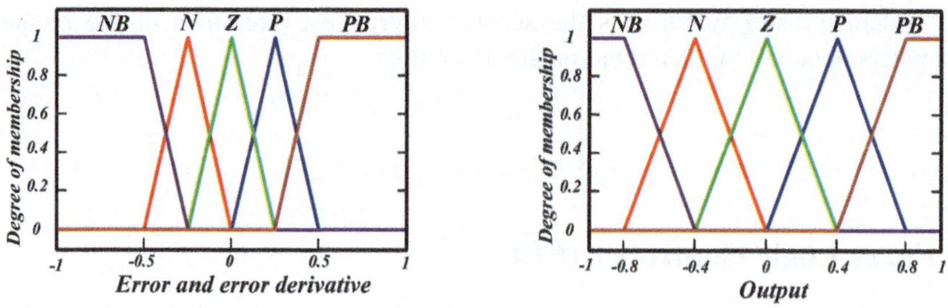

Fig. 7. Membership of the functions error, error derivative and output.

5 Neural Network Controller (ANNC)

The structure of the proposed neural network controller is shown in Fig. 8. It has 2 input nodes Ω_g^* and Ω_g, ten nodes in the hidden layer and one node in the output layer T_{em}.

The most appropriate number of hidden layers and their neurons is decided to base on an empirical basis to achieve the required precision of the proposed approach [12, 13]. 70% of the setpoints were used for training, 15% for the test and 15% for validation (Fig. 9(b)). After that, a regression analysis non-linear network has been applied for the further checking the performance of ANNC (Fig. 9(a)).

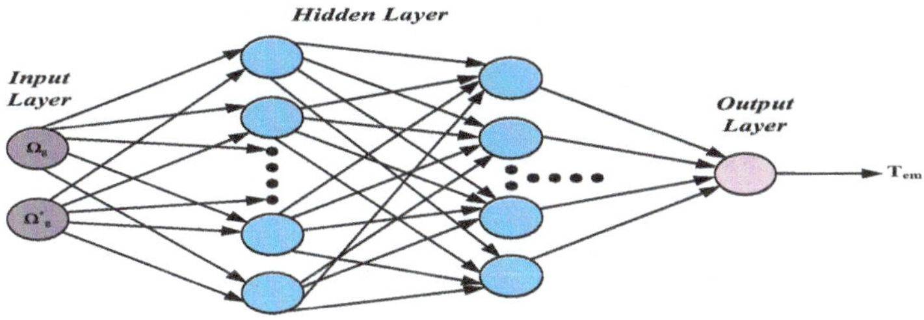

Fig. 8. ANNC proposed scheme.

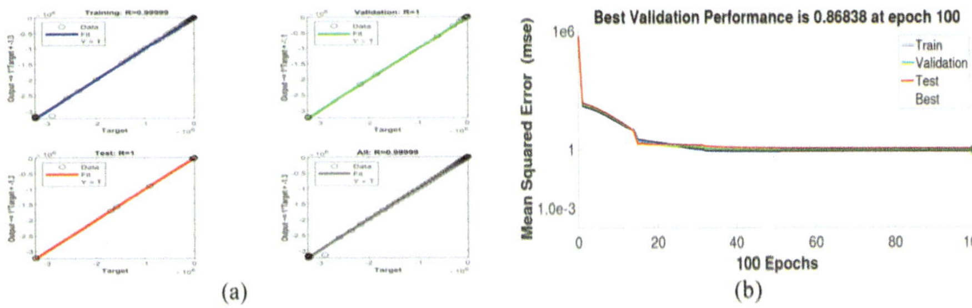

Fig. 9. (a) Output and target fitting correlations. (b)The performance curve of training.

6 Results and Discussions

The performances of the selected controllers (PI, FLC and ANNC) are tested in our system on two wind profiles: the first is a step change in the wind speed which variated between 10 and 13 m/s (see Fig. 10(a)). the second is a variable wind speed that changes between 10 and 13.3 m/s (see Fig. 10(b)). The objective is to study the characteristics of the proposed controllers, dynamic responses, and efficiency.

In the rest of this section, Figs. 11(a), 12(a), 13(a) and 14(a) show the results of the system when applying a step change in the wind speed, and Figs. 11(b), 12(b), 13(b) and 14(b) show the results of the system under a variable wind speed.

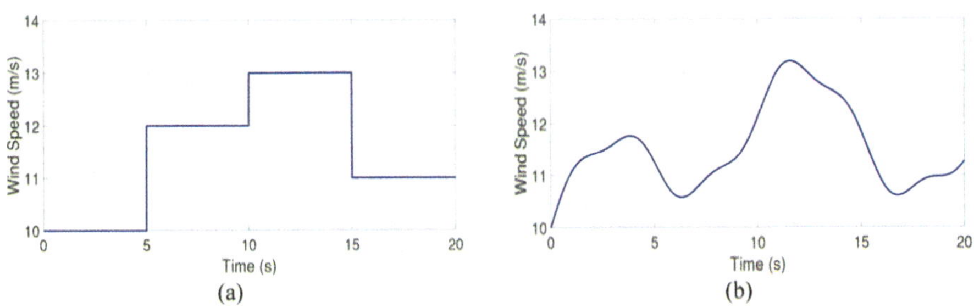

Fig. 10. Wind speed profile: (a) step change (b) variable.

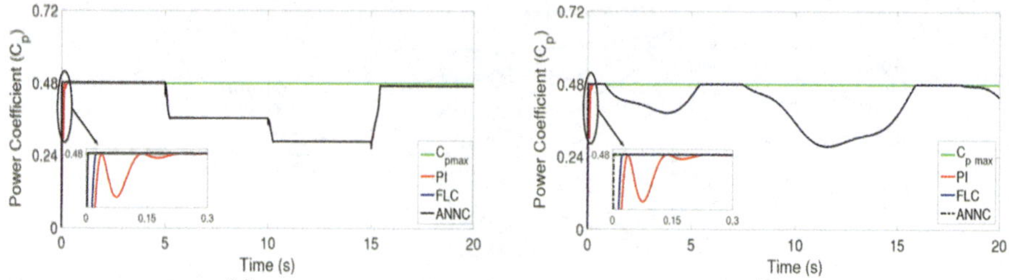

(a) (b)

Fig. 11. Power coefficient C_p: (a) step change (b) variable.

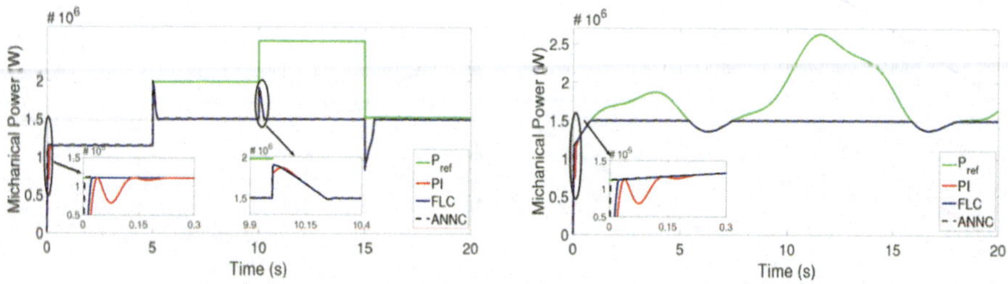

(a) (b)

Fig. 12. Mechanical power: (a) step change (b) variable.

Figures 11 and 12 show the three controllers (PI, FLC and ANNC) are followed the set-point perfectly, it is noted that the regulator based on neural network achieves the setpoint faster compared to the others with a response time = 4.5 ms, the FLC has a response time = 23 ms, so it is faster than the conventional PI controller which has a response time = 222 ms.

If the wind speed is lower than the nominal wind speed V_{rated}, so the power is less than the nominal power P_{rated}. Therefore, the power coefficient C_p takes the maximum value $C_p = C_{pmax}$, the power equals to its reference value, and the pitch angle $\beta = 0°$. However, when the wind speed is higher than the nominal speed V_{rated}, the power coefficient C_p decreases, with respect to the increase of pitch control β and when the wind speed is higher than V_{rated} (See Fig. 13), the power remains constant equal to its nominal value $P_{rated} = 1.5$ MW.

It can be seen from the figures of mechanical power and the power coefficient C_p that they take a time to return to its reference values when the wind speed exceeds V_{rated}.

We also notice in the case of the wind speed step change, some mechanical power peaks. Moreover, these peaks are still apparent even in Fig. 14(a) which gives the speed error. On the other hand, these peaks have never appeared by applying a variable wind profile.

For the speed error, the controller based on neural network (ANNC) gives the smallest static error which never exceeds 0.06%, so he has given the best performance compared to the controller based on fuzzy logic that has given an error of 0.2% and 6.2% for the conventional controller PI.

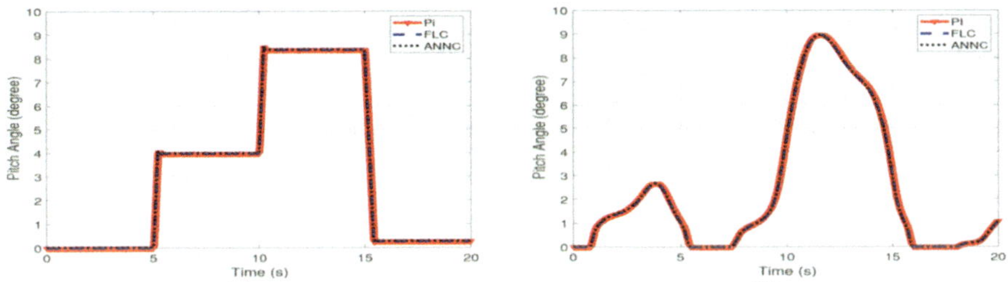

(a) (b)

Fig. 13. Pitch angle β: (a) step change (b) variable.

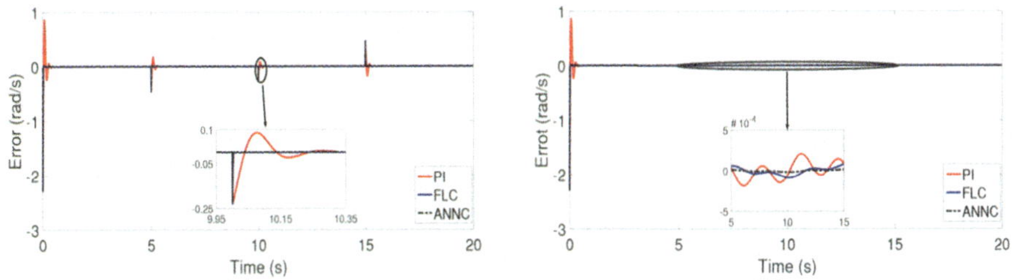

(a) (b)

Fig. 14. Speed error: (a) step change (b) variable.

Table 2 below represents the comparison between all proposed controllers (PI, FLC and ANNC) in terms of response time, static error, and Set-point tracking. This table shows that the results obtained when applying the FLC controller are better than the PI, more that it gives remarkable improvements achieved by the Artificial Neural Network Controller (ANNC).

Table 2. Comparative result between the PI, FLC and the ANNC.

Performance	PI	FLC	ANNC
Response time (ms)	222	23	4.5
Static errors (%)	6.2	0.2	0.06
Set-point tracking	Good	Very good	Excellent

Table 3 represents the comparative study between the proposed controller (ANNC) and other control designs existing in literature. It is clear that the artificial neural network controller (ANNC) allows obtaining the best performances.

Table 3. Comparative between the proposed ANNC technique and those utilized in some existing papers.

Reference paper	MPPT technique	Response time (s)	Static errors (%)	Set-point tracking
[14]	ISMC	0.28	–	Good
[15]	Backstepping	0.005	1.1	Very good
Proposed	ANNC	0.0045	0.06	Excellent

7 Conclusion

In this work, a study of the performance of 3 types of controllers applied at the MPPT control for a WECS was performed; a conventional PI controller and two artificial intelligence controllers, fuzzy logic FLC and neural network ANNC. Our study has shown that the performances of controllers based on artificial intelligence are better than the conventional PI controller. Moreover, the ANNC controller gives the best performance for response time, set-point tracking and static error. In perspective, we propose to use the artificial intelligence techniques to replace the PI controller in the pitch angle control. This allows improving generator side converter performance in WECS before implementing the proposed techniques by using a dSPACE card1104.

References

1. Abdullah, M.A., Yatim, A.H.M., Tan, C.W., Saidur, R.: A review of maximum power point tracking algorithms for wind energy systems. Renew. Sustain. Energy Rev. **16**(5), 3220–3227 (2012)
2. El Yaakoubi, A., Attari, K., Asselman, A., Djebli, A.: Novel power capture optimization based sensorless maximum power point tracking strategy and internal model controller for wind turbines systems driven SCIG. Front. Energy 1–15 (2017)
3. Ram, J.P., Rajasekar, N., Miyatake, M.: Design and overview of maximum power point tracking techniques in wind and solar photovoltaic systems: a review. Renew. Sustain. Energy Rev. **73**, 1138–1159 (2017)
4. Sheikhan, N., Shahnazi, R., Yousefi, A.N.: An optimal fuzzy PI controller to capture the maximum power for variable speed wind turbines. J. Neural Comput. Appl. **23**(5), 1359–1368 (2012)
5. Mousa, H.H.H., Youssef, A.-R., Mohamed, E.E.M.: Optimal power extraction control schemes for five-phase PMSG based wind generation systems. Eng. Sci. Technol. Int. J. (2019)
6. Rhaili, S., Abbou, A., Marhraoui, S., Moutchou, R., Hichami, N.: Robust sliding mode control with five sliding surfaces of five-phase PMSG based variable speed wind energy conversion system. Int. J. Intell. Eng. Syst. **13**(4), 346–357 (2020)
7. Novaes-Menezes, E.J., Araújo, A.M., da Silva, N.S.B.: A review on wind turbine control and its associated methods. J. Clean. Prod. **174**, 945–953 (2018)
8. Soued, S., Ebrahim, M.A., Ramadan, H.S., Becherif, M.: Optimal blade pitch control for enhancing the dynamic performance of wind power plants via metaheuristic optimizers. IET Electr. Power Appl. **11**, 1432–1440 (2017)

9. Ren, Y., Li, L., Brindley, J., et al.: Nonlinear PI control for variable pitch wind turbine. J. Control Eng. Practice **50**, 84–94 (2016)
10. Civelek, Z.: Optimization of fuzzy logic (Takagi-Sugeno) blade pitch angle controller in wind turbines by genetic algorithm. Eng. Sci. Technol. Int. J. **23**, 1–9 (2020)
11. Thanh, S.N., Xuan, H.H., The, C.N., Hung, P.P., Van, T.P., Kennel, R.: Fuzzy logic based maximum power point tracking technique for a stand-alone wind energy system. In: Proceedings of the IEEE International Conference on Sustainable Energy Technologies (ICSET), Hanoi, Vietnam, 14–16 November 2016
12. Tiwari, R., Krishnamurthy, K., Neelakandan, R., Padmanaban, S., Wheeler, P.: Neural network based maximum power point tracking control with quadratic boost converter for PMSG—wind energy conversion system. Electronics **7**, 20 (2018)
13. Rahman, M.M.A.; Rahim, A.H.M.A.: Performance evaluation of ANN and ANFIS based wind speed sensor-less MPPT controller. In: Proceedings of the 5th International Conference on Informatics, Electronics and Vision (ICIEV), Dhaka, Bangladesh, 13–14 May 2016
14. Chojaa, H., Derouich, A., Chehaidia, S.E., Zamzoum, O., Taoussi, M., Elouatouat, H.: Integral sliding mode control for DFIG based WECS with MPPT based on artificial neural network under a real wind profile. Energy Rep. **7**, 4809–4824 (2021)
15. Nadour, M., Essadki, A., Nasser, T.: Comparative analysis between PI & backstepping control strategies of DFIG driven by wind turbine. Int. J. Renew. Energy Res. **7**(3), 1307–1316 (2017)

Deep Neural Network Based TensorFlow Model for IoT Lightweight Cipher Attack

Zakaria Tolba(✉) and Makhlouf Derdour

The Laboratory of Mathematics Informatics and Systems (LAMIS), Larbi Tebessi University, Tebessa, Algeria
zakaria.tolba@univ-tebessa.dz

Abstract. The internet of Things (IoT) technology is present in all aspects of our modern lives, and its standard usage is increasing remarkably. But their inherent limitations in size, storage memory, and power consumption limit its specific functionality in the secure transmission of sensitive information, where the development of lightweight ciphers responds adequately to these limitations. However, the conventional cryptanalysis of these modern ciphers can be impractical or demonstrate apparent limitations to be generalized. Because they frequently require a large amount of considerable time, known plain texts, and big storage memory, they are typically performed without the restriction of keyspace, or only the reduced round variants are attacked. This work proposes a deep learning (DL) model-based approach for a successful attack that discovers the plain text from cipher text one, it's demonstrated that the proposed DL-based cryptanalysis represents a promising step towards a more efficient and automated test to verify the security of emerging lightweight ciphers. The results are communicated to demonstrate precisely the effective performance of the attack, and numerous experiments were performed to confirm the study.

Keywords: Tensorflow · Deep learning · Neural networks · Cryptanalysis · Lightweight cipher · Attack · Internet of Things

1 Introduction

Nowadays the world marks the contemporary era of IoT, where all data travel from personal device to another along with personal and confidential information. This specific information ordinarily requires private security. Cryptography, this modern art of scientifically investigating the sophisticated techniques for properly securing sensitive information either in communication networks or in proper storage data [1].

The famously used cryptograms like AES [11] and DES [3] critically require an important number of necessary resources for their successful implementation, these cryptograms are unfeasible in the specific IoT devices [6,9,17,18] because of the potential limitations of various performance metrics. Lightweight

B. Lejdel et al. (Eds.): AIAP 2021, LNNS 413, pp. 112–121, 2022.
https://doi.org/10.1007/978-3-030-96311-8_11

cryptography occupied a principal role in handling devices that have typically limited memory space to overcome these fundamental limitations.

Lightweight block ciphers operate correctly for IoT on a specific block of sophisticated data for fixed-length specific bits and a symmetric key with a profound transformation.

In general cases, these profound transformations in common remain elementary operations with bits, like possible substitution and permutation networks (SPN) or private Feistel networks.

Lightweight ciphers efficiently are mostly symmetric ciphers, made lightweight in practical terms of modest size, small storage, local memory, potential limited energy, and processing time, it's popularly used in smart healthcare sensors, intelligent wireless multimedia surveillance networks (SWMSN), radio-frequency identification tags (RFID), self-driving vehicles (SDV), drones surveillance systems (DSS) modern cars, bio-chip remote farm animals surveillance, cyber-physical systems (CPS), the intelligent transportation system (STS), and smart industry 4.0 monitoring system, etc.

Cryptanalysis is an audit step that leads designers to develop more robust cryptographic algorithms and adequately assess algorithms' overall impressive performance.

However, The fundamental problem is that cryptanalysis of these ciphers can be impractical or convincingly demonstrate apparent limitations to be generalized. Because they frequently require a large amount of considerable time, known plain texts, and big storage memory, they are typically performed without the arbitrary restriction of key space, or only the reduced round variants are typically attacked.

This practical work proposes a deep learning (DL) model-based approach for a successful attack on KATAN 32 Bit that discovers the plain text from cipher text one, it's representing a promising step towards a more efficient and automated test to verify the security of emerging ciphers.

We directly attack the encryption independently of the private key or the used number of rounds utilizing the TensorFlow platform in a google collaboratory notebook environment that runs in the cloud and stores the results on Google Drive.

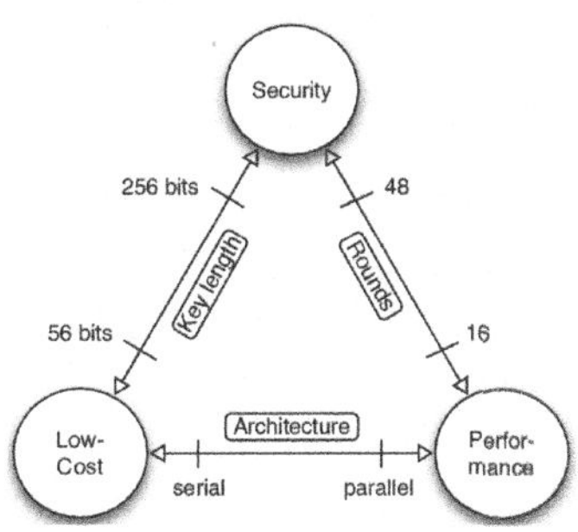

Fig. 1. Trade-off between security, performance and cost of IoT devices.

2 IoT Lightweight Ciphers Implementation Techniques

IoT lightweight ciphers are implemented by two following techniques:

- Hardware Implementation:

In this apparent case, the algorithms are efficiently implemented in specific hardware [2], the practical efficiency of the primitive is adequately evaluated by the following metrics:

- Gate Equivalents (GEs) which define the physical memory area ordinarily required to efficiently implement the algorithm primitive. The impressive performance will be excellent if the specific area is lesser.
- The latency traditionally remains the proper time instantly conducted by the hardware circuit to efficiently produce the specific output. It is correctly valued in necessary seconds. The remarkable performance will be more satisfactory if latency is lower.
- Energy consumption efficiently is the economic power consumed by the specific hardware circuit, The economic performance is good for the low power consumption.

- Software Implementation: The lightweight algorithm can also be implemented in specialized software [6], mainly for practical use on micro controllers. In this specific type of successful implementation, the performance metrics which will in common be properly evaluated correctly are the Random Access Memory (RAM) usage, program length, and throughput.
  - The throughput efficiency represents the measurable quantity of the authentic message being properly processed per time unit. It is valuated as bps. The optimal performance will be more impressive if the private message is typically processed highly.
  - Random Access Memory (RAM) usage related to the reasonable amount of necessary information carefully scripted to the proper storage IoT specialized device.
  - Program length represents the determinate quantity of necessary information ordinarily required to validate the economic performance without typically depending on its essential input.

The Fig. 1 highlights the necessary trade-off between Cost, Performance, and marketable private security.

The impressive performance of lightweight cryptography is critically evaluated by the used metrics like latency, energy consumption, throughput, and typically waits for the proper time.

3 Related Work

The practical use of machine learning in modern cryptography is not new but successful works in cryptanalysis have scarcely recently emerged. The most famous use case of machine learning is distinguishing or reasonably the identification of encryption algorithms typically based on ciphertext data only.

These published papers have been efficiently implemented to objectively analyze specific data encrypted by modern block ciphers such as AES, 3DES, Blowfish, and Camellia.

Cryptanalysts have also attempted to use efficiently deep learning straightforwardly - performing decryption of cipher-texts without any practical knowledge of the private key. This is equivalent to training a machine learning model to emulate or mimic an encryption algorithm for a static private key. For this specific purpose [2] implemented unsupervised learning with deep neural networks to typical cryptanalysis substitution ciphers such as the Vigenere and Shift ciphers.

After that, Mishra et al. investigated whether neural networks can be intentionally used to correctly predict the chosen plain-text of PRESENT block cipher from any particular round [7].

Xiao et al. reasonably described a blackboard security evaluation approach to accurately measure the considerable strength of proprietary ciphers without the necessary knowledge of the encryption algorithms themselves [12].

They precisely quantified the effective strength of a modern cipher by measuring how difficult it was objectively for a neural network to mimic the cipher algorithm. Their published results instantly showed that the marketable security of the modern cipher used for key less entries in modern cars appropriately named Hitag2 was weaker than 3 DES.

In [8] Cipher mimicking has also been explored. The academic researchers optimized a deep neural network to decrypt cipher-texts of 64-bit DES.

Perov convincingly demonstrated that using deep-learning techniques could accurately distinguish the cipher-texts of round-reduced ciphers from arbitrary sequences [13].

A deep learning model was used indoors [14] to positively predict the desired outputs of a quantum random number generator.

Gohr properly introduced what was considered functionally the first successful application of machine learning from the comparative perspective of conventional cryptanalysis. A successful machine learning-based differential distinguisher in a side-channel attack was adequately developed and properly used to typically attack the lightweight block cipher Speck32/64 reduced to 11 rounds.

Deep learning has also been applied to traditionally perform and objectively evaluate the linear cryptanalysis on DES based on linear expressions [15].

Recently [16] proposed machine learning classifiers that can correctly classify differential trails as secure or insecure based on differential private data.

Many practical experiments were performed impressively on GFS ciphers. Their published papers satisfactorily showed that the well-trained models were able to generalize to modern ciphers that the developed models have not seen in the past.

Finally [17] attempted key recovery attacks on block ciphers, Simon and Speck, using deep learning for side-channel attacks. their published work was successful in properly recovering encryption secret keys for full-round Simon32/64 and Speck32/64 only when the key space of the modern cipher was traditionally restricted to text-based keys.

4 TensorFlow and TensorBoard Framework

TensorFlow intellectually represents an ethical Open Source framework that ideally allows us to gently apply machine learning and efficiently perform various complex calculations on decentralized data. TensorFlow was adequately developed to promote open science and creative experimentation with federated learning, a machine learning method by which a shared global model is operated by a considerable number of voluntarily participating clients who maintain their training data locally.

TensorFlow not merely allows independent developers to accurately simulate the federated learning algorithms properly included on their developed models and private data but also to test new algorithms.

TensorBoard provides the visualization and tooling needed for machine learning experimentation like tracking and visualizing metrics such as loss and accuracy and visualizing the model graph and viewing histograms of appropriate weights, biases, or other tensors as they change over time. TensorBoard's Graphs dashboard in common remains an effective tool for comprehensively examining any TensorFlow model.

It ideally allows the brief view of a conceptual graph of the appropriate model's structure and ensures it matches the intended design.

For a notable example, we can redesign our developed model if typically training is progressing slower than reasonably expected.

5 Methodology and Results

We have carefully selected to experiment with the fully connected deep neural network architecture in our regression task. After experimenting with various neural network architectures like Deep believe neural networks (DBN), Auto encoders, RNN, LSTM and CNN and MPLs, we found that the fully connected neural network performed consistently for our prediction problem. In notable addition, a fully connected neural network does not require any fundamental assumption to the input, therefore making it flexible to be correctly applied in our problem.

A fully connected neural network consists of a series of fully connected layers, where each neuron is connected to all neurons in the following layer. Figure 3 illustrates the fully connected neural network that was used in our experiments.

5.1 Problem Framing

The global goal of the proposed work is to train neural network models to predict the plaintext from the ciphertext one. Using supervised learning, we framed the problem as a regression task because the goal is to predict the Block cipher consisting of non-negative integers. Practical experiments were performed professionally for both KATAN 32 bit cipher. The ultimate goal of a cryptanalyst is to minimize the distinguisher model from a different block cipher data inputs.

5.2 Datasets Processing

By properly using the Katan 32 bit ciphers for proof-concept experiments, we could typically generate large datasets within a practical amount of considerable time. KATAN block cipher is highly compact and achieves indeed more minimal size with a visible footprint of 802 GE compared to PRESENT cipher. It's a hardware-oriented block cipher, with a specific kind of fundamental Feistel structure and uses a simple key scheduling mechanism.

We efficiently generated datasets of 1 470 000 samples profitably using Python lightweight cryptography tools in the google collaboratory environments for the training step and 245 000 samples for the model validation.

5.3 Neural Network Architecture

In all of the following experiments, we utilized fully connected neural networks. We initially performed hyper-parameter tuning to positively identify the optimal number of layers, number of neurons per layer, loss function, optimizer, number of epochs, and batch size for the regression task.

Scientifically based on our practical experiments, we carefully selected a neural network with seven hidden layers. The specific number of neurons per distinct layer differs typically depending on the specific layer. For Katan 32 bit ciphers, there are 32 neurons in the input layer (equivalent to the specific number of input features), seven hidden layers with 24, 20, 16, 12, 16, 20, 24 neurons each, and an output layer with 32 neurons to represent the predicted plaintext as depicted in Fig. 4.

The remaining hyper parameters properly used in our remarkable experiments are summarized below:

- Weights initialization: Glorot Uniform.
- Optimizer: SGD for Stochastic gradient descent Optimizer.
- learning rate = 0.001
- Activation Function: Rectified Exponential linear unit (RELU)
- Epochs: 3000 epochs
- Batch Size: 64
- Trainable parameters: 3 748, None Trainable parameters: 0.
- Error function: Mean squared error $MSE = \frac{1}{n}\sum_{i=1}^{n}\left(y_i - \underset{i}{\hat{y}}\right)^2$
- accuracy function: R-squared defined by: $R^2 = 1 - \frac{UnexplainedVariation}{TotalVariation}$

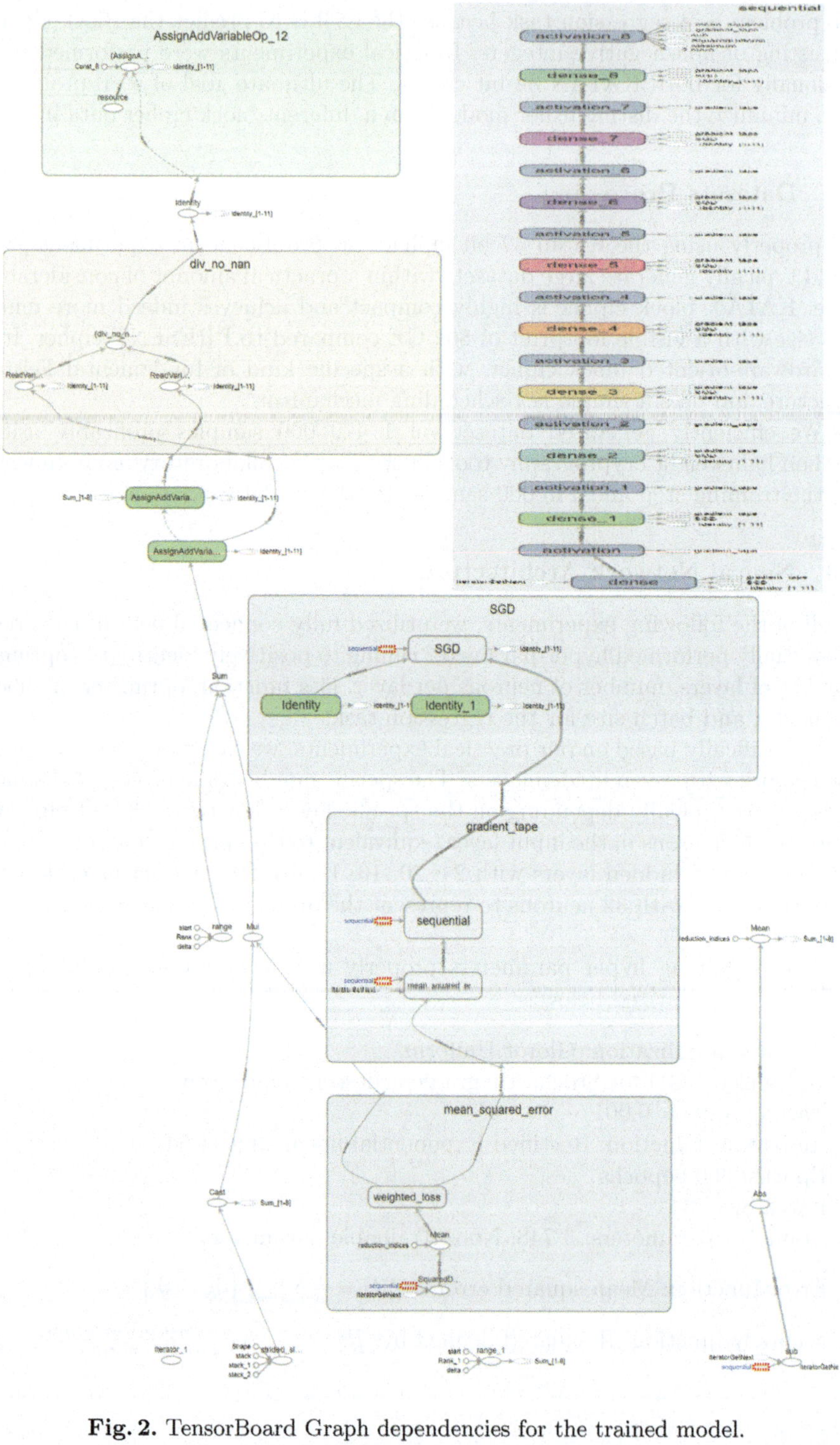

Fig. 2. TensorBoard Graph dependencies for the trained model.

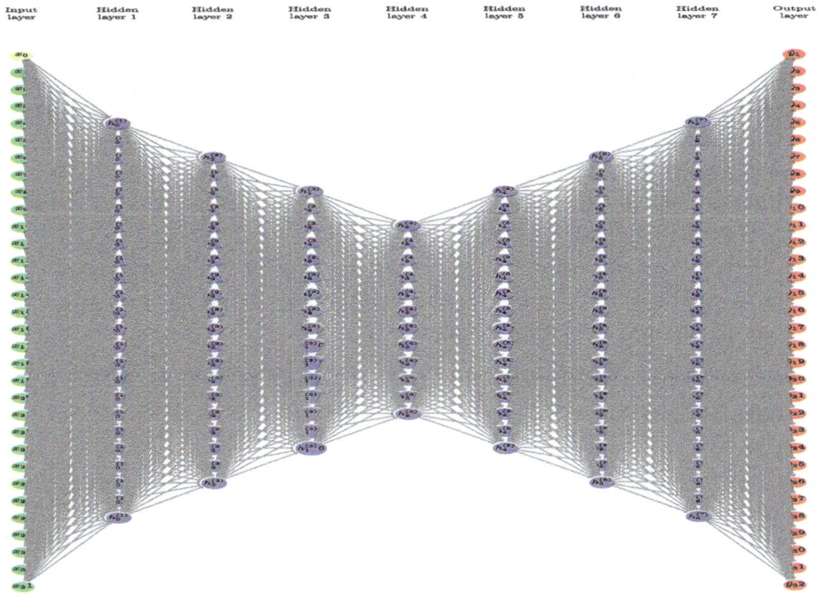

Fig. 3. The fully connected deep neural network Trained model

The R-squared value R^2 is always between 0 and 1 inclusive, An R^2 of 1 indicates that the regression predictions perfectly fit the data.

Our work is a regression problem the accuracy of the model is defined with these additional metrics:

- Cosine proximity, Root Mean squared error, Absolute squared error. Figure 2 illustrates the Graph dependencies for the trained model.

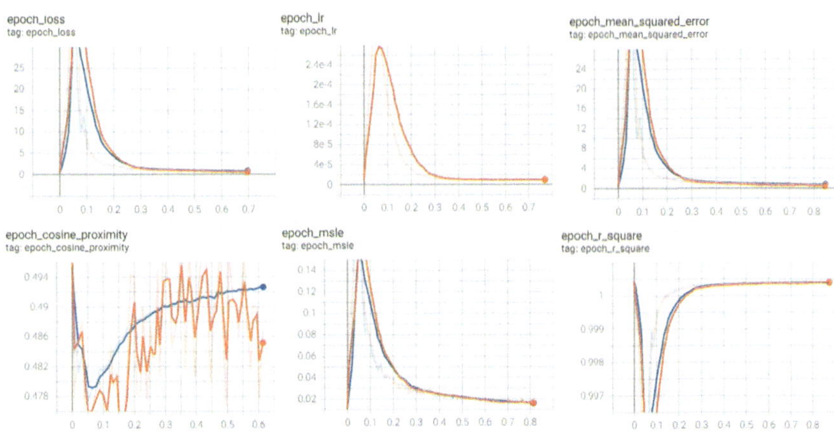

Fig. 4. The experimental results for training and validation.

5.4 Experimental Results

We have operated the Keras checkpoint named Call Backs to typically save the most satisfactory results after every iteration and to saving the weight and bias of our trained model, After 3000 epoch iterations in 11 h and 43 min, the most impressive results are properly obtained in the 2759 epoch.
Error function: Mean squared error = 0.0087.
accuracy function = 0.89.

The high R-squared value typically ranging from 0.89 also indicates there is an effective relationship between predicted plain-text and the cipher-text. The results are illustrated in the following Fig. 4.

6 Conclusion

In this illustrated paper, Our work is properly established as a regression task where we tentatively proposed a deep learning approach to KATAN 32 bit lightweight block cipher security analysis. Exclusively, we typically train fully connected deep neural network models to accurately predict the plain-text from the chosen cipher-text one.

The fully connected deep neural networks are improved using TensorFlow Framework in a Google Cloud environment. We properly investigate the economic feasibility of the proposed attack using cloud tools.

In the future works, we plan to extend this work for more larger Block cipher size like KATAN 48 bit and 64 bit, and for more promising step towards a more efficient and automated test to verify the security of emerging lightweight ciphers such as: RECTANGLE, Humming Bird, SIMON, GRAIN, WG-8, ESPRESSO, and TRIVIUM.

References

1. Burnside, R.S.: The electronic communications privacy act of 1986: the challenge of applying ambiguous statutory language to intricate telecommunication technologies. Rutgers Comput. Tech. L.J. **13**, 451 (1987)
2. Gomez, A.N., Huang, S., Zhang, I., Li, B.M., Osama, M., Kaiser, L.: Unsupervised cipher cracking using discrete GANs. In: International Conference on Learning Representations (2018)
3. Wu, W., Zhang, L.: LBlock: a lightweight block cipher. In: Lopez, J., Tsudik, G. (eds.) ACNS 2011. LNCS, vol. 6715, pp. 327–344. Springer, Heidelberg (2011). https://doi.org/10.1007/978-3-642-21554-4_19
4. Pradeepthi, K.V., Tiwari, V., Saxena, A.: Machine learning approach for analysing encrypted data. In: 2018 Tenth International Conference on Advanced Computing (ICoAC). IEEE (December 2018)
5. Zhang, W., Zhao, Y., Fan, S.: Cryptosystem identification scheme based on ASCII code statistics. Secur. Commun. Netw. **2020**, 1–10 (2020)
6. Yu, F., Gong, X., Li, H., Wang, S.: Differential cryptanalysis of image cipher using block-based scrambling and image filtering. Inf. Sci. **554**, 145–156 (2021)

7. Mishra, G., Krishna Murthy, S.V.S.S.N.V.G., Pal, S.K.: Neural network based analysis of lightweight block cipher present. In: Yadav, N., Yadav, A., Bansal, J.C., Deep, K., Kim, J.H. (eds.) Harmony Search and Nature Inspired Optimization Algorithms. AISC, vol. 741, pp. 969–978. Springer, Singapore (2019). https://doi.org/10.1007/978-981-13-0761-4_91
8. Mundra, A., Mundra, S., Srivastava, J.S., Gupta, P.: Optimized deep neural network for cryptanalysis of DES. J. Intell. Fuzzy Syst. **38**, 5921–5931 (2020)
9. Bansod, G., Raval, N., Pisharoty, N.: Implementation of a new lightweight encryption design for embedded security. IEEE Trans. Inf. Forensics Secur. **10**(1), 142–151 (2015)
10. Jain, A., Mishra, G.: Analysis of lightweight block cipher few on the basis of neural network. In: Yadav, N., Yadav, A., Bansal, J.C., Deep, K., Kim, J.H. (eds.) Harmony Search and Nature Inspired Optimization Algorithms. AISC, vol. 741, pp. 1041–1047. Springer, Singapore (2019). https://doi.org/10.1007/978-981-13-0761-4_97
11. Bogdanov, A., et al.: PRESENT: an ultra-lightweight block cipher. In: Paillier, P., Verbauwhede, I. (eds.) CHES 2007. LNCS, vol. 4727, pp. 450–466. Springer, Heidelberg (2007). https://doi.org/10.1007/978-3-540-74735-2_31
12. Xiao, Y., Hao, Q., Yao, D.D.: Neural cryptanalysis: metrics, methodology, and applications in CPS ciphers. In: Proceedings of the 2019 IEEE Conference on Dependable and Secure Computing (DSC). IEEE (November 2019)
13. Perov, A.: Using machine learning technologies for carrying out statistical analysis of block ciphers. In: Proceedings of the 2019 International Multi-Conference on Engineering, Computer and Information Sciences (SIBIRCON). IEEE (October 2019)
14. Truong, N.D., Haw, J.Y., Assad, S.M., Lam, P.K., Kavehei, O.: Machine learning cryptanalysis of a quantum random number generator. IEEE Trans. Inf. Forensics Secur. **14**(2), 403–414 (2019)
15. Hou, B., Li, Y., Zhao, H., Wu, B.: Linear attack on round-reduced des using deep learning. In: Chen, L., Li, N., Liang, K., Schneider, S. (eds.) ESORICS 2020. LNCS, vol. 12309, pp. 131–145. Springer, Cham (2020). https://doi.org/10.1007/978-3-030-59013-0_7
16. Lee, T.R., Teh, J.S., Yan, J.L.S., Jamil, N., Yeoh, W.Z.: A machine learning approach to predicting block cipher security. In: Cryptology and Information Security Conference. Universiti Putra Malaysia (2020)
17. So, J.: Deep learning-based cryptanalysis of lightweight block ciphers. Secur. Commun. Netw. **2020**, 1–11 (2020)
18. Biham, E., Shamir, A.: Differential cryptanalysis of DES-like cryptosystems. J. Cryptol. **4**(1), 3–72 (1991). https://doi.org/10.1007/BF00630563

Sentiment Analysis of Algerian Dialect Using a Deep Learning Approach

Badia Klouche(✉), Sidi Mohamed Benslimane, and Nadir Mahammed

LabRi Laboratory, Ecole Supérieure en Informatique Sidi Bel Abbès, Sidi Bel Abbès, Algeria
{b.klouche,s.benslimane,n.mahammed}@esi-sba.dz

Abstract. Nowadays the Internet has become an essential tool for exchanging information, both on a personal and professional level. Today, the analysis of sentiment offers us a great interest for research, marketing and industry. With millions of comments and tweeting published every day, the information available on the Internet and in social media has become a gold mine for companies developing in their production, management and distribution. In this article, we propose a novel approach to analyze the sentiments of the Algerian dialect for the benefit of the Algerian Telephone Operator Ooredoo. The proposed approach is based on a CNN deep learning model and the SVM machine learning algorithm, which provides state-of-the-art results on a dataset written in Algerian dialect. In this study, the Facebook comments shared in Modern Standard Arabic (MSA) and Algerian dialect of the customers of the Algerian telephone operator Ooredoo are analyzed in order to allow the operator to retain and satisfy its customers to the maximum. Experimental results show that deep learning approaches outperformed traditional methods of sentiment analysis.

Keywords: Sentiment analysis · Deep learning · CNN · Algerian dialect · NLP

1 Introduction

Today E-commerce allows users to express their opinions, views and sentiments through comments on products/services, in different social media platforms such as Facebook, Tweeter and Instagram. The information derived from the comments of Internet users is very important; it influences everyone's decision to take action to opt for a given article, based on the experience and opinions of other users. Thus, sentiment analysis (SA), also called opinion mining, is the field of study that exploits the opinions, sentiments, evaluations, assessments, attitudes and emotions of individuals towards entities, such as these products, services, organizations, individuals, problems, events and subjects [1]. SA is becoming a very active field of research, its objective being to analyse people's opinions, sentiments, attitudes and emotions on different topics with different languages from texts shared in different social networks [2].

B. Lejdel et al. (Eds.): AIAP 2021, LNNS 413, pp. 122–131, 2022.
https://doi.org/10.1007/978-3-030-96311-8_12

Several approaches to sentiment analysis have been proposed and introduced by several authors [3,4] and [5] with good results. Consequently, the researchers realized that finding the sentiment generated by current user data required a thorough understanding and effective methods of learning text without resorting to manual feature engineering [6,7] and [8]. In recent years, it has been demonstrated that deep learning models are a promising solution to the challenges of Natural Language Processing (NLP). Indeed, Deep Learning approaches have proven to be more effective than traditional methods of sentiment analysis [9]. The objective of our work is to analyze the polarity of customer comments of the Algerian telephone operator Ooredoo; published in different forms (MSA, Algerian Dialect) using Deep Learning approaches.

The remaining part of this paper is organized as follows: Sect. 2 presents a current state-of-the-art on deep learning based sentiment analysis. Section 3 explains our approach. Section 4 summarizes the experiences and analyzes the results. Finally, Sect. 5 presents the conclusions of this paper and highlights future work.

2 Related Work

In this section, we present related work on sentiment analysis, using the deep learning of Arabic texts (Modern Standard Arabic and Dialect Arabic).

[10] Proposed a deep learning model for sentiment analysis in Arabic, based on a CNN architecture layer for extracting local features and two LSTM layers for maintaining long-term dependencies. The feature maps learned by CNN and LSTM are passed to the SVM classifier for final classification. Their model reaches an accuracy of 90.75%. [11] Proposed a deep learning (DL) method for the analysis of sentiments in dialectal Arabic, which combines long-term and short-term memory (LSTM) with convolutional neural networks (CNN) memory. Their model achieved an accuracy of 81% to 93% for binary classification and 66% to 76% accuracy for three-way classification. [12] Presented a deep learning study to classify sentiments from texts in the Saudi dialect. They applied two deep learning techniques to perform sentiment analysis: Long-Short-Term Memory (LSTM) and Bi-Directional Long-Short-Term Memory (Bi-LSTM). The experimental results of Bi-LSTM were 94% higher than those of LSTM 92%, while SVM had the lowest performance at 86.4%. [13] Used an ensemble model combining the CNN (Convolutional Neural Network) and LSTM (Long Short-Term Memory) models to predict the sentiment of Arabic tweets. Their model scored 64.46% higher than the F1 Advanced Deep Learning Model score of 53.6% of the Arabic tweets dataset. [14] Proposed to combine convolutional and recurrent layer methods into a single model, in addition to the preformed word vectors, to capture long-term dependencies in short texts more efficiently. They proved that the CNN and RNN models can fill the gaps in short texts in deep learning models. [15] Discussed a neural network (CNN) model that integrates user compartmental information into an Arabic tweet document. They presented the "Mazajak" tool, the first online sentiment analysis tool in Arabic.

3 Proposed Approach

In this section, we illustrate the main steps of our approach (see Fig. 1). First comments are collected on the social networks Facebook and tweeter of the telephone operator Ooredoo. Next, the comments go through the cleaning and pre-treatment step in order to eliminate unwanted symbols and tokens. Finally, the comments are listed for preparation for the sentiment analysis step.

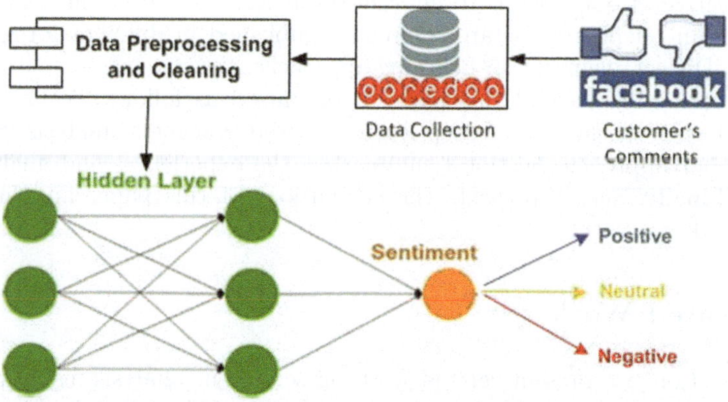

Fig. 1. Main steps of the proposed approach

3.1 Data Collection

Sentiment analysis in Deep Learning requires the collection of a large dataset. All comments used in our research are mainly extracted from the social networks Facebook and Tweeter. All the data processed in this work were collected and annotated using the Facepager and Tweepy APIs, a Python wrapper for the Twitter API.

Indeed, our corpus contains 65,125 comments, where the Algerian dialect is widely used with 75% of the collected data, while the rest of the dataset consists of other languages (Fig. 2).

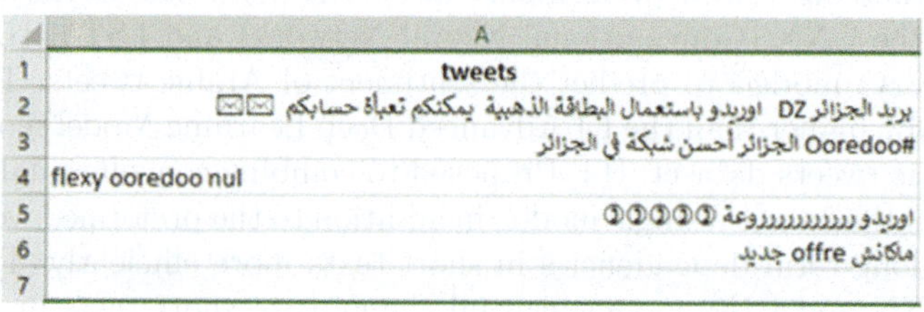

Fig. 2. Example of Tweets Ooredoo

3.2 Language Detection

Algerian Internet users communicate in the majority of cases in a multilingual language, written in French, Arabic and English, while we often find other informal languages written by users, such as the Algerian dialect and Arabizi (Arabic chat alphabet that uses Latin script to write Arabic text).

In this context, our research is interested in the classification of the sentiments of the comments and posts written in Algerian dialect of a sample of customers of the Algerian Telephone Operator Ooredoo. To do this, we use the Python Alphabet Detector 11 library to detect Latin and Arabic characters in our Dataset, in order to create an Arabic-specific corpus, where we then translate the Dataset into the other languages used with the Google Translation API. In our study, we used a Python library allowing the detection of the alphabet, in order to keep an Arabic corpus.

3.3 Cleaning and Pre-treatment

Before starting the sentiment analysis stage, a preliminary cleaning and preprocessing phase of the comments and posts is necessary in order to remove unwanted noises and symbols, empty words, URLs, etc.

In this framework, the following steps are listed for cleaning and preprocessing:

- Tokenization.
- Removal of empty words.
- Removal of special characters, punctuation marks and all diacritics
- Deletion of all non-Arabic characters.
- Removal of URLs.
- emmatization.
- Removal of repeated letters.
- Lexical normalization.
- Removal of hashtags.

Figure 3 shows an example of Ooredoo Tweeets after the preprocessing phase:

1	tweets
2	بريد الجزائر دز اوريدو باستعمال البطاقة الذهبية يمكنكم تعبأة حسابكم
3	اوريدو الجزائر أحسن شبكة في الجزائر
4	فليكسي اوريدو نول
5	اوريدو روعة
6	ماكانش عرض جديد
7	

Fig. 3. Example of Tweets Ooredoo after the preprocessing

3.4 Sentiment Analysis

In this phase, we proposed to use the deep learning model Convolutional Neural Network CNN, in order to allow the analysis of the feelings of the Algerian dialect on a set of data collected from Internet users from the official pages of the Telephone Operator Ooredoo.

With the aim of extracting morphological information, we started a deep character representation with the use of the CNN model inspired by the model proposed by the authors [16]. Indeed, it is a matter of generating a new vector representative of an input word by using a convolution layer followed by a max-grouping layer. It should be noted that for the preparation of the CNN model, we used the python API TensorFlow and Sklearn open source libraries for sentiment analysis. In this context, note that an SVM classifier from the machine learning approach was also used to classify the polarity of the data into positive, negative and neutral classes.

4 Expermentation and Evaluation

In this section, we present and discuss the results of applying SA using convolutional neural network (CNN) and support vector machine (SVM) for the data set.

4.1 Sentiment Analysis Results

The objective of this research is to study and explore the improvement of sentiment analysis by deep learning of the Algerian dialect DAlg, where we compared the CNN model with the SVM classifier, to the effect of classifying the polarity according to the classes: positive, negative or neutral.

In this step, several experiments were conducted, where the results obtained are illustrated in the following tables using the three measures namely: precision, recall and F-measure. Table 1 represents the results of the precision values for the classes: positive, negative and neutral of the data set, from which it is noted in this context that the positive class obtained the highest precision compared to the other two classes.

The Fig. 4 shows the precision values for each of the three classes: positive, negative and neutral. Thus, we notice that the accuracy values for the positive class obtained the best results compared to each of the other two classes, with a rate of 76% for the CNN model and 72% for the SVM classifier. For the negative class, the accuracy of the CNN model is 72% and 71% for the SVM classifier. As for the neutral class, the results obtained are different from the two previous classes, with an accuracy value of 70% for the CNN model and only 64% for the SVM classifier.

Table 1. Precision of positive, negative and neutral classes

Classifiers	Positive	Negative	Neutral
CNN	0.76	0.72	0.70
SVM	0.72	0.71	0.64

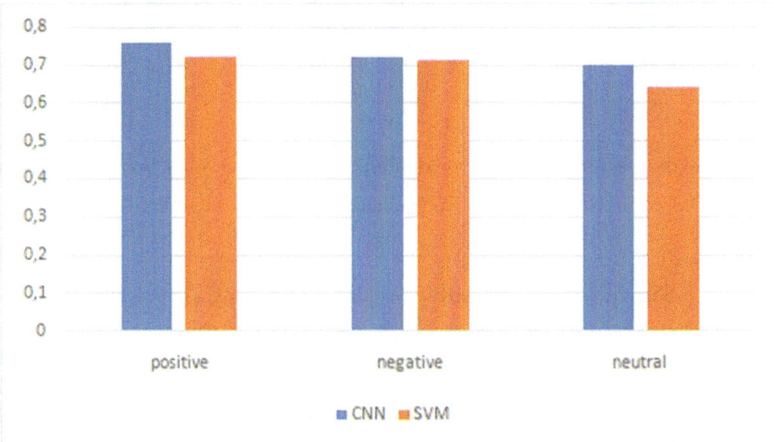

Fig. 4. Precision of positive, negative and neutral classes

Table 2 describes the recall values for each of the three classes.

Table 2. Recall of positive, negative and neutral classes

Classifiers	Positive	Negative	Neutral
CNN	0.37	0.81	0.73
SVM	0.24	0.77	0.74

Figure 5 demonstrates the recall values for the three classes: positive, negative and neutral of the SVM classifier and the CNN model.

We can remark that the negative class obtained the best recall compared to each of the two other classes: positive and neutral with a rate of 81% for the CNN and 77% for the SVM. However, for the neutral class, the recall value is 73% for the CNN and 74% for the SVM classifier. Figure 7 shows that the positive class obtained the lowest recall with a rate of only 37% for the CNN model and 24% for the SVM classifier.

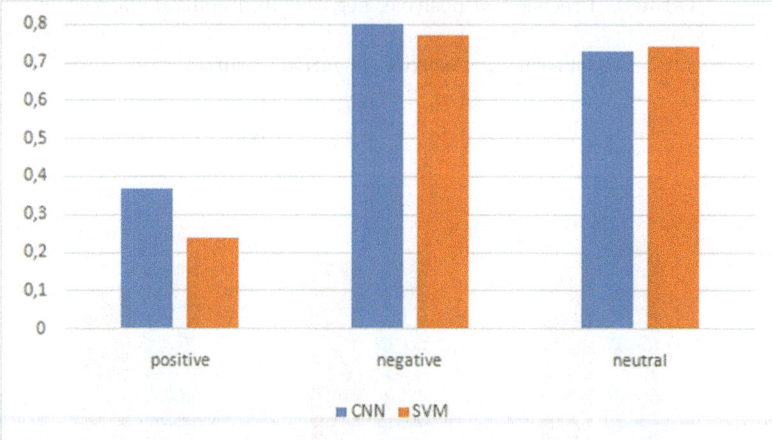

Fig. 5. Recall of positive, negative and neutral classes

Table 3 and Fig. 6 represent the F-measure values for each of the three classes. We see that the F-measure results for the negative class are the best performing with a rate of 73% for the CNN model and 72% for the SVM classifier, compared to the other two classes: positive and neutral. Concerning the F-measure rates of the neutral class, these are close to each other, evaluated at 68% for the CNN and 69% for the SVM.

For the positive class, Fig. 6 illustrates that the CNN model obtained a rate of only 40% and only 29% for the SVM classifier, rates that are significantly lower than those obtained for the other two classes.

Table 3. F-measure of positive, negative and neutral classes

Classifiers	Positive	Negative	Neutral
CNN	0.40	0.73	0.68
SVM	0.29	0.72	0.69

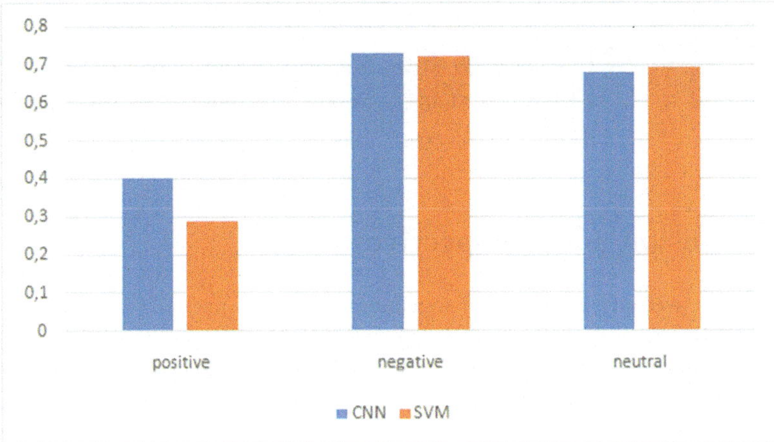

Fig. 6. F-measure of positive, negative and neutral classes

Table 4 and Fig. 7 illustrate the experimental results obtained from the SVM classifier and the CNN model.

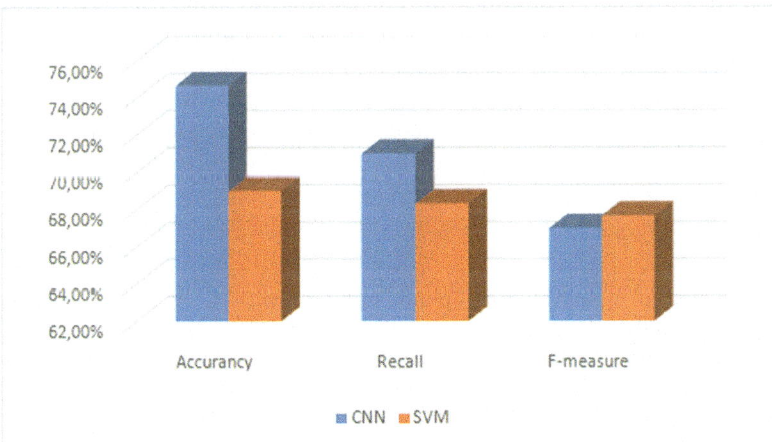

Fig. 7. The experimental results obtained

From the results obtained in Table 4 and Fig. 7, we deduce that the CNN model achieved high accuracy compared to that obtained by the SVM. The CNN model achieved 74.66% accuracy, while the SVM achieved only 69.00%. This shows that deep learning handles the large amount of data better compared to machine learning algorithms, such as the SVM classifier.

Table 4. The obtained experimental results

Classifiers	Positive	Negative	Neutral
CNN	74.66%	71.00%	67.00%
SVM	69.00%	68.33%	67.66%

5 Conclusion and Perspectives

In this paper, we have presented a sentiment analysis approach using deep learning related to the comments of the Algerian telephone operator Ooredoo's customers on different social networks. The objective of this work is to address the concerns of the Algerian telephone operator, whose main concern is to best satisfy its customers who subscribe to the social network Facebook among others. Ooredoo in this context and in its new strategy seeks to improve the quality of its services to its customers to retain them and encourage them to a long-term subscription. In this paper, we presented a deep learning approach for sentiment analysis of Arabic comments, as we were able to classify and analyze the polarity of comments from customers of the Algerian telephone operator Ooredoo, written in Algerian dialect and in MSA. In this regard, we conducted various experiments using several algorithms such as CNN and SVM to enable sentiment analysis. As a result, we obtained an precision score using the CNN and SVM models of 76% and 72% for the positive class, 72% and 71% for the negative class, and 70% and 64% for the neutral class, respectively.

In our future work, we plan to conduct a comparative study of sentiment analysis using deep learning models, introducing multilingualism, in addition to Arabic.

References

1. Liu, B.: Sentiment analysis and opinion mining. Synth. Lect. Hum. Lang. Technol. **5**(1), 1–167 (2012)
2. Klouche, B., Benslimane, S.M.: Multilingual sentiments analysis to improve the quality of services provided by algerian telephone operator. In: JERI (2019)
3. Pang, B., Lee, L., Vaithyanathan, S.: Thumbs up? sentiment classification using machine learning techniques. arXiv preprint cs/0205070 (2002)
4. Hu, M., Liu, B.: Mining and summarizing customer reviews. In: Proceedings of the tenth ACM SIGKDD International Conference on Knowledge Discovery and Data Mining, pp. 168–177 (2004)
5. Stone, P.J., Bales, R.F., Namenwirth, J.Z., Ogilvie, D.M.: The general inquirer: a computer system for content analysis and retrieval based on the sentence as a unit of information. Behav. Sci. **7**(4), 484 (1962)
6. Abandah, G.A., Graves, A., Al-Shagoor, B., Arabiyat, A., Jamour, F., Al-Taee, M.: Automatic diacritization of Arabic text using recurrent neural networks. Int. J. Doc. Anal. Recognit. (IJDAR) **18**(2), 183–197 (2015)
7. Lulu, L., Elnagar, A.: Automatic Arabic dialect classification using deep learning models. Procedia Comput. Sci. **142**, 262–269 (2018)

8. Klouche, B., Benslimane, S.M., Bennabi, S.R.: Ooredoo rayek: a business decision support system based on multi-language sentiment analysis of algerian operator telephones. Int. J. Technol. Diffus. (IJTD) **11**(2), 66–81 (2020)
9. Elaraby, M., Abdul-Mageed, M.: Deep models for arabic dialect identification on benchmarked data. In: Proceedings of the Fifth Workshop on NLP for Similar Languages, Varieties and Dialects (VarDial 2018), pp. 263–274 (2018)
10. Ombabi, A.H., Ouarda, W., Alimi, A.M.: Deep learning cnn-lstm framework for Arabic sentiment analysis using textual information shared in social networks. Soc. Netw. Anal. Min. **10**(1), 1–13 (2020)
11. Abu Kwaik, K., Saad, M., Chatzikyriakidis, S., Dobnik, S.: LSTM-CNN deep learning model for sentiment analysis of dialectal Arabic. In: Smaïli, K. (ed.) ICALP 2019. CCIS, vol. 1108, pp. 108–121. Springer, Cham (2019). https://doi.org/10.1007/978-3-030-32959-4_8
12. Alahmary, R.M., Al-Dossari, H.Z., Emam, A.Z.: Sentiment analysis of Saudi dialect using deep learning techniques. In: 2019 International Conference on Electronics, Information, and Communication (ICEIC), pp. 1–6. IEEE (2019)
13. Heikal, M., Torki, M., El-Makky, N.: Sentiment analysis of Arabic tweets using deep learning. Procedia Comput. Sci. **142**, 114–122 (2018)
14. Hassan, A., Mahmood, A.: Deep learning approach for sentiment analysis of short texts. In: 2017 3rd International Conference on Control, Automation and Robotics (ICCAR), pp. 705–710. IEEE (2017)
15. Farha, I.A., Magdy, W.: Mazajak: an online Arabic sentiment analyser. In: Proceedings of the Fourth Arabic Natural Language Processing Workshop, pp. 192–198 (2019)
16. Chiu, J.P., Nichols, E.: Named entity recognition with bidirectional lstm-cnns. Trans. Assoc. Comput. Linguist. **4**, 357–370 (2016)

Do We Need Change Detection for Dynamic Optimization Problems?: A Survey

Abdennour Boulesnane[1(✉)] and Souham Meshoul[2]

[1] Faculty of Medicine, Salah Boubenider-Constantine 3 University, Constantine, Algeria
aboulesnane@univ-constantine3.dz

[2] Princess Nourah Bint Abdulrahman University RC-CCIS, Riyadh, Saudi Arabia
sbmeshoul@pnu.edu.sa

Abstract. Solving dynamic optimization problems is more challenging than static ones. When a change in the objective landscape occurs, the search process may not be powerful enough to track new optima. For population based algorithms this is referred to as diversity loss problem. Furthermore, the memory of old optima becomes outdated and if not correctly dealt with, the evolution of the search process may be misguided. Recently, a new interesting trend in dealing with optimization in dynamic environments has emerged toward developing new algorithms that are able to effectively handle changes without using any change detection scheme, and hence no extra computational cost is needed. There exist several works in the literature that attempt to maintain diversity without change detection. However, not that much work has been devoted to studies that investigate the possibility to overcome the outdated memory problem without expensive change detection. This study presents a comprehensive survey of the various change detection based methods. As part of this survey, we include a classification of the change detection schemes and we identify the main features of each method.

Keywords: Dynamic optimization problems · Change detection · Diversity loss problem · Outdated memory problem · Memory schemes

1 Introduction

Many real world problems require optimization over time because of the dynamic nature of the environments. Typical fields where such problems need to be solved include economics, engineering, communication systems, machine learning, bioinformatics to name just a few. Time dependent optimization problems are most commonly known as dynamic optimization problems (DOPs). Solving DOPs is not only a matter of locating global optima as in static optimization but of being able to track such optima in changing objective landscapes as well. Hence, a DOP can be viewed as a sequence of static optimization problems

B. Lejdel et al. (Eds.): AIAP 2021, LNNS 413, pp. 132–142, 2022.
https://doi.org/10.1007/978-3-030-96311-8_13

over time [26]. DOPs have been defined in different ways. Over the past decade, Swarm Intelligence (SI) [16] and Evolutionary Algorithms (EAs) [21] are considered to be a good choice for solving DOPs. However, two main problems are encountered when using traditional methods to solve DOPs. The first one is the diversity loss caused by the convergence of these approaches preventing them from tracking new optima in an efficient manner. The second one is the outdated memory caused by changes in the environment and that may misguide the evolution of the search process.

On the other hand, dynamic change detection is an integral part of dynamic evolutionary algorithms design. Following the change detection, the diversity loss and the outdated memory problems can be efficiently resolved by reacting properly to the environmental change using mechanism such as [32]: reevaluation of population to update memory, clearing the memory, introducing diversity, re-initialization of parameters of the algorithm, etc. For that purpose, different kinds of change detection schemes have been proposed in the literature [23]. However, difficulties in detection schemes include proper choice of memory solutions that represent the whole search space, increased computational evaluation cost, detecting partial change in the landscape, noisy environments, etc.

To avoid these drawbacks, recently, a new interesting trend in dealing with dynamic environments has emerged toward developing new algorithms that are able to effectively handle dynamics without any change detection schemes by focusing on the optimization process rather than spending computational resources on change detection. Therefore, several studies in the literature have been carried out in an attempt to maintain diversity without change detection [13]. On the other hand, very little work has been done to investigate the possibility to overcome the outdated memory problem without expensive change detection.

In this paper, we analyze the existing state of the art change detection based methods proposed for the dynamic environments in the literature. Moreover, we present for first time a classification of these schemes and highlight their advantages and limitations. Furthermore, we discuss the importance of strategies that do not require the knowledge of the change point time to handle future changes and what kind of factors must be taken into consideration to move towards this new dynamic optimization design framework.

The rest of the paper is organized as follows. Firstly, problem statement and challenges are presented in Sect. 2. Section 3 and 4 describe change detection and without change detection based methods respectively. Finally, conclusions and plans for future work are given in Sect. 5.

2 Dynamic Optimization Problems (DOPs): Problem Statement and Challenges

Optimization in dynamic environments, popularly known as *dynamic optimization*, was, and still is, an active open research area, due to its obvious and direct relation to real-world problems. The main challenge in the field of optimization is the dynamic nature of these problems which imposes to find solutions subject to time constraints. We can formally describe a dynamic optimization problem as the task that aims to find the sequence $(x_1^*, x_2^*, ..., x_n^*)$ that:

$$\begin{array}{l} \text{Optimize } f(x,t) \\ \text{subject to. } h_j(x,t) = 0 \text{ for } j = 1,2,...,u \\ g_k(x,t) \leq 0 \text{ for } k = 1,2,...,v. \\ \text{with } x \in \mathbb{R}^n \end{array} \quad (1)$$

Where $f(x,t)$ is a time dependent objective function, $(x_1^*, x_2^*, ..., x_n^*)$ is the sequence of n optima found as the fitness landscape changes. In other ways, it depicts optima tracking, $h_j(x,t)$ denotes the jth equality constraint and $g_k(x,t)$ denotes the kth inequality constraint. All these functions may change dynamically at any time, as shown by the dependence on the time parameter t. When applied to DOPs, the objective of an optimization algorithm is no longer to simply find the optimal solution, but also to track its movement in a dynamic search space. Hence, traditional optimizers have been revisited and significant improvements have been introduced to ensure appropriate behavior when dealing with dynamic environments [32].

2.1 Diversity Loss Problem

One of the most important challenge posed by the dynamic behavior of DOPs is the diversity loss issue [32]. The diversity loss occurs when all solution candidates converge to the same region in the search space and lose their ability to find new optimum as required after any change in the environment. In the literature, several approaches have been developed to address this limitation. Indeed, at the origin most of them were initially designed to handle static optimization problems, including approaches based on swarm intelligence and evolutionary algorithms. To fit the purpose, they have been intelligently adapted to cope with dynamic environments by using some strategies such as [32]: maintaining diversity during the run, increasing diversity after a change, using several populations (multi-population approaches), using memory schemes, using prediction mechanisms and hybrid approaches.

Memory schemes to tackle diversity loss is a connecting link between all these strategies, where almost in every study at least one memory-based method was used. Despite the large variety of memory schemes, they all rely on a core and common feature which is the reuse of information accumulated during the optimization process to enhance their adaptability to the environmental change. According to a comprehensive study on various memory schemes [4], the most popular memory approach is the use of an explicit memory to store the good solutions collected over the successive environments, known as *global memory scheme*.

2.2 Outdated Memory Problem

Whatever the adopted memory pattern is (explicit, implicit, global, local, associative, direct, etc.), the issue that makes the dynamic optimization task more challenging is the outdated memory problem. This problem refers to the condition in which all existing information that the dynamic optimization algorithm has accumulated during the search process (i.e. stored personal best and/or global best positions and/or their corresponding function values, etc.) may no longer be useful or even valid after a dynamic change. Therefore, incorporating directly this stored knowledge into the search process has the potential to negatively affect and mislead the dynamic optimizer in its quest to follow the global optimum in the changing environment. In the literature, the outdated memory problem has been solved in two ways:

- Using change detection approaches by either re-evaluating or forgetting the memory when changes in the environment occur.
- Without change detection by iteratively re-evaluating the memory contents [12].

Although the efficiency of using change detection approaches or iterative function re-evaluation to overcome the outdated memory problem was well proved, the optimization task becomes time and effort intensive. Therefore, a new research direction has been followed [12] which consists of designing new dynamic optimization algorithms able to cope with DOPs adaptively without using any change detection process while saving a large amount of computational resources. In this context and motivated by these facts, the goal of this paper is to refer to the need to think not only of the diversity problem, but also of solving the outdated memory problem by using new methods with the aim of laying a solid foundation for a new dynamic optimization design framework.

The aforementioned issues have been tackled in different ways in the literature. Related methods can be broadly classified into different categories according to whether they make use of a change detection mechanism or not.

In the rest of this section, we present a review of methods in both classes.

3 Change Detection Based Methods

For solving the change detection task, a change point $t_{cp} \in \mathbb{N}_0$ can be defined as follows [23]:

$$f(x, t_{cp}) \neq f(x, t_{cp} + 1) \tag{2}$$

where $x \in M$ is an element of a fixed bounded search space $M \subset \mathbb{R}^n$. The change point definition in Eq. (2) says that a change in the function landscape has taken place no matter how small and irrelevant the alteration in the environment is. Besides, approaches based on change detection need to know this moment to react properly to the dynamic environments. Hence, a change detection mechanism should be implemented.

Once the change is detected, the algorithm takes explicit mechanism (change reaction schemes) to respond to changes and to increase or introduce diversity in population, therefore, tracking moving optima becomes easy. The following mechanisms are widely utilized : a) employing memory [24]; b) hyper mutating

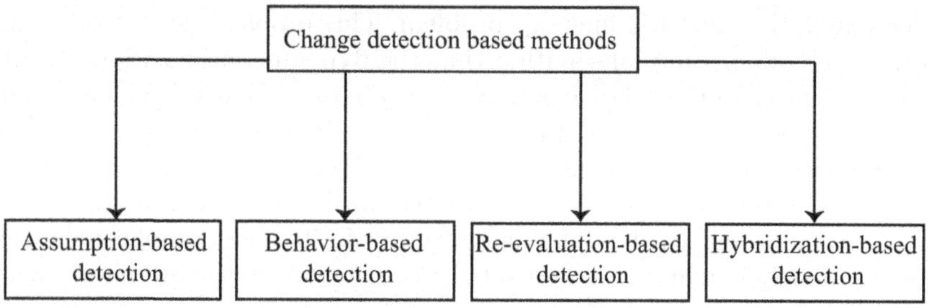

Fig. 1. Different schemes in change detection based methods.

the previous population [17]; c) randomly generating new solutions [30]; and d) anticipating and predicting [24], etc.

In this section, and as shown in Fig. 1, we present a classification and an overview on the state of the art of change detection policies that have been used to deal with dynamic environment.

3.1 Assumption-Based Detection

Many dynamic optimization approaches take an explicit action to detect changes in the environment. The simplest way to detect the environmental changes is to assume that changes in the environment are either known a priori to the algorithm or can be easily predicted. Assumption based approaches do not spend any additional function evaluations to detect changes. Usually, in the state-of-the-art experimental studies, this method allows to fairly evaluate and analyze the performance of the proposed techniques offering an ideal situation where all changes are 100% detected [27]. Furthermore, this method can be used to handle with periodic DOPs [29] in which the changes are deterministic and predictable, hence the system can easily calculate the time of the next change.

This method makes sense for solving the academic benchmark problems. However, it may be inefficient, especially in cases where the changes are often random or unpredictable (i.e. real-world applications).

3.2 Behavior-Based Detection

Changes that influence the search space, can influence the algorithm's behavior as well. Based on this idea, behavior based approaches try to exploit the irregularities in the algorithm behavior in order to confirm the occurrence of changes in the environment.

A change detection step is needed when the algorithm has no earlier information of the time of next change. In many studies, monitoring the quality of the best solutions in the population can be used as a change detection mechanism. Either by observing the drop in the average of best stored solutions during a certain number of iterations [20] or if those archived solutions do not change over a number of generations [10], then it is inferred that the change has taken place.

[8] proposed Partitioned Hierarchical Particle Swarm Optimizer (PH-PSO) for dynamic and noisy function optimization. PH-PSO divides the swarm into a tree-based hierarchy of sub-swarms for a certain number of iterations after the occurrence of a change. A new approach to detect environmental changes under the presence of noise is introduced, called hierarchy monitoring change detection. In this method, change detection is performed by observing changes in the hierarchy itself. Doing so does not involve any additional evaluation of the fitness function and allows working in noisy environment.

Another interesting idea has been investigated in [23] in order to detect changes in a systematic way by analyzing the behavior of population over generations. The key idea is to find difference between fitness distributions of the populations through non-parametric statistical hypothesis tests. Performance degradation over consecutives generations is taken as an indicator of change detection. However, the used non-parametric tests involve using independent samples which is not satisfied given fitness distributions [25]. To address this issue, an immunological approach using negative selection has been proposed in [22].

In [18], a detection mechanism is proposed by which the deviation amount of the locally mutated Differential Evolution (DE) individuals is tracked to detect changes. The basic idea relies on the fact that locally mutated DE offsprings differ slightly from their parents resulting in small fitness difference. However, the deviation amount is expected to be significant with dynamic change because offspring are produced on a thoroughly revised landscape whereas parents keep the fitness obtained in the previous generation. Hence, tracking such deviation would be good change indicator. As a consequence, neither extra functional evaluations are involved nor high complexity computations.

By using this schemes, there is no need for additional function evaluations. However, since behavior based approaches are mainly related to the algorithms used, they can be defined as specific approaches, which means it is impossible to directly incorporate these approaches into other algorithms.

3.3 Reevaluation-Based Detection

Re-evaluating solutions is the most commonly used change detection approach [21]. Some specific solutions in the search space (named detectors or sensors) are employed and then their objective values are evaluated after each iteration. If the present and past objective values are different, then, indeed, the environment has changed.

In the literature, numerous methods have been developed with the integration of this change detection scheme. They differ by each other by the strategy they use to place sensors in the landscape, to determine the required number of sensors and the kind of solutions that these sensors should represent. Accordingly, the type of the used detectors influences the performance of the detection. Solution used for change detection can be part of the population as in [9], where a percentage of solution from the current population selected randomly is re-evaluated. Otherwise, in [11], change detection is done using a random point called test solution. The fitness value of this test solution is calculated at every generation. If a significant difference is recorded, this can be an indication of a change in the environment. Also in [6], the change detection problem is addressed by using a fixed set of detectors that are not part of the swarm and distributed in the search space in a way to cover its parts more evenly.

On the other hand, stored points from the memory can be used as change detectors as in [5] where some detectors are re-evaluated to state whether a change has taken place or not. For instance in [14], the top 5 best solutions at each iteration are considered in the re-evaluation process. In [19], the current best solution is taken as a detector. This may cover only some part of the landscape. Therefore, a multi-population approach has the advantage to cover more parts of the landscape. Also in [15], where cooperative methods for DOPs that are based on a set of cooperating agents are used. an agent re-evaluates its best performance recorded so far and checks the extent to which it changes.

Furthermore, in [1], an empirical study has been conducted to thoroughly study the impact of how to distribute a set of detectors in the search space. The performance of four different sensor-based detection schemes has been investigated, which are: random selection of individuals from the population, best-so-far individuals, randomly selected individuals that do not belong to the population and distributing sensors. Each placement strategy has been measured by using two commonly known dynamic optimization problems.

It is worth noting that this approach is easy to implement, as it does not require elaborated statistical analysis and it is more favorable when the change detection is more difficult [23]. However, it assumes that there is no noise in function evaluations, which means it is not robust with all problems instances. Moreover, it spends a large amount of computation throughout the search space on detecting the changes.

3.4 Hybridization-Based Detection

All existing change detection methods have limitations, an efficient way to overcome these drawbacks is the hybridization between two or more change detection approaches. As in [2], a hybrid change detection schemes for dynamic optimization problems is proposed. In this study, the well known statistical hypothesis testing approaches (Behavior-based detection) are combined with three sensor-based detection schemes (Reevaluation-based detection) in order to increase detection capability.

Recently, in [28], a new hybrid scheme is proposed that incorporates sensor-based schemes with the population-based ones for detecting changes in dynamic environments. The results of the experimental study demonstrate the effectiveness of the proposed hybrid scheme compared to other change detection schemes.

Despite the effectiveness of the hybrid change detection scheme, it is always suffering from the problem of high computational costs.

4 Without Change Detection Based Methods

Some dynamic environments can pose challenges such as noise or partial landscape changes, which will be hard or even impossible to properly detect change in. Therefore, dynamic optimization algorithms in such environments depend for their success on an efficient change detection method and an adequate change reaction strategy. However, and as has been explained above, the change detection schemes in the literature are also suffering from several performance issues especially the high computational costs.

To avoid these drawbacks, a new initiative has been recently undertaken in [12]. This new idea consists of developing new algorithms that will be able to deal with dynamic environments adaptively without using any change detection schemes. In other words, solving DOPs without using any change detection methods to deal with the diversity loss problem and the outdated memory problem.

In fact, there has been much literature on the subject of the diversity loss problem. Generally, this problem has been tackled with the ultimate goal to maintain population diversity without using any change detection mechanism throughout the entire run of the algorithm [31]. However, these methods show the convergence and focusing constantly on diversity may affect the optimization process in an undesirable way. Multipopulation approaches have been proposed as an effective way to address the problem of diversity loss [12,13]. For example in [12], when the population diversity drops below a certain level, a strategy that selects random immigrants is performed to inject them into the population and boost its diversity.

On the other hand, the outdated memory problem has been always solved using iterative re-evaluation which is computationally expensive. This problem may prevent the algorithm finding new optima and misguide the optimization process. Therefore, solving this issue is of crucial importance. However, very little research effort has been devoted to it compared to diversity loss problem solving. One example is the evaporation mechanism proposed in [7]. This mechanism proposes to reduce the fitness value of the best position found by each particle during the search process. Therefore, better performance can be achieved by spending the efforts avoided in re-evaluation stored solutions, on the optimization process.

In the same context, a recent new Q-learning RL model has been proposed to solve DOPs [3]. This RL model does not require any change detector to know the occurrence of future changes in the environment. Rather, it is the task of the learner agent to learn how to deal with different complex situations.

5 Conclusion

Solving DOPs without using any change detection scheme is an open research issue. Whatever the problem's characteristics and whenever the changes occur, this kind of algorithms is not based on the knowledge of the change point time to handle future changes. Therefore, they will be effective for solving hard problems with recurrent changes and fast changes due to the fact that these algorithms do not need to spend any unnecessary effort on detecting changes.

In this paper we have tried to provide a survey of the literature on various change detection based schemes. We have also proposed a new classification of change detection based methods.

For future work, it would be interesting to design a new optimization algorithm that can dynamically adapt to changes by maintaining diversity and remedy the problem of outdated information without change detection.

References

1. Altin, L., Topcuoglu, H.R.: Impact of sensor-based change detection schemes on the performance of evolutionary dynamic optimization techniques. Soft Comput. **22**(14), 4741–4762 (2017). https://doi.org/10.1007/s00500-017-2660-1
2. Altin, L., Topcuoglu, H.R., Ermis, M.: Hybridizing change detection schemes for dynamic optimization problems. In: 2017 IEEE Congress on Evolutionary Computation (CEC), pp. 2086–2093. San Sebastian (2017)
3. Boulesnane, A., Meshoul, S.: Reinforcement learning for dynamic optimization problems. In: Proceedings of the Genetic and Evolutionary Computation Conference Companion, GECCO 2021, pp. 201–202. Association for Computing Machinery, New York, NY, USA (2021)
4. Bravo, Y., Luque, G., Alba, E.: Global memory schemes for dynamic optimization. Nat. Comput. **15**(2), 319–333 (2015). https://doi.org/10.1007/s11047-015-9497-2
5. Bu, C., Luo, W., Yue, L.: Continuous dynamic constrained optimization with ensemble of locating and tracking feasible regions strategies. IEEE Trans. Evol. Comput. **21**, 14–33 (2017)
6. Campos, M., Krohling, R.A.: Entropy-based bare bones particle swarm for dynamic constrained optimization. Knowl. Based Syst. **97**, 203–223 (2016)
7. Fernandez-Marquez, J.L., Arcos, J.L.: An evaporation mechanism for dynamic and noisy multimodal optimization. In: Proceedings of the 11th Annual Conference on Genetic and Evolutionary Computation, pp. 17–24. Montreal, Québec, Canada (2009)
8. Janson, S., Middendorf, M.: A hierarchical particle swarm optimizer for noisy and dynamic environments. Genet. Program. Evolvable Mach. **7**, 329–354 (2006)
9. Jiang, S., Yang, S.: A steady-state and generational evolutionary algorithm for dynamic multiobjective optimization. IEEE Trans. Evol. Comput. **21**, 65–82 (2017)
10. Jordehi, A.R.: Particle swarm optimisation for dynamic optimisation problems: a review. Neural Comput. Appl. **25**, 1507–1516 (2014)
11. Kundu, S., Biswas, S., Das, S., Suganthan, P.N.: Crowding-based local differential evolution with speciation-based memory archive for dynamic multimodal optimization. In: Proceedings of the 15th Annual Conference on Genetic and Evolutionary Computation, pp. 33–40. Amsterdam, The Netherlands (2013)
12. Li, C., Yang, S.: A general framework of multipopulation methods with clustering in undetectable dynamic environments. IEEE Trans. Evol. Comput. **16**, 556–577 (2012)
13. Li, C., Yang, S., Yang, M.: An adaptive multi-swarm optimizer for dynamic optimization problems. Evol. Comput. **22**, 559–594 (2014)

14. Li, X., Branke, J., Blackwell, T.: Particle swarm with speciation and adaptation in a dynamic environment. In: Proceedings of the 8th Annual Conference on Genetic and Evolutionary Computation, pp. 51–58. Seattle, Washington, USA (2006)
15. Masegosa, A.D., Pelta, D., Amo, I.G.D.: The role of cardinality and neighborhood sampling strategy in agent-based cooperative strategies for dynamic optimization problems. Appl. Soft Comput. **14**, 577–593 (2014)
16. Mavrovouniotis, M., Li, C., Yang, S.: A survey of swarm intelligence for dynamic optimization: algorithms and applications. Swarm Evol. Comput. **33**, 1–17 (2017)
17. Morrison, R.W., Jong, K.A.D.: Triggered hypermutation revisited. In: Proceedings of the 2000 Congress on Evolutionary Computation. CEC00 (Cat. No.00TH8512), vol. 2, pp. 1025–1032. La Jolla, CA (2000)
18. Mukherjee, R., Debchoudhury, S., Das, S.: Modified differential evolution with locality induced genetic operators for dynamic optimization. Eur. J. Oper. Res. **253**, 337–355 (2016)
19. Mukherjee, R., Patra, G.R., Kundu, R., Das, S.: Cluster-based differential evolution with crowding archive for niching in dynamic environments. Inf. Sci. (Ny) **267**, 58–82 (2014)
20. Nguyen, T.T.: Continuous dynamic optimisation using evolutionary algorithms. Ph.D. thesis, University of Birmingham, Birmingham, U.K. (2011). http://etheses.bham.ac.uk/1296
21. Nguyen, T.T., Yang, S., Branke, J.: Evolutionary dynamic optimization: a survey of the state of the art. Swarm Evol. Comput. **6**, 1–24 (2012)
22. Richter, H.: Change detection in dynamic fitness landscapes: an immunological approach. In: 2009 World Congress on Nature Biologically Inspired Computing (NaBIC), pp. 719–724. Coimbatore (2009)
23. Richter, H.: Detecting change in dynamic fitness landscapes. In: 2009 IEEE Congress on Evolutionary Computation, pp. 1613–1620. Trondheim (2009)
24. Richter, H., Yang, S.: Learning behavior in abstract memory schemes for dynamic optimization problems. Soft Comput. **13**, 1163–1173 (2009)
25. Richter, H., Yang, S.: Dynamic optimization using analytic and evolutionary approaches: a comparative review. In: Zelinka, I., Snášel, V., Abraham, A. (eds.) Handbook of Optimization. Intelligent Systems Reference Library, vol. 38, pp. 1–28. Springer, Berlin, Heidelberg (2013). https://doi.org/10.1007/978-3-642-30504-7_1
26. Rohlfshagen, P., Yao, X.: Dynamic combinatorial optimisation problems: an analysis of the subset sum problem. Soft Comput. **15**, 1723–1734 (2011)
27. Sahmoud, S., Topcuoglu, H.R.: A memory-based NSGA-II algorithm for dynamic multi-objective optimization problems. In: Squillero, G., Burelli, P. (eds.) EvoApplications 2016. LNCS, vol. 9598, pp. 296–310. Springer, Cham (2016). https://doi.org/10.1007/978-3-319-31153-1_20
28. Sahmoud, S., Topcuoglu, H.R.: Hybrid techniques for detecting changes in less detectable dynamic multiobjective optimization problems. In: Proceedings of the Genetic and Evolutionary Computation Conference Companion. ACM (2019)
29. Tinós, R., Yang, S.: Analyzing evolutionary algorithms for dynamic optimization problems based on the dynamical systems approach. In: Yang, S., Yao, X. (eds.) Evolutionary Computation for Dynamic Optimization Problems. Studies in Computational Intelligence, vol. 490, pp. 241–267. Springer, Berlin, Heidelberg (2013). https://doi.org/10.1007/978-3-642-38416-5_10
30. Tinós, R., Yang, S.: A self-organizing random immigrants genetic algorithm for dynamic optimization problems. Genet. Program. Evolvable Mach. **8**, 255–286 (2007)

31. Yang, S., Yao, X.: Experimental study on population-based incremental learning algorithms for dynamic optimization problems. Soft Comput. **9**, 815–834 (2005)
32. Yazdani, D., Cheng, R., Yazdani, D., Branke, J., Jin, Y., Yao, X.: A survey of evolutionary continuous dynamic optimization over two decades–Part A. IEEE Trans. Evolut. Comput. **25**, 1 (2021)

GPS/IMU in Direct Configuration Based on Extended Kalman Filter Controlled by Degree of Observability

Bendehiba Dahmane[1,5](✉), Brahim Lejdel[2], Fayssal Harrats[3], Sameh Nassar[4], and Lahcene Hadj Abderrahmane[5]

[1] Department of Technology, Faculty of Technology, El-Oued University, El-Oued, Algeria
[2] Department of Computer Sciences, Faculty of Exact Sciences, El-Oued University, El-Oued, Algeria
[3] Department of Electronics, Faculty of Electrical Engineering, Science and Technology University (USTO-MB), Oran, Algeria
[4] Mobile Multi-Sensor Systems (MMSS) Research Group, Department of Geomatics Engineering, University of Calgary, Calgary, Alberta, Canada
[5] Department of Space Instrumentation, Satellite Development Center (CDS), Space Agency (ASAL), Oran, Algeria

Abstract. In this paper a practical method for estimating the full kinematic state of a land-vehicle, along with sensors, low-cost inertial measuring unit (IMU), and Global Positioning System (GPS). However, this INS-GPS system requires in generally a robust architecture such as an Extended Kalman Filter (EKF) approach in direct configuration, by reason of its properties of extensive evaluations of non-linear equations. In addition, a practical approach for controlling the Degree of Observability (DoO) in GPS-INS integrated systems is used in these tests. Other than that, traditional observability analysis is inadequate for a long navigation trajectories matrix that becomes very large, such that it rises computational difficulties. Two datasets are used to verify the efficacy of the proposed approach against the existing GPS-INS integration scheme. The first set is real road data collected from a higher grade IMU at each (0.01 s) that was combined with DGPS data at each (1 s) in order to obtain the assumed true solution for the trajectory. The second one is real test data collected during land-vehicle trajectory. The implementation consists of three main algorithms that as well namely: Strapdown (Dead Reckoning DR), DoO, and EKF algorithms. The results are shown, implementation of the both approaches based on EKF and concept of DoO in GPS/INS Integrated systems are enough robust for its use along with low-cost sensors.

Keywords: Kalman filter · IMU · EKF · INS-GPS · DoO

1 Introduction

The autonomous navigation is an important ability for both manned and unmanned vehicles. This property of «autonomous» permit the system for estimating the state of

B. Lejdel et al. (Eds.): AIAP 2021, LNNS 413, pp. 143–152, 2022.
https://doi.org/10.1007/978-3-030-96311-8_14

the vehicle without the aid of a human operator. In many situations the autonomous navigation is a prerequisite for control tasks.

Low-cost inertial measuring unit (IMU) is an autonomous navigation [1, 2]. This an important capability contributed to the emergence of many research and industrial development [3, 4]. IMUs low-cost systems have become in along time proven a principal part of vehicle navigation systems [5, 6]. However, the precision of an INS error model with a Kalman Filter (KF) of 54 states is given in [8, 9]. INS-GPS is less perfect over time by the accumulation of the errors such as IMU alignment [7] model is obtained by employing a first-order model of the IMU. Besides, the complex integration system based in both method: estimate and absolute. That are justified by important complementarity of proprioceptive sensors and exteroceptive sensors [10].

Kalman Filter estimates position, velocity and attitude errors based on an INS error model and GPS updates [11, 12]. GPS has acceptable long-term accuracy; it is used to update the position and velocity in output of IMUs. Hence, it limits the long-term improve of INS errors. Besides, the short-term, precise data provided by the IMU is used to defeat of GPS outages and multipath errors. If a GPS outage happens, the Kalman Filter operates in the prediction mode, correcting the IMU data based on the system error model.

The Concept of the Degree of Observability (DoO) with regarding GPS/INS integrated systems is investigated in this paper. The traditional observability analysis is incompetent for exceptional navigation scenarios matrix that becomes very large with the passage of time, such that it rises computational difficulties. However, an unobservable system would not product an accurate estimation [13] and is apt to deviation [14], even if the noise level is negligible. Hence, the observability imposes a lower limit for the estimation error, for more details are given in [15].

On that report, and based on the above discussion, the paper objectives are:

- Pushing the low-cost IMU systems to be used as autonomous navigation system during long GPS outages for general land-vehicle navigation. Then, the fusion of IMU and GPS sensors is assured by proposed EKF that used as an estimator technique.
- Apply a practical approach for the observability, especially in dynamic analysis system, which to define the KF efficiency in the estimated states.

This paper is organized as follows: Sect. 2 illustrates the used methodology. Summaries of our tests and discussions of the results are demonstrated in Sect. 3. Finally, the conclusions are presented in Sect. 4.

2 Research Method

2.1 Extended Kalman Filter

Kalman Filter in direct configuration combine two estimators' values IMU and GPS data, which each contains values PVA (position, velocity, and attitude) [16, 17].

In our test, the first estimation is provided directly from IMU and the second estimation is the measurement provided from GPS receiver. In [18], linear equations in the continuous state are presented the Dynamic system.

$$\dot{x}(t) = F(t)x(t) + G(t)u(t) \tag{1}$$

Where:
F(t) is the dynamic matrix (partial derivatives)
x(t) is the state vector
G(t) is the design matrix
u(t) is the forcing function
$u(t) = [\delta f^b, \delta w_{ib}^b]^T$ are white noise whose covariance matrix is given by

$$Q = diag\left(\sigma_{ax}^2 \sigma_{ay}^2 \sigma_{az}^2 \sigma_{wx}^2 \sigma_{wy}^2 \sigma_{wz}^2\right) \tag{2}$$

The measurement model is given by:

$$z(t) = H(t)x(t) + v(t) \tag{3}$$

Table 1. Discrete-time KF equations

System model Measurement model	$x_k = \Phi_{k-1}x_{k-1} + w_{k-1}, w_k \sim N(0, Q_k)$ $z_k = H_k x_k + v_k, v_k \sim N(0, R_k)$
Initialization	$\hat{x}_0^- = E[x_0], P_0^- = var\left(x_0^-\right)$
Gain calculation	$K_k = P_k^- H_k^T \left(R_k + H_k P_k^- H_k^T\right)^{-1}$
Measurement update	$\hat{x}_0^+ = \hat{x}_0^- + K_k(\tilde{y}_k - \hat{y}_k)$
Covariance matrix update	$P_k^+ = [I - K_k H_k]P_k^-$
Time propagation	$\hat{x}_{k+1}^- = \Phi_k \hat{x}_k^- + G_k u_k$ $P_{k+1}^- = \Phi_k P_k^+ \Phi_k^T + \Phi_{dk}$ $\left(\Phi_{dk} = G_k \Phi_k G_k^T \Delta T\right)$
Where:	x^-, x^+: are, respectively, the Priori and Posteriori state vector, P^-, P^+: are, respectively, the Priori and Posteriori error covariance matrix $G = \begin{bmatrix} 0 & 0 \\ -R_b^n & 0 \\ 0 & R_b^n \end{bmatrix}$ *and* $Q = diag\left(\sigma_{ax}^2, \sigma_{ay}^2, \sigma_{az}^2, \sigma_{wx}^2, \sigma_{wy}^2, \sigma_{wz}^2\right)$ are, respectively, the standard deviation of the accelerometers and gyro-meters

Where:

$z(t)$ is the measurement at time t, H is the observation matrix and $v(t)$ is the white noise with $v(t) \sim N(0, R)$. Implementation of IMU information be based on very small sampling time interval $\Delta t = t_k - t_{k-1}$ (update each IMU = 100 Hz), the position (vehicle's movement: PVA variation vector) and the measurement model are given in Table 1 [16, 19]. Table 1 summarizes the discrete-time KF equations [3, 18].

2.2 Observability Analysis

A method that allows us to know the degree the health of internal system if it's good or not. Which, be based on measuring the external information output detects. This Method is named 'Observability' [20]. Here, in our non-linear system, H(t) = I3 × 3 is time constants; thus, implementation of Observability based on Boolean conditions

$$rank(O_\upsilon) \stackrel{?}{=} rank(O_{\upsilon+1}) \tag{4}$$

$$\text{let } O = \begin{bmatrix} N_0(t) \\ N_1(t) \\ \vdots \\ N_{\upsilon-1}(t) \end{bmatrix} \Rightarrow \begin{bmatrix} H_0(t) \\ H_0(t)F(t) + \frac{\partial}{\partial x} H_0(t) \\ \vdots \\ H_{\upsilon-1}(t)F(t) + \frac{\partial}{\partial x} H_{\upsilon-1}(t) \end{bmatrix} \Rightarrow \begin{bmatrix} H \\ HF \\ \cdots \\ HF^{\upsilon-1} \end{bmatrix} \tag{5}$$

In 1983 [21], the innovation of the concept of the "Degree of Observability" was based on a quantitative approach. Covariance error (P) is obtained from many iterations in the Extended Kalman filter process in which error describes the difference between an estimate and true states values. In addition, the common mathematical analysis has been used for description Normalized error (P′)

$$P'(k) = \left(\sqrt{P(0)}\right)^{-1} P(k) \left(\sqrt{P(0)}\right)^{-1} \tag{6}$$

Where:

P(0) is the initial error covariance matrix, P(k) is the current error covariance matrix. In bellow the acquired matrix can be presented:

$$P'(k) \Rightarrow \begin{bmatrix} \frac{P_{11}}{P_{11}(0)} & \frac{P_{12}}{\sqrt{P_{11}(0)P_{22}(0)}} & \cdots & \frac{P_{12}}{\sqrt{P_{11}(0)P_{nn}(0)}} \\ \frac{P_{21}}{\sqrt{P_{22}(0)P_{11}(0)}} & \frac{P_{11}}{P_{22}(0)} & \cdots & \frac{P_{12}}{\sqrt{P_{22}(0)P_{nn}(0)}} \\ \vdots & \vdots & \ddots & \vdots \\ \frac{P_{n1}}{\sqrt{P_{nn}(0)P_{11}(0)}} & \frac{P_{n2}}{\sqrt{P_{nn}(0)P_{22}(0)}} & \cdots & \frac{P_{nn}}{P_{nn}(0)} \end{bmatrix} \tag{7}$$

P_{ij} and $P_{ij}(0)$ are the error covariance matrix elements. Where, the pursuit is obtained by the sum of all of the eigenvalues, after that we obtain the normalized error covariance in (8). The eigenvalues of P″(k) are without dimension and limited between $0 < \lambda_i \leq n$, such that the DoO is defined better, as the error turns smaller.

$$P''(k) = \frac{n}{tr(P'(k))} P'(k) \tag{8}$$

2.3 INS-GPS Integration Methods

The integration of inertial sensors with GPS is general classified as follows [22]

- Loosely coupled system.
- Tightly coupled system.
- Ultra-tightly coupled system.

In our work, loosely coupled systems [23–26] and [27] scheme is investigated, the GPS data (e.g. position velocity, etc.) are fused explicitly with IMU data. This kind of systems is safely dependent on the availability of GPS data in Fig. 1.

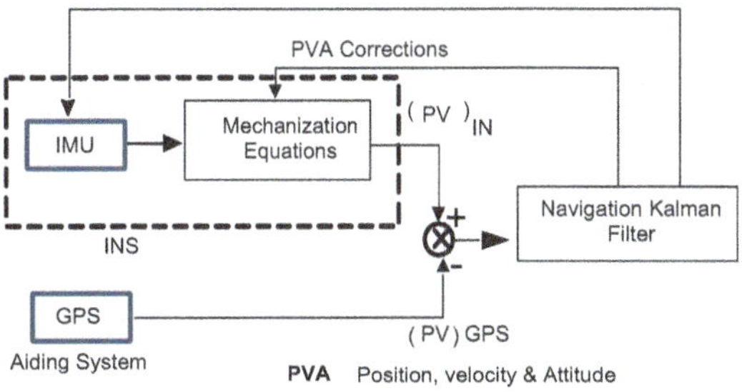

Fig. 1. Loosely coupled integration

Fig. 2. Reference trajectory

3 Results and Discussion

3.1 Field-Test Data Description

A Micro-Electro-Mechanical System (MEMS) based IMU was used. The Motion Pak II is a solid-state sensor cluster used for measuring linear accelerations and angular rates in instrumentation and control applications (Dead Reckoning Aiding GPS, Robotics, and Flight testing etc.).

This IMU (Motion Pak II) unit was mounted on the roof of the vehicle, with the NovAtel OEM4 GPS receiver [27]. Using this setup, a test trajectory was generated. In addition, for comparison purposes, a higher grade IMU (CIMU) was combined with DGPS data to obtain the assumed true solution for the trajectory Fig. 2.

3.2 IMU (Motion Pak II) Properties

Measurement frequency at 100 Hz for the IMU and 1 Hz for the GPS are use in our tests in Table 2.

Following are the results of IMU measurements

It's clearly shown in Fig. 3, that the input of f_b^y and w_z are noises. which, as described in Fig. 4, that the land-vehicle is not tracking the reference trajectory in all directions.

This reason is due to the absence of fusion GPS/IMU.

Table 2. Accelerometer and gyro-meter properties.

		x	y	z
Bias Factory Set (mg^2, /s)	Accelerometer	±125		
	Gyrometer	±5.0		
Scale Factor (mg^2, °/s)	Accelerometer	6.66 V/g		
	Gyrometer	0.133 V/°/s		
Input Axis Alignment (°typical)	Accelerometer	1		
	Gyrometer			

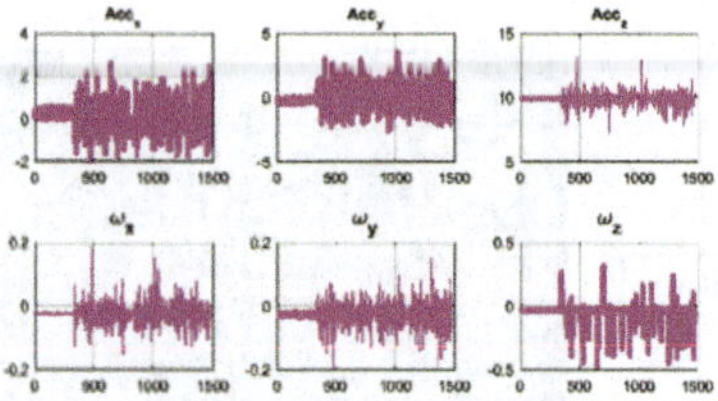

Fig. 3. IMU measurements along run

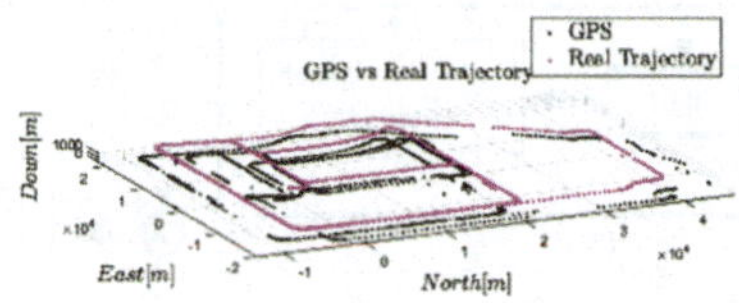

Fig. 4. True trajectory vs GPS measurements

3.3 Error State Results

In Fig. 5, As shown the decay from the initial value along time and the representation of the standard of deviation is the magenta dashed line. As seen the geodetic angle errors (φ, λ) show small changes that oncoming to zero.

Contrarily in Fig. 6, the position that is being corrected by the observation (GPS measurement) itself, here the errors increase over time and develop in a random walk. v_N and v_E show noisy style while v_D succeeds to stabilize, after a period, as it can be clarified our strap down model's inexactness.

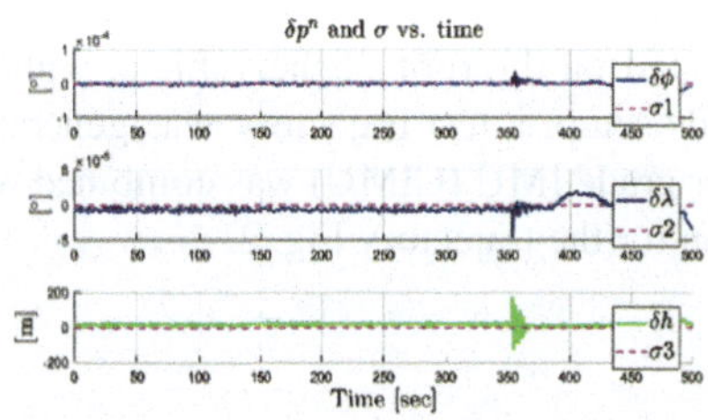

Fig. 5. Position error vs time

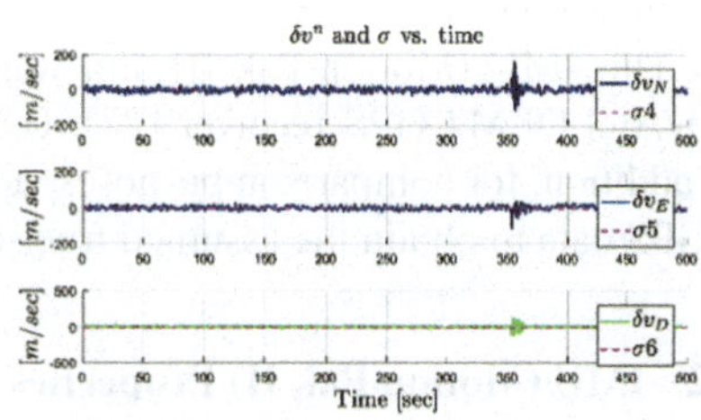

Fig. 6. Velocity error vs time

As shown in Fig. 7, variation of Euler angles error during the time of test time. Small variations occurred on the axis φ (roll) and θ (pitch). However, a great variation in azimuthal (δψ). That means, variation in the planar motions is related to the orientation of the land vehicle.

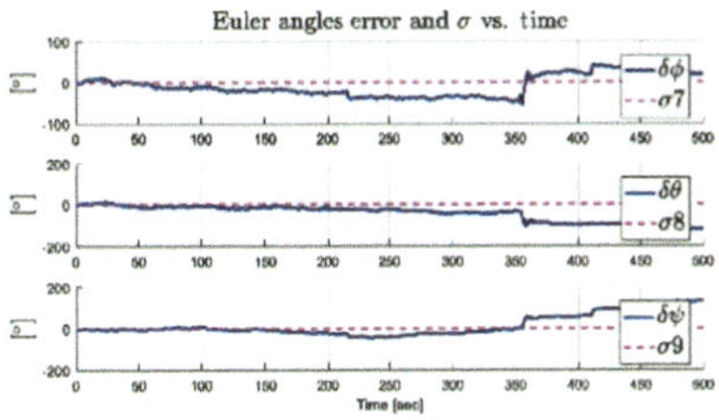

Fig. 7. Euler angles error vs time

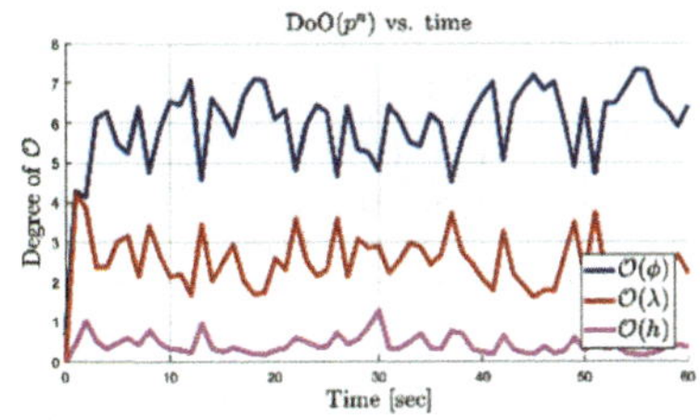

Fig. 8. DoO of position error vs time

3.4 Degree of Observability (DoO) Results

Depending on Sect. 2.2, we can analyze the DoO, this opportunity gives us a way to know the best states estimated (↓DoO), and which are most weak (↑DoO).

It's clearly shown in Fig. 8, there are cambers in starting of the land vehicle in (φ, λ and h) Latitude, longitude and altitude. Afterward, continues with many drifts and outages caused by two sensors IMU and GPS respectively.

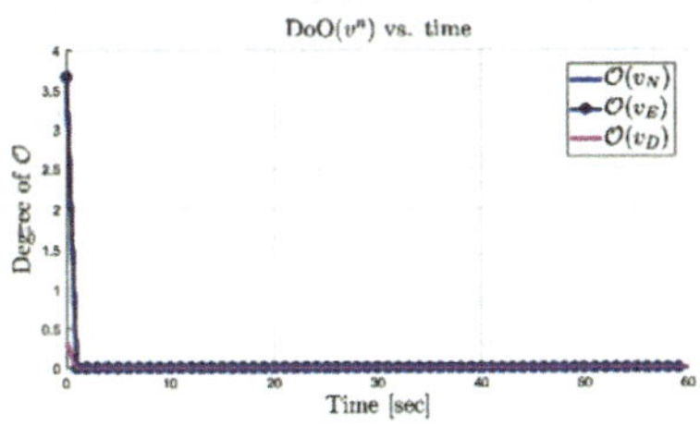

Fig. 9. DoO of velocity error vs time

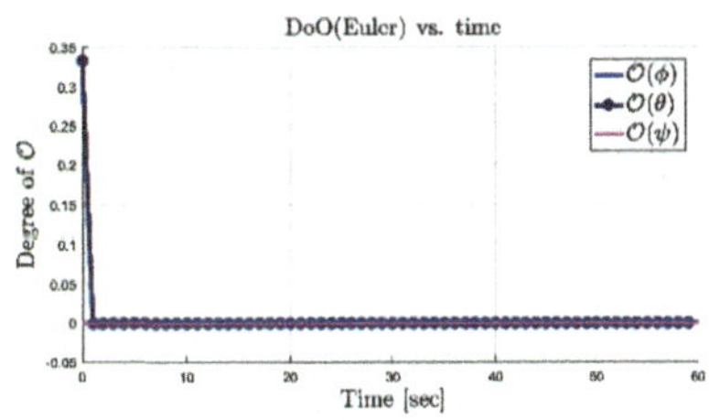

Fig. 10. DoO of euler angle error vs time

In Fig. 9, it appears the degree of observability for v_N eerror experiences a lag, while the degree of observability for v_D and v_D error quickly converges. because the system is stationary at first, then it starts in motion. that the v_D error and v_E error have good observability, with regard to the v_E error has bad observability when the system is stationary.

From Fig. 10, it shown the degree of observability for roll and pitch error quickly converges, while the degree of observability for yaw error is very late, and then vastly converges to small values. It appears also, the observability of an integrated GPS/INS system is better when the motion variation of land vehicles increases.

3.5 Comparison GPS Measurements, Reference and Estimate Trajectories

As shown in Figs. 11 and 12, the proposed approach shows comparative results for the movable vehicle in all directions.

It can be comprehended from the below figures that the vehicle is pursuit the true trajectory, particularly after applying the fusion IMU-GPS. which confirms the strength of the proposed approach in the assurance of vehicle tracking the reference trajectory with almost minimal errors in all directions, when GPS signal outages happen.

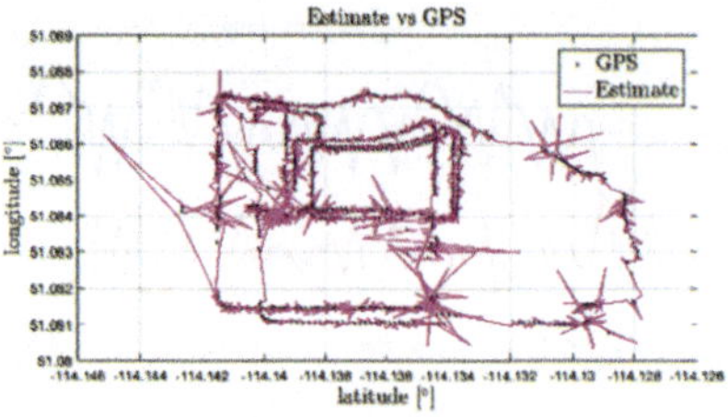

Fig. 11. Estimate vis Real trajectory

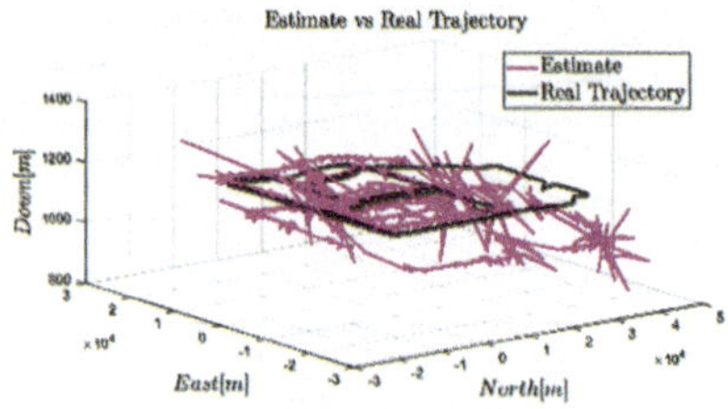

Fig. 12. Estimate vs GPS measurements

4 Conclusion

This work presents a practical method for estimating the full kinematic state of a land-vehicle navigation application, along with noised Inertial Measuring Unit (IMU) and Global Positioning System (GPS) sensors based on a loosely coupled approach. In addition, the concept of the Degree of Observability (DoO) is investigated in GPS/INS integrated system in order to control and obtain more accuracy.

The architecture of the system is based in an Extended Kalman filtering approach in direct configuration. The EKF is still the standard estimation technique for this kind of systems. One of the motivations of the proposed approach is due to the clarity and simplicity associated with the EKF in direct configuration. this method could be easily modified for its use with another kind of vehicle or craft (e.g. aerial robots).

In general, the results of our tests showed, the position and velocity errors converge to zero while the orientation errors remain small during the run. There are three possibilities of the reasons that orientation doesn't converge to zero since could be:

- the indirect connection between them and the GPS measurements.
- Error from the strap-down algorithm (model error).

Afterward, we examined the filter with several different fusion ratios between GPS and IMU rates and saw that as the ratio gets higher the accuracy, but also the calculation's complexity increases.

Finally, it is considered that the both approaches of GPS/INS Integrated systems Based on Extended Kalman Filter (EKF) and the concept of the Degree of Observability (DoO) are enough robust for its use along with low-cost sensors.

References

1. Han, H., Wang, J., Du, M.: A fast SINS initial alignment method based on RTS forward and backward resolution. J. Sens. **2017**, Article ID 7161858 (2017)
2. Wang, X., Ni, W.: An improved particle filter and its application to an INS/GPS integrated navigation system in a serious noisy scenario. Meas. Sci. Technol. **27**(9), article 095005 (2016)
3. Farrell, J.: Aided Navigation GPS with High-Rate Sensors. McGraw-Hill, New York, NY, USA (2008)

4. Rezaei, S., Sengupta, R.: Kalman filter-based integration of DGPS and vehicle sensors for localization. IEEE Trans. Control Syst. Technol. **15**, 1080–1088 (2007)
5. Skog, I.: A low-cost aided inertial navigation system for vehicle applications. MSc, Royal Institute of Technology, Stockholm, Sweden (2005)
6. Li, Y., Mumford, P., Rizos, C.: Performance of a low-cost field re-configurable real-time GPS/INS integrated system in urban navigation. In: IEEE/ION Position, Location and Navigation Symposium, pp. 878–885. Monterey, CA, USA, 5–8 May 2008
7. Titterton, D., Weston, J.L., Weston, J.: Strapdown Inertial Navigation Technology, 2nd edn. The American Institute of Aeronautics and Astronautics, VA, USA, Reston (2004)
8. Grewal, M.S., Andrews, A.P.: Kalman, Filtering: Theory and Practice Using MATLAB. Wiley, USA (2001)
9. Grewal, M.S., et al.: Global Positioning Systems, Inertial Navigation, and Integration. Wiley-Interscience, USA (2007)
10. El Hadji Amadou, G.: Localisation garantie d'automobiles. Contribution aux techniques de satisfaction de contraintes sur les intervalles. In: Doctorat, Université de Technologie de Compiègne, France (2006)
11. Quinchia, A.G., Ferrer, C.: A low-cost GPS & INS integrated system based on an FPGA platform. In: International Conference on Localization and GNSS, pp. 152–157. Tampere, Finland, 29–30 June 2011
12. Mohinder, G.S., Andrews, A.P.: Kalman Filtering: Theory and Practice Using MATLAB, 3rd edn. Wiley & Sons, New York, NY, USA (2008)
13. Goshen-Meskin, D., Bar-Itzhack, I.Y.: Observability analysis of piece-wise constant systems-Part II: application to inertial navigation in-flight alignment. IEEE Trans. Aerosp. Electron. Syst. **28**, 1068–1075 (1992)
14. Wang, J., Lee, H.K., Hewitson, S., Lee, H.K.: Influence of dynamics and trajectory on integrated GPS/INS navigation performance. J. Glob. Position. Syst. **2**, 109–116 (2003)
15. Shin, E., El-sheimy, N.: Report on the Innovate Calgary Aided Inertial Navigation System (AINSTM) Toolbox. Calgary, Canada (2004)
16. Tran, D.T., Luu, M.H., Nguyen, T.L., Nguyen, D.D., Nguyen, P.T.: Land-vehicle MEMS INS/GPS positioning during GPS signal blockage periods. VNU J. Sci. Math. Phys. 243–251 (2007)
17. Zhao, Y.: Key technologies in low-cost integrated vehicle navigation systems. Ph.D. Royal Institute of Technology, Stockholm, Sweden (2013)
18. Mohinder, G.S., Andrews, A.P.: Kalman Filtering Theory and Practice Using MATLAB, 3rd edn. Wiley & Sons, New York, NY, USA (2008)
19. Titteron, D.H., Weston, J.L.: Strapdown Inertial Navigation Technology, 2nd edn. IEEE, New York, NY, USA (2004)
20. Rhere, I., Abdel-Hafez, M., Speyer, J.: Observability of an integrated GPS/INS during maneuvers. IEEE Trans. Aerosp. Electron. Syst. 526–535 (2004)
21. Ham, F., Brown, R.: Observability, eigenvalues, and Kalman filtering. IEEE Trans. Aerosp. Electron. Syst. 269–273 (1983)
22. Zhou, J., et al.: INS/GPS tightly-coupled integration using adaptive unscented particle filter. J. Navig. **63**, 491–511 (2010)
23. Syed, Z.F., et al.: Civilian vehicle navigation: required alignment of the inertial sensors for acceptable navigation accuracies. IEEE Trans. Veh. Technol. **57**(6), 3402–3412 (2008)
24. Bruggemann, T.S., et al.: GPS fault detection with IMU and aircraft dynamics. IEEE Trans. Aerosp. Electron. Syst. **47**(1), 305–316 (2011)
25. Crassidis, J.L.: Sigma-point Kalman filtering for integrated GPS and inertial navigation. IEEE Trans. Aerosp. Electron. Syst. **42**(2), 750–756 (2006)
26. Georgy, J., et al.: Low-cost three-dimensional navigation solution for RISS/GPS integration using mixture particle filter. IEEE Trans. Veh. Technol. **59**(2), 599–615 (2010)

27. El-Sheimy, N., et al.: The utilization of artificial neural networks for multisensor system integration in navigation and positioning instruments. IEEE Trans. Instrum. Meas. **55**(5), 1606–1615 (2006)
28. Aggarwal, P., Gu, D., Nassar, S., Syed, Z., El-Sheimy, N.: Extended particle filter (EPF) for ins/GPS land vehicle navigation applications. In: ICON GNSS 20th International Technical Meeting of the Satellite Division, pp. 25–28 (September 2007)

Recognizing Arabic Handwritten Literal Amount Using Convolutional Neural Networks

Aicha Korichi[1(✉)], Sihem Slatnia[1], Najiba Tagougui[2], Ramzi Zouari[3], Monji Kherallah[3], and Oussama Aiadi[4]

[1] LINFI Laboratory, University of Mohammed Khider Biskra, Biskra, Algeria
{aicha.korichi,sihem.slatnia}@univ-biskra.dz
[2] The Higher Institute of Computer Science and Multimedia of Sfax, Sakiet Ezzit, Tunisia
[3] Faculty of Science of Sfax, Sfax, Tunisia
monji.kherallah@fss.usf.tn
[4] Department of Computer Science, University of Kasdi Merbah Ouargla, Ouargla, Algeria

Abstract. Currently, deep learning techniques have become the core of recent research in pattern recognition domain and especially for the handwriting recognition field where the challenges for the Arabic language are stilling. Despite their high importance and performances, for the best of our acknowledge, deep learning techniques have not been investigated in the context of Arabic handwritten literal amount recognition. The main aim of this paper is to investigate the effect of several Convolutional Neural Networks CNNs based on the proposed architecture with regularization parameters for such context. To achieve this aim, the AHDB database was used where very promising results were obtained outperforming the previous works on this database.

Keywords: Arabic handwriting · Literal amount recognition · Offline recognition · Deep learning · Resnet · VGG

1 Introduction

Handwriting recognition has received a growing interest by researchers and it has become a very active field of research in recent years due to its important applications including automatic postal mail sorting, historical handwritten documents digitization, automatic checks recognition... etc. The handwriting recognition systems are divided into online and offline branches according to the data acquisition mode [1–3]. In the online mode, the input data is acquired from a digitized tactile screen and both static and dynamic information about the handwriting trajectory are available like the trajectory coordinates, temporal order, speed and acceleration [4]. In the offline mode, the input data is captured from a scanned image of the text, and therefore only static information representing

B. Lejdel et al. (Eds.): AIAP 2021, LNNS 413, pp. 153–165, 2022.
https://doi.org/10.1007/978-3-030-96311-8_15

the pixel values (0 or 1) is available. The lack of dynamic information makes offline handwriting recognition a very challenging task compared to the online one. Despite the many solutions proposed to deal with offline issues, it is still a challenging task because of several inherent characteristics that are related to the writing style and the writing language itself. In fact, for the Arabic script which is written from right to left, it contains 28 letters where each letter has more than three shapes according to its position within the world. Furthermore, additional strokes can be written above and below the letters like dots, chadda, fatha, etc. Moreover, Arabic is the 4th spoken language with more than 400 million speakers [5], it is one of the six United Nations official languages.

In the literature, existing approaches can roughly be classified into two categories, namely holistic and segmentation approach [6]. As its name indicates, the segmentation approach consists of the segmentation of the word into characters or sub words, meanwhile, the holistic approach considers the whole image as an entity that must be recognized without decomposition in smaller units and it is generally used with limited lexical datasets.

For the actual research purpose, we proposed in this paper to deal with the recognition of the Arabic handwriting literal amount that is still highly used in checks. Although this is not a new research area especially that many digitization attempts where realized in the early 1990s, our motivation to revive this domain has been consolidated by two major facts, the first one is that unlike it was expected in the digitization area, online electronic payment methods haven't replaced the use of checks and these ones are still spread used nowadays. The second reason is that according to recent statistical studies, about 100 billion checks are treated manually each year [7,8]. Therefore, checks are a fundamental payment tool used in many countries and there are still processed thanks to the effort of human agents. There is no real need to stop using them, especially that they offer a trusted biometric measure within the personal handwritten style that differs from one person to another and which is more secure than laying on magnetic cards where the security property could be violated.

For all those reasons, and convinced by the power of the deep learning for the field of handwriting recognition, we tried to investigate in a real solution that enhances the actual automating checks reading methods so as to bring several benefits to ensure a better solution that is efficient, robust, and fast.

Indeed, interest for deep learning has emerged in the last few years [9] due to its effectiveness with the ability to learn high-level abstractions automatically in the data, it has replaced the handcrafted ways where the features were extracted manually. Convolutional Neural Networks (CNN) are among the most famous neural network models. It has proved its efficiency for many pattern recognition tasks. CNNs are present in the literature with several architectures that we decide to test in this experimental study to choose the most relevant one that matches our main motivation of designing a robust and efficient automatic checks reading system using the recognition of Arabic handwriting literal amounts.

To achieve this goal, several CNNs architectures starting from simple architecture and then going to complex ones namely Visual Geometry Group (VGG) and Residual Network (Resnet) were used with regularization parameters for the recognition of Arabic handwriting literal amounts.

The rest of the paper is organized as follows: In Sect. 2, we present some related works, which have dealt with the Arabic handwritten recognition issue. Section 3 gives details of the proposed CNNs architectures and an overview of the system. Section 4 presents the AHDB database of Arabic handwritten literal amounts. The experimental evaluation results are given in Sect. 5. Finally, we finalize the paper by outlining some conclusions.

2 Related Work

Nowadays, Deep Learning techniques become the state of the art of the majority of research, they proved their efficiency for many pattern recognition systems [10,11]. Despite their good performances, little attention has been devoted to dealing with them in the context of Arabic handwriting recognition. Almaageg et al. [12] proposed a new system for Arabic handwriting recognition based on two deep neural network techniques. The first one is CNN, and it is used for feature extraction, the second one is bidirectional Long Short-Term Memory (BLSTM) followed by Connectionist Temporal Classification layer (CTC) for classification purposes. In [13], the proposed model is based on the combination of CNN with Support Vector Machines (SVM) classifier and using raw pixel data. This system was tested on both HACDB and IFN/ENIT Arabic handwriting letters databases. The same system was reproduced in [14] with the application of the Dropout technique. Authors in [15] proposed a handwriting recognition system based on hybrid CNN architectures applied on several databases.

El-Melegy et al. [16] were the first that apply deep learning for the recognition of complete literal amount words. The proposed system is based on VGG architecture composed of 16 hidden convolution layers and 1 fully-connected layer by using data augmentation.

On the other hand, most researchers have devoted to using handcrafted features. They are divided into three categories. In the first sub-category, the Arabic handwriting was considered as a series of statistical characteristics. Assayony et al. [17] proposed a new system of Arabic handwritten literal amount recognition based on a holistic approach using Gabor filters with Bag of Features (BoF). Gabor filter was applied with different scales and orientations to extract local features which will be arranged and fed to BoF frameworks. Hassen et al. [18] proposed a Multi statistical features system for Arabic handwriting literal amount recognition. They used a set of statistical features including Invariant Moments (IV), Histogram of Oriented Gradients (HOG), and Gabor filters. Thereafter, Sequential Minimal Optimization (SMO) classifier was applied.

In the second sub-category, the Arabic handwriting literal amount is considered as a series of structural features. Al-Nuzaili et al. [19] presented an improvement of the Perceptual Feature Extraction Model (PFM) by considering the shapes of loops and dots. In another work [20], the handwriting is considered as a set of distance, angle, vertical and horizontal span features. In the classification stage, three ELM classifiers were combined using the majority vote technique.

The third category considers the handwriting as a mixture of statistical and structural features. In [21], the Arabic handwriting literal amounts were represented using statistical features like Zernike moment invariants (ZMI), local chain code histograms (CCH), zoning, and the density profile histograms (DPH), and some other structural features extracted from the different parts of the image. In the classification stage, SVM was applied based on the extracted features. In [22], the proposed method proceeds by applying Discrete Cosine Transform (DCT) and Histogram of Oriented Gradient (HOG) to extract structural features merged with some other statistical features. An artificial neural network was used in the classification stage.

3 Deep Learning for Arabic Literal Amount Recognition

Deep learning has become the core of recent pattern recognition research and it has taken an incredible growth for many computer vision tasks due to its high performance in automatically capturing complex characteristics from the low level to high level. In other words, it aims to model high-level abstractions based on a set of traditional machine learning algorithms using several nonlinear transformations.

From many deep learning techniques, Convolutional Neural Networks CNNs [23] are the most commonly used and popular ones. They proved their efficiency, especially for handwriting recognition and they considered as top solutions in such issues [14,24–27].

Referring to the aforementioned criteria, we are presenting several CNN architectures for Arabic handwriting literal amount recognition, starting from a simple CNN, until using complex ones like Resnet and VGG. In Spite of the major advantages of deep CNNs that can learn automatically more abstract information by constructing deep architecture, the huge quantity of parameters used could lead to other problems known as over fitting. To deal with such situations and protect the CNN against such cases, we opted to add a dropout layer. Moreover, CNNs need a huge quantity of data to be more efficient, we increase the number of training data by applying data augmentation.

3.1 Convolutional Neural Networks

As we have mentioned above, the Convolutional Neural Network is a type of deep learning technique that proved its efficiency for handwriting character recognition due to its ability to learn the visual patterns from image pixels [28]. Generally, CNN comprises three main layers which are convolutional layers, pooling layer, and fully connected layer (FCL) [29]. CNN performs the nonlinearity by the activation functions and the pooling layers. As an activation function, we have exploited the ReLU function (Rectified Linear Units).

$$f(x) = \max(0, x) \tag{1}$$

FCL is performed as follows:

$$y_j^l = \max\left(0, \sum_i y_i^{l-1}.w_{i,j}^l + b_j^l\right) \tag{2}$$

Where: y_i^l represents the jth node in the lth layer, $w_{i,j}^l$ represents the weights between y_i^l and y_i^{l-1}. b_j^l is the bias.

The last step of predicting a distribution p(y_i)is to handle a softmax over the outputs Z_i:

$$y_i = \frac{\exp(Z_i)}{\sum_k \exp(Z_k)} \tag{3}$$

$$Z_i = \sum_j y_i^{l-1}.w_{i,j} + b_i \tag{4}$$

The proposed CNN network scheme based on the dropout layer is illustrated in Fig. 1. At first, the network receives the image in the input layer as a sequence of pixels and passed them through convolution layers where the image is convolved with a set of filters. Thereafter, the obtained activation maps are passed to an activation function layer, followed by a MaxPooling layer in order to preserve the pixels of higher values. To protect the network against the over fitting, a dropout layer is added just before the fully connected layers where the classification task is done.

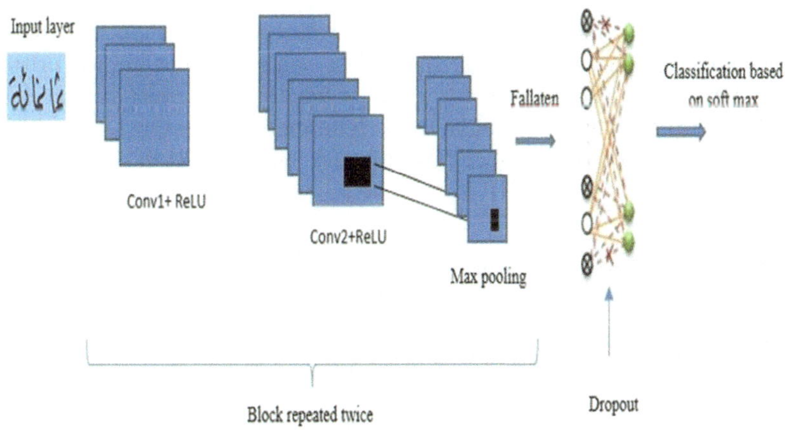

Fig. 1. General scheme of our proposed network based on the dropout layer.

3.2 Visual Geometry Group (VGG)

Starting from the common idea that says going deeper through a network will give better accuracies, Simonyan et al. [30] introduced the Visual Geometry Group (VGG) architecture which includes mainly two convolutional layers with

ReLU as an activation function followed by max-pooling layer. The final layer of VGG architecture is the softmax for classification purposes. One of the advantages of VGG architecture is its simplicity, unlike previous networks by minimizing the kernel-size filters into 3 × 3 which allows learning more complex features.

3.3 Residual Network (ResNet)

Going deeper into a network by increasing the number of layers, does not mean that the network is more efficient, sometimes going deeper could harm the network efficiency during the training data because of vanishing gradient problem. Recently, Residual Network ResNet network architecture, which was proposed by [31], has becomes very popular tools in the community of deep learning due to solving such a vanishing problem. The main idea behind the residual network is skipping one or more layers by a residual mapping. In another term, adding a parameter to the output from the previous layer to the layer ahead

4 Database Presentation

For many Arabic countries, checks with the handwritten format are stilling the fundamental tool for financial transactions where about one hundred billion checks are treated over the world and the majority of them are treated manually basing on human agents. Automatic check reading has become an active area of research. AHDB benchmark database [32] is a publicly available database that contains the 63 different classes representing the Arabic handwritten literal amounts normally used on checks. Each class contains 105 samples written by different writers. As it is using for many researchers, a cross-Validation with three folds (two folds for training and the remainder for testing) is used for this study where each fold contains 2205 samples. A sample of each class is shown in (Table 1):

Table 1. Arabic words used to express amounts on checks extracted from AHDB database

N°	Class	N°	Class	N°	Class	N°	Class	N°	Class
1	أحد	14	خمسة	27	تسعة	40	ستة	53	ثلاثمئة
2	احدى	15	خمسمئة	28	تسعمئة	41	ستمئة	54	ثلاثمائة
3	واحد	16	خمسمائة	29	تسعمائة	42	ستمائة	55	عشرون
4	ثمان	17	أربع	30	تسعون	43	ستون	56	عشرين
5	ثمانية	18	أربعة	31	تسعين	44	ستين	57	اثنان
6	ثمانمئة	19	أربعمئة	32	لا	45	عشر	58	اثنين
7	ثمانمائة	20	أربعمائة	33	سبع	46	عشرة	59	مئتين
8	ثمانون	21	أربعون	34	سبعة	47	ثلاثون	60	مائتين
9	ثمانين	22	أربعين	35	سبعمئة	48	ثلاثين	61	ألفان
10	اثنى	23	مئة	36	سبعمائة	49	ألف	62	ألفين
11	خمسون	24	مائة	37	سبعون	50	آلاف	63	غير
12	خمسين	25	مليون	38	سبعين	51	ثلاث		
13	خمس	26	تسع	39	ست	52	ثلاثة		

5 Experimental Results

For our case, we are dealing with 63 different classes as previously described. The limited lexicon that we used argues the use of the holistic approach where the images were fed directly to the network without any segmentation. Since the CNNs require a huge amount of data to be efficient which is not available in the AHDB database, for all experiments done in this section, we have used data augmentation technique for training images with criteria that are related and interpreted by the Arabic language orientation, zoom, and writing width). Moreover, the batch size was selected to be 32 batches. All the experiments done in this section were obtained using Python with Keras and Tensorflow installed on a computer with a Core i7 "7th generation" processor, 16 GB of RAM and AMDA Radeon graphical card. As a metric of evaluation, we have used the accuracy metric, which is the quotient of the total number of correctly classified words corrected over the total number of words.

5.1 Results of the Proposed Architecture

First, we tested the effect of CNNs on the recognition of the literal amounts based on a simple CNN composed of four layers. After every two layers, a max-pooling layer is added. The aim of using a simple CNN is to study the effect of adding the dropout layer to a network for several epochs where the experiments can be easily done. To achieve such objective, several experiments have been carried out without using a dropout layer, and by adding the dropout layer in both extraction and classification parts, and only adding the dropout layer before the classification layer. The obtained results are illustrated in the following figures (Fig. 2, Fig. 3, and Fig. 4):

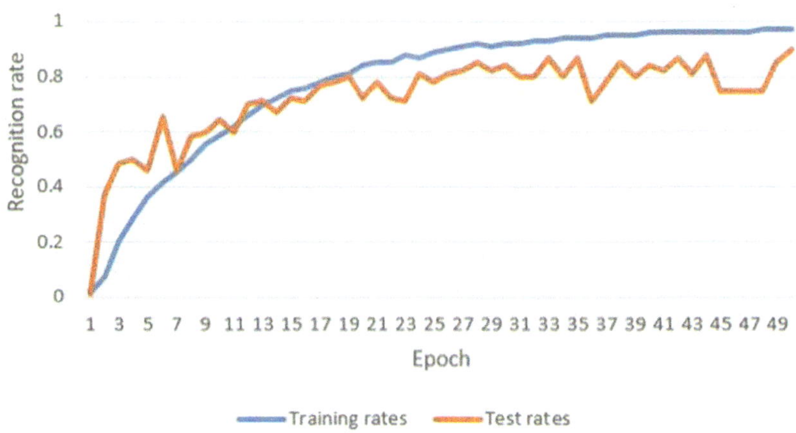

Fig. 2. Results without using dropout.

Fig. 3. Results by adding the dropout layer in both stages of feature extraction and classification.

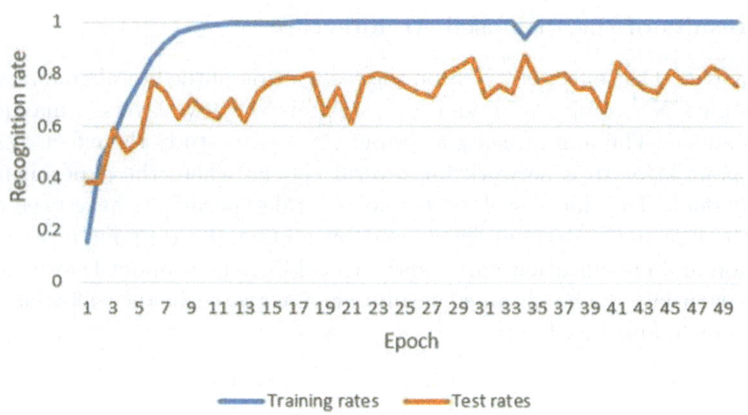

Fig. 4. Results by adding the dropout layer only before the classification layer.

Based on the above figures, it is clearly shown the positive effect of adding a dropout layer on the test data regardless of its position, as shown in Fig. 3 and Fig. 4. Moreover, it is obvious that using the dropout layer on just before the fully connected layer gave in some epoch's very high performance compared with those that have been obtained by using dropout out in both feature extraction and classification parts. Whilst, the average recognition rate obtained by adding the dropout layer on both parts is better than which is achieved by adding it just once before the classification stage. It can be caused by the negative influence of the irrelevant characteristics without removing them by using dropout just on the classification.

5.2 VGG Results

As we have mentioned in the previous sections, the Visual Geometry Group (VGG) network is one of the simplest and efficient networks. For this study, we used a VGG network architecture based on 13 layers to investigate the impact of adding more layers in the recognition rate. During the first fifth epochs, the recognition rates obtained by the VGG16 network are very low compared to those obtained by the previously proposed architectures. However, the architecture faithfully success in increasing the recognition rate of training samples to be stabilized on 100%. Despite that VGG-16 contains more layers than the previously proposed ones; this hasn't affected positively the recognition rates especially for the test samples (Fig. 5). This can be interpreted by the nature of Arabic writing and the existing features that could be captured just by a few layers.

Fig. 5. VGG16 results.

5.3 Resnet Results

In order to investigate the effect of more complex architectures by going deeper, we opted to apply Resnet architecture using always a dropout layer, with a depth of 17 convolutional layers with randomly initialized weight. According to the obtained results (Fig. 6), it is clearly shown the positive effect of increasing the number of layers to go more depth basing on Resnet network. This allows weights to be updated correctly through the backpropagation of the gradient error. Resnet tries to find an optimized number of layers to eliminate the vanishing gradient problem. Unlike the previous architectures, the recognition rate is starting with high performance in training and test samples then decrease on some epochs of the test images, and finally stabilizes from the 14th epoch for both stages. The recognition rate is about 100% to 98% for training and test samples respectively, after a number of 100 epochs.

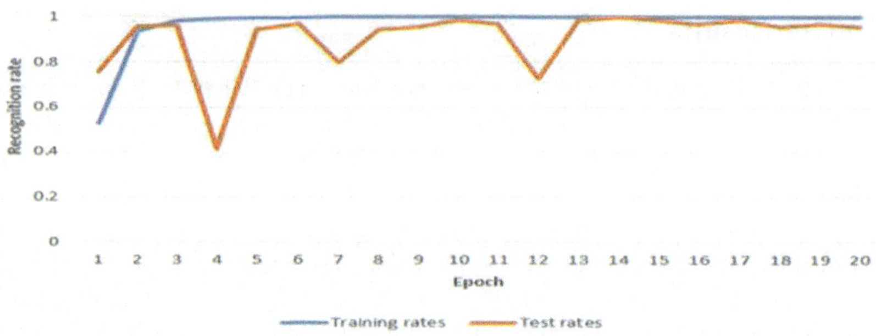

Fig. 6. Resnet results.)

5.4 Results Discussion

In order to investigate the impact of deep learning algorithms on the recognition of Arabic handwriting, in this paper have tested three architectures with different parameters. Table 2 summarizes all experiments done for this study.

Table 2. Proposed architectures results.

Architecture	Accuracy %
Proposed architecture with dropout	95.7
VGG16	97.14
Resnet	98.57

It is clearly shows the high performance of deep architectures for the recognition with even simple or complex ones. First, we have tested the performance of the proposed architecture with several positions of dropout layer where the best recognition rate was 95.71% by adding a dropout layer just before the classification stage. In spite of the low number of layers used in our proposed architectures, it gives very good results which are very close to them obtained with VGG and Resnet architectures where we are going deep on the network by increasing the number of layers by adding always the dropout layer in the same position. VGG and Resnet architectures are given results close to each other. However, the nature of Resnet eliminating the vanishing problem by trying to find the optimized number of layers allows it to outperform the other architecture with the best average recognition rate of 98.57%.

5.5 Comparison with State of the Art Systems

In this work, we studied the effectiveness of several CNN architectures in such a context, where the best recognition was 98.57%, and it was attained when using

Resnet architecture. The comparisons of the obtained results with other recent and relevant works made on the AHDB database are summarized in Table 3:

Table 3. Proposed architectures results.

Authors	Accuracy '%
Menasria et al. [21]	89.13
Assayony and Mahmoud [17]	86.44
Hassan et al. [18]	95
Al-Nuzaili et al. [19]	92.13
El-Melegy et al. [16]	97.8
Amani Ali et al. [15]	96.8
Our proposed system	98.57

Based on the above table, it is clear that the implemented CNN architectures have proven their efficiency against handcraft based methods, in the context of Arabic handwriting literal amount images from AHDB database.

6 Conclusion and Perspectives

Arabic handwriting recognition is still a challenging task due to several inherent characteristics of Arabic script and it is still in the level of experiments. Arabic handwritten literal amount recognition is a typical application of the handwriting recognition domain since the checks are considered as the fundamental financial transactions tool for several Arab countries. In this paper, several Convolutional Neural Networks CNNs were implemented like simple CNN, VGG-16 and ResNet, with the application of regularization methods including dropout and data augmentation techniques. The obtained results proved the efficiency of the CNNs architectures that outperform the existing methods based on handcrafted features. As a perspective, we intend to evaluate an enhanced version using transfer learning of the implemented architectures for recognizing Arabic words from other vocabularies and databases.

References

1. Lorigo, L.M., Govindaraju, V.: Offline Arabic handwriting recognition: a survey. IEEE Trans. Pattern Anal. Mach. Intell. **28**(5), 712–724 (2006)
2. Korichi, A., et al.: Off-line Arabic handwriting recognition system based on ML-LPQ and classifiers combination. In: 2018 International Conference on Signal, Image, Vision and their Applications (SIVA), pp. 1–6. IEEE (2018)
3. Korichi, A., et al.: Arabic handwriting recognition: Between handcrafted methods and deep learning techniques. In: 2020 21st International Arab Conference on Information Technology (ACIT), pp. 1–6. IEEE (2020)

4. Zouari, R., Boubaker, H., Kherallah, M.: A time delay neural network for online Arabic handwriting recognition. In: Madureira, A.M., Abraham, A., Gamboa, D., Novais, P. (eds.) ISDA 2016. AISC, vol. 557, pp. 1005–1014. Springer, Cham (2017). https://doi.org/10.1007/978-3-319-53480-0_99
5. Gary, F.S., Fennig, C.D. Ethnologue: languages of Asia. sil International Dallas (2017)
6. Khorsheed, M.S.: Off-line Arabic character recognition-a review. Pattern Anal. Appl. **5**(1), 31–45 (2002)
7. Ahmad, I., Mahmoud, S.A.: Arabic bank check analysis and zone extraction. In: Campilho, A., Kamel, M. (eds.) ICIAR 2012. LNCS, vol. 7324, pp. 141–148. Springer, Heidelberg (2012). https://doi.org/10.1007/978-3-642-31295-3_17
8. Ahmad, I., Mahmoud, S.A.: Arabic bank check processing: state of the art. J. Comput. Sci. Technol. **28**(2), 285–299 (2013)
9. LeCun, Y., Bengio, Y., Hinton, G.: Deep learning. Nature **521**(7553), 436–444 (2015)
10. Granet, A., et al.: Transfer learning for handwriting recognition on historical documents. In: ICPRAM, pp. 432–439 (2018)
11. Altwaijry, N., Al-Turaiki, I.: Arabic handwriting recognition system using convolutional neural network. Neural Comput. Appl. **33**(7), 2249–2261 (2020). https://doi.org/10.1007/s00521-020-05070-8
12. Maalej, R., Kherallah, M.: Convolutional neural network and BLSTM for offline Arabic handwriting recognition. In: 2018 International Arab Conference on Information Technology (ACIT), pp. 1–6. IEEE (2018)
13. Elleuch, M., Maalej, R., Kherallah, M.: A new design based-SVM of the CNN classifier architecture with dropout for offline Arabic handwritten recognition. Procedia Comput. Sci. **80**, 1712–1723 (2016)
14. Elleuch, M., Tagougui, N., Kherallah, M.: A novel architecture of CNN based on SVM classifier for recognising Arabic handwritten script. Int. J. Intell. Syst. Technol. Appl. **15**(4), 323–340 (2016)
15. Ali, A.A.A., Mallaiah, S.: Intelligent handwritten recognition using hybrid CNN architectures based-SVM classifier with dropout. J. King Saud Univ. Comput. Inf. Sci. (2021)
16. El-Melegy, M., Abdelbaset, A., Abdel-Hakim, A., El-Sayed, G.: Recognition of Arabic handwritten literal amounts using deep convolutional neural networks. In: Morales, A., Fierrez, J., Sánchez, J.S., Ribeiro, B. (eds.) IbPRIA 2019. LNCS, vol. 11868, pp. 169–176. Springer, Cham (2019). https://doi.org/10.1007/978-3-030-31321-0_15
17. Assayony, M.O., Mahmoud, S.A.: Recognition of Arabic handwritten words using Gabor-based bag-of-features framework. Int. J. Comput. Digit. Syst. **7**(01), 35–42 (2018)
18. Hassen, H., Al-Maadeed, S.: Arabic handwriting recognition using sequential minimal optimization. In: 2017 1st International Workshop on Arabic Script Analysis and Recognition (ASAR), pp. 79–84. IEEE (2017)
19. Al-Nuzaili, Q., et al.: Enhanced structural perceptual feature extraction model for Arabic literal amount recognition. Int. J. Intell. Syst. Technol. Appl. **15**(3), 240–254 (2016)
20. Al-Nuzaili, Q.A., et al.: Pixel distribution-based features for offline Arabic handwritten word recognition. Int. J. Comput. Vis. Robot. **7**(1-2), 99–122 (2017)
21. Menasria, A., et al.: Multiclassifiers system for handwritten Arabic literal amounts recognition based on enhanced feature extraction model. J. Electron. Imaging **27**(3), 033024 (2018)

22. Hassan, A.K.A., Kadhm, M.S.: Handwriting word recognition based on neural networks. Int. J. Appl. Eng. Res. **10**(22), 43120–43124 (2015)
23. Fukushima, K.: A hierarchical neural network capable of visual pattern recognition. In: Neural Network, p. 1 (1989)
24. Guo, Y., et al.: Deep learning for visual understanding: a review. Neurocomputing **187**, 27–48 (2016)
25. Ahmed, R., Al-Khatib, W.G., Mahmoud, S.: A survey on handwritten documents word spotting. Int. J. Multimed. Inf. Retr. **6**(1), 31–47 (2017)
26. Hafemann, L.G., Sabourin, R., Oliveira, L.S.: Learning features for offline handwritten signature verification using deep convolutional neural networks. Pattern Recognit. **70**, 163–176 (2017)
27. Jin, L., et al.: Online handwritten Chinese character recognition: from a bayesian approach to deep learning. In: Advances in Chinese Document and Text Processing. World Scientific, pp. 79–126 (2017)
28. LeCun, Y., et al.: Convolutional networks for images, speech, and time series. Handb. Brain Theory Neural Netw. **3361**(10), 1995 (1995)
29. Krizhevsky, A., Sutskever, I., Hinton, G.E.: Imagenet classification with deep convolutional neural networks. In: Advances in Neural Information Processing Systems, pp. 1097–1105 (2012)
30. Simonyan, K., Zisserman, A.: Very deep convolutional networks for large-scale image recognition. In: arXiv preprint arXiv:1409.1556 (2014)
31. He, K., et al.: Deep residual learning for image recognition. In: Proceedings of the IEEE Conference on Computer Vision and Pattern Recognition, pp. 770–778 (2016)
32. Al-Ma'adeed, S., Elliman, D., Higgins, C.A.: A data base for Arabic handwritten text recognition research. In: Proceedings Eighth International Workshop on Frontiers in Handwriting Recognition, pp. 485–489. IEEE (2002)

A Novel Separable Convolutional Neural Network for Human Activity Recognition

Ali Boudjema(✉) and Faiza Titouna

LaSTIC Laboratory of Computer Sciences, University of Batna2, Batna, Algeria
{a.boudjema,f.titouna}@univ-batna2.dz

Abstract. The issue with the time series classification arises in several human applications such as healthcare, industrial monitoring and cyber-security. Recently, various methods have been developed in order to deal with this matter. In this paper, a novel deep learning-based model for human activity recognition is developed. The proposal examines deeply the training phase in which the acceleration metric is considered by exploring all components of the model. To this end, the architecture of the Convolutional Neural Network (CNN) is studied: a) first, we employ a separable CNN, where we integrate a particular filter model for the depthwise convolution; b) second, we combine the extracted features with the handcrafted features. The proposed classifier is evaluated using a human activity recognition dataset and compared to a set of recent works. The obtained results show that our model outperforms the compared methods under various metrics.

Keywords: Time series classification · Human activity recognition · Convolutional Neural Network (CNN) · Separable CNN

1 Introduction

Human activity recognition (HAR) aims to analyze and recognize activities obtained from a sequence of observations. The classification problem based on time series appears in many real applications such as health monitoring, medical care, human-computer interaction, etc. There are two families of time series, video-based, where data are collected using cameras (videos) and sensor-based using smartphones, smartwatches, tablets, MP3 players, or any other digital devices that could detect body movement just by adding specific sensors. The first family requires the recording of body movements with the help of cameras, which presents a significant risk of violating personal data. The quality of the data collected may also be influenced by external conditions (climate, camera quality, lighting, etc.); also, the preprocessing of video requires enormous resources (RAM, CPU, GPU, etc.). Meanwhile, sensors are portable, low cost and their data are not influenced by external conditions. Human activities recognition includes four primary applications based on activities [3], covering gestures recognition which aims to recognize hand or face movements. We also

B. Lejdel et al. (Eds.): AIAP 2021, LNNS 413, pp. 166–176, 2022.
https://doi.org/10.1007/978-3-030-96311-8_16

cite action recognition that comprises movements and actions of a single person; another human activity recognition application is interaction recognition that tries to identify actions executed while interacting with an object or another person. The last category regroups the previous classes. It collects data such as wrist-worn accelerometers, gyroscopes and magnetometers. Many other examples of data from real-life applications are represented as a time series, such as biomedical signals (e.g. EEG[1] and ECG[2]), industrial devices (e.g. gas sensors and laser excitation), etc.

In the meantime, exploiting recent methods such as deep learning (DL) has been applied in automatic feature-extraction [29], and achieves a high rate in fields such as computer vision, speech recognition and natural language processing.

The rest of the paper is structured as follows. In the next section, some recent works dealing with the classification of time series are presented. Section 3 covers a wide range of preliminary concepts such as time series, convolutional neural network and feature extraction, In Sect. 4, we describe the proposed architecture based mainly on the separable convolutional neural network model. The experimental results are presented and discussed in Sect. 5. We conclude this work in the last section by relating some perspectives.

2 Related Works

There are significant major categories of time series. The first one is frequency-domain, which includes methods as spectral analysis and wavelet analysis. In contrast, the second is time-domain which contains auto-regression, cross-correlation analysis and auto-correlation methods. Time series classification (TSC) problems are classically solved using model-based, instance-based and feature-based strategies. The first one used algorithms such as the hidden Markov model (HMM) and Auto-regression (AR), in which a model is built for each class by adapting its parameters to this class. The weakness of this approach emerges when it deals with stationary and symbolic non-stationary time series. The second category is based on similarity (dissimilarity) measurement (distance), such as the Euclidean distance-based1-Nearest Neighbor (1-NN) and Dynamic Time Wrapping (DTW) [22]. This solution is known as computationally expensive. Finally, the feature-based family aims to extract essential features; it includes methods such as the discrete Fourier transform (DFT) [23], the discrete wavelet transform (DWT) [5], singular value decomposition (SVD) [12], and sparse coding [4]. Another family of classification combines a set of classifiers known as ensemble-based. For example, we can cite the flat collective of transform-based ensembles (COTE) [17]. These methods need a massive work on preprocessing and feature engineering.

In the last years, CNNs have been exploited to solve the problems of time series classification. Two main approaches are proposed; the first is based on existing (the well-known) CNN architecture [9] that uses 1D time-series signals.

[1] Electroencephalography.

[2] Electrocardiogram.

Meanwhile, the second approach reshaped 1D time series' signals into 2D matrices then the CNN is applied. The authors in [10] proposed a time-delay neural network (TDNN) adapted to EEG classification. They used one single hidden layer, which was not able to learn hierarchical features. The convolutional Deep Belief Network(CDBN) was also exploited in [16] to classify audios using the frequency domain. In [31], the authors proposed a multichannel CNN (MCCNN) to deal with multivariate TS. The end-to-end neural network method applies multiple transformations of different scales, sampling rates and frequencies. Then, the authors used convolution operations followed by traditional MLP (Multi layer Perceptron) to classify obtained feature maps. The authors also proposed a pretrained version of MCCNN. This model achieves high accuracy on several real-world data sets. Furthermore, The CNN is also applied to speech recognition within the framework of hybrid NN-HMM mode in [2]. The Multi-scale convolutional neural network for time series classification is presented in [6]. Other papers proposed models such as a Fully Convolutional Neural Network (FCN), a deep multi-layer perceptron network (Dense Neural Network, DNN) and a deep Residual Neural Network on univariate Time series [27].

Recurrent neural networks (LSTM) get involved in human activity recognition and achieves good results. Authors in [28] used bidirectional LSTM by incorporating temporal dependencies. Authors of [30] proposed a deep residual Bidir-LSTM, while later in 2019 [25], another model is created based on LSTM and named it Stacked LSTM network by making a network with two parts. The first one contains a single layer neural network followed by a stack of LSTM cells. In 2020, the authors in [26] evaluate the performance of a set of models (SVM, MLP, CNN, LSTM and BLSTM) and compare the results, while authors in [18] optimized set of models (1D regular CNN, 1D separable CNN, GRU and LSTM) and proposed an edge-based IoT system.

Before describing our proposed model, we first need to give some background to different concepts on time series classification.

3 Preliminaries and Methodology

3.1 Time Series

Time series form a sequence of data (measurements) naturally ordered over cycles of time [1]. This kind of data is characterized by high dimensionality and updating continuously. It is divided into two families, univariate and multivariate series. The multivariate time series contain more than one observation. On the other hand, univariate time series is characterized by a single observation. Formally, time series can be represented as:

$$X = \{X_1, X_2, X_3, ..., X_l\}^n \tag{1}$$

where ℓ is the length of time series and n its rank.

Time series classification is a learning procedure. This task consists of training a model over a set of samples of times series to each one is associated a label that is the probability distribution over the class values and it is represented as follows:

$$\{X_1, X_2, X_3, ..., X_l\}^n \rightarrow Y^n \tag{2}$$

Where Y^n is the label of time series of rank n.

3.2 Convolutional Neural Networks

The Convolutional Neural Networks (CNNs) are a powerful family of neural networks. In the simplest neural network known as a multi-layer neural network (MLP), information is propagated through different layers of interconnecting nodes. Through these layers, a non-linear transformation is applied to compute the output of each layer that is expressed in the following equation:

$$y_l = \phi(\sum w_{i_l}.x_{i_l} + b_l) \tag{3}$$

where ϕ corresponds to the non-linearity function applied to the neurons of layer l. The weights and the bias are denoted by w_{i_l} and b_{i_l}. The x_{i_l} is the input time series.

The training process is then performed according to the feed-forward step and the back-propagation step minimizing the global error based on the gradient descent algorithm and adjusting the parameters (randomly initialized weights) of the model. Different loss functions can be used during this process.

In the CNN model, there are two main parts. The first one is considered an extractor of features and it consists of multiple convolutional and pooling layers [15]. The second part is a discriminative layer known as a fully connected layer and defined by a multi-layer neural network.

Feature Extraction Process. In the CNN model, the convolution layer creates a feature map by applying a filter or a kernel to an input. This operation is performed by sliding a filter on the data. Performing several convolutions on the input data leads to different feature maps. Moreover, the padding operation which consists of adding zeros to data, is critical since it avoids shrinking the feature map [15].

Classification Process. The latent features issued from each layer are fed into an MLP to perform classification. It takes the feature map of the previous part and mapped it into the output classes. A flatten layer precedes this phase in order to turn the multidimensional feature map into 1D data. All layers usually use the "Relu" activation function (Restricted Linear Unit) defined by $\max(0, \sum w_i.y_i)$ where y_i is the input of each layer. This function allows the model to overcome the vanishing gradient problem and make it learning more efficient [20].

3.3 Separable Convolutional Neural Networks

Some neural net architectures such as MobileNet [11] use the separable CNN, which allows performing separable convolution spatially or depthwise.

Spatial Separable Convolutions. This family of neural networks deals with the spatial dimensions of the input and the kernel by decomposing the latter into two small sub-Kernels. A $N \times N$ kernel will be divided into two kernels of sizes $N \times 1$ and $1 \times N$, respectively. In other terms, it consists of factorizing the initial matrix defined by the kernel as the product of two rectangular matrices having lower dimensions. More formally, we have:

$$K(i,j) = \sum_i \sum_j K_1(i,1).K_2(1,j) \quad (4)$$

with K_1 and K_2 two smaller kernels which, when multiplied, the original Kernel K is found.

Many works used this strategy such as Flattened networks [13] and Inception models [24].

Depthwise Separable Convolution. It is not always obvious to factorize a matrix into two matrices such as defined in Eq. 4. So, using spatial separable CNN causes troubles during training the model. A Depthwise separable convolution consists of splitting the kernel into two separate kernels known as depthwise and pointwise convolutions. Meanwhile, the conventional CNN applies the kernel on all N channels; the depthwise procedure applies different kernels on each channel individually; as a result, we obtain smaller N outputs with smaller sizes. In the pointwise Convolution, the process consists of applying a kernel of 1×1 and combining the result with the N output obtained by the depthwise phase [14].

3.4 HAR-Model Architecture

In this section, we present our proposed method of learning for time series classification. To categorize activities based on raw data collected using a wearable sensors, we use a separable convolutional network architecture which includes a set of 4 separable convolution layers obtained after testing and evaluating same architectures with different number of separable layers on the test set using cross validation. We organized those 4 layers in serial structure such as the output from each previous layer is transmitted as an input to the next layer. A separating layer called max pooling is added between every two convolution layers. Figure 1 illustrates this architecture well.

To achieve our goal, we start the first convolution layer with $1 \times 128 \times 9$ time series sequence and we use 64 filters of size $1 \times 11 \times 9$. A stride of 1 is used. The result in the first layer is $1 \times 128 \times 64$. A Relu as activation function is computed at each separable layer. The performance of our model is improved by incorporating a specific filter of dimension $(11, 1)$. This kernel operates at the level of the depthwise convolution layer while the kernel of dimension (1, 1) is applied at the level of pointwise convolution, as explained in the previous section.

Compared to the standard convolution operation, the depthwise separable convolution network with a kernel of dimension (11, 1) consumes fewer parameters and its computational cost is much lower.

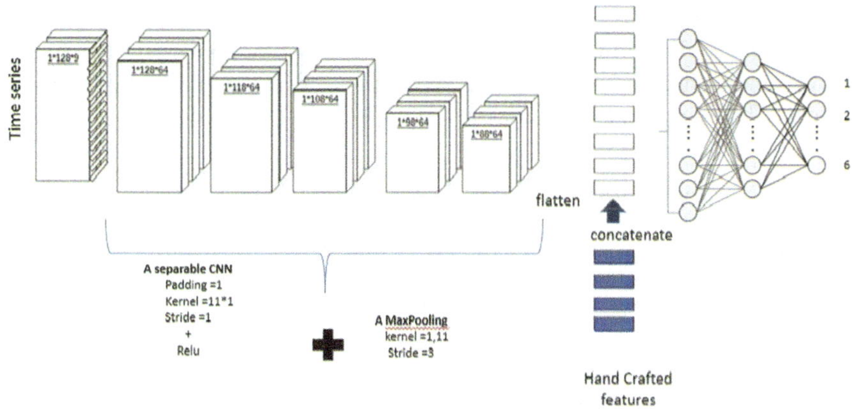

Fig. 1. The proposed architecture

Before introducing the last phase of our architecture, we add a layer block that contains the result of the concatenation operation between the output of the previous layer with handcrafted features. Finally, we use Softmax as an activation function in the last fully connected layer since it is a multiclass classification (see Eq. 5). The output is normalized and corresponds to the probability distribution of learned activity classes.

$$Class = argmax_k(\frac{e^k}{\sum_j^m e^j}) \tag{5}$$

To optimize the error during the learning procedure, we use the loss function defined by categorical cross-entropy and expressed as follows [8]:

$$Loss(x) = -\sum_i^m y_i \log y_i \tag{6}$$

with m is the number of classes given in the dataset.

4 Experimental

4.1 HAR Dataset Description

In our experiments, the data set used is the UCI-HAR provided by [7]; it contains activities performed by volunteers in the age range from 19 to 48, those persons wear a sensor set (Samsung Galaxy S II) on their waist to find out their state (WALKING, WALKING UPSTAIRS, WALKING DOWNSTAIRS, SITTING, STANDING, LAYING). It contains 3-axial linear acceleration and 3-axial angular velocity at a constant rate 50 Hz captured using the embedded accelerometer and gyroscope. It includes nine files representing accelerometer and gyroscope signals. The data were labeled manually after being video-recorded and fragmented both part training and testing. The data set provides another file containing handcrafted data where each row represents a sample and contains 561 features. This hand crafted is considered entirely in our model.

4.2 Experiments Setup

To train this model, we used Keras API[3] and Scikit-learn library[4] on Google Colab environment[5]. The batch size used was 128, the number of epochs was 100, the training dataset contains 7352 experiences for the training part and 2947 for the testing part. As an optimizer, we have used Adam optimizer with its default learning rate equal to 0.001.

For comparison purposes, some evaluation metrics, such as accuracy, precision, recall and F1-score [21], are computed for each model. All these metrics are obtained from the confusion matrix which is a powerful tool for measuring the performance of a machine learning model. Each classification model tries to achieve high performance by correctly predicting the appropriate class for each activity; by testing it, we get four outputs. For example, accuracy counts the ratio of correctly classified data and the error rate that represents the ratio of misclassified data.

5 Results and Discussion

To evaluate the results of our classifier, we used the measurements mentioned in the previous subsection. Indeed, our model aims to classify an activity among six possible activities given in the UCI-HAR dataset. In Table 1, we show the results obtained from two models. So, we remark that the presence or the absence of handcrafted features significantly affect the classifiers' performance score. Indeed, the accuracy of the best model reaches a value of 94.77.

The confusion matrix depicted in Fig. 2 shows clearly the low classification error and a high performance for correct class labels.

As can be seen from the Fig. 3, as the number of epochs increases, the values of both train and test losses decrease. Nevertheless, the accuracy increases and reaches a high performance. In Table 2, we report our results of classification reports for each activity using both models (classical and separable CNN) and we compare a various works in terms of accuracy measure.

Furthermore, the success of classical CNN appears clearly when we handle applications of computer vision, on the other hand, performance is less attractive when we use time series data.

[3] https://keras.io/.
[4] https://scikit-learn.org/.
[5] https://colab.research.google.com/notebooks/intro.ipynb.

Table 1. Classification report model 1 vs model 2

Metrics	Model 1 (without handcrafted)			Model 2 (with handcrafted)			Support
	Precision	Recall	F1-score	Precision	Recall	F1-score	
WALKING	0.74	0.83	0.78	0.92	0.99	0.95	496
WALKING_UPSTAIRS	0.93	0.94	0.82	0.91	0.94	0.92	471
WALKING_DOWNSTAIRS	0.79	0.78	0.78	0.99	0.88	0.93	420
SITTING	0.79	0.87	0.83	0.94	0.93	0.93	491
STANDING	0.80	0.83	0.82	0.94	0.94	0.94	532
LAYING	1	0.95	0.97	1.00	1.00	1.00	537
Accuracy	83.64%			94.77%			2947

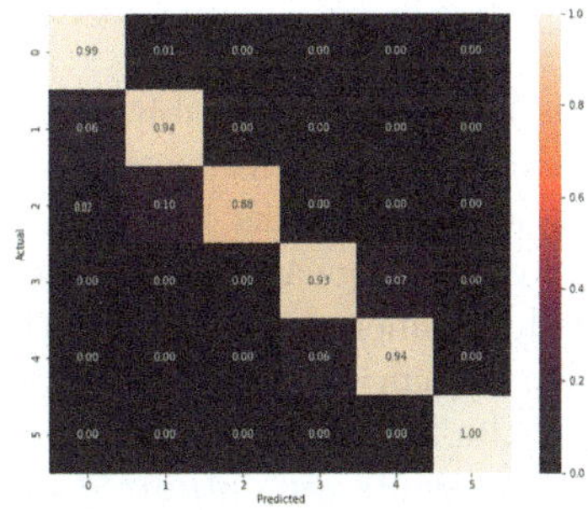

Fig. 2. Confusion matrix

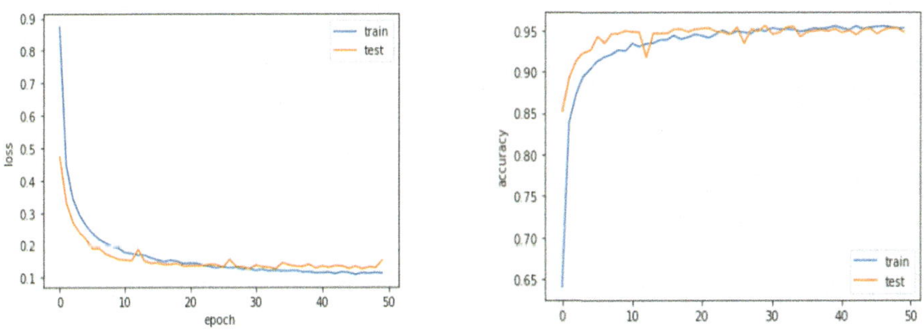

Fig. 3. Loss vs Accuracy curve

Only one proposed model achieves better performance compared to the other existing models. Indeed, the model 1 which used the CNN architecture within a kernel of dimension 11×11, gives an accuracy of about 92.67%, In contrast, for the second model, which incorporated the kernel of 11×1 in depthwise separable convolution and took in consideration handcrafted features, provides a more exciting result that is 94.77%. Moreover, we can see clearly the number of reduced parameters in the model 2.

Table 2. Accuracy comparing of the models for UCI Dataset

Model	Accuracy
CNN [26]	92.71
Stacked Lstm [25]	93.13
Bidir Lstm [28]	93.79
Res Lstm [30]	91.6
Res Bidir Lstm [30]	93.6
Cnn Lstm [19]	92.14

Model	Accuracy	No.Parameters
Model1(cnn modified)	92.67%	6,490,566
Model2(separable)	**94.77** %	6,364,137

6 Conclusion

Time series classification is a challenging problem in particular when we handle activities applications. In this paper, we have proposed a novel architecture of separable convolutional neural networks based on a specific kernel and followed by the handcrafted features concatenation process. Experimental results showed that the elaborated classifier outperformed the state-of-art models on the UCI-HAR dataset. The human activity recognition is then achieved with better accuracy. Although, the hyper-parameters were selected basis on trial and error process and further results are optimized to achieve better accuracy. In the future the tuning of parameters may be carried out in detail for better performance. Also, it will be interesting to eliminate irrelevant and redundant features, which will enable the network to learn more effectively and perform in a robust manner. Another future work consists to evaluate the model on other HAR datasets and working on recognizing composite activities and concurrent activities.

References

1. Wang, J., Liu, P., She, M.F., Nahavandi, S., Kouzani, A.: Bag-of-words representation for biomedical time series classification. Biomed. Signal Process. Control **8**(6), 634–644 (2013)
2. Abdel-Hamid, O., Mohamed, A.R., Jiang, H., Penn, G.: Applying convolutional neural networks concepts to hybrid nn-hmm model for speech recognition. In: 2012 IEEE International Conference on Acoustics, Speech and Signal Processing (ICASSP), pp. 4277–4280. IEEE (2012)
3. Aggarwal, J.K., Ryoo, M.S.: Human activity analysis: a review. ACM Comput. Surv. (Csur) **43**(3), 1–43 (2011)
4. Bahrampour, S., Nasrabad, N.M., Ray, A.: Sparse representation for time-series classification. In: Pattern Recognition And Big Data, pp. 199–215. World Scientific (2017)
5. Chaovalit, P., Gangopadhyay, A., Karabatis, G., Chen, Z.: Discrete wavelet transform-based time series analysis and mining. ACM Comput. Surv. (CSUR) **43**(2), 1–37 (2011)
6. Cui, Z., Chen, W., Chen, Y.: Multi-scale convolutional neural networks for time series classification. arXiv preprint arXiv:1603.06995 (2016)

7. Dua, D., Graff, C.: UCI machine learning repository (2017)
8. Ismail Fawaz, H., Forestier, G., Weber, J., Idoumghar, L., Muller, P.-A.: Deep learning for time series classification: a review. Data Min. Knowl. Discov. **33**(4), 917–963 (2019). https://doi.org/10.1007/s10618-019-00619-1
9. Fawaz, H.I., et al.: Inceptiontime: finding alexnet for time series classification. Data Min. Knowl. Discov. **34**(6), 1936–1962 (2020)
10. Haselsteiner, E., Pfurtscheller, G.: Using time-dependent neural networks for eeg classification. IEEE Trans. Rehabil. Eng. **8**(4), 457–463 (2000)
11. Howard, A.G., et al.: Mobilenets: Efficient convolutional neural networks for mobile vision applications. CoRR **abs/1704.04861** (2017)
12. Hui, Z., Tu, B.H., Kawasaki, S.: Wrapper feature extraction for time series classification using singular value decomposition (2005)
13. Jin, J., Dundar, A., Culurciello, E.: Flattened convolutional neural networks for feed forward acceleration. arXiv preprint arXiv:1412.5474 (2014)
14. Kaiser, L., Gomez, A.N., Chollet, F.: Depthwise separable convolutions for neural machine translation. arXiv preprint arXiv:1706.03059 (2017)
15. Krizhevsky, A., Sutskever, I., Hinton, G.E.: Imagenet classification with deep convolutional neural networks. Adv. Neural Inf. Process. Syst. **25** (2012)
16. Lee, H., Pham, P., Largman, Y., Ng, A.: Unsupervised feature learning for audio classification using convolutional deep belief networks. Adv. Neural Inf. Process. Syst. **22**, 1096–1104 (2009)
17. Lines, J., Taylor, S., Bagnall, A.: Hive-cote: the hierarchical vote collective of transformation-based ensembles for time series classification. In: 2016 IEEE 16th International Conference on Data Mining (ICDM), pp. 1041–1046. IEEE (2016)
18. Mukherjee, A., et al.: Edge-based human activity recognition system for smart healthcare. J. Inst. Eng. (India): Ser. B, 1–7 (2021)
19. Mutegeki, R., Han, D.S.: A cnn-lstm approach to human activity recognition. In: 2020 International Conference on Artificial Intelligence in Information and Communication (ICAIIC), pp. 362–366. IEEE (2020)
20. Nwankpa, C., Ijomah, W., Gachagan, A., Marshall, S.: Activation functions: comparison of trends in practice and research for deep learning. CoRR **abs/1811.03378** (2018)
21. Powers, D.M.W.: Evaluation: from precision, recall and f-measure to roc, informedness, markedness and correlation (2020)
22. Rakthanmanon, T., et al.: Addressing big data time series: mining trillions of time series subsequences under dynamic time warping. ACM Trans. Knowl. Discov. Data (TKDD) **7**(3), 1–31 (2013)
23. Schäfer, P.: The boss is concerned with time series classification in the presence of noise. Data Min. Knowl. Discov. **29**(6), 1505–1530 (2015)
24. Szegedy, C., Ioffe, S., Vanhoucke, V., Alemi, A.: Inception-v4, inception-resnet and the impact of residual connections on learning (2016)
25. Ullah, M., Ullah, H., Khan, S.D., Cheikh, F.A.: Stacked lstm network for human activity recognition using smartphone data. In: 2019 8th European Workshop on Visual Information Processing (EUVIP), pp. 175–180. IEEE (2019)
26. Wan, S., Qi, L., Xu, X., Tong, C., Gu, Z.: Deep learning models for real-time human activity recognition with smartphones. Mob. Netw. Appl. **25**(2), 743–755 (2020)
27. Wang, Z., Yan, W., Oates, T.: Time series classification from scratch with deep neural networks: a strong baseline. In: 2017 International Joint Conference on Neural Networks (IJCNN), pp. 1578–1585. IEEE (2017)

28. Yu, S., Qin, L.: Human activity recognition with smartphone inertial sensors using bidir-lstm networks. In: 2018 3rd International Conference on Mechanical, Control and Computer Engineering (icmcce), pp. 219–224. IEEE (2018)
29. Zhao, B., Lu, H., Chen, S., Liu, J., Wu, D.: Convolutional neural networks for time series classification. J. Syst. Eng. Electron. **28**(1), 162–169 (2017)
30. Zhao, Y., Yang, R., Chevalier, G., Xu, X., Zhang, Z.: Deep residual bidir-lstm for human activity recognition using wearable sensors. Math. Probl. Eng. **2018** (2018)
31. Zheng, Y., Liu, Q., Chen, E., Ge, Y., Zhao, J.L.: Time series classification using multi-channels deep convolutional neural networks. In: Li, F., Li, G., Hwang, S., Yao, B., Zhang, Z. (eds.) WAIM 2014. LNCS, vol. 8485, pp. 298–310. Springer, Cham (2014). https://doi.org/10.1007/978-3-319-08010-9_33

Deep Approach Based on User's Profile Analysis for Capturing User's Interests

Randa Benkhelifa[1(✉)] and Nasria Bouhyaoui[1,2]

[1] Laboratoire de l'intelligence artificielle et des technologies de l'information, Université Kasdi Merbah Ouargla, BP 511, Route de Ghardaia, 30 000 Ouargla, Algérie
{randa.benkhelifa,nasria.bouhyaoui}@univ-ouargla.dz

[2] Ecole normale supérieure de Ouargla, BP 511, Route de Ghardaia, 30 000 Ouargla, Algérie

Abstract. Capturing user's interests and preferences by analyzing and interpreting the daily-shared contents in online social networks offer a unique information source for several domains such as business, marketing and politics.

User's profile describes its owner's characteristics, where it contains several important personal information such as (age, sex, job title, level of education, etc.), which can help to improve the process of user's interests identification. This information can typically represent a range of values representing only one user profile. Hence, the shared posts, the reactions on other posts and their circle of friends can help to reflect their interests. However, exploiting all this information through the analysis of user profiles can help to enhance user's interests identification performances.

In this paper, we propose a deep learning user's profile analysis based approach CNN-PA and RNN-PA that relies on users' personal information and textual content for detecting user's interests and preferences. We experimented our approach using a large Facebook dataset, and show how the deep learning approaches perform significantly better than the classical algorithms such as SVM.

Keywords: Deep learning · Personal information attributes · Profiling · Text classification · Social text · Online Social Networks · User's interests · CNN-PA · RNN-PA

1 Introduction

Online social network (OSN) has become a part of the daily routine of a huge number of persons. OSNs users can fill information that represents their demographic attributes, posts, and comments covering several topics. A large number of fields including: trend-setting, future prediction, recommendation systems, community detection, business marketing, and sentiment classification are interested in the automatic use of this information. User's personal attributes such as gender, age, location, marital status, education, career, etc. are the personal static information that describes user's profile in OSN. Some of the previous researches have interested to the categorization of author's characteristics relying on textual content (textual features) generated by this OSNs user, a part of them

B. Lejdel et al. (Eds.): AIAP 2021, LNNS 413, pp. 177–186, 2022.
https://doi.org/10.1007/978-3-030-96311-8_17

is interested in the detection of the author's gender [17, 18, 23], and [19]. Other works focused on the identification of the author's age [12] and [13]. The mentioned works relied on textual content generated by the user to detect what are their demographic attributes. The previous works prove the existence of strong relation, which led as consequence, to use the text generated by users in order to identify user's personal attributes. Hence, we suggest exploiting personal information attributes textual content generated by the user for identifying user's interests.

We notice from the existing studies on OSNs text classification, that only user-generated content was taken in concern for the aim of improving social text classification performances. In the work in [8] and others have relied on sentiment bags, and some others on embedding models [22]. Whereas, there are a studies that have proposed approaches which deal with the specific characteristics of the social short text [22] and [6]. In The research proposed by (use the name of author here) [2], the first order Markov model for hierarchical Arabic text classification has been used. In addition, [1] has relied on the singular value decomposition method in order to identify textual feature, where [4] has presented an improved method of Chi-square feature selection method. The research in [19] has explored the use of Natural Language Processing techniques in a gender classification system, where the results have determined that word embedding models have significantly performed better using multiple machine learning techniques in opposite of the traditional Bag of Words model. Both of demographic information and user-generated content are available in online social networks. All the existing studies on OSNs text classification depend only on textual content generated by the user. We suggest including the demographic attributes in addition of user-generated content. That step can play a crucial role in improving the classification performances. Hence, it is feasible to leverage these attributes to build a smarter classifier and achieve better performance. In this work, we investigate how both user-generated content (textual data) and personal attributes are exploited to categorize textual content by topics of interests.

Inspired by the recent success of deep learning techniques in many NLP tasks, we further propose a deep demographic-content-based approach relied on both of textual features and demographic features for the classification of user textual content by topics of interests. To evaluate our approach, we propose to compare between the well-known classical classifier SVM that got the best results in [5] and the deep learning classifiers results.

We discussed the previous related works. In Sect. 3, we present our methodology including our proposed approach. In Sect. 4, we represent the details of the used dataset. In Sect. 5, we detailed the results applied to the dataset extracted from Facebook, and finally in Sect. 6, we present a conclusion and perspectives.

2 Related Works

This section emphasizes previous studies relevant to this research, and pinpoints the considered features in each work for improving the performance of the social text classification. Usually, the previous studies have been classified the social text considering only the textual content. The authors in [14] have introduced a text-based hidden Markov models, which utilizes word orders without the need of sentiment lexicons. A posts classification model based on neural network with incorporating user tastes, topic tastes, and user comments has been developed by [9]. The authors in [22], have described a Twitter election classification task that aims to detect election-related tweets, this work is based on embedding models for improving the classification performance. [3] has proposed an approach which deals with the limitations of the social short texts for improving the classification performance. In addition, [6] have introduced a novel preprocessing methods adapted to the special characteristics of the social text in order to improve the classification performances. A real-time system has been developed by [8] to extract and classify the YouTube cooking recipes reviews automatically in order to improve the performance of this system some sentiment bags, based on emoticons and injections have been constructed. The authors in [2] have proposed a text classification method which is a space efficient method that utilizes the first order Markov model for hierarchical Arabic text classification. The work in [1] has used the singular value decomposition method to extract textual features; they compare between some of the well-known classification methods.

The research [4] has presented an improved method of Chi-square feature selection method to minimize the data and produce higher classification accuracy. The authors in [16] have focused on the evaluation of feature selection methods for improving text classification performance. The authors in [20] have used the structure of social media opinions for enhancing the sentiment classification performance.

All these previous works mentioned above have focused only on content-based methods and especially, the textual features for analyzing and classifying the social text. They totally neglect the demographic aspect of the user, which can play a crucial role toward the content classification. Moreover, we remarked a strong link between user-generated content and his/her demographic attributes trough several previous studies that have exploited this text shared in OSNs to detect one or more demographic attributes of the user (author's text).

Many studies in OSNs have focused on the task of capturing author's gender. The authors [18] have considered a set of text features such as function words and part of speech n-grams for providing a system of gender classification. The authors [23] have presented an efficient gender classification model to predict the gender values of specified users crawled from a Chinese micro-blog service. The problem of gender classification has been treated by [17] which have proposed a typical surface-level text classification approach by identifying differences between genders in the ways they use the same words. Other works have focused on the age detection, the authors [12] have proposed an approach considering the writing style and both users' history and profile for determining the age groups of Twitter users. Other research have focused on several other characteristics, where a hybrid text-based and community-based method for the demographic estimation of Twitter users has been proposed by [13].The authors in [19] have achieved a significantly better results using word embedding models with multiple machine learning techniques than the traditional Bag of Words model in a gender classification system.

The authors in [5] have considered the demographic attributes with proposing a content-demographic-based approach which use not only the textual features, but both the textual content shared by the user and his/her demographic attributes in orders to improve the classification of the textual content by a topic of interests. Here, the authors have used only the classical algorithm SVM.

In this paper, we propose a deep content-demographic-based approach which uses both of textual content and user's demographic attributes. This approach is based on Deep Learning algorithms. To evaluate the proposed approach, we propose to compare between the well-known classical classifier SVM and the deep learning classifiers results.

3 Proposed Methodology

We consider the interests of a user *u* in a specific period *d*, as a set of interests *I* where each interest is a category *c* with a score *s* associated with that category.

$$I_u = \{(c, s) | c \in C\} \quad (1)$$

Where, C is the set of categories c_j, and *P* is the set of messages p_i shared by user *u* in a specific period.

$$z(p_i, c_j) = \begin{cases} 1, & \text{if } p_i \text{ is classified at } c_j \\ 0, & \text{else} \end{cases}$$

where, $p_i \in P$ and $c_j \in C$

$$S_{u,c_j}(d) = \sum_i z(p_i, c_j) \quad (2)$$

Where *S* is a score calculated for a category c_j based on posts (user-generated text) shared by a user *u* in period.

User-generated text in OSNs is exploited in order to discover user's preferences and interests, which represent an important piece of information. Hence, in the work [7] the authors have been discover that even users' demographic attributes can be useful in predicting users' interests, and improving the performance of the social text categorization. To understand the link between users demographic attributes and user's generated text, they have proposed a quantitative analysis by computing the number of users with similar demographic attribute in each topic of interest. Hence, they have deduced that the link between users demographic attributes and user-generated text is strong or weak relatively to the topic of interest in question. "Each attribute has a strong impact on some categories (topic of interest) and weak impact on others" [7]. Basing on this interesting study proposed by the authors in [7] and due to the power of deep learning in NLP and text classification, we propose a new approach based on deep learning and the users' demographic attributes.

3.1 Traditional Classification Approach

Exploiting both content and demographic information for social text classification have been studied in [5] using traditional machine learning algorithms such as support vector machine (SVM). In this work, we use SVM model, which achieves the best performance in [5] for our social text classification task. The SVM model is based on the following objective function for classification: for each post p_i. The vector v_i is the corresponding content and demographic attributes vector.

$$f(p_i) = w.v_i + b \tag{3}$$

where w is the normal vector to the hyperplane, and v_i is the input vector. The post class is attributed by the sign of f:

$$h = sing(f(p_i)) = \begin{cases} +1, \, if \; f(pi) > 0 \\ -1, \; else \end{cases}$$

For multi class SVM model, a multi class categorization can be obtained by combining a set of binary classifiers $f_1, f_2, \ldots f_m$ for M classes, and each classifier is trained to differentiate class from the rest. The combination is carried out according to the maximal output before applying the sign function [21].

3.2 Deep Learning Approach

Due to the recent success of deep learning techniques in text classification [15], we propose a content-demographic-based approach, which use deep neural networks models to classify social text. We have used TensorFlow2 Python library for text pre-processing (Fig. 1).

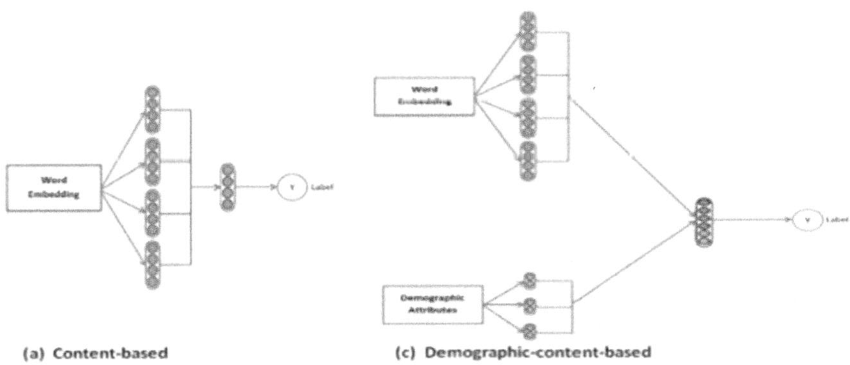

Fig. 1. Deep learning content-based approach vs deep learning demographic-content-based approach for the classification process.

Convolutional Neural Network based on Personal Attributes (CNN-PA). Inspired by the good results of convolutional neural network model [15] in text classification task, we further propose to use CNN model in our proposed approach based on the demographic information (CNN-PA). In the training phase, in order to produce a new feature f_i, a convolution operation uses a filter w which has a window h of words sequence $x_{i:i+h-1}$. Usually, the feature fi in the content-based CNN is defined as follows:

$$f_i = g(w.x_{i:i+h-1} + b) \tag{4}$$

where b is a bias term and g is a non-linear function such as hyperbolic tangent. This filter is applied to each possible window of words in the sentence $\{x_{1:h}, x_{2:h+1}, x_{n-h+1:n}\}$ to produce a feature map $f = \{f_1, f_2, \ldots, f_{n-h+1}\}$. Then a max-pooling operation [10] is applied on the feature map to capture the most important feature by taking the maximum value $\hat{f} = max\{f\}$. Finally, we use a softmax function to generate the probability distribution over labels. Now, we introduce the demographic-content based approach CNN-PA, for each post p_i, v is the combination between words sequence $x_{i:i+h-1}$ and personal attributes A, $V = (x_{i:i+h-1}, A)$.

$$f_i = g(w.v + b) \tag{5}$$

Recurrent Neural Network based on Personal Attributes (RNN-PA): Recurrent neural networks (RNNs) are a rich class of dynamic models. RNNs can be trained for sequence generation by processing real data sequences one step at a time and predicting what comes next [11]. An input vector sequence $x = (x_1, \ldots, x_T)$ is passed through weighted connections to a stack of N recurrently connected hidden layers to compute first the hidden vector sequences $h^n = (h_1^n, \ldots, h_T^n)$ and then the output vector sequence $y = (y_1, \ldots, y_T)$. Each output vector y_t is used to parameterize a predictive distribution $Pr(x_{t+1}|yt)$ over the possible next inputs x_{t+1}.

$$h_t^1 = H(W_{ih}^n x_t + W_{h\,h}^{1\,1} h_{t-1}^1 + b_t^1) \tag{6}$$

$$h_t^n = \mathrm{H}(\mathrm{W}_{\mathrm{ih}}^{\mathrm{n}} \mathrm{x}_\mathrm{t} + \mathrm{W}_{\mathrm{h}\;\;\mathrm{h}}^{\mathrm{n-1}\,\mathrm{n}} h_t^n + \mathrm{W}_{\mathrm{h}\;\mathrm{h\,n}}^{\mathrm{n}} h_{t-1}^n + h_t^n) \tag{7}$$

where the W denote weights, the b denote bias vectors and H is the hidden layer function. Given the hidden sequences, the output sequence is computed as follow:

$$\hat{y} = b_y + \sum\nolimits_{n+1}^{N} w_{h^n y} h_t^n \tag{8}$$

$$y_t = Y(\hat{y}_t) \tag{9}$$

where Y is the output layer function. The output vectors y_t are used to parameterize the predictive distribution $Pr(x_{t+1}|y_t)$ for the next input. The probability given by the network to the input sequence x is:

$$\Pr(\mathrm{x}) = \prod\nolimits_{t=1}^{T} \Pr(x_{t+1}|y_t) \tag{10}$$

Usually, for classifying a textual post p the input vector sequence x is constructed only from the content d of this post. Here, in our model called Recurrent Neural Network based on Demographic information (RNN-PA) the input vector sequence x corresponds to both the post-content d and the user's personal attributes A.

4 Dataset Collection and Construction

The dataset provided in [7] consists of 72900 Facebook posts, these posts contain only the textual content, which are sufficient for training our models.The posts are collected from July to September 2016 using Facebook API. They are collected from 300 specific users' profiles, which are public, active and real. This dataset are made of the posts accompanied by their authors' demographic attributes. These posts are annotated to eight (8) different classes, which are (art, fashion, sport, technology, business, news, science and education, and other). Considering the privacy reasons, the users' IDs and names are deleted. Table 1 summarizes the basic statistics of our data (i.e., the average number of posts and the average number of users in each category).

Table 1. Statistics of our data.

Data	Statistics
Posts	72,900
Users	300
Categories	News (8748), Business (8311), Technology (13413), Art (9112), Sport (6561), Mode& Fashion (6780), Science & Education (8165), Other (11810)
Gender	Female (42.66%), Male (57.34%)
Age	13–17 (19.4%), 18–27 (36.8%), 28–37 (27.8%), 38–60 (16%)
Work	Worker (60.3%), Unempoloyed (20.3%), Not concerned (19.4%)
Edudcation	University (yes, 44.1%), (No, 55.9%)
Marital status	Married (46.1%), Not married (34.5%), Not concerned (19.4%)

5 Results and Discussion

In this section, we compare between the results obtained applying the different algorithms based only on textual content and the results obtained by these different algorithms based on both textual content and personal attributes (gender, age, marital status, work, and education). These results are summarized in Table 2. Here, we evaluate all our ranking models using 10-fold cross validation to reliably compute e statistical significance values. All the deep learning classifiers were build using Python.

By observing the results showed in Table 2 we remark that the personal attributes have an impact on the classification performances. Especially, deep neural network models have shown a more powerful classification a model using both content and demographic attributes features. Comparison between classifiers results: As we remark from the Table 2, the highest results we achieved are: 94,9% in accuracy, 0.949 in precision, 0.949 in recall and 0.949 in F-measure using the textual content and users characteristics (users demographic attributes) by CNN-PA classifier. Firstly, using only the textual content, we got the following results for the accuracies, 75.55% using SVM, 78, 6% using CNN, 75,7% using RNN, the best accuracy obtained here are given by CNN classifier. In term of precision, CNN got the best result with 0.786, SVM with 0.774, and RNN with 0.778. In terms of recall and F-Measure, we got these results respectively, CNN with (0.786 and 0.787) which represent the best results followed by RNN with (0.757

Table 2. Comparison between classifiers results, Precision, Recall, F-Measure, and Accuracy, using 10- folds cross validation.

Method	Content	Demographic	Accuracy (%)	Precision	Recall	F-Measure
SVM [7]	*		75.55	0.774	0.755	0.761
CNN	*		78.6	0.786	0.786	0.787
RNN	*		75.7	0.778	0.757	0.757
SVM [7]	*	*	90.01	0.896	0.9	0.897
CNN-PA (ours)	*	*	90.7	0.909	0.907	0.908
RNN-PA (ours)	*	*	94.9	0.949	0.949	0.949

and 0.757) and SVM with (0.755 and 0.761). Now, we discuss the results obtained by adding demographic information of users in the classification process. Where, we got the following results, accuracies with 90.01% using SVM, 94.9% using CNN-PA and 90.7% using RNN-PA, here we notice that the best accuracy is achieved by CNN-PA classifier. In term of precision, CNN-PA got the best result with 0.949, SVM with 0.896, and RNN-PA with 0.909. In terms of recall and F-Measure, we got these results respectively; CNN-PA classifier with (0.949 and 0.949) represents the best results followed by RNN-PA with (0.907 and 0.908) and SVM with (0.9 and 0.897).

After presenting, the global results obtained using only the textual content and using both textual content and the prsonal attributes, and showing the positive impact of the demographic attributes on the classification performances.

6 Conclusion and Future Works

In this article, we have presented a deep learning-based approach to sequential social short-text classification. We demonstrate that adding user's personal information improves the performance of the user's interests identification, and the use of deep learning model performs significantly better than the classical algorithms used in the stat of the art, such as the SVM classifier.

Limitations and future work: The main idea of the proposed approach is to include users' personal attributes in the classification process of the social text in order to improve its performances. However, two major limitations are presented here. Firstly, only the explicit attribute are taken. As such, our approach does not work well when user's Social information is not available, or users restrict their personal information. In addition, this work is limited to the textual data where one of the OSNs specificity is the data heterogeneity (posts can be images, videos, text, etc.).

In our future works, we will spread out our dataset and mainly focus on solving the posed limitations: firstly, proposing an automatic model, which can extract the implicit information about the user, existed in his social content. Secondly, we will analyze other multimedia content (such as images). Thirdly, we will exploit other information available on OSNs to build a more powerful classification model, for example, using different interaction information such as the reactions (Like, lovely, etc.), and friendship.

Acknowledgment. The authors gratefully acknowledge financial support from "La Direction Générale de la Recherche Scientifique et du Développement Technologique (DGRSDT)" of Algeria.

References

1. Al-Anzi, F.S., AbuZeina, D.: Toward an enhanced Arabic text classification using cosine similarity and latent semantic indexing. J. King Saud Univ. Comput. Inf. Sci. **29**(2), 189–195 (2017). https://doi.org/10.1016/j.jksuci.2016.04.00
2. Al-Anzi, F.S., AbuZeina, D.: Beyond vector space model for hierarchical Arabic text classification: a Markov chain approach. Inf. Process. Manag. **54**(1), 105–115 (2018). https://doi.org/10.1016/j.ipm.2017.10.003
3. Alsmadi, I., Hoon, G.K.: Term weighting scheme for short-text classification: twitter corpuses. Neural Comput. Appl. **31**(8), 3819–3831 (2018). https://doi.org/10.1007/s00521-017-3298-8
4. Bahassine, S., Madani, A., Al-Sarem, M., Kissi, M.: Feature selection using an improved Chi-square for Arabic text classification. J. King Saud Univ. Comput. Inf. Sci. (2018). https://doi.org/10.1016/j.jksuci.2018.05.010
5. Benkhelifa, R., Bouhyaoui, N., Laallam, F.Z.: A demographic-based approach for improved content categorization in social networking. In: Natural Language and Speech Processing (ICNLSP), 2018 2nd International Conference on, pp. 1–5. IEEE (2018)
6. Benkhelifa, R., Laallam, F.Z.: Facebook posts text classification to improve information filtering. In: Proceedings of the 12th International Conference on Web Information Systems and Technologies, 2016, pp. 202–207. Rome, Italy (2016). https://doi.org/10.5220/0005907702020207. 8
7. Benkhelifa, R., Laallam, F.Z.: Exploring demographic information in online social networks for improving content classification. J. King Saud Univ. Comput. Inf. Sci. **32**(9), 1034 1044 (2020)
8. Benkhelifa, R., Laallam, F.Z.: Opinion extraction and classification of real-time youtube cooking recipes comments. In: Hassanien, A., Tolba, M., Elhoseny, M., Mostafa, M. (eds.) The International Conference on Advanced Machine Learning Technologies and Applications (AMLTA2018). AMLTA 2018. Advances in Intelligent Systems and Computing, vol. 723, pp. 395–404. Springer, Cham (2018). https://doi.org/10.1007/978-3-319-74690-6_39
9. Chen, W.F., Ku, L.W.: UTCNN: a deep learning model of stance classification on social media text. In: Proceedings of the 26th International Conference on Computational Linguistics, pp.1635–1645 (2016)
10. Collobert, R., Weston, J., Bottou, L., Karlen, M., Kavukcuoglu, K., Kuksa, P.: Natural language processing (almost) from scratch. J. Mach. Learn. Res. **12**, 2493–2537 (2011)
11. Graves, A.: Generating sequences with recurrent neural networks. arXiv preprint arXiv:1308.0850 (2013)
12. Guimaraes, R.G., Rosa, R.L., De Gaetano, D., Rodriguez, D.Z., Bressan, G.: Age groups classification in social network using deep learning. IEEE Access **5**, 10805–10816 (2017). https://doi.org/10.1109/ACCESS.2017.2706674
13. Ikeda, K., Hattori, G., Ono, C., et al.: Twitter user profiling based on text and community mining for market analysis. Knowl. Based Syst. **51**, 3547 (2013). https://doi.org/10.1016/j.knosys.2013.06.020
14. Kang, M., Ahn, J., Lee, K.: Opinion mining using ensemble text hidden Markov models for text classification. Expert Syst. Appl. **94**, 218–227 (2018). https://doi.org/10.1016/j.eswa.2017.07.019

15. Kim, Y.: Convolutional neural networks for sentence classification. In: Proceedings of the Empirical Methods in Natural Language Processing, October 2014, pp.1746–1751 (2014)
16. Kou, G., Yang, P., Peng, Y., Xiao, F., Chen, Y., Alsaadi, F.E.: Evaluation of feature selection methods for text classification with small datasets using multiple criteria decision-making methods. Appl. Soft Comput. **86**, 105836 (2020)
17. Mihalcea, R., Garimella, A.: What men say, what women hear: finding gender-specific meaning shades. IEEE Intell. Syst. **31**(4), 62–67 (2016). https://doi.org/10.1109/MIS.2016.71
18. Mukherjee, S., Bala, P.K.: Gender classification of microblog text based on authorial style. IseB **15**(1), 117–138 (2016). https://doi.org/10.1007/s10257-016-0312-0
19. Vashisth, P., Meehan, K.: Gender classification using twitter text data. In: 2020 31st Irish Signals and Systems Conference (ISSC), pp. 1–6. IEEE (2020)
20. Vairetti, C., Martínez-Cámara, E., Maldonado, S., Luzon, V., Herrera, F.: Enhancing the classification of social media opinions by optimizing the structural information. Future Gener. Comput. Syst. **102**, 838–846 (2020)
21. Weston, J., Watkins, C.: Multi-class support vector machines. Technical report CSD-TR98-04, Department of Computer Science, Royal Holloway, University of London, May 1998
22. Yang, X., Macdonald, C., Ounis, I.: Using word embeddings in Twitter election classification. Inf. Retr. J. **21**(2–3), 183–207 (2017). https://doi.org/10.1007/s10791-017-9319-5
23. Yu, Y., Yao, T.: Gender classification of Chinese weibo users. In: Proceedings of the 2017 International Conference on E-commerce, E-Business and E-Government, pp. 5–8. ACM (June 2017). https://doi.org/10.1145/3108421.3108423

Multi Agent Systems Based CPPS – An Industry 4.0 Test Case

Abdelhamid Bendjelloul[1,2(✉)], Mehdi Gaham[1], Brahim Bouzouia[1], Mansour Moufid[2], and Bachir Mihoubi[1]

[1] Center of Development of Advanced Technologies, CDTA, SRP Team, Baba Hassan, Algiers, Algeria
abendjelloul@cdta.dz

[2] LRPE Laboratory, USTHB University, BP 32, El Alia, Bab Ezzouar, Algiers, Algeria

Abstract. With the rise of the Industry 4.0 Revolution, Artificial Intelligence, digitalization, and connectivity have been more than ever; adopted in the industrial world. This adoption is leading to the transformation of the mechatronic systems used in production into Cyber-Physical Production Systems. Such a concept is taking industrial Automation and computer integrated manufacturing to the next level. The massive migration of traditional production systems into Cyber-Physical Production Systems, including the MAS-based CPPS, made the reviewing of the traditional methods of Engineering and Commissioning a must. Which explains the increase in the number of research works during recent years about the application of these Architectures on practical cases. In the present paper, we propose a way of developing and implementing MAS-based CPPS on an Industry 4.0 Assembly Platform. Moreover, we test the behavior of the Multi-Agent systems with interaction with SIEMENS Programmable Logic Controllers via OPC UA Protocol, during a Software-In-the-Loop "SIL" Test on a 3D Model of the Platform running on a separate Computer. The test assesses the behavior of the components of a typical Cyber-Physical production module during the treatment of a given operation on the product, to extract the vulnerabilities in the treatment of the operation and search for appropriate improvements.

Keywords: Multi agent systems · Cyber physical production systems · Intelligent manufacturing systems · SIL-testing

1 Introduction

Nowadays, the rise of competitiveness in several industries like electronics, cars, accessories or even clothes, lead to the fast development of products into better versions and the demand for more and more new features which lead to the mass personalization on the one hand. On the other hand, the need for better competitiveness in the market. That means the need for more reliable plants with less downtime due to unpredicted breakdowns, with faster response time of maintenance staff and logistics to react to every new situation, which make the use of software more important than ever, a generalization of software use in every aspect of the production, leading to its digitalization. Without

B. Lejdel et al. (Eds.): AIAP 2021, LNNS 413, pp. 187–196, 2022.
https://doi.org/10.1007/978-3-030-96311-8_18

omitting the importance of more data availability at all levels of the factory and control systems, which is only possible by more connectivity.

Lot of initiatives were made to meet these requirements, making the Industry migrating to an entire digitalization and connectivity, from where was born the Industry 4.0 [12]. The fourth industrial revolution is based mainly on cyber-physical production systems, which include smart machines and production facilities that have been developed digitally, and have their logistics, production, marketing, and service entirely integrable based on ICT [13]. In fact, the transformation of mechatronic systems into cyber-physical systems (CPS) is the source of some of the main Industry 4.0 objectives [14].

Being a founding brick of the industry 4.0 [1], Cyber-Physical Production systems are defined by *Monostori et al.* [17] as following "CPPS consist of autonomous and cooperative elements and sub-systems that are getting into connection with each other in situation-dependent ways, on and across all levels of production, from processes through machines up to production and logistics networks." Moreover, the cyber-physical systems architecture is divided into 5 Levels [7, 18], and [2] as follows: The connection level, the conversion level, the cyber level, the cognition level, and the configuration level. Several papers highlight the requirements that have to be met for Cyber-Physical Production Systems [3–5], and [6]. In [3] CPPS characteristics are categorized into four groups. The first group is Architectural models which could be based on SOA or MAS due to their openness. The second group is Communication and data consistency. The third group is intelligent products and production facilities inside a CPPS, which are able to flexibly adapt to change in customer requirements, variation in the demands, and breakdowns during production. The fourth group is Data Preparation for Humans, about the CPPS, its architecture, products, and production as long as the concepts support the CPPS engineering and capability to pre-process production data.

In this context, several practical applications have already emerged. Among them, we can cite the work of [9], where the authors developed a method for the systematic engineering of industrial CPS. They applied modularity under consideration of a smart factory, smart data, smart products, and smart services. In [10], the authors proposed a modular MAS-based CPPS architecture where software agents are running on the fog level. In [11] the authors developed an efficient MAS based CPPS for a discrete flexible manufacturing system. In the same direction, we propose in the present paper a MAS-based CPPS architecture, for the Management and control of an Industry 4.0 Assembly Platform situated in SRP Lab "Robotized Systems for Production" at the "Robotics and Integrated Manufacturing" division in Algerian Center of Development of Advanced Technologies CDTA. This case study covers the development and the virtual commissioning of the proposed CPPS Architecture.

The rest of this paper is organized in the following way. Section 2 describes the use case platform subject of the study. In Sect. 3, we describe the CPPS Architecture developed in this paper. Section 4 is dedicated to the Software-In-the-Loop "SIL" Testing procedure. Finally, Sect. 5 concludes the paper.

2 Laboratory Assembly Cell Use Case

The use case studied in this paper is an Industry 4.0 platform made by the Robotized Production systems team at the Robotics and Industrial Automation Division of CDTA

(Centre de Développement des Technologies Avancées) in Algeria with the collaboration of SIEMENS Algeria.

The cyber-physical system in the case of the study is a robotized cell made up of four stations "One pic and place station for Entry/Exit of shuttles to the platform, and two pic and place workstations for part feeding the product shuttles with spare items. One Assembly Station, and a closed-loop conveying system (Fig. 1). Each station is equipped with a photoelectric sensor to detect the presence of a product shuttle in front of the Workstation, and an RFID reader; to read/write the product-specific data on the RFID TAG fixed to the product Shuttle.

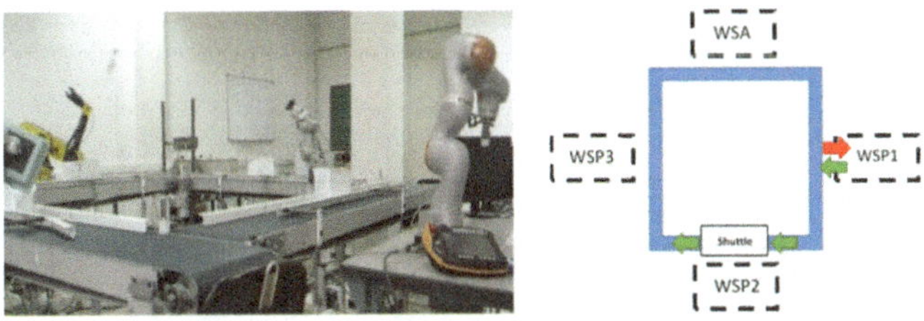

Fig. 1. General overview of the assembly platform at SRP laboratory

The production system represented by the robotic platform is a synchronous flow shop. During the rest of this paper, we will call the pic and place station for Entry/Exit of shuttles as "WSP1". The two pic and place workstations for part feeding as "WSP2" and "WSP3" respectively. The Assembly Station will be called "WSA".

The product Shuttle is put in the system at the first station "WSP1". After that, the RFID reader on this station writes the initial information to the corresponding TAG. The conveying system transports then the Shuttle to the next station "WSP2". The RFID reader of this station reads the Shuttle TAG, to determine what to do on this product. In this case, it is a part feeding of the product shuttle. The Robotic station executes then the operation, and the conveying system transports the Shuttle to the third station "WSP3" where another part feeding operation is executed. The product shuttle is then transported to the Assembly Station "WSA", where the spare items are extracted from the shuttle, and assembled by a collaborative robot with the help of a human operator. Then the assembled product is put in the shuttle, and placed by the robot on the conveyor. Finally, the product is returned to the first station "WSP1" where it is verified and extracted from the production system.

As there is not yet installed Robot in the "WSP1", the results obtained in the virtual commissioning of the MAS-based CPPS architecture presented in the next section will partly contribute to the successful integration of a future installed Robot in this Workstation to the production cell.

3 The Developed CPPS Architecture

In the present work, a cyber-physical system Architecture is presented, where the control tasks are divided between PLCs and Multi-Agent systems.

One of the benefits of the proposed MAS-based CPPS solution is a smooth integration in existing Manufacturing Plants based on traditional control systems. By keeping everything controlled by the PLC. The list of tasks that can be performed by the physical resource is transmitted to its corresponding Software Agent. This one can combine these tasks into a set of different operations, dynamically modifiable without the need to modify the PLC logic. So the Production becomes partly controllable by the Software Agents via the PLC inside each individual Cyber-Physical Module CPPM. This is useful in the case of plants where it is not allowed to give full control of the production by the Software Agents for different reasons including safety or security requirements.

The division of control functions between PLC and the different software agents is described below.

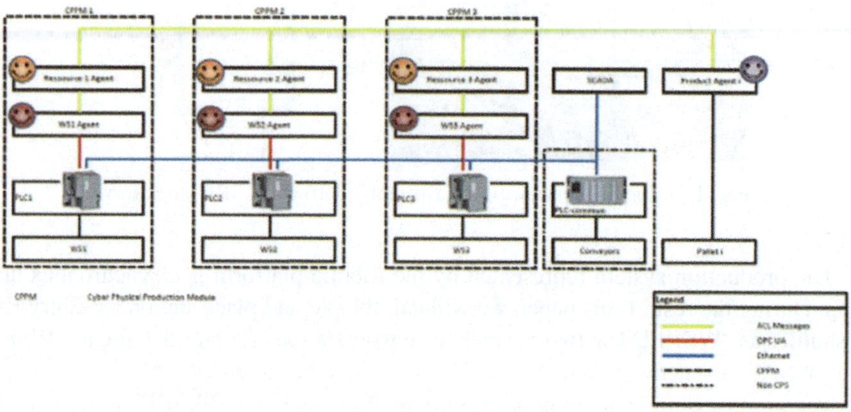

Fig. 2. Developed MAS based CPPS architecture

3.1 Description of the CPPS Architecture

Since the assembly cell is constituted of four workstations and a conveying system, among them three workstations are concerned by this work. This work aims to design a CPPS Architecture of the cell, the workstations are considered to be the cyber-physical modules, and the conveying system will be considered in the present as a Non-Cyber-Physical Entity. In this section, we present the system architecture. At the beginning we will describe the different types of Agents and their functions, then the interaction between them, inside the CPPM. Among different MAS Architectures proposed for CPPS, there are MAS Architectures, where they may have also different degrees of control of the process, along with the Edge controller. In this work, the Software Agent of each WorkStation controls high-level operations, and only supervises the low-level operations done by the respective PLC of this Workstation.

In this Architecture, there are three types of Software Agents: Product Agents, Resource Agents responsible for the resource (like Machine or Robot…), and the Workstation Agent. This later is responsible for the corresponding Workstation and ensuring the abstraction of the PLC Data to usable information by the Resource Agent, and the transfer of instructions in the opposite direction.

The Cyber Physical Modules

Each station is controlled by a distributed controller SIEMENS ET200SP with an S7 1500 CPU and a couple of agents. The ET200SP is responsible for the low-level control functions of the robotic station. A software agent is assigned to the resource, and it is responsible for the decision and/or the personalization of the products by the workstation. Then there is a Workstation Software Agent for the interface between the PLC and the resource agent. It transforms the data produced by the PLC to semantics utilizable by the resource agent, sent to this one via Agent Communication Language (FIPA-ACL) [19]. The workstation Agent is connected to the PLC via OPC UA (Open Platform Communications Unified Architecture) [20]. In the normal case, the operations to do on the product by the workstation are stored on the product RFID Tag. In this study, the RFID TAG does not contain the information on the operations; they are stored in a PLC data bloc, which is modifiable by the product Software Agent. Upon the arrival of the product Shuttle to the Workstation, the RFID reader reads the Tag on the Shuttle; then sends it to the PLC. This later reads the operations to be done on the product from the product data bloc. The operations Data is provided by the Product Software Agent. This gives the possibility to modify the operations list by the software agent at any time. This is very useful in the case of personalization or correction that has to be done on the product.

Different Layers Inside the CPPM

Description of the different layers of the CPPS Architecture and the responsibilities of each layer. The first layer includes the physical parts of the system, represented by the workstations with their different components (Robot, part feeding system) as physical modules. The second layer is composed of the Programmable Logic Controllers, responsible for the low-level control functions, at the operational level. The third layer is an abstraction layer, represented by the Workstation Agent, also called PLC Agent. It is responsible for the abstraction of the data collected from the PLC to understandable information by the software Agents in the upper layer, and the translation of their instructions sent via ACL messages into precise commands to send by the Interface Agent to PLC via OPC UA. The Final layer contains the Resource Agents, and the product Agents, responsible for the High-level control operations, at the information level.

Communication between the CPPMs and with Non-CP Entities

The communication between the CPPM is performed at both the operational level (see Fig. 2) between the PLCs via Ethernet communication, and at the information level between the software Agents at the Multi-Agent System.

The communication between the CPPS and the non-Cyber-Physical “non-CP” entities is ensured by the Programmable Logic Controllers in the control network, while the pertinent information is transferred to the Multi-Agent system, creating an indirect interaction with the rest of the plant.

3.2 Implementation of the Multi Agent System

As mentioned in the previous sub-section, the MAS Architecture is composed of two kinds of Software Agents: The first is the cognitive Agent responsible for Management and control of the Cyber-Physical Production System. This Agent represents the Resource Agents and Product Agents. The Resource Agent is responsible for the Management of the affected physical resource (Machine, Robot…). It can inform the other Agents about the state of the resource (Free, occupied, out of service…etc.) and decide to accept or not perform a given operation of a product, during the negotiation process. The Products Agent represents the software part of its associated physical product on the shop floor. It can give the user and the other Agents the state and the progress of the scheduled operations on the product, and it can ask a resource Agent to allocate its physical resource to perform a given operation at a given time according to its production schedule. The second one is a reactive Agent called "Workstation Agent", situated at the abstraction layer of each Cyber-Physical Module. It is responsible for its respective workstation "WS". It also ensures both the abstraction of the Workstation PLC data to usable information by the cognitive Agents and the transfer of the requests in the opposite direction. The Multi-Agent System is implemented on JADE (Java Agent Development Framework) with JAVA. There are three Agent Classes: The Workstation Agent Class called "Agent_PLC", the Resource Agent Class called "Agent_Ressource", the Product Agent Class called "Agent_Produit". The Agents of type "PLC Agent" are declared using each corresponding PLC OPC UA Server URI as shown in the Jade Container UML Diagram below (Fig. 3).

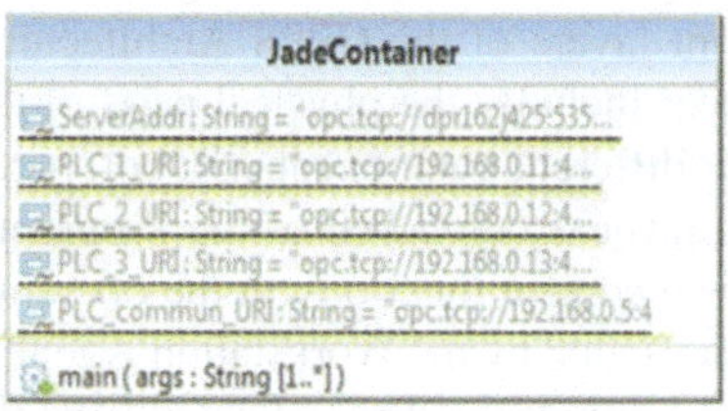

Fig. 3. JADE container diagram

The communication between Agents is based on FIPA Contract Net in two cases: The first is the negotiation between Resource Agents and Product Agents for the Allocation of the Resource to a given Product. The second is between two or more Product Agents to determine the order of Entry of the products to the production process. If an agent requests the execution of a given action from another agent, FIPA Request is used.

4 The Virtual Commissioning

4.1 General Description

Virtual commissioning is usually used to reduce the costs of validating solutions on real equipment [8, 15, 16].

In our case, three reasons motivated the choice of virtual commissioning approach:

- Seize the time while the platform was under assembly
- Faster tuning of the PLC logic compared to the traditional commissioning
- Drastically reduce the risk to damage equipment later, at real commissioning

In order to commission virtually the Robotized manufacturing cell. We first created the virtual model of the cell in 3D environment using FLEXSIM, by emulating the robotic workstations, in each workstation; we have replicated the robot and spare parts buffers. The RFID readers and photoelectric sensors have been replicated also in the virtual model, as well as the product Shuttles and the conveying system Fig. 4.

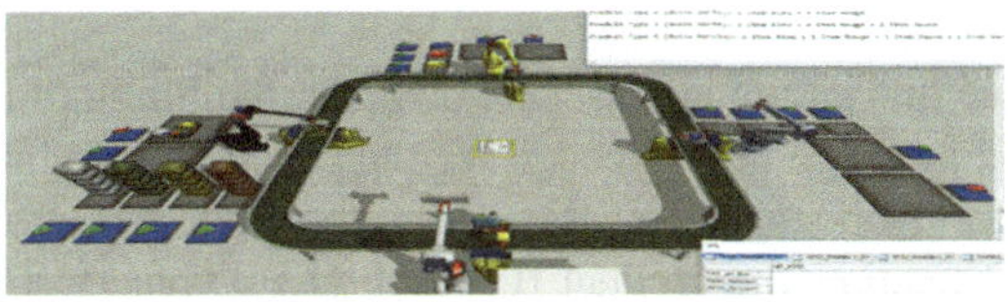

Fig. 4. The 3D model of the Robotized cell in FLEXSIM environment

After that, we connected the Virtual PLCs to the emulated model of the platform, and performed an overall testing of Input/Output signals, with respecting the following order: The PLC 1 is controlling the Workstation WSP1, the PLC 2 for WSP2 and PLC 3 for WSP3. The Assembly workstation WSA containing the collaborative robot KUKA IRB IIWA is not included in the present work, as its virtual commissioning has been treated in a separate work.

Therefore, the complete system is available to implement and validate the solution developed in the previous section.

4.2 SIL Testing Validation

The complete software in the loop (SIL) Validation test includes the four Virtual PLCs and the Multi-Agent System, for controlling the virtual Model of the production cell.

During the SIL test, the Simulation of the four programmable logic controllers is performed using SIEMENS PLCSIM ADVANCED software. The Multi-Agent System is developed under JAVA using the JADE platform, with Eclipse Editor. Both are executed on the same Personal Computer communicating via OPC UA protocol, while the FLEXSIM 3D Model of the production system is running on another PC. On the same Local Area Network of the first PC, and communicating with the Virtual PLCs via OPC UA Protocol (Fig. 5).

Test Case

In order to validate the developed MAS-based CPPS Architecture, we use the SIL validation technique cited above in practical cases during production. The first case is the entry of the products to the system and the way their request is handled and executed. The second case is the execution of an operation on a given product in a workstation, and how the detailed list of tasks is loaded to the PLC, and executed by the Resource

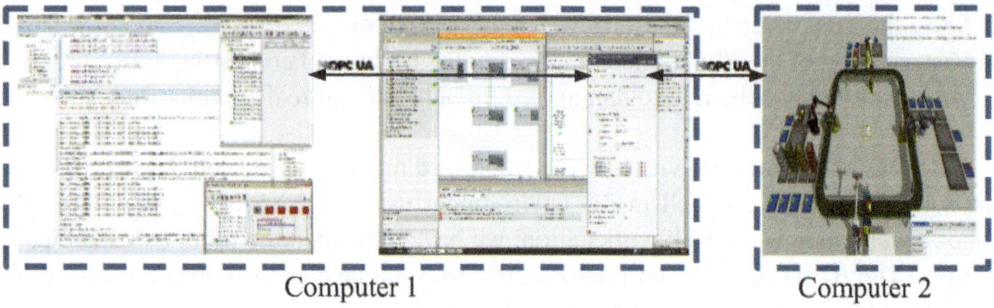

Fig. 5. SIL-testing validation

(Robot). To illustrate this particular case, we consider the filling of the Product shuttle with the corresponding spare items needed for the product assembly, this operation is done by the Robot2 (Workstation WSP2). Upon the detection of the product shuttle at WSP2. The PLC acquires the product ID via the RFID reader. The WSP2 Agent is notified of the presence of the product shuttle. This later reads the Product ID from the PLC via its embedded OPC UA Client. Then it requests the list of tasks to execute from the corresponding PA using the Product ID. Once it receives the response, the WSP2 Agent extracts the corresponding tasks and then writes them on a dedicated Data bloc in the PLC. This DB is used in the PLC logic to set the tasks to the Robot controller. The WS Agent informs the PA and RA once all tasks are over (see Fig. 6).

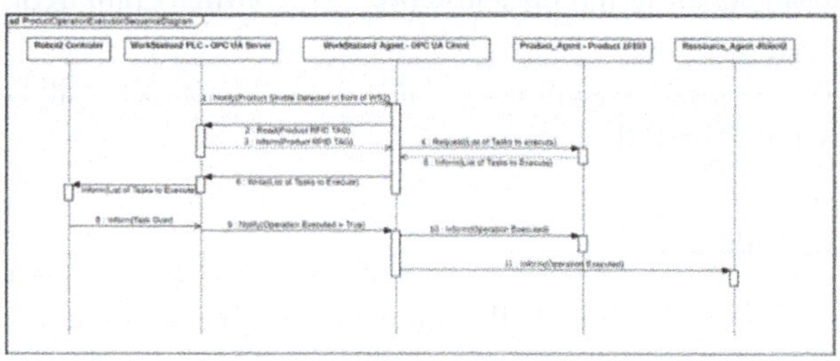

Fig. 6. Sequence diagram of interaction among agents and PLC for operation execution on a product with TAG 10103 at workstation N°2

4.3 Discussions

We observed a lack of synchronism between the Workstation Agent and the PLC while treating the scheduled Operations on the product. The PLC did not get the Task list on time, which disturbed the proper functioning of the latter. A possible explanation is that the Agents processing time (Soft Real-Time) is greater than the cycle time of the PLC (Hard Real-Time). Furthermore, the inter-Agents communication consumes an additional time to treat the Product operations information. We solved the problem by adding a watchdog and a Fail-Safe routine on the PLC side and sending another Status feedback of the Workstation Agent to the PLC.

5 Conclusion

The existence of several patterns and standardization works regarding MAS-based CPPS makes their development much easier than before. However, many aspects regarding the adoption of Multi-Agent systems in the development of CPPS are continuously evolving. Indeed, MAS are well adapted to develop distributed intelligent systems, featuring flexibility, agility, and self-configuration. In this paper, we proposed MAS-based CPPS Architecture in an Industry 4.0 context production system. We assessed the behavior of the components of a typical Cyber-Physical production module during the treatment of a given operation on the product during a SIL test. We noticed the importance of supervising the MAS status by the PLC besides adding Fail-Safe routines regarding Software Agents during the elaboration of the PLC logic. We used a SIL virtual commissioning validation approach in a practical case during production in Industry 4.0 context. The SIL test has proven its adequacy for PLC logic tuning and assessing the interaction between PLC logic and the Multi-Agent system inside the CPPS. As a perspective, we will complete the Commissioning of the system by a Hardware-in-The-Loop HIL validation, followed by an implementation in the real Assembly platform at SRP Lab in CDTA.

References

1. Leitao, P., Karnouskos, S., Ribeiro, L., Lee, J., Strasser, T., Colombo, A.W.: Smart agents in industrial cyber–physical systems. Proc. IEEE **104**(5), 1086–1101 (2016)
2. Monostori, L., et al.: Cyber-physical systems in manufacturing. CIRP Ann. **65**(2), 621–641 (2016)
3. Vogel-Heuser, B., Diedrich, C., Pantförder, D., Göhner, P.: Coupling heterogeneous production systems by a multi-agent based cyber-physical production system. In 2014 12th IEEE International Conference on Industrial Informatics (INDIN), pp. 713–719. IEEE (July 2014)
4. Cruz, S.L.A.,Vogel-Heuser, B.: Comparison of agent oriented software methodologies to apply in cyber physical production systems. In 2017 IEEE 15th International Conference on Industrial Informatics (INDIN), pp. 65–71. IEEE (July 2017)
5. Salazar, L.A.C., Mayer, F., Schütz, D., Vogel-Heuser, B.: Platform independent multi-agent system for robust networks of production systems. IFAC-PapersOnLine **51**(11), 1261–1268 (2018)
6. Salazar, L.A.C., Ryashentseva, D., Lüder, A., Vogel-Heuser, B.: Cyber-physical production systems architecture based on multi-agent's design pattern—comparison of selected approaches mapping four agent patterns. Int. J. Adv. Manuf. Technol. **105**(9), 4005–4034 (2019)
7. Lee, J., Bagheri, B., Kao, H.A.: A cyber-physical systems architecture for industry 4.0-based manufacturing systems. Manuf. Lett. **3**, 18–23 (2015)
8. Drath, R., Weber, P., Mauser, N.: An evolutionary approach for the industrial introduction of virtual commissioning. In: 2008 IEEE International Conference on Emerging Technologies and Factory Automation, pp. 5–8. IEEE (September 2008)
9. Oks, S.J., Fritzsche, A., Möslein, K.M.: Engineering industrial cyber-physical systems: an application map based method. Procedia CIRP **72**, 456–461 (2018)
10. Rocha, A.D., Tripa, J., Alemão, D., Peres, R.S., Barata, J.: Agent-based Plug and Produce Cyber-Physical Production System–Test Case. In 2019 IEEE 17th International Conference on Industrial Informatics (INDIN), vol. 1, pp. 1545–1551. IEEE (July 2019)

11. Mihoubi, B., Bouzouia, B., Tebani, K., Gaham, M.: Hardware in the loop simulation for product driven control of a cyber-physical manufacturing system. Prod. Eng. **14**(3), 329–343 (2020)
12. Rasche, C., Tinkleman, M.: Industry 4.0-a discussion of qualifications and skills in the factory of the future: a German and American perspective. VDI, ASME, Düsseldorf, Germany (April 2015)
13. Kagermann, H., Wahlster, W., Helbig, J.: Recommendations for implementing the strategic initiative INDUSTRIE 4.0-Final report of the Industrie 4.0 Working Group. acatech–National Academy of Science and Engineering, Germany (April 2013)
14. DIN and DKE ROADMAP: German Standardization Roadmap Industrie 4.0. Standardization council Industrie 4.0. Germany (2016)
15. Berger, T., Deneux, D., Bonte, T., Cocquebert, E., Trentesaux, D.: Arezzo-flexible manufacturing system: a generic flexible manufacturing system shop floor emulator approach for high-level control virtual commissioning. Concurr. Eng. **23**(4), 333–342 (2015)
16. Quintanilla, F.G., Cardin, O., L'Anton, A., Castagna, P.: Virtual commissioning-based development and implementation of a service-oriented holonic control for retrofit manufacturing systems. In: Borangiu, T., Trentesaux, D., Thomas, A., McFarlane, D. (eds.) Service Orientation in Holonic and Multi-Agent Manufacturing. Studies in Computational Intelligence, vol. 640, pp. 233–242. Springer, Cham (2016). https://doi.org/10.1007/978-3-319-30337-6_22
17. Monostori, L.: Cyber-physical production systems: roots, expectations and R&D challenges. Procedia Cirp **17**, 9–13 (2014)
18. Vogel-Heuser, B., Lee, J., Leitão, P.: Agents enabling cyber-physical production systems. at-Automatisierungstechnik **63**(10), 777–789 (2015)
19. FIPA Homepage. http://www.fipa.org/specs/fipa00061/index.html
20. Mahnke, W., Leitner, S.H., Damm, M.: OPC Unified Architecture. Springer Science & Business Media, Heidelberg (2009)

Ranking Social Media News Feeds: A Comparative Study of Personalized and Non-personalized Prediction Models

Sami Belkacem[1(✉)], Kamel Boukhalfa[1], and Omar Boussaid[2]

[1] LSI Laboratory, USTHB, Algiers, Algeria
{s.belkacem,kboukhalfa}@usthb.dz
[2] ERIC Laboratory, University of Lyon 2, Lyon, France
omar.boussaid@univ-lyon2.fr

Abstract. Ranking news feed updates by relevance has been proposed to help social media users catch up with the content they may find interesting. For this matter, a single non-personalized model has been used to predict the relevance for all users. However, as user interests and preferences are different, we believe that using a personalized model for each user is crucial to refine the ranking. In this work, to predict the relevance of news feed updates and improve user experience, we use the *random forest* algorithm to train and introduce a personalized prediction model for each user. Then, we compare personalized and non-personalized models according to six criteria: (1) the overall prediction performance; (2) the amount of data in the training set; (3) the cold-start problem; (4) the incorporation of user preferences over time; (5) the model fine-tuning; and (6) the personalization of feature importance for users. Experimental results on *Twitter* show that a single non-personalized model for all users is easy to manage and fine-tune, is less likely to overfit, and it addresses the problem of cold-start and inactive users. On the other hand, the personalized models we introduce allow personalized feature importance, take into consideration the preferences of each user, and allow to track changes in user preferences over time. Furthermore, personalized models give a higher prediction accuracy than non-personalized models.

Keywords: Social media · News feed · Relevance · Personalization

1 Introduction

In several research approaches, ranking news feed updates in descending relevance order has been proposed to help social media users quickly catch up with the content they may find interesting in the news feed [1]. For this matter, supervised prediction models have been commonly used to predict the relevance of updates using labeled training data [2]. These models analyze past user behaviors to predict whether they will find an update relevant or not in the future [2]. However, in related work, to train a prediction model and predict the relevance, data of all users were first merged as if there is only one user. Then, a

B. Lejdel et al. (Eds.): AIAP 2021, LNNS 413, pp. 197–209, 2022.
https://doi.org/10.1007/978-3-030-96311-8_19

single non-personalized model has been trained on all data for all users. Indeed, according to Vougioukas et al. [1], in non-personalized models, a global model is typically trained on a large collection of updates received by multiple users and the interaction of each user with each update, e.g. retweets. The trained model is then used to predict a user-independent relevance score to each new update. By contrast, personalized models should be trained only on updates received by a particular user and the interactions of the particular user, e.g. whether the user retweeted each tweet. Hence, a separate model should be trained per user and then employed to provide user-specific relevance scores for each new tweet or, generally, social update. We believe that non-personalized models are useful to learn the overall interests of the majority of users (e.g., in general, users are likely to find relevant tweets that are similar to their own tweets), but generalize such unrealistic assumptions to all users makes it difficult to predict their individual preferences. For example, a given user might be more interested in new content that is different from his own tweets. Indeed, Paek et al. [3] noticed in their study 44 cases in which several participants had rated the same news feed post and found out that 82% of the cases differ in ratings suggesting that the relevance judgment can be subjective depending on the preferences of each user.

In this paper, we first provide background on ranking news feed updates according to a typical approach and a reminder of the non-personalized models used in related work. Then, to predict the relevance of news feed updates given that user preferences are different, we introduce a personalized prediction model for each user based on the *random forest* algorithm. Finally, we conduct a comparative study of personalized and non-personalized models according to six criteria: (1) the overall prediction performance of both approaches to get a global overview of the most effective model; (2) the amount of data in the training set to investigate the robustness of each model; (3) the cold-start problem, which is a common problem in recommender systems; (4) the incorporation of user preferences over time; (5) the model fine-tuning to investigate the manageability of each model; and (6) the personalization of feature importance for users.

The paper is structured as follows: Sect. 2 presents background on ranking news feed updates on *Twitter*, Sect. 3 provides a reminder of non-personalized prediction models, Sect. 4 introduces our personalized model, Sect. 5 discusses the experiments we performed to compare both models and highlight the need for personalization, and finally, Sect. 6 concludes the paper.

2 Background

In this work, we focus like most related work on *Twitter*. Note, however, that it is possible to use this work on other social media platforms with some adaptations. Figure 1 describes the primary non-personalized technique used to predict the relevance score $\boldsymbol{R(t,u)}$ of a tweet $\boldsymbol{t} \in \boldsymbol{F(u)}$, where $\boldsymbol{F(u)}$ denotes tweets unread by the recipient user $\boldsymbol{u}$ that can potentially be included in the news feed. This technique is based on a supervised prediction model that analyzes labeled training data of tweets users read in the past to predict if the recipient user u will find the tweet t relevant in the future. Let $D(u)$ denotes a subset of tweets previously read by the user u and D the overall labeled training data of all users. The training data of a user u is a set of input-out pairs such that an input represents a vector of features that may influence the relevance of a tweet $t' \in D$ to u, and the output represents the relevance score $R(t',u)$. The primary technique involves three steps: (1) label tweets by relevance scores; (2) extract the features that may influence relevance; and (3) train the prediction model. In this section, we describe each step according to a typical approach [4].

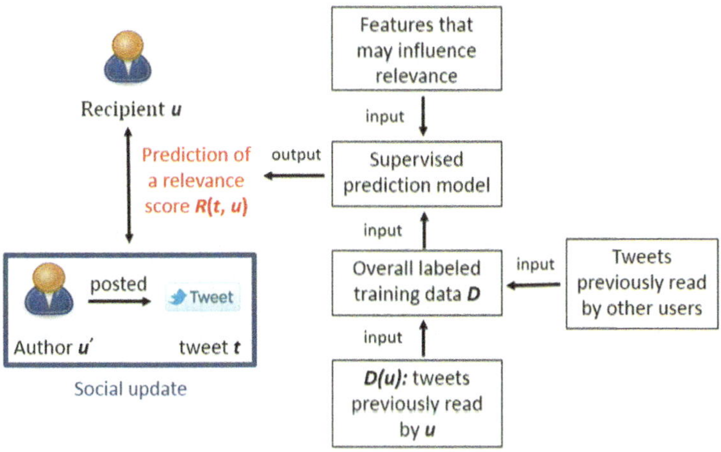

Fig. 1. Non-personalized prediction of a relevance score

First, to label tweets by relevance scores, we use the implicit method used by most related work [4]. It assumes that a previously read tweet $t' \in D(u)$ is relevant to a user u if u interacted with t' (retweet, reply, like). Predicting relevance scores results in a binary classification problem. Note that some machine learning models such as *random forest* allow to predict the probability of classes and rank tweets by relevance according to the probability of having class 1.

Second, according to related work [4], we use 13 most relevant features that may influence the relevance of a tweet t, posted by an author u', to the recipient u. The features are divided in four categories, while more details are given in [4]:

- Features that match between the content of t and the interests of u.
- Features that measure social tie strength between u and u'. The assumption is that t could be relevant to u if u and u' are close friends.
- Features that measure the authority of u'. The assumption is that t could be relevant to u if u' is important and has authority on the platform.
- Features that measure the quality of t: length, popularity, if it has a photo, etc. The assumption is that t could be relevant to u if t is of high quality.

Finally, the prediction model aims to analyze labeled training data of tweets users read in the past to predict if they will find a tweet relevant in the future. Let $\boldsymbol{S}$ denotes the set of recipient users. First, we generate training data instances for each recipient user $\boldsymbol{u} \in \boldsymbol{S}$ in the form of input-output pairs considering each previously read tweet $\boldsymbol{t'} \in \boldsymbol{D(u)}$. An input represents a vector of features that may influence the relevance of $\boldsymbol{t'}$ to $\boldsymbol{u}$, and the output represents the implicit relevance score $\boldsymbol{R(t',u)}$. Then, we can either train a personalized prediction model for each user $\boldsymbol{u} \in \boldsymbol{S}$, or merge all data as if there was only one user to train a single non-personalized model for all users. The aim of both approaches is to map new input features of a tweet unread by a user $\boldsymbol{u}$ to a relevance score using a binary classifier learned from previously read tweets in the training set. In the next section, we provide a reminder of non-personalized models.

3 Non-personalized models

In non-personalized models, a single model is trained on a large collection of tweets received by multiple users and the interactions of all users with each tweet [1]. The trained model is then used to assign a user-independent relevance score to a new incoming tweet. Figure 2 describes the primary technique used in related work to train a non-personalized prediction model. First, historical user data, which consists of previously read tweets $\boldsymbol{D}_i$, are merged and scaled to have feature values within the same range. Then, the overall data $\boldsymbol{D}$ is shuffled as if there is only one user and no chronological order of tweets. Finally, data is split into two sets: a training set to train the prediction model with 70% of the data and a test set to evaluate the performance with the 30% remaining data.

Table 1 indicates the non-personalized models used in related work. The table shows that different supervised algorithms were used: *logistic regression* [1,5–7], *Support Vector Machines* [3], *artificial neural networks* [7–10], etc. In each work, a single algorithm was used for either: (1) all users [1,8–13]; (2) each fold/partition of data with five folds in [3] and three partitions in [5]; or (3) each demographic subset of users [7]. In other words, no related work has used a single model for each user, such that in the best of cases, five models were used for 24 users in [3] and n models in [7], where n is the number of demographic subsets of users. The research work state that non-personalized models benefit from a large collection of tweets in the training set. Each tweet is represented as a feature vector that includes user-specific features. If two users receive the same tweet, it will be represented by two different feature vectors, which allows the model to produce different predictions per user for the same incoming tweet.

Nonetheless, since non-personalized models are trained on all data as if there is only one user, the models may learn and generalize unrealistic assumptions (e.g., all users are likely to find relevant the tweets that are similar to their own tweets). The importance/weight of the features learned by non-personalized models is assumed to be the same for all users, but such assumptions may not apply to some users. For example, a given user might be more interested in new content that is different from his own. Indeed, Paek et al. [3] asked 24

Table 1. Non-personalized models in related work

Research work	Data	Supervised algorithm	A prediction model for
[11]	665 tweets	Coordinate ascent algorithm	All users
[12]	816 users	*Gradient Boosting*	
[13]	675 users	*Naive Bayes*	
[1]	122 users	*logistic regression*	
[10]	2 users	*artificial neural networks*	
[9]	307 users		
[8]	1000 users		
[3]	24 users	*Support Vector Machines*	Each fold of data (5 folds)
[5]	LinkedIn users	*logistic regression*	Each partition of data (3 partitions)
[6]			
Facebook [7]	Trillions of examples	*logistic regression, artificial neural networks, Gradient Boosting, etc.*	Each demographic subset of users

participants to rate news feed posts and noticed that 82% of ratings that concern the same tweets are different. This study indicates that the relevance judgment is subjective as user preferences and interests are different. Therefore, we believe that using a personalized user-dependent model is crucial to enhance the news feed content. In the next section, we introduce our approach that uses the *random forest* algorithm to train a personalized prediction model for each user.

4 A Personalized Prediction Model

In contrast to non-personalized models, personalized models should be trained on tweets received by a particular user and the interactions of the particular user with each tweet. Hence, a separate model should be trained per user and then employed to provide user-specific relevance scores to new incoming tweets. Figure 2 describes the technique we use to train a personalized prediction model for each user and assign user-specific relevance scores to tweets. First, we sort tweets by time and divide the training data $\boldsymbol{D}_i$ of each user $\boldsymbol{u}_i \in \boldsymbol{S}$ into two sets: a training set of the prediction model for the 70% least recent instances and a test set for the remaining 30% most recent instances. The purpose is to keep a chronological track of the relevance judgment of tweets by users over time. Then, we use the training set of each user $\boldsymbol{u}_i \in \boldsymbol{S}$ to train the corresponding *random forest* model $\boldsymbol{M}_i$. *Random Forest* [14] is a popular *ensemble learning* method[1] for

[1] A method that uses multiple machine learning algorithms to obtain better predictive performance than could be obtained by any of the constituent learning algorithms.

classification and regression problems that operate by constructing a multitude of *decision trees*. In our previous work [2], we compared several machine learning algorithms used in related work and found that ensemble learning models are the most suitable to predict the relevance of news feed updates.

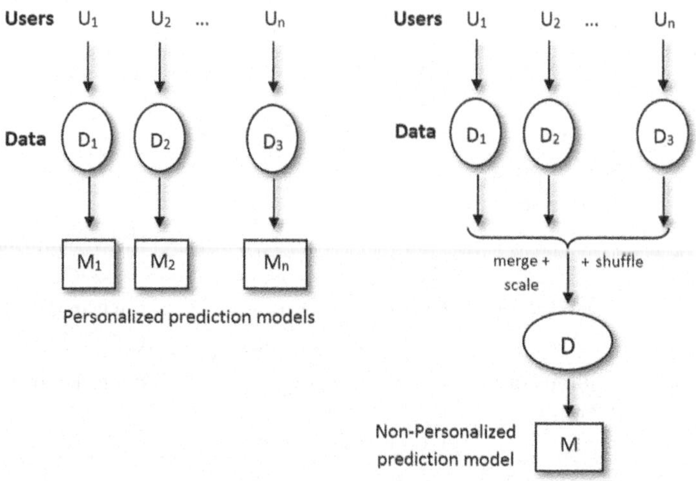

Fig. 2. Personalized and Non-personalized models

The aim of using a personalized *random forest* model for each user is to make tailored recommendations, which may not coincide with the interests of the majority of users that non-personalized models are trained to predict. Indeed, unlike non-personalized models, not only the feature vector is different for each user-tweet pair, but also the feature importance/weight for each user. In other words, as a model is trained on the data of a given user independently of the other users, the model learns the individual user preferences and interests (e.g., a user interested in art is more likely to find tweets with a multimedia content relevant). Another reason to use a personalized model for each user is to sort and split the corresponding train and test data by time. Train the model on recent data allows to track changes in user preferences over time and make time-sensitive recommendations accordingly. In the next section, we describe the experiments we used to compare personalized and non-personalized models.

5 Experiments and Comparison Results

To compare personalized and non-personalized models and highlight the need for personalization, we describe in this section: (1) the dataset used in the experiments we performed; (2) the measures we used to evaluate the performance; (3) the methodology we used in the comparison, and (4) the obtained results.

5.1 Dataset

First, we randomly selected a set $\boldsymbol{S}$ of 46 recipient users and collected data over ten months using *Twitter Rest API*[2]. Then, to simulate the news feed of each user $\boldsymbol{u} \in \boldsymbol{S}$, we used the following principle to select, $\boldsymbol{D(u)}$, tweets posted by the followings of $\boldsymbol{u}$ that he may have read: (1) sort the tweets posted by the followings of $\boldsymbol{u}$ from least recent to most recent; (2) for each tweet **t'** with which $\boldsymbol{u}$ interacted, keep the chronological session defined by the tweet **t'**, the tweet before **t'**, and the tweet after **t'**. This resulted in 26180 tweets, a 35% interaction rate with tweets and 569 tweets on average as training data for each user.

5.2 Measures

First, we train *random forest* classifiers for both personalized and non-personalized models using 70% of the data. Then, we define the following concepts to evaluate the models using the corresponding test set with 30% of the data [15]:

- True Positive (TP): # of relevant tweets correctly predicted relevant
- True Negative (TN): # of irrelevant tweets correctly predicted irrelevant
- False Positive (FP): # of irrelevant tweets incorrectly predicted relevant
- False Negative (FN): # of relevant tweets incorrectly predicted irrelevant

After that, we use the weighted *F1 score* measure given by Eq. 1 [15], which is a popular measure for binary and unbalanced classification problems.

$$F = \frac{(F_r \times (TP + FN)) + (F_i \times (TN + FP))}{TP + TN + FP + FN} \tag{1}$$

Where:

- F_r is the standard *F1 score* for the class of relevant tweets
- F_i is the standard *F1 score* for the class of irrelevant tweets

5.3 Methodology

In the experiments, we first selected the best *random forest* parameters (number of trees, maximum three depth, splitting criterion, etc.) for a fair comparison between non-personalized and personalized models. Hence, a *random search* was run over different parameter values so that the parameters are optimized by a cross-validated search [16]. Indeed, we used a cross-validation for the non-personalized model and a time-series cross-validation for the personalized model as the latter preserves the chronological order of tweets [15], unlike the non-personalized model where data is shuffled. Then, to study the model stability with several runs and small changes to training data, we retrained each model on 30 different random state[3] values and evaluated it on the test set. Finally, we select the average *F score* for personalized and non-personalized approaches.

[2] https://dev.twitter.com/rest/public.

[3] A variable used in randomized machine learning algorithms to determine the random seed of the pseudo-random number generator.

5.4 Results

The comparison and evaluation results are presented and discussed according to six criteria: (1) the overall prediction performance of both approaches to get a global overview of the most effective model; (2) the amount of data in the training set to investigate the robustness of each model; (3) the cold-start problem, which is a common problem in recommender systems; (4) the incorporation of user preferences over time; (5) the model fine-tuning to investigate the manageability of each model; and (6) the personalization of feature importance for users.

First, the results show that introducing a personalized model for each user has improved the average *F score* by +3.12%, from 77.73% with the non-personalized model to 80.85% with the personalized model. Therefore, to make refined predictions and select the tweets that might be relevant to a given user, it is more convenient to train a model on tweets the user has found relevant in the past rather than including tweets and behaviors about other users in the training process. Undoubtedly, tweets that are relevant to one user are not necessarily relevant to another user, which illustrates the importance of the personalized model we introduce to capture individual user needs and improve the prediction accuracy. Time-aware user preferences are another advantage of personalized models that makes them more accurate. Indeed, train the model on recent data allows time-sensitive recommendations. The personalized models capture the chronological evolution of user relevance judgment of tweets, which may change with time (e.g., a user may over time give less importance to popular tweets and more importance to tweets related to his interests). In contrast, the non-personalized model cannot predict such behaviors since data of all users are merged and shuffled as if there is only one user and no chronological order of tweets.

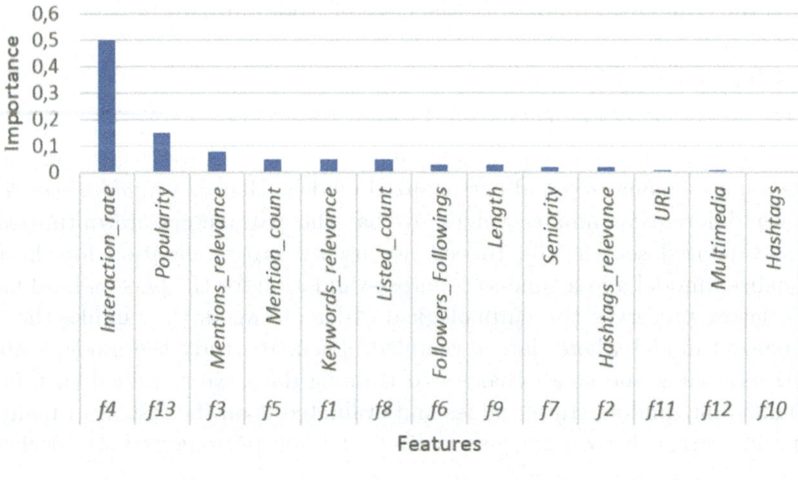

Fig. 3. Non-personalized feature importance

Second, we computed feature importance values[4] [14] in both personalized and non-personalized models, which are presented in Table 2 and Fig. 3 respectively. Figure 3 gives the average feature importance for all users. The figure shows that non-personalized models can learn and provide an overview of the features that influence the relevance judgment of tweets by users, which is useful to understand user behaviors and the assessment of relevance in general. For example, the top feature is the feature f_4 (0.5), the interaction rate of $\boldsymbol{u}$ with tweets posted by $\boldsymbol{u'}$. In contrast to non-personalized feature importance, Table 2 gives the personalized feature importance for each recipient user. First, we note that feature importance differs according to users, i.e. features that are important to one user are not necessarily important to another user, e.g. the feature f_9 which stands for the tweet length is very important to the user *Red_or_MC1R* when judging the relevance of tweets (0.22) but not to the user *Medium* (0.02). Certainly, user preferences are different, and this illustrates the gain brought by a personalized prediction model for each user, which takes into consideration individual interests. Furthermore, we note that the features learned as highly important by the non-personalized model are in fact, not important to all users. For example, the feature f_4 (0.5), the interaction rate of $\boldsymbol{u}$ with tweets posted by $\boldsymbol{u'}$, is important to many users when judging the relevance of tweets, but not to some users, e.g. the users *TheMuslimReform* (0.01), *LKrauss1* (0.02), and *bamwxcom* (0.03). This proves that non-personalized models generalize unrealistic assumptions to all users. In opposite, personalized models allow tailored recommendations that are different from the preferences of the majority of users.

Despite all the improvements the personalized models have brought in, we observe from the evaluation results that the proposed approach has some limitations. Figure 4 presents the learning curve of the non-personalized model for all users along with the learning curve of the personalized model for the user *ch402*. Note that the learning curves of the 46 users in the dataset are quite similar; hence we randomly selected a single user as a case study due to lack of space.

First, Fig. 4 shows that the non-personalized model benefits from a large collection of tweets in the training set compared to the personalized model (20000 against 400 tweets). Indeed, unlike personalized models which are trained on the individual data of each user, the non-personalized model merges the data of all users, which allows it to be trained on a large collection of tweets received by different users. Note that in the dataset, there is an average of 569 tweets in the training database of each recipient user and a median of 343 tweets.

Second, the training and cross-validation curves in Fig. 4 indicate that both personalized and non-personalized models converge, suggesting that the models are able to learn to classify tweets according to their relevance. However, we notice that train a non-personalized model on a larger training set makes it more robust and less likely to overfit comparing to the personalized model. In other words, the training and cross-validation curves of the non-personalized model converge to the same *F score* value (76%), indicating that the model can

[4] *Random Forest* computes the importance of a feature as the normalized total reduction of the criterion brought by that feature, also known as the *Gini importance*.

Table 2. Personalized feature importance

User	f1	f2	f3	f4	f5	f6	f7	f8	f9	f10	f11	f12	f13
Astro0Glen	0.02	0.04	0.0	0.39	0.06	0.13	0.07	0.1	0.04	0.01	0.01	0.0	0.14
Astro_Pam	0.05	0.06	0.03	0.23	0.06	0.09	0.05	0.1	0.09	0.01	0.03	0.02	0.2
bamwxcom	0.0	0.01	0.13	0.03	0.0	0.1	0.06	0.08	0.08	0.01	0.02	0.08	0.39
Baronatrix	0.03	0.01	0.0	0.21	0.0	0.11	0.09	0.15	0.12	0.01	0.04	0.01	0.23
BethStamper6	0.04	0.0	0.0	0.3	0.0	0.06	0.06	0.1	0.14	0.02	0.04	0.02	0.22
byudkowsky	0.03	0.0	0.0	0.14	0.12	0.21	0.05	0.12	0.17	0.0	0.01	0.0	0.14
ch402	0.03	0.0	0.01	0.19	0.02	0.11	0.07	0.15	0.13	0.01	0.02	0.01	0.25
demishassabis	0.06	0.0	0.02	0.25	0.0	0.09	0.23	0.09	0.08	0.01	0.01	0.0	0.17
eevil_abby	0.2	0.0	0.07	0.06	0.05	0.06	0.05	0.09	0.19	0.02	0.01	0.03	0.18
elonmusk	0.04	0.0	0.0	0.23	0.02	0.1	0.06	0.07	0.06	0.01	0.02	0.01	0.39
GeorgeHarrison	0.0	0.01	0.0	0.2	0.22	0.07	0.03	0.15	0.02	0.0	0.0	0.0	0.29
GilmoreGuysShow	0.03	0.0	0.0	0.1	0.0	0.14	0.08	0.10	0.13	0.02	0.07	0.01	0.22
gwern	0.03	0.0	0.0	0.21	0.01	0.16	0.07	0.11	0.08	0.04	0.01	0.0	0.29
homebrew	0.02	0.01	0.01	0.2	0.09	0.07	0.03	0.1	0.08	0.01	0.01	0.01	0.37
HybridZizi	0.04	0.0	0.0	0.3	0.0	0.11	0.06	0.13	0.1	0.02	0.03	0.02	0.21
jadelgador	0.08	0.01	0.0	0.29	0.04	0.07	0.05	0.09	0.03	0.01	0.0	0.01	0.33
JHUBME	0.05	0.04	0.11	0.17	0.13	0.08	0.04	0.11	0.07	0.01	0.0	0.0	0.19
JohnDawsonFox26	0.04	0.02	0.0	0.26	0.12	0.07	0.04	0.06	0.05	0.01	0.02	0.08	0.25
john_walsh	0.0	0.21	0.02	0.13	0.03	0.14	0.03	0.17	0.05	0.03	0.0	0.0	0.18
kilcherfrontier	0.01	0.12	0.12	0.27	0.07	0.07	0.04	0.19	0.03	0.0	0.02	0.0	0.06
LKrauss1	0.05	0.0	0.08	0.02	0.06	0.11	0.06	0.15	0.17	0.04	0.14	0.01	0.12
mastenspace	0.14	0.0	0.3	0.23	0.01	0.05	0.02	0.17	0.02	0.0	0.0	0.0	0.05
Medium	0.4	0.0	0.21	0.06	0.0	0.03	0.01	0.16	0.02	0.02	0.0	0.0	0.09
microphilosophy	0.13	0.01	0.0	0.11	0.01	0.12	0.16	0.08	0.1	0.02	0.01	0.01	0.25
MIRIBerkeley	0.23	0.0	0.0	0.09	0.04	0.09	0.18	0.12	0.08	0.01	0.01	0.0	0.15
NASAKepler	0.12	0.07	0.31	0.03	0.06	0.04	0.01	0.15	0.1	0.0	0.0	0.0	0.1
NASA_Wallops	0.08	0.05	0.03	0.11	0.02	0.07	0.0	0.18	0.03	0.0	0.05	0.0	0.38
newscientist	0.26	0.0	0.19	0.07	0.0	0.08	0.03	0.1	0.07	0.01	0.0	0.02	0.18
PattiPiatt	0.03	0.01	0.0	0.34	0.01	0.07	0.11	0.12	0.06	0.02	0.06	0.01	0.17
peterboghossian	0.05	0.01	0.06	0.23	0.03	0.08	0.05	0.13	0.1	0.01	0.02	0.01	0.22
rafat	0.03	0.0	0.03	0.18	0.0	0.11	0.04	0.11	0.08	0.01	0.03	0.04	0.33
realDonaldTrump	0.02	0.03	0.0	0.1	0.0	0.12	0.02	0.07	0.03	0.0	0.0	0.01	0.58
Red_or_MC1R	0.03	0.0	0.0	0.11	0.0	0.13	0.06	0.13	0.22	0.02	0.05	0.01	0.24
renormalized	0.03	0.0	0.0	0.16	0.0	0.11	0.07	0.13	0.13	0.0	0.06	0.01	0.3
RossTuckerNFL	0.03	0.0	0.22	0.2	0.05	0.09	0.09	0.13	0.04	0.01	0.02	0.02	0.11
RoxanneDawn	0.02	0.04	0.0	0.28	0.03	0.14	0.07	0.16	0.07	0.01	0.02	0.02	0.15
scimichael	0.06	0.03	0.0	0.32	0.0	0.14	0.04	0.12	0.05	0.01	0.0	0.0	0.24
SfNtweets	0.04	0.18	0.02	0.13	0.02	0.06	0.13	0.12	0.08	0.01	0.01	0.0	0.2
slatestarcodex	0.01	0.0	0.0	0.14	0.0	0.22	0.09	0.13	0.13	0.0	0.03	0.0	0.25
SLSingh	0.06	0.0	0.0	0.23	0.02	0.11	0.11	0.2	0.03	0.0	0.02	0.0	0.21
sxbegle	0.04	0.0	0.14	0.17	0.02	0.09	0.05	0.11	0.1	0.01	0.02	0.0	0.23
TeslaRoadTrip	0.05	0.01	0.0	0.23	0.0	0.12	0.06	0.1	0.04	0.01	0.02	0.03	0.34
TheMuslimReform	0.04	0.0	0.0	0.01	0.24	0.13	0.13	0.1	0.11	0.02	0.0	0.0	0.22
TheRickDore	0.01	0.02	0.22	0.1	0.0	0.12	0.03	0.11	0.08	0.02	0.05	0.02	0.2
USDISA	0.07	0.01	0.12	0.16	0.03	0.12	0.04	0.13	0.12	0.02	0.01	0.01	0.16
WestWingWeekly	0.04	0.02	0.14	0.21	0.11	0.09	0.04	0.07	0.08	0.01	0.01	0.01	0.18

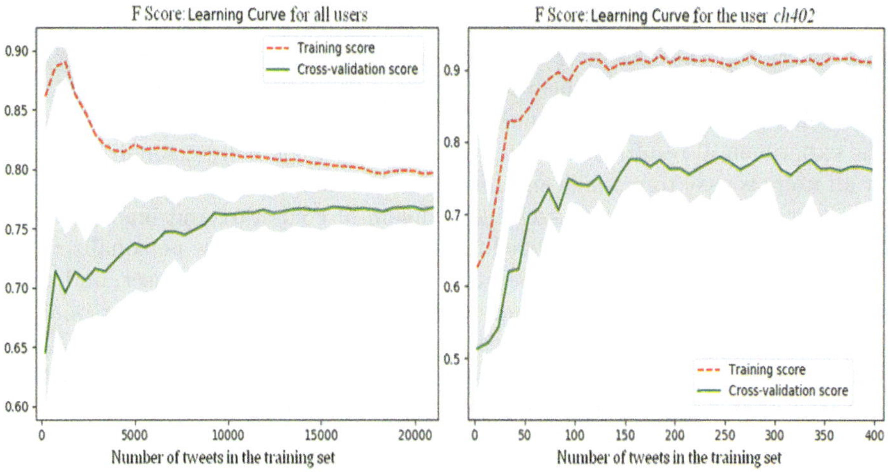

Fig. 4. Learning curves: Non-Personalized (left) vs. Personalized (right)

generalize relevance predictions to unseen tweets. As to the personalized model, which is trained on a smaller training set, we observe that the model fits the training dataset too well with a high *F score* value (90%) and loses some of its ability to generalize to the cross-validation set with a lower *F score* value (78%). Therefore, to make more accurate predictions to new and unseen tweets, it would be advisable to use one of the many machine learning techniques to prevent overfitting: regularization, early stopping, data augmentation, etc. [15].

Finally, another notable difference is that non-personalized models may work better with new or inactive users, for which personalized models may have very few training instances. Indeed, in such cases, the personalized model does not have information about user preferences and interests in order to make specific recommendations. Hence, it is important to suggest alternatives to address this common problem in recommender systems known as the *cold-start problem*. Non-personalized models address this issue by default since the same model can be used to any user on the social media, even new or inactive users. Lastly, it is easier for social media administrators/developers to fine-tune and manage a single non-personalized model than fine-tuning a personalized model for each user. For example, in our case, it was somewhat possible to look at each of the 46 prediction models corresponding to the 46 recipient users, but this may become more challenging as the number of users increases. In such a situation, it is necessary to provide reliable automatic techniques to validate user models.

6 Conclusion

In this paper, to predict the relevance of news feed updates and improve user experience, we used the *random forest* algorithm to train and introduce a personalized prediction model for each user. Then, we conducted a comprative study of personalized and non-personalized models according to six criteria: (1) the overall prediction performance; (2) the amount of data in the training set; (3) the cold-start problem; (4) the incorporation of user preferences over time; (5) the model fine-tuning; and (6) the personalization of feature importance for users. The experimental results on *Twitter* show that a single non-personalized model for all users is easy to manage and fine-tune, is less likely to overfit as it benefits from more data, and it addresses the problem of cold-start and inactive users. On the other hand, the personalized models we introduce allow personalized feature importance, take into consideration the preferences of each user, and allow to track changes in user preferences over time. Furthermore, the personalized models give a higher prediction accuracy than non-personalized models. These findings highlight the need for personalization to effectively rank the news feed.

Despite the advantages that personalized models have brought over the classical non-personalized models, we observed that non-personalized models may still work better with new or inactive users, for which personalized models may have very few training instances. Hence, it is important to suggest alternatives to address this common problem in recommender systems known as the *cold-start problem*. Non-personalized models address this issue by default since the same model can be used for any user, even new or inactive users. To address this problem, for example, it would be interesting to propose a hybrid method that takes the advantages of both personalized and non-personalized models.

References

1. Vougioukas, M., Androutsopoulos, I., Paliouras, G.: Identifying retweetable tweets with a personalized global classifier. In: Proceedings of the 10th Hellenic Conference on Artificial Intelligence–SETN 2018, pp. 1–8. ACM Press, Patras, Greece (2018)
2. Belkacem, S., Boussaid, O., Boukhalfa, K.: Ranking news feed updates on social media: a comparative study of supervised models. In: EGC–Conference on Knowledge Extraction and Management, vol. 36, pp. 499–506. Revue des Nouvelles Technologies de l'Information (2020)
3. Paek, T., Gamon, M., Counts, S., Chickering, D.M., Dhesi, A.: Predicting the importance of newsfeed posts and social network friends. In: AAAI, vol. 10, pp. 1419–1424 (2010)
4. Belkacem, S., Boukhalfa, K., Boussaid, O.: Expertise-aware news feed updates recommendation: a random forest approach. Clust. Comput. **23**(3), 2375–2388 (2019). https://doi.org/10.1007/s10586-019-03009-w
5. Agarwal, D., et al.: Activity ranking in LinkedIn feed. In: Proceedings of the 20th ACM SIGKDD International Conference on Knowledge Discovery and Data Mining, pp. 1603–1612 (2014)
6. Agarwal, D., et al.: Personalizing linkedin feed. In: Proceedings of the 21th ACM SIGKDD International Conference on Knowledge Discovery and Data Mining, pp. 1651–1660 (2015)
7. Backstrom, L.: Serving a billion personalized news feeds. In: Proceedings of the Ninth ACM International Conference on Web Search and Data Mining–WSDM 2016, p. 469. ACM Press, San Francisco, California, USA (2016)

8. Zhang, Q., Gong, Y., Wu, J., Huang, H., Huang, X.: Retweet prediction with attention-based deep neural network. In: Proceedings of the 25th ACM International on Conference on Information and Knowledge Management, pp. 75–84. ACM, Indianapolis Indiana USA, October 2016
9. Piao, G., Breslin, J.G.: Learning to rank tweets with author-based long short-term memory networks. In: Mikkonen, T., Klamma, R., Hernández, J. (eds.) ICWE 2018. LNCS, vol. 10845, pp. 288–295. Springer, Cham (2018). https://doi.org/10.1007/978-3-319-91662-0_22
10. De Maio, C., Fenza, G., Gallo, M., Loia, V., Parente, M.: Time-aware adaptive tweets ranking through deep learning. Future Gener. Comput. Syst. **93**, 924–932 (2019)
11. Uysal, I., Croft, W.B.: User oriented tweet ranking: a filtering approach to microblogs. In: Proceedings of the 20th ACM International Conference on Information and Knowledge Management–CIKM 2011, pp. 2261. ACM Press, Glasgow, Scotland, UK (2011)
12. Shen, K., et al.: Reorder user's tweets. ACM Trans. Intell. Syst. Technol. **4**(1), 1–17 (2013)
13. Song, K., Wang, D., Feng, S., Zhang, Y., Qu, W., Yu, G.: CTROF: a collaborative tweet ranking framework for online personalized recommendation. In: Tseng, V.S., Ho, T.B., Zhou, Z.-H., Chen, A.L.P., Kao, H.-Y. (eds.) PAKDD 2014. LNCS (LNAI), vol. 8444, pp. 1–12. Springer, Cham (2014). https://doi.org/10.1007/978-3-319-06605-9_1
14. Breiman, L.: Random forests. Mach. Learn. **45**(1), 5–32 (2001)
15. Sammut, C., Webb, G.I.: Encyclopedia of Machine Learning. Springer Science & Business Media, Heidelberg (2011)
16. Bergstra, J., Bengio, Y.: Random search for hyper-parameter optimization. J. Mach. Learn. Res. **13**(Feb), 281–305 (2012)

A Social Media Approach for Improving Decision-Making Systems

Islam Sadat(✉) and Kamel Boukhalfa

Computer Science Department, University of Science and Technology
Houari Boumediene, Algiers, Algeria
{isadat,kboukhalfa}@usthb.dz

Abstract. The appearance of social networks has revolutionized the way of interaction and exchange of information on the internet, and has led to an explosion of data on these networks. The availability of this data via a set of APIs triggered a wave of initiatives and work aiming to make use of this massive amount of information in order to extract knowledge that will facilitate the decision-making process. In fact, due to growing business competition, leaders from different sectors have already started using reporting tools, making it possible to provide reliable and relevant analyses at the right time. In this paper we present an approach for improving an existing decision-making system by combining its core datawarehouse with a social one. This social datawarehouse built using extracted tweets will expand the analyzes possibilities of the existing system, and will allow carrying out various analyses, facilitating the decision-making process, and this for any sector.

Keywords: Datawarehousing · Social media · Decision making · Data analysis

1 Introduction

Contemporary information and decision support systems have been essential to the proper functioning and growth of successful businesses around the world for more than two decades. Data warehousing and OLAP technology are at the center of these systems and have been instrumental in analyzing data in multiple areas such as manufacturing, retail, transport, health care, education, research and government. The data warehousing technology as well as its underlying techniques have been extended to provide better performance, by taking advantage of the emergence of new data types and sources, especially data of the public shared on the phenomena of social media. Indeed, social media have shaped the last two decades of the 21st century, and is considered as a revolution that sparked almost all ways of life. The users content on these sites represents huge volumes of data that is generated at a high rate and attracts a lot of research and interest. Since then, companies and organizations with decision-making systems have sought to make this continuous flow of information concerning them beneficial, and to use it as an asset to facilitate decision-making. Table 1 represents the number of users by social network for the year 2020.

B. Lejdel et al. (Eds.): AIAP 2021, LNNS 413, pp. 210–230, 2022.
https://doi.org/10.1007/978-3-030-96311-8_20

Table 1. Number of users by social network

Facebook	2 740 000
Youtube	2 291 000
Instagram	1 221 000
Twitter	353 000
Quora	300 000

Datawarehouses as well as multidimensional analyzes are essential for any decision-making process; however, the absence of tools allowing the use of social networks in decision support constitutes a niche research field that aroused the interest of a good number of researchers, what resulted in various works aiming to carry out multidimensional analyzes on social data. We aim through this study to improve an existing decision-making system by transforming its datawarehouse into a more robust one, composed of its original dimensions and aggregations, as well as new social activity tables and dimensions built using social media data (Twitter). This hybrid datawarehouse will offer the possibility of performing more consistent analyses, and will facilitate the decision-making process. In order to build this system we encountered the following challenges:

- The extracted social data (tweets) is multidimensional, time stamped, geographical and composed of many attributes. Designing a database architecture capable of storing this type of social data was a challenging subject.
- Choosing the most appropriate social dimensions that allow the fusion of social and non-social data.
- Summing the social data and finding the correct aggregations with their respective dimensions was a major step in order to answer the users queries in the most efficient way.

In this paper, we first explain the sources of the social data, the extraction process as well as the preparation of the social corpus. Then we introduce the architecture of the activity and dimensions tables which will store the aggregated social information. Next we discuss the different multidimensional analyses possibilities offered by our system. Finally, we conclude the paper and discuss future research.

2 Literature Background

2.1 Datawarehouse

A data warehouse is a type of data management system designed to enable and support business intelligence activities, particularly analytics [18]. Data warehouses are intended only for performing queries and analyzes. They often contain large amounts of historical data. The data in a data warehouse usually comes from a wide variety of sources, such as application log files and transaction applications. According to the firm Gartner [19], the business intelligence market reached a worldwide turnover of 27 billion dollars in 2019. Business intelligence is targeting to improve business decision-making on the basis of established facts and offers non-IT decision-makers a transversal vision of strategic information.

2.2 OLAP Technology

OLAP (online analytical processing) is the application of complex queries to large amounts of historical data, aggregated from OLTP databases and other sources, for data mining, analysis and business projects intelligence. In an OLAP database, data is integrated through a process called extract, transform, load (ETL). This pre-aggregation is what dramatically reduces response time to complex queries. Each query involves one or more columns of data aggregated from multiple rows. Examples include year-over-year financial performance or lead generation marketing trends [21]. OLAP databases and data warehouses allow analysts and decision makers to use custom reporting tools to turn data into information. Figure 1 shows an example of a multidimensional cube.

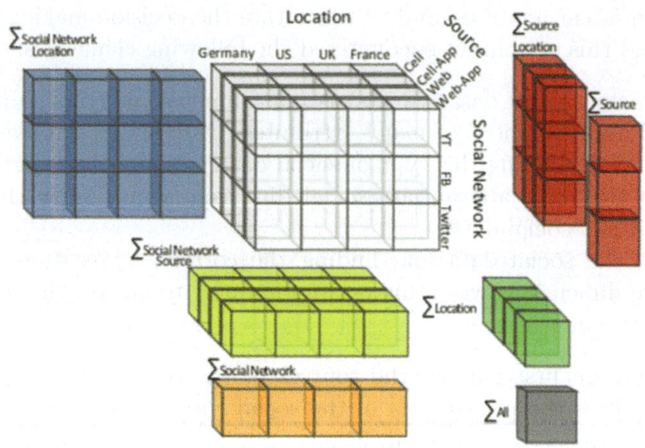

Fig. 1. A three-dimensional cube (gray) with its projections (colored).

2.3 Decision-Making System

A BI (business intelligence or decision-making system) is designed to answer questions like "What happened?", "Why did it happen?" And "What's going to happen?" and this with the aim of enabling organizations to understand their internal and external environment, through the acquisition, collation, analysis, interpretation and systematic use of information [22]. A BI system typically performs customer profiling, customer support, market research, market segmentation, product profitability, and inventory and distribution analysis. These functions can be accomplished through data warehousing, OLAP, data mining, descriptive and predictive analytics. BI systems have played a key role in running a successful business, and today it's hard to find a successful business that hasn't taken advantage of BI technology. Figure 1 shows the process of entering and leaving data in a BI framework.

2.4 Twitter

Microblogging is a network service with which users can share messages, links to external websites, images, or videos that are visible to users subscribed to the service. Messages that are posted on microblogs are short in contrast to traditional blogs. Like It or Not: A Survey of Twitter Sentiment Analysis Methods 28:5 Currently, a number of different microblogging platforms are available, including Twitter, Tumblr, FourSquare, Google+ and LinkedIn. One of the most popular microblogs is Twitter, which was launched in 2006 and since then has attracted a large number of users. Currently, Twitter has 284 million users who post 500 million messages per day. Due to the fact that it provides an easy way to access and download published posts, Twitter is considered one of the largest datasets of user generated content [3].

2.5 Big Data

The rapid development of the Internet of Things (IoT) results in a massive explosion of data generated from ubiquitous wearable devices and sensors [5]. The unprecedented increase of data volumes associated with advances of analytic techniques empowered from AI has led to the emergence of a big data era [5,6]. Big data has been employed in a wide range of industrial application domains, including healthcare where electronic healthcare records (EHRs) are exploited by using intelligent analytics for facilitating medical services. For example, health big data potentially supports patient health analysis, diagnosis assistance, and drug manufacturing [7]. Big data can be generated from a number of sources which may include online social graphs, mobile devices (i.e. smartphones), IoT devices (i.e. sensors), and public data [8] in various formats such as text or video. In the context of COVID-19, big data refers to the patient care data such as physician notes, X-Ray reports, case history, list of doctors and nurses, and information of outbreak areas. In general, big data is the information asset characterized by such a high volume, velocity and variety to acquire specific technology and analytical methods for its transformation into useful information to serve the end users.

2.6 Sentiment Analysis

Sentiment analysis is a natural language processing techniques to quantify an expressed opinion or sentiment within a selection of tweets. There are two main approaches for extracting sentiment automatically which are the lexicon-based approach and machine-learning-based approach [9] The sentiment can be found in comments or tweets to provide useful indicators for many different purposes, and its mainly categorized into 3 groups: negative, positive and neutral sentiment words.

3 Related work

Many researchers are working on analysing data on Twitter social media, some key contributions provide support for finding users behaviors and situations in the different cases while happening around the world, for this some of the essential papers are included in this section.

T. Yigitcanlar and al [9], proposed a social media data analyzing study, by collecting 100 k tweets generated by Australian users only, in order to capture the attitudes and perceptions of the Australian community during the covid-19 pandemic. The main objective of their study was to exploit social media in guiding interventions and decisions of the authorities, and to identify community needs and demands in a pandemic situation.

K. H. Manguri and al [10], been pulled out using python programming language Tweepy library a dataset of Tweets. The extraction of the tweets was based on two specific hashtag keywords, which are ("COVID-19, coronavirus"). The date of searching data is seven days from 09-04-2020 to 15-04-2020. Then they used python TextBlob library for analysing the users emitted sentiments. The obtained measures were after that represented graphically.

Vijay and al [11], gathered tweets regarding COVID-19 from November, 2019 to May, 2020 in India. Multiple datasets were created Month-wise, then combined to analyze the people's reactions towards Lockdown in June, 2020 and about everything related to COVID-19. The general feeling was negative at first shifted towards positive and neutral comments. In April, 2020 most comments were Positive and about winning against Corona virus.

R. Lamsal [12], prsented a large-scale Twitter dataset with more than 310 million COVID-19 specific English language tweets and their sentiment score, as well as GeoCOV19Tweets, The dataset's geo version. Lamsal's paper discussed the datasets design in detail, and the tweets in both the datasets were analyzed, giving a better understanding of spatial and temporal dimensions of the public shared tweets related to the ongoing pandemic.

Ben Kraiem and al [22] presented a multidimensional modeling tool. Contrary to the work cited above, the tool named OLAP4tweets exploits the association of OLAP technology with data mining techniques to allow an analysis centered on the aggregation of metadata on Twitter users and their web activity. messaging.

In [23] the authors present a tool called SocialCube. The latter helps organize social media data into multiple dimensions and hierarchies for efficient viewing and visualization of information from multiple perspectives, through the application of multidimensional manipulations of social data cubes into different dimensions.

The research papers cited previously encounter many limits. Most of these studies expressed awareness of the importance of using social data. Some of them created informative visuals using a social corpus, but didn't explore the multidimensional aspect of the social data. The others built datawarehouses with data from twitter, allowing multidimensional analyses, but did not make use of already non social datawarehouses. Our study exceeds the limits of other studies by mixing social and non social data in a single hybrid datawarehouse, meaning that our contribution covers the user of social media, it's multidimentional aspect improves the decision making process.

4 Data and Methodology

The process of improving the performances of a decision making system is presented in this section. The hybrid datawarehouse resulting in our approach is mainly composed of two parts:

1) A non social part (activity and dimension tables): It represents the datawarehouse behind the decision-making system concerned by the social aspect improvements. We have chosen a covid-19 decision-making system as a study case. Its datawarehouse is composed of activity tables and dimension related to the pandemic, allowing visualisation of the variation of number of deaths, recoveries and cases due to covid-19.
2) A social part (activity and dimension tables): This part represent the new activitiy and dimension tables established using attributes and aggregations from a twitter social corpus set up beforehand. This social corpus is composed of an important number of social data (tweets), treated, analyzed and enriched.

The merging of these two parts will result in what we named a "hybrid datawarehouse". We will in what follows explain the steps for setting up the two parts, as well as the fusion process.

4.1 Non Social Datawarehouse Part

In our case of study we will use "covid.sql", a file comprising a multidimensional cube having as a core unit aggregations related to coronavirus, and that allow to follow the evolution of this pandemic. This cube has two dimensions that are time and location. We have chosen this issue as a case study given the impact this pandemic had globally and the availability of information on social networks, more specifically on TWITTER. As shown in Fig. 2, the covid-19 datawarehouse is composed of a single activiy tables with 3 aggregates:

- **number of deaths**
- **number of recoveries**
- **number of cases**

This activity table is related to two dimension tables: **Location** and **Time**, which will allow temporal and geographic analyses. It is clear that the scope of analysis of this datawarehouse is very limited, and we will show in what follows how to extend this scope by setting up a datawarehouse of social data related to coronavirus.

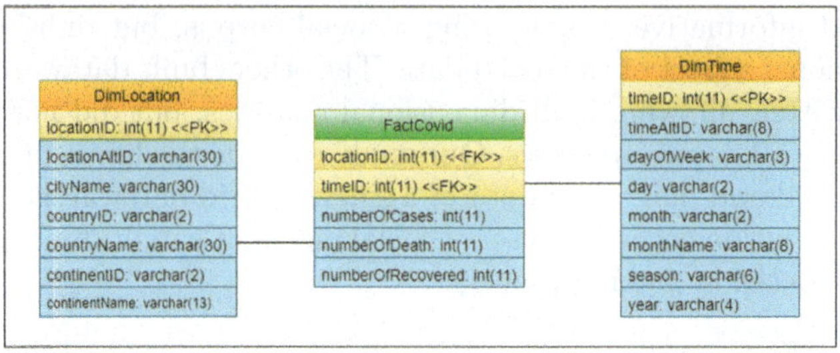

Fig. 2. Schema of the covid-19 datawarehouse (non social datawarehouse)

4.2 Social Datawarehouse Part

This part represents the added value that our approach brings to the decision-making system. This social part is built thanks to a social corpus composed of tweets mentioning the coronavirus. We will detail in what follows the stages of construction of the corpus, as well as the definition of the architecture of the social darawarehouse

1) Building the social corpus. Twitter social media network provides across Twitter Developer Platform a set of free APIs that allow retrieval of tweets. The most used API is the Twitter REST API. It basically takes key words as input in order to extract relevant tweets. After trying different open source tools, we concluded that the python library "Tweepy" is the most convenient one for extracting data from Twitter API. "Tweepy" offers the possibility of fetching tweets by location, hashtags, keywords and date. In spite of that, the built dataset for this study is general and doesn't concern a specific country or continent, considering that coronavirus is a world wide pandemic. Despite of having many limitations from Twitter API, 126,000 tweets relative to coronavirus were successfully withdrawn from 29/02/2020 to 31/05/2020. The gathered data was at first saved as csv files. Table 2 shows indicators relative to the dataset.

Table 2. Collected data indicators

Indicator	Value
No. of tweets	126,000
No. of continents	6
No. of countries	159
No. of languages	52

The text of tweets very often contains elements that do not interest us in our case study, such as URLs, Hashtags, Mentions, Emojis, Smileys, JSON and even words specific to Twitter, such as "RT" for example which means "Retweet".

Therefore, text filtering is extremely important to keep only representative content. It is essential to enrich the data extracted in the previous step, and this in order to offer different possibilities of analysis. Enrichment is done as follows:

- Language detection: The language used in tweets allows to calculate the number of tweets for each language.
- Detection of the country and the continent: Geo-location will for example make it possible to know the global distribution of tweets according to the continent.
- Detection of the feeling expressed: To classify the publications contained in the corpus according to the emotion released, we use a lexical technique of sentiment analysis.

2) Social datawarehouse schema. The production of a social data-warehouse is at the center of our approach, it goes through two phases:

- Choice of the modelization scheme: The modeling scheme we have chosen to model our social data warehouse is the constellation scheme. The constellation diagram is the amalgamation of several star diagrams that use common dimensions. For this, we must model our different datamarts based on the star diagram.
- Dimensional modeling process: After studying the Twitter meta-model as well as the format of a "tweet", we found that it is possible to have two processes to model. The activities that we are going to model are: "Average sentiment of tweets" And "Distribution of tweets".
  Figure 3 represents the schema of "Average Tweets Sentiment" activity.

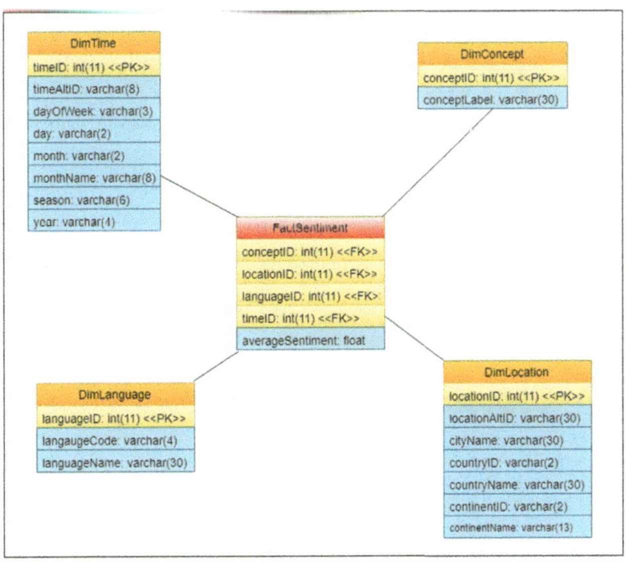

Fig. 3. Star model of the "Average Tweets Sentiment" activity

Figure 4 represents the schema of "Distribution of tweets" activity.

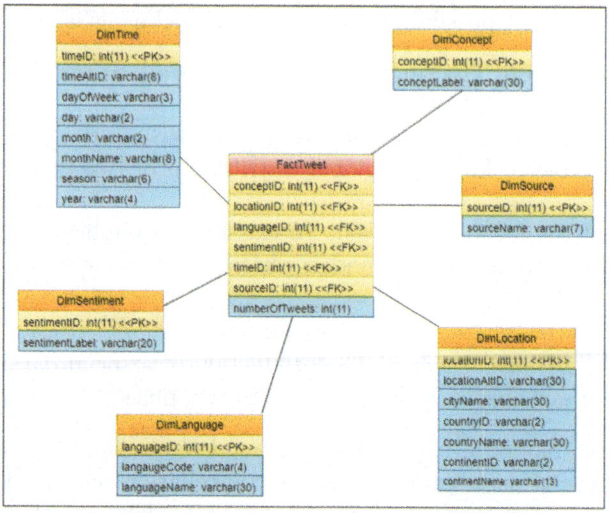

Fig. 4. Star model of the "Distribution of tweets" activity

The two previous schemes are merged into a single constellation schema as shown in Fig. 5.

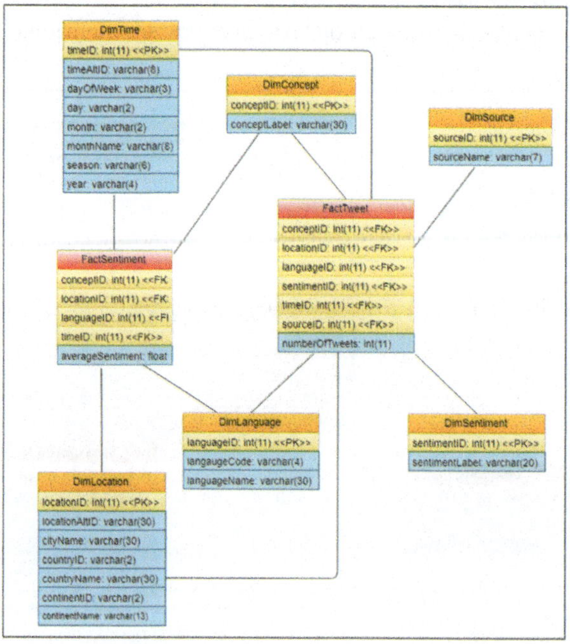

Fig. 5. Constellation scheme of the social datawarehous

3) Building the hybrid datawarehouse. Merging the two datawarehouses presented above will result in a datawarehouse structured according to a constellation scheme. This datawarehouse will be composed of social and non-social data, and will offer axes of analysis, and the possibility of much richer OLAP operations. The merge between the two datawarehouses is done based of the common dimensions between these two. In our case the shared dimensions are:

- Time
- Location

the hybrid datawarehouse scheme is shown in Fig. 6.

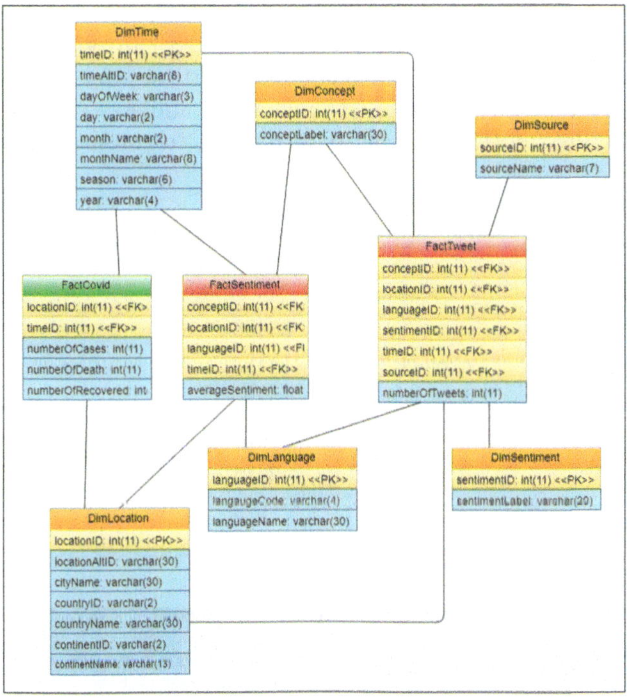

Fig. 6. Constellation scheme of the hybrid datawarehouse

5 Multidimensional Analysis

It is in this step that the importance of our work is highlighted. Indeed, at this level we run various OLAP queries on the data warehouse modeled by the set of multidimensional cubes. The results of OLAP queries will be made available in the form of graphical visualizations. In this section of the paper we present the different analyses performed of the social coronavirus datawarehouse. The procedure of each analysis will be explained, as well as the obtained results.

5.1 Geographical Analysis

In this phase we analysed the tweets by based on the Location dimension, using the Tweets Distribution cube. Table 3 show the number of tweets related to coronavirus by continent.

By analysing the report [14] of the ECDC (European Centre for Disease Prevention and Control) regarding the number of deaths due to covid-19 by continent at the end of Mai 2020, we noticed that the top three continents in number of deaths are Europe, America and Asia, the same continents with the more number of tweets relative to coronavirus in our study's dataset.

5.2 Linguistic Analysis

Analyzing in which languages social data is shared can be very beneficial. It may indicate to different entities which languages to use for communicating on social networks in order to have great impact and audience. Figure 7 shows the analysis of the distribution of social data relating to coronavirus as a function of language.

Table 3. No. of tweets by continent

Continent	No. of tweets
Asia	23438
Europe	18487
North America	10370
South America	8075
Africa	2614
Australia	529
Unknown	62487

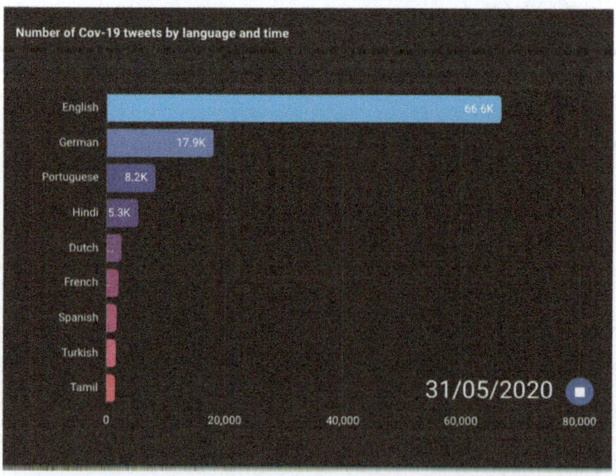

Fig. 7. Distribution of tweets by language

5.3 Sources Type Analysis

This analysis consists of distributing the tweets according to the emission sources (Android, iOS, Web). The results of this analysis could prove to be of great interest for choosing the computer software format (Mobile application or website) for having the maximum audience. The numbers of tweets by source are presented in Table 4.

It is clear from the obtained above that more than 70% of tweets were generated from mobile devices, since most internet users today communicate through smartphones.

Figure 8 shows the distribution of social data related to coronavirus by emission source for each continent.

5.4 Sentiment Analysis

This type of analysis allows the visualisation of the variation of sentiments and opinions of the public regarding covid-19 pandemic. The Tweets sentiment activity present in our social datawarehouse is very useful for analysing displaying the number of tweets by sentiment. Figure 9 presents the distribution of social data on "coronavirus" as a function of sentiment.

Table 4. No. of tweets by source type

Source type	No. of tweets
Android	47196
Iphone	23958
Web	29172
Unknown	25674

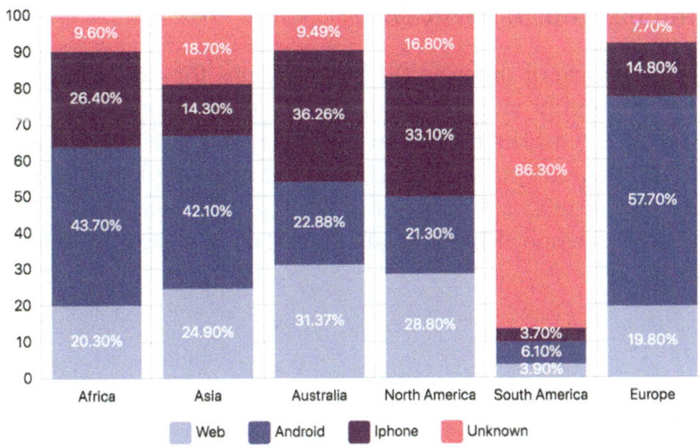

Fig. 8. Distribution of tweets by source for each continent

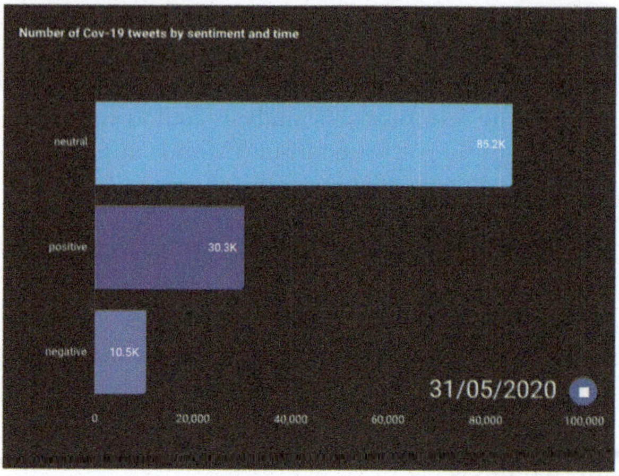

Fig. 9. Distribution of tweets by sentiment

In this study, the bar chart demonstrates emitted sentiment over Twitter starting from February 29th, 2020 to Mai 31st, 2020. It is clear that the total number of tweets is 126 K. Overall, more than 30.3 k of people published optimistic views, while only around 10.5 k of the tweets were negative. However, the No. of neutral tweets was significantly high (85.2 k). Such a Large quantity of neutral tweets might have happened because most of the tweets contained were facts rather than opinions, or because of the presence of a lot of prayers phrases that do not express negative nor positive emotions. The different analyses and indicators presented above were obtained using a tool that we developed as part of our study. In the following section, we will explain in detail the different stages of the implementation of our tool as well as the technologies used.

5.5 Redundant Topics Analysis

Although the existence of different NLP techniques for extracting efficient keywords, we have chosen in this work to analyse define relative subjects based on the words redundancy in the tweets texts. This redundant items will most likely offer significant new perceptions, especially after cleaning the data from non significant words like stop words. The more the topic is mentioned, the bigger the node is in the topics cloud. Figure 10 shows the node corresponding to the most redundant topic.

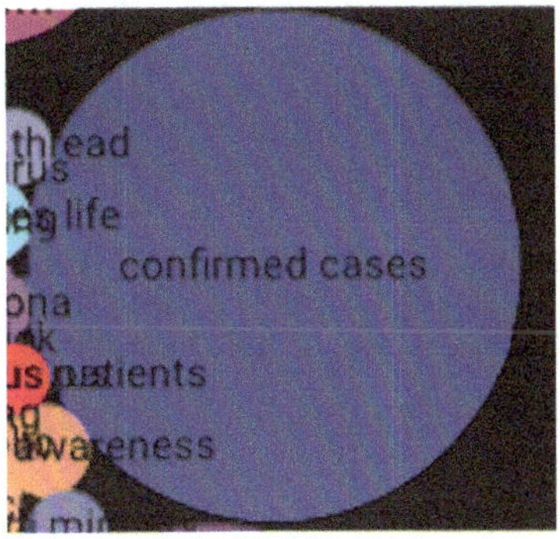

Fig. 10. Most redundant topic node in covid-19 dataset

Table 5 shows the four most redundant topics in our study data-set and their No. of occurrences.

Table 5. Four most redundant topics in data-set

Topic	No. of occurrences
confirmed cases	5668
corona virus	5173
vote amp	2758
cases deaths	1547

6 Tool Implementation

This section is devoted to the detailed description of the implementation of our application which we named "Cubes Creator". We will expose the different produced modules as well as the interfaces of our application.

6.1 Development Architecture

In order to provide a tool available on different types of devices and operating systems, we have chosen to develop a web application based on the RESTFUL architecture [38], and this using two very popular technologies which are the VUE.JS framework for the development of the front-end part of the application and the FLASK framework for the back-end part.

6.2 Developed Modules

In what follows, we will detail the different components of our tool by presenting the graphical interfaces relating to the features offered. Our tool essentially consists of (04) modules:

- External data integration module
- Social corpus construction module
- Social data cube construction module
- Social data cube manipulation module

The values of the "Consumer key", "Consumer secret", "access token" and "access secret" necessary for the user of the Twitter API are previously integrated at the application level, so the user does not have to. only have to enter his username and password to authenticate. Figure 11 shows the authentication interface.

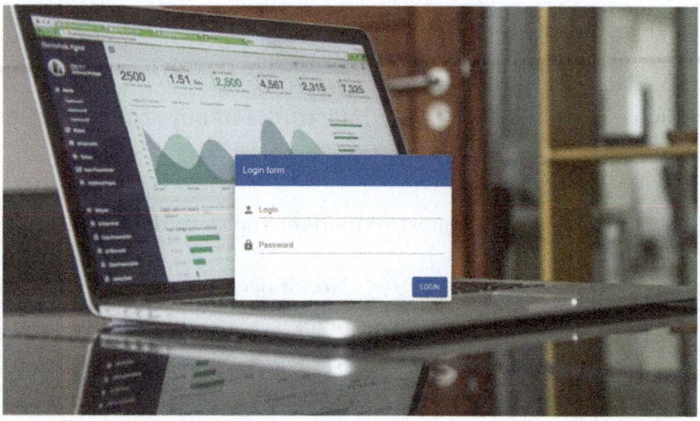

Fig. 11. Authentication interface

6.3 External Data Integration Module

We allowed the user of our tool to upload external data relative to covid-19 (number of deaths, number of cases). The uploaded information will be displayed in a relevant dashboard as shown in Fig. 12.

6.4 Social Corpus Construction Module

The construction of the social data corpus is a consistent step in our project. This is the first step in the ETL (Extract-transform-load) process, it aims to extract useful data from specific sources for use in next steps. Our tool offers the possibility of extraction data from the build dataset presented in previous sections, or real time extractions from Twitter using Twitter's REST API. Figure 13 represents the interface for constructing the social corpus 13.

Fig. 12. Global visualization interface of the evolution of the convid-19 pandemic

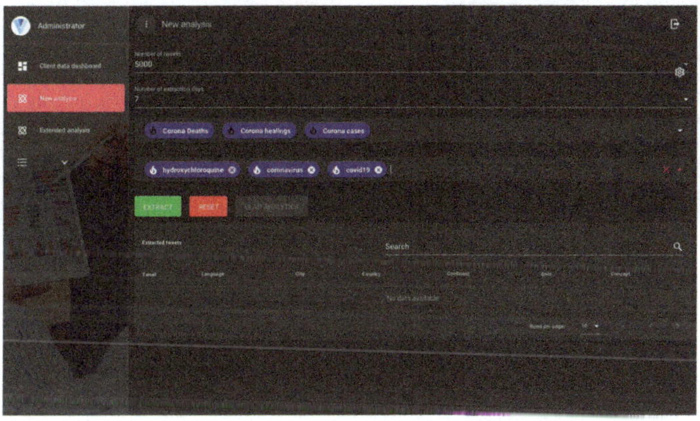

Fig. 13. Social corpus construction interface

6.5 Social Data Cube Construction Module

To apply OLAP technology we calculate from the social corpus the values of the measures according to the dimensions. These calculated values will be inserted at the level of the social part of the datawarehouse, completing the non social part representing in our case of study data related to the covid-19 pandemic.

6.6 Social Data Cube Manipulation Module

Multidimensional cubes produced from the global data warehouse (social data and external data) will allow the user to get a better analysis experience, considering the reduced execution time of OLAP queries, which is due the aggregations made during the creation of the cubes. We will detail in what follows the different analyzes that our application offers.

1) Sentiment analysis of social data Sentiment analysis is very interesting for decision making. It gives an idea of the opinions expressed by a population concerning a specific subject. Thanks to the multidimensional cube "Cube-Tweet", we can for example execute an OLAP query manipulating the measure "numberOfT-weets", to know the number of positive, negative or neutral tweets dealing with the subject "coronavirus", and this at a location and at a specific date, as shown in the Fig. 14.

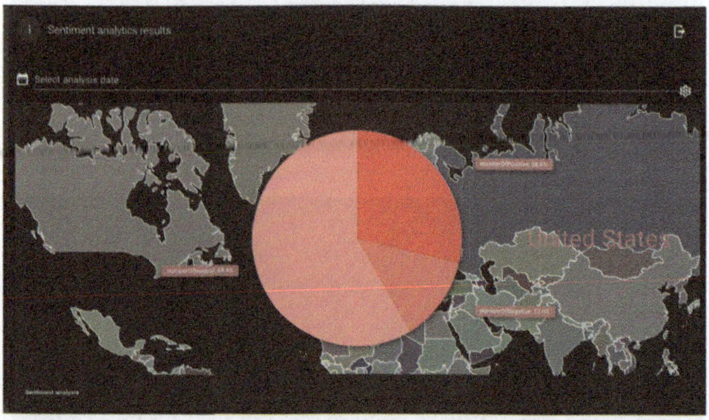

Fig. 14. Analysis of the sentiments of tweets in the USA talking about the "coronavirus" according to time and place

2) Analysis of the distribution of social data Having multiple dimensions linked to the "CubeTweet" fact table allows us to perform distribution analyzes against multiple axes. That is, get different distributions of social data. Figure 15 represents the interface for visualizing the distribution graphs of social data.

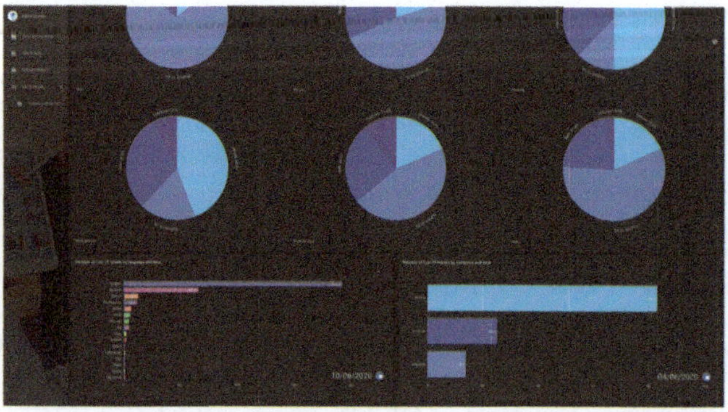

Fig. 15. Social data distribution analysis interface

3) Correlation analysis This aspect of the application offers the user the possibility of comparing two different graphs, to be able to verify the existence of correlation between external covid-19 data uploaded by user and social data. The following figure represents the graphs of a comparative analysis between the variation of the global sentiment emitted on Twitter concerning the "coronavirus", obtained by exploiting the social cube "CubeSentiment", and the variation of the number of deaths, cures and active cases, obtained by exploiting data provided by the users. Figure 16 shows the interface for comparing the variation in customer metric values with the variation in sentiment on Twitter.

6.7 Other Features of Our Tool

1) Analysis of related topics. The analysis of relative subjects is the extraction of all the most redundant terms in the social corpus, and which are considered relevant to the concepts on which the social corpus was built. This functionality does not depend on OLAP technology, but we considered it relevant to integrate it, because it offers the user the opportunity to broaden his field of analysis to new aspects. Figure 17 shows the interface for viewing topics related to a concept extracted from social data using our application.

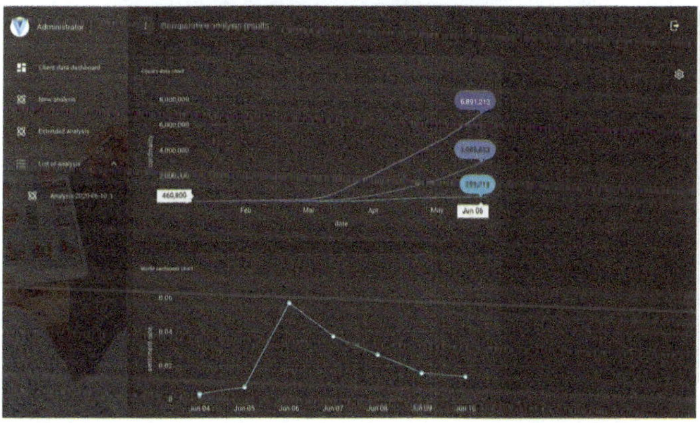

Fig. 16. Analysis interface between the measurements of the user provided data and the social cubes

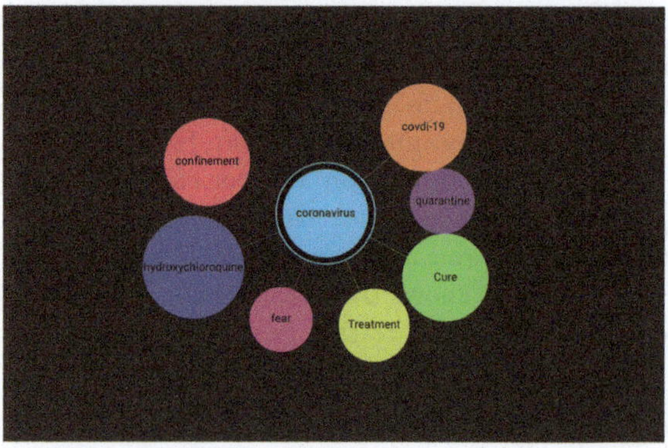

Fig. 17. Analysis interface for related topics extraction

2) Enrichment of an analysis. It is possible at any time for the user to enrich the social corpus of an analysis by performing a new data extraction. The tweets will be processed and added to the set of tweets for the relevant analysis. Figure 18 shows the interface for enriching an analysis.

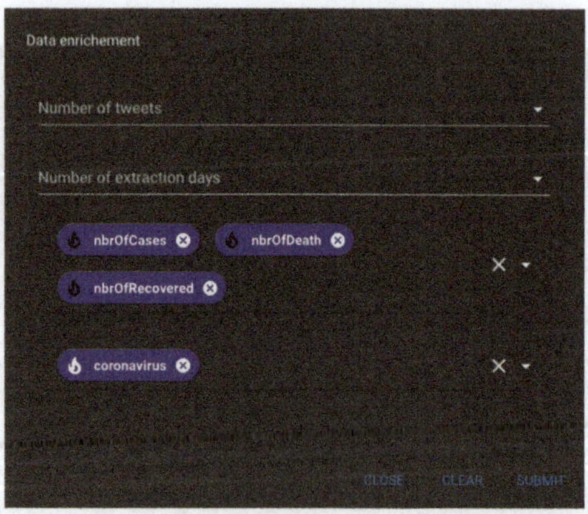

Fig. 18. Data enrichment interface for an analysis

7 Conclusion

The huge share of opinions on social media is undoubtedly an important draw for both businesses and world wide organizations. In an increasingly complex and competitive environment, various institutions have become aware of the importance of exploiting this social data, in order to obtain more powerful decision-making and knowledge extraction tools, allowing them to make strategic decisions. timely. If the stakes largely justify this desire, we must not neglect the technical constraints to be overcome. In this modest work, we have focused on proposing an approach to exploit social data in order to improve decision-making systems. More precisely, data from the social network Twitter. We concluded that providing the ability to apply comprehensive analytics to social data is a critical asset in facilitating decision making, and our tool was designed for that purpose. Indeed, it allows you to perform multidimensional OLAP analyzes on data from Twitter, in order to offer the user several axes of analysis, and thus offer him new perspectives not available in a traditional decision-making system.

The tool we have developed allows the user to perform a social data extraction, either directly from the Twitter API for instantaneous analyses, or from a manually prepared corpus of data that can be enriched by the tool at any time. This second possibility is intended for analyzes of liabilities. Subsequently, the previously constructed social corpus will be used to calculate the values of different measures according to several dimensions, which will be saved at the application data warehouse according to a previously defined scheme. Finally, the data warehouse set up will be used to produce multidimensional social data cubes, which our "Cubes Creator" tool will use to perform a set of several OLAP operations.

Our approach is always subject to improvement, so as perspectives of our work, we propose:

- Implementing a sentiment analyzer using machine learning techniques for better classification.
- Developing an android version of the tool given the emergence of smartphones and tablets
- Using a professional Twitter account that can extract an unlimited number of tweets for more consistent analyses.

References

1. Qiu, W., Rutherford, S., Mao, A., Chu, C.: The pandemic and its impacts. Health Cult. Soc. **9**, 1–11 (2017)
2. Alsaeedi, A., Khan, M.Z.: A study on sentiment analysis techniques of twitter data. Int. J. Adv. Comput. Sci. Appl. **10**, 361–374 (2019). https://doi.org/10.14569/IJACSA.2019.0100248
3. Li, H., Liu, S.-M., Yu, X.-H., Tang, S.-L., Tang, C.-K.: Coronavirus disease 2019 (COVID-19): current status and future perspectives. Int. J. Antimicrob. Agents **55**(5), 105951 (2020)
4. Vinayakumar, R., Alazab, M., Srinivasan, S., Pham, Q.V., Padannayil, S.K., Simran, K.: A visualized botnet detection system based deep learning for the internet of things networks of smart cities. IEEE Trans. Ind. Appl. (2020). in press
5. Tsai, C.-W., Lai, C.-F., Chao, H.-C., Vasilakos, A.V.: Big data analytics: a survey. J. Big Data **2**(1), 1–32 (2015). https://doi.org/10.1186/s40537-015-0030-3
6. Priyanka, K., Kulennavar, N.: A survey on big data analytics in health care. Int. J. Comput. Sci. Inf. Technol. **5**(4), 5865–5868 (2014)
7. Cottle, M., Hoover, W., Kanwal, S., Kohn, M., Strome, T., Treister, N.: Transforming health care through big data strategies for leveraging big data in the health care industry. Inst. Health Technol. Transform. (2013). http://ihealthtran.com/big-data-in-healthcare

8. Sarlan, A., Nadam, C., Basri, S.: Twitter sentiment analysis. Computer Information Science PETRONAS University of Technology, Perak, Malaysia
9. Yigitcanlar, T., et al.: How can social media analytics assist authorities in pandemic-related policy decisions? Insights from Australian states and territories. Health Inf. Sci. Syst. **8**(1), 1–21 (2020). https://doi.org/10.1007/s13755-020-00121-9
10. Manguri, K.H., Ramadhan, R.N., Amin, P.R.M.: Twitter sentiment analysis on worldwide COVID-19 outbreaks. Kurdistan J. Appl. Res. **5**(3), 54–65 (2020)
11. Vijay, T., Chawla, A., Dhanka, B., Karmakar, P.: Sentiment analysis on COVID-19 twitter data. In: 2020 5th IEEE International Conference on Recent Advances and Innovations in Engineering (ICRAIE), pp. 1–7 (2020). https://doi.org/10.1109/ICRAIE51050.2020.9358301
12. Lamsal, R.: Design and analysis of a large-scale COVID-19 tweets dataset. Appl. Intell. **51**(5), 2790–2804 (2020). https://doi.org/10.1007/s10489-020-02029-z
13. Ahmed, W., Vidal-Alaball, J., Downing, J., Seguí, F.L. (2020)
14. COVID-19: Situation update worldwide, as of 5 June 2020 [archive], sur European Centre for Disease Prevention and Control
15. TextBlob: Simplified Text Processing. https://textblob.readthedocs.io/
16. Coronavirus disease (COVID-19): World Health Organization, Situation Report-132 Data as received by WHO from national authorities by 10:00 CEST, 31 May 2020
17. Oracle: WHAT IS A DATA WAREHOUSE?. www.oracle.com/ca-fr/database/what-is-a-data-warehouse
18. Analytics and Business Intelligence. https://www.gartner.com/en
19. Darmont, J., Marcel, P.: Data warehouses and OLAP, analysis and decision in the company. In: CNRS Editions. Big Data Uncovered, pp. 132–133 (2017). 978-2-271-11464-8. ffhal-01493948f
20. Tournier, R.: Online document analysis (OLAP), PhD thesis in computer science, under the direction of Gilles Zurfluh, Toulouse, University of Toulouse (2007)
21. Ben Kraiem, M., Feki, J., Khrouf, K., Ravat, F., Teste, O.: OLAP4Tweets: multidimensional modeling of tweets. In: Morzy, T., Valduriez, P., Bellatreche, L. (eds.) ADBIS 2015. CCIS, vol. 539, pp. 68–75. Springer, Cham (2015). https://doi.org/10.1007/978-3-319-23201-0_9. hal-01343054
22. Liu, X., et al.: SocialCube: a text cube framework for analyzing social media data. In: 2012 International Conference on Social Informatics, Lausanne, 2012, pp. 252–259 (2012). https://doi.org/10.1109/SocialInformatics.2012.87. emergence and transmission. Pathogens. 2016;5(4):66

Applying Artificial Intelligence Techniques for Predicting Amount of CO_2 Emissions from Calcined Cement Raw Materials

Yakoub Boukhari(✉)

Chemistry Department, Ziane Achour University, Djelfa, Algeria
jacoubchimie@yahoo.fr

Abstract. This paper aims to predict the amount of carbon dioxide CO_2 emissions from raw material used in cement clinker production during calcination process. The amount of CO_2 emissions is mainly from the decarbonisation thermal process that is directly related to chemical composition, distribution of particle size and time exposed at high temperature. These influencing factors interact with each other, making the calculation of the amount of CO_2 emissions with conventional techniques more difficult. For this reason, several artificial intelligence techniques are applied to predict the amount of CO_2 emissions. The key advantage of the proposed techniques is its ability to learn and to generalise without any prior knowledge of an explicit relationship between target and its influencing parameters. The intelligence techniques applied are deep neural network (DNN), artificial neural networks (ANN) optimised using ant colony optimization (ACO-ANN) and genetic algorithm (GA-ANN).

The results obtained are promising and show that all intelligence techniques can provide excellent accuracy with high R^2 and low error. DNN predicts the amount of CO_2 emissions very accurately when comparing to other techniques. Overall, the performance accuracy of ACO-ANN technique is higher than the GA-ANN. According to R_2 values, there are more than 99%, 98.5% and 98% of experimental data in testing phases can be explained by DNN, ACO-ANN and GA-ANN respectively with average relative error less than 1.04%. As conclusion, all intelligence techniques can be employed as an excellent alternative to predict the amount of CO_2 emissions.

Keywords: CO_2 emissions · Calcination process · Deep neural network · Artificial neural networks · Ant colony optimization · Genetic algorithm

1 Introduction

The cement industry is one of the most important heavy industry sectors in the world. It is a major source of CO_2 emissions. In terms of greenhouse gas emissions, it contributes approximately 5 to 7% of global CO_2 emissions [1]. Total emissions of carbon dioxide CO_2 from cement manufacturing is mainly due to the combustions of fuels in a rotary kiln and the calcinations of raw materials at high temperature [2] during cement production. During cement production, about 62% of CO_2 is released from calcination process [3]

B. Lejdel et al. (Eds.): AIAP 2021, LNNS 413, pp. 231–243, 2022.
https://doi.org/10.1007/978-3-030-96311-8_21

of raw materials. The raw materials used to produce cement are generally composed of the following main oxides CaO, SiO_2, MgO, Al_2O_3 and Fe_2O_3. These oxides play an important role in determining the quality of cement products [4]. The amount of CO_2 emissions is an important indicator of control in cement production. It is affected by many factors which are mainly on chemical composition, powder fineness and burning time.

The emission process of CO_2 is still not fully clarified and explained. It also is complicated phenomena [5] which is difficult to express analytically. Moreover, the mathematic relationships between the influencing parameters and the amount of CO_2 emissions is not precisely known. It is well known that the laboratory experiments are generally difficult, complicated, costly and require number of chemical reagents and equipments. Therefore, using artificial intelligence techniques are required to predict the amount of CO_2 emissions and are important now more than ever. These techniques do not need to specify mathematical relationships or prior knowledge about the relationship between a different influencing factors and the amount of CO_2 emissions.

For decades, artificial intelligence techniques are extensively used in many fields to predict the behavior of complex systems [6, 7]. In this study, several artificial intelligence tools are used to predict the amount of CO_2 emissions from cement raw materials with different chemical composition and various fineness in the range of time. The different artificial intelligence techniques proposed in the present paper are deep neural networks (DNN), artificial neural network optimised using ant colony optimization (ACO-ANN) and ANN optimised using genetic algorithm (GA-ANN).

Currently, artificial intelligence techniques are exploited in various fields because they help to transform traditional industry towards the real intelligent industry. There are plenty of successful application of DNN in many fields such as prediction of reservoir production [8] and for predicting drug-target interaction [9]. Hybrid ACO-ANN is utilised as useful tool to model the diesel engine emissions [10] and to predict the capital cost of mining projects [11]. Hybrid GA-ANN is successfully applied to predict turbidity and chlorophyll-a variations [12] and sheet metal forming [13].

The present paper is arranged as follows: brief description of artificial intelligence techniques are presented in Sect. 2; materials and methods are established in Sect. 3; the Sect. 4 discuss the results obtained; Sect. 5 is reserved for our conclusions.

2 Brief Description of Artificial Intelligence Techniques

2.1 Deep Neural Networks (DNN)

Deep neural network which imitates the human neural system is powerful predictive technique [14]. It is characterized by a more complicated structure than conventional artificial neural networks. The DNN performance is significantly influenced by the pre-training and training stages. One amongst the several pre-training algorithms and straining are Autoencoder AE [15] and Softmax function [16] respectively. DNN proposed is created by stacking autoencoder with a softmax layer (SL). Autoencoder and softmax are first trained individually and separately, one layer at a time. The AE is trained using the initial inputs where softmax layer is trained with the output of the AE layer. Then AE and SL are joined together to form DNN. Finally, to improve the accuracy, DNN is fine-tuned using back propagation learning algorithm.

Autoencoder AE proposed for pre-training of DNN is a kind of an unsupervised learning ANN with three layers: input layer, hidden layer and output layer. It is used to extract hidden features and reconstruct the input of DNN in order to build suitable representation from [17]. The learning process of AE is self-supervised by raw inputs without a class label or outputs. It consists of two processes of encoding and decoding. The encoder is used to compress input vector and a decoder is used to extend the compressed input for reconstructing and extracting the essential features. The new reconstruction data are built according the activation function chosen by both encoder and decoder. The reconstruction error is calculated through cost functions. The network weights and bias are iteratively updates to converge cost function in the AE towards a minimum. The cost function used for training AE is MSE adjusted by adding a regularization term (L_2 regularization term and sparsity regularization) [18] to prevent over-fitting phenomena during learning process. In order to ensure a perfect accuracy and good reconstruction data, scaled conjugate gradient descent is used to train AE [19].

Softmax layer which applies a softmax function is added to autoencoder as supervised learning to produce good solutions for some nonlinear problems and especially to solve multiple classification problems. It is characterised by simple structure and easy integration. Similarly to AE, the scaled conjugate gradient algorithm is used by SL for training. The goal of supervised learning is minimise the cost function by updating weight and bias parameters iteratively. The loss function used by softmax layer is cross entropy (log loss) [20].

2.2 Artificial Neural Network Optimised by Ant Colony Optimization (ACO-ANN)

The possibility of falling into local minimum problems without obtain a globally optimal solution and slow convergence speed to optimal points of ANN is inevitable. To overcome shortcomings of ANN, ant colony optimization (ACO) is combined with ANN to optimise its parameters for producing powerful technique called ACO-ANN.

Ant Colony Optimization (ACO) algorithm is an optimization tool that is mainly used to optimise and to solve different optimisation problems. It is inspired by the behavior of ants. In nature, ants is able to search for an optimal and shortest trajectory (edge) between their nest and food source. Each ant moves randomly through all to find the food source. When ant find a food leave behind a chemical substance on the ground, called pheromone. The quantity, the quality and the distance of the food source are related to the quantity of the laid pheromone. The pheromone evaporation of pheromone [21] over the time plays an important role to fall into the local optimum and to avoid the rapid convergence to a local optimal. Without evaporation of pheromone, the first paths selected by the first ant is extremely attractive to the other ants, and consequently, the research space could be limited. The path which contained more quantity of pheromone is becoming the main path. In each iteration with sufficient number of ants, the quantity of pheromone is updated and reinforced which help to attain more precise and optimal solutions.

The main advantage of ACO algorithm is to converge rapidly to optimal solutions without falling into a local optimum. The optimal performance of the ACO mainly depends on suitable parameters settings. The main parameters of ACO are number of ants per iteration, number iterations and evaporation rate of pheromone.

Ant colony optimization algorithm is mainly used to train and optimise the weights and biases of ANN. The risk of getting stuck in local optima is sharply reduced by ACO algorithm [22]. The ACO combined with ANN on the basis of considering that BP neural networks have strong predictive ability and that the ant colony optimization has good ability of searching for optimisation.

2.3 Artificial Neural Network Optimised by Genetic Algorithm (GA-ANN)

Like the ACO-ANN technique, genetic algorithms GA is used as powerful optimizing tool to optimise and select the optimal parameters of ANN, called GA-ANN technique [23].

Genetic algorithm is one of the most important and popular optimization algorithms. It is much more frequently employed for solving the most optimization problems by finding the optimal values of a function [24]. The main idea of GA is to imitate natural ways of evolution. The three evolutionary principles of genetic operators used by GA algorithms are crossover, mutations, natural selection.

The initial population of n individuals (solutions) is randomly generated. Each individual is characterised their genes. The crossover operator is used for combining genetic information (solutions) of two parents to generate two new offspring. Then, the genetic operators continued with the mutation operator to introduce and maintain diversity from one generation of a population to the next one. The idea of selection is to choose two parents from the previous generation and let them pass their genes to the next population. The process is repeated to produce better solutions in each new iteration until reachs the termination condition [25]. The main parameters of GA are number of population size per generation, number generations (iterations), crossover probability and mutation probability.

3 Material and Methods

3.1 Materials and Experiments

In the present study, the amount of CO_2 emission is considered as a function of chemical composition, grain size and time exposed. The raw materials are blended and preheated to round 300 °C to remove water combined in the hydration products and then up to 850 °C to remove impurities, which can affect the cement quality.

The grain size selected are set in four sizes 71 μm, 125 μm, 250 μm and 350 μm. The chemical composition and mix proportions of four raw materials used are summarised as shown in Table 1. Finally, each mixture of raw materials with gain size are burned in the laboratory furnace at 1000 °C for different time 5, 10, 15, 20, 30 min. The amount of CO_2 emissions from each mixture of raw materials is calculated before and after burning at 1000 °C.

Table 1. Chemical composition (% by weight) for each raw

Raw materials	Oxides				
	SiO_2	CaO	MgO	Fe_2O_3	Al_2O_3
Material 1	12,38	80,28	1,38	1,69	4,27
Material 2	3,96	92,62	0,99	0,65	1,78
Material 3	14,06	78,69	1,35	1,68	4,22
Material 4	14,16	78,04	1,36	2,21	4,23

3.2 Data Collection

The data extracted from the experimentation are collected in table of 80 rows and 8 columns. Each row in table presents experiment. From 1 to 7 columns are inputs where the last column is output. The size particle, time exposed, SiO_2(%), CaO (%), MgO (%), Fe_2O_3 (%), Al_2O_3 (%) are inputs while the amount of CO_2 emissions is the output.

The total data are randomly divided into two sets: training and testing. For each intelligent techniques, 75% of data are used for training while the remaining 25% (not unseen data) of data are kept out to evaluate the generalisation ability. The most common performance criterion used to evaluate the accuracy of each intelligent techniques are the coefficient of determination R^2 and the mean absolute percentage error (MAPE). The technique performance is perfect when values of R^2 and MAPE are very close to 1 and 0, respectively.

4 Results and Discussion

First of all, finding the appropriate parameter settings values of each models play a key role in achieving highest prediction accuracy [26] and in avoiding over-fitting and under-fitting. Often, it is not easy to select these parameters. Consequently, the appropriate parameters of each models are obtained after several tests during training process.

4.1 DNN Results

The main appropriate parameters values of the autoencoder and softmax layers are predefined as listed in Table 2.

Table 2. DNN parameters

Parameters	Values
L_2 weight regularization	0.01
Sparsity regularization	4

(continued)

Table 2. (*continued*)

Parameters	Values
Sparsity proportion	0.05
Number of hidden layer	1
Neuron number	18
Encoder transfer function	logsig
Decoder transfer function	purelin
Iteration number of AE	1050
Loss function of softmax layer	Crossentropy
Iteration number of softmax layer	1250

Figure 1 illustrates comparison between experimental and predicted amount of CO_2 emissions obtained by DNN.

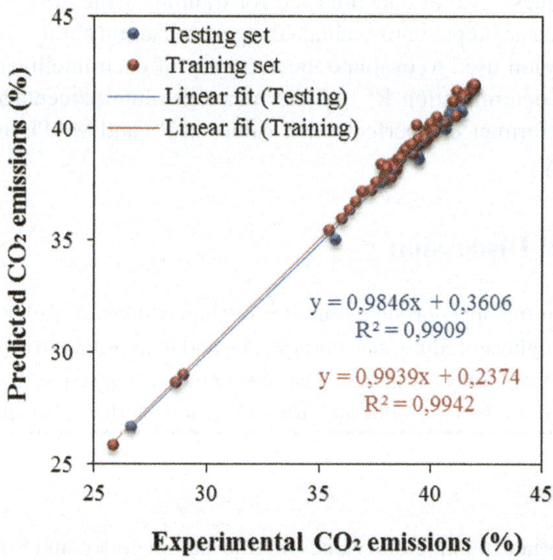

Fig. 1. The comparison between the predicted and experimental CO_2 emissions.

It is observed that almost of points for both phases fall exactly along the linear fit which express perfect equality between experimental and predicted CO_2 emissions. From Fig. 1, the values of R^2 calculated for training and testing phases are 0.9942 and 0.9909 respectively. The According to values of R^2, there is a high degree of association between experimental and predicted values without overfitting or underfitting problems.

Distribution of the relative error of CO_2 emissions for DNN during training and testing phases are given in Fig. 2. This figure illustrates clearly that almost of points are

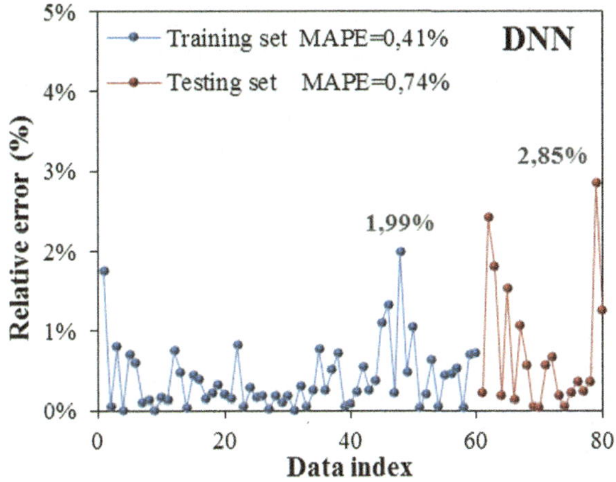

Fig. 2. Distribution of the relative error

closest to zero line. The maximum relative error obtained in training and testing phases are 1.99% and 2.85%, respectively. The average relative error for training and testing phases are 0.41% and 0.74% respectively. These values are indicative of high accuracy in prediction and generalisation.

It is clear that DNN is able to provide excellent performance in both phases, since the values of MAPE and R^2 are near to 0 and 1, respectively. The perfect performance of DNN is due to AE layer which is able to extract the essential information efficiently from data and excellent predicted probability distribution via softmax layer. From the results obtained, it can be concluded that DNN is useful technique.

4.2 ACO-ANN Results

The optimal parameters obtained which leads to good results is summarised in Table 3.

Table 3. ACO-ANN parameters

Parameters	Values
Number of hidden layer	1
Neuron number in hidden layer	12
Transfer function of the hidden layer	transig
Transfer function of the output layer	purelin
Numbers of ants	18
Number of iterations	12
Evaporation rate	0.2
Pheromone concentration	0.8

The performance of the ACO-ANN is shown in Fig. 3 presenting the comparison between the predicted and experimental amount of CO_2 emissions. As a result, the R^2 value obtained in training phase is 0.9890 while the R^2 value in testing phase is 0.9876. The predicted and experimental CO_2 emissions are highly correlated with R^2 closer to unity. It is evident that, ACO-ANN is capable to predict most of the testing and training points closely.

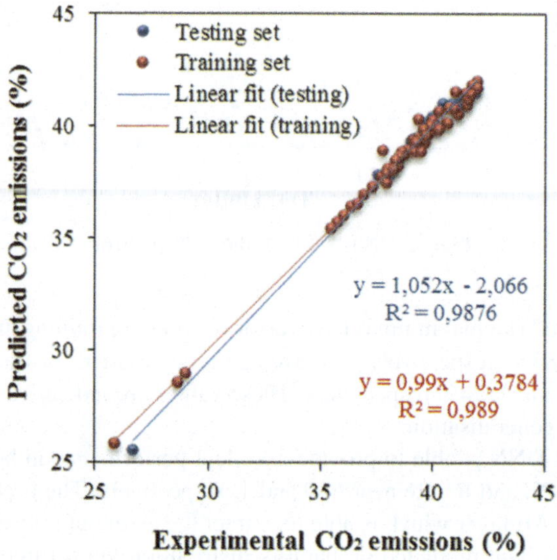

Fig. 3. The comparison between the predicted and experimental CO_2 emissions

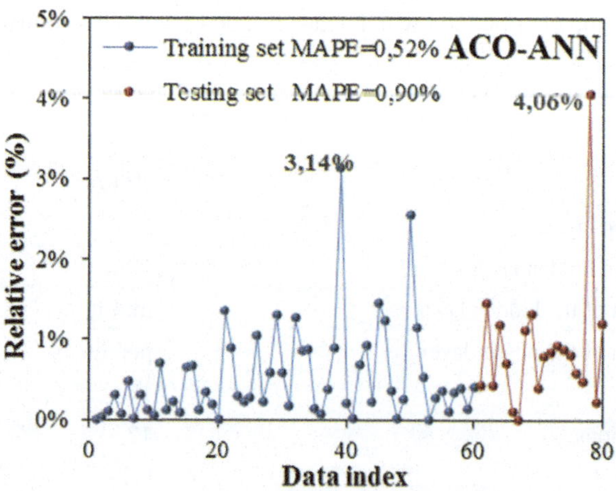

Fig. 4. Distribution of the relative error

Figure 4 show the distribution of the relative error of amount of CO_2 emissions for ACO-ANN during training and testing phases.

As is clearly seen in the Fig. 4, the average relative error for training and testing phases are 0.52% and 0.9%, respectively indicating the ability of ACO-ANN to find suitable predicted values. Additionally, the relative error does not exceed an error of 3.14% for training phase and 4.06% for testing phase. The performance obtained is reasonable due to the flexibility of ANN in solving complex function and ability of ACO algorithms to optimise initial parameters of ANN.

4.3 GA-ANN Results

It should be noted that optimal architecture of ANN used by ACO-ANN and GA-ANN are the same, which is used mainly for comparisons purposes. The optimal parameters of GA-ANN is listed in Table 4.

Table 4. GA-ANN parameters

Parameters	Values
Number of hidden layer	1
Neuron number in hidden layer	12
Transfer function of the hidden layer	transig
Transfer function of the output layer	purelin
Population size	15
Number of iterations	20
Probability of selecting the best	0.09

Figure 5 presents the comparison between the predicted and experimental amount of CO_2 emissions for two phases. As shown in Fig. 5, the GA-ANN can offer good prediction for experimental amount of CO_2 emissions with high values of R^2. However, the R^2 value of the training data is equal to 0.9865. While the R^2 value of the testing data is 0.9834. These values of R^2 indicate a strong linear relationship between the experimental and predicted amount CO^2 emissions during both phases.

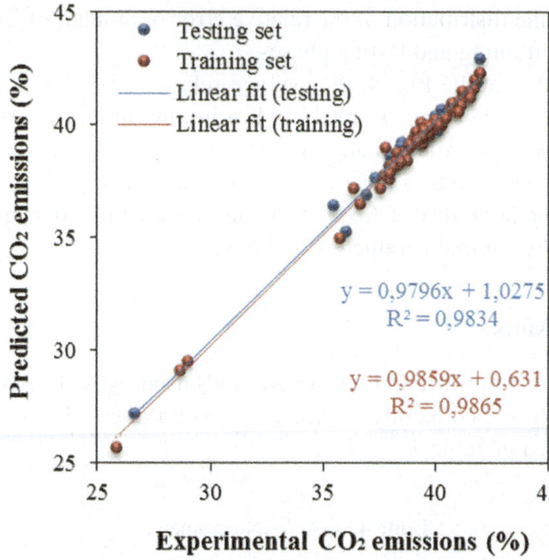

Fig. 5. The comparison between the predicted and experimental CO_2 emissions

Figure 6 a shows the distribution of the relative error between experimental and predicted amount of CO_2 emissions obtained by GA-ANN during training and testing phases. From Fig. 6, highest error values of 3.21% and 2.83% are observed in training and testing phases, respectively. The values of MAPE for training and testing phases are 0.71% and 1.04%, respectively. The result obtained from the figure, it still reflects a suitable performance of GA-ANN. The high accuracy of GA-ANN is mostly due to great optimisation of GA and the particularity of ANN in solving non-linear system.

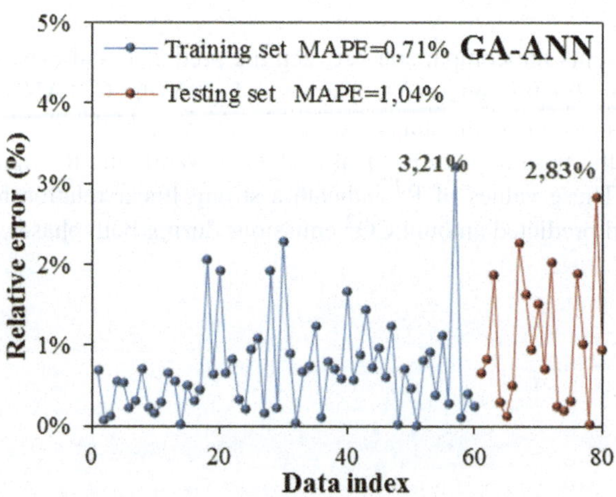

Fig. 6. Distribution of the relative error

4.4 Comparison Between Different Techniques

The results obtained by DNN, ACO-ANN and GA-ANN during testing phase are compared based on generalisation accuracy. The common performance criterion used to determine the performance are R^2 and MAPE. A comparison of the results obtained from DNN, ACO-ANN and GA-ANN techniques briefly shows in Fig. 7.

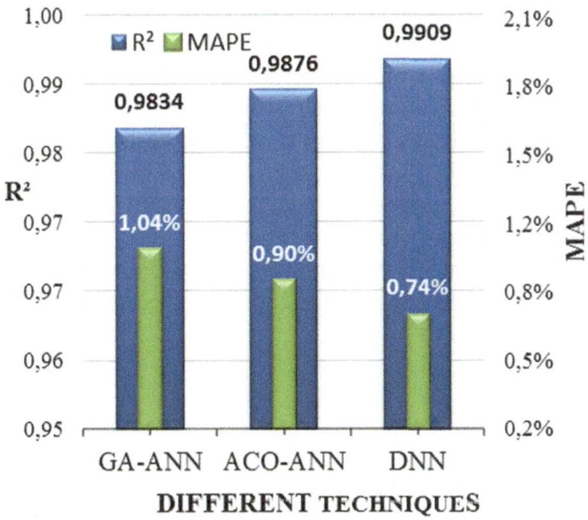

Fig. 7. The comparison between DNN, ACO-ANN and GA-ANN results

The Fig. 7 clearly illustrate that all techniques proposed could provide the average relative error less than 1.04% and R^2 more than 0.98. The DNN technique has best generalisation accuracy in a comparison of the other techniques with highest R^2 and lowest MAPE. Whereas, lowest average accuracies at 1.04% is provided by GA-ANN. in addition, the results obtained demonstrate that the DNN is more advanced than ACO-ANN in terms of generalisation accuracy followed by GA-ANN (Table 5).

Table 5. Other performance indices used

Parameters	GA-ANN	ACO-ANN	DNN
R^2 adjusted	98,25%	98,70%	99,04%
RMSE	0,4857	0,4080	0,3835

The R_2 adjusted values indicates that 99.04%, 98.7 and 98.25% of data can be explained by DNN, ACO-ANN and GA-ANN respectively. Moreover, values of root mean square error RMSE are very close to 0 mean a higher performance accuracy and the predicted amount of CO_2 emissions are very close to the real data.

5 Conclusion

In the current study, the percentage amount of CO_2 emissions from raw materials used in cement clinker production is predicted by the DNN, ACO-ANN, GA-ANN techniques. These techniques are trained and tested using 75% and 25% of experimental data, respectively. The performance of intelligence techniques is judged by R^2 and MAPE as performance criterion.

As a conclusion, the DNN is stamped as superior technique for the prediction of amount of CO_2 emissions with lowest MAPE and highest R^2. The performance accuracy of ACO-ANN technique is higher than the GA-ANN. Each artificial intelligence techniques can provide an useful and interesting techniques to predict the amount of CO_2 emissions produced without needing to actually do the experiment and conventional computational methods. Moreover, it can be used as an alternative technique to calculate the amount of CO_2 emissions. The average relative error obtained by DNN, ACO-ANN AND GA-ANN in testing phase are 0.74%, 0.90% and 1.04%, respectively. According to R^2 adjusted values, there are 99.04%, 98.7 and 98.25% of experimental data can be explained by DNN, ACO-ANN and GA-ANN respectively with very small error.

References

1. Possan, E., Thomaz, W.A., Aleandri, G.A., Felix, E.F., dos Santos, A.C.P.: CO_2 uptake potential due to concrete carbonation: a case study. Case Stud. Constr. Mater. **6**, 147–161 (2017)
2. Ali, M.B., Saidur, R., Hossain, M.S.: A review on emission analysis in cement industries. Renew. Sust. Energ. Rev. **15**, 2252–2261 (2011)
3. Deja, J., Bochenczyk, A.U., Mokrzyck, E.: CO_2 emissions from Polish cement industry. Int. J. Greenh. Gas Control **4**, 583–588 (2010)
4. Cao, Z., Shen, L., Zhao, J., Liu, L., Zhong, S., Yang, Y.: Modeling the dynamic mechanism between cement CO_2 emissions and clinker quality to realize low-carbon cement. Resour. Conserv. Recycl. **113**, 116–126 (2016)
5. Summerbell, D.L., Barlow, C.Y., Cullen, J.M.: Potential reduction of carbon emissions by performance improvement: a cement industry case study. J. Clean. Prod. **135**, 1327–1339 (2016)
6. Boukhari, Y.: Using intelligent models to predict weight loss of raw materials during cement clinker production. Rev. d'Intelligence Artif. **34**, 101–110 (2020)
7. Abubakar, A.M., Behravesh, E., Rezapouraghdam, H., BahaYildiz, S.: Applying artificial intelligence technique to predict knowledge hiding behavior. Int. J. Inf. Manag. Sci. **49**, 45–57 (2019)
8. Kim, J., Kim, S., Park, C., Lee, K.: Construction of prior models for ES-MDA by a deep neural network with a stacked autoencoder for predicting reservoir production. J. Pet. Sci. Eng. **187**, 106800 (2020)
9. You, J., McLeod, R.D., Hu, P.: Predicting drug-target interaction network using deep learning model. Comput. Biol. Chem. **80**, 90–101 (2019)
10. Mohammadhassani, J., Dadvand, A., Khalilarya, S., Solimanpur, M.: Prediction and reduction of diesel engine emissions using a combined ANN–ACO method. Appl. Soft Comput. **34**, 139–150 (2015)

11. Zhang, H., et al.: Developing a novel artificial intelligence model to estimate the capital cost of mining projects using deep neural network-based ant colony optimization algorithm. Resour. Policy **66**, 101604 (2020)
12. Mulia, I.E., Tay, H., Roopsekhar, K., Tkalich, P.: Hybrid ANN–GA model for predicting turbidity and chlorophyll-a concentrations. J. Hydro-Environ. Res. **7**, 279–299 (2013)
13. Liu, W., Liu, Q., Ruan, F., Liang, Z., Qiu, H.: Springback prediction for sheet metal forming based on GA-ANN technology. J. Mater. Process. Technol. **187–188**, 227–231 (2007)
14. Katuwal, R., Suganthan, P.N.: Stacked autoencoder based deep random vector functional link neural network for classification. Appl. Soft. Comput. **85**, 105854 (2019)
15. Feng, S., Zhou, H., Dong, H.: Using deep neural network with small dataset to predict material defects. Mater. Des. **162**, 300–310 (2019)
16. Kannadasa, K., Damodar, R.E., Venkatanareshbabu, K.: Type 2 diabetes data classification using stacked autoencoders in deep neural networks. Clin. Epidemiology. Glob. Health **7**, 530–535 (2019)
17. Vincent, P., Larochelle, H., Lajoie, I., Bengio, Y., Manzagol, P.-A.: Stacked denoising autoencoders: learning useful representations in a deep network with a local denoising criterion. J. Mach. Learn. Res. **11**, 3371–3408 (2010)
18. Olshausen, B.A., Field, D.J.: Sparse coding with an overcomplete basis set: a strategy employed by V1? Vis. Res. **37**, 3311–3325 (1997)
19. Moller, M.F.: A scaled conjugate gradient algorithm for fast supervised learning. Neural Netw. **6**, 525–533 (1993)
20. Angelob, G.D., Ficcoa, M., Palmierib, F.: Malware detection in mobile environments based on Autoencoders and API-images. J. Parallel Distrib. Comput. **137**, 26–33 (2020)
21. Jovanovica, R., Tubab, M., Voß, S.: An efficient ant colony optimization algorithm for the blocks relocation problem. Eur. J. Oper. Res. **274**, 78–90 (2019)
22. Liu, Y.P., Wu, M.G., Qian, J.X.: Predicting coal ash fusion temperature based on its chemical composition using ACO-BP neural network. Thermochim. Acta **454**, 64–68 (2007)
23. Beltramoa, T., Klockeb, M., Hitzmanna, B.: Prediction of the biogas production using GA and ACO input features selection method for ANN model. Inf. Process. Agric. **6**, 349–356 (2019)
24. Samira, G., Jarek, K., Yousef, M.: A hybrid Genetic Algorithm and Monte Carlo simulation approach to predict hourly energy consumption and generation by a cluster of Net Zero Energy Buildings. Appl. Energy **179**, 626–637 (2016)
25. Ghorbani, B., Arulrajah, A., Narsilio, G., Horpibulsuk, S., Bo, M.W.: Development of genetic-based models for predicting the resilient modulus of cohesive pavement subgrade soils. Soils Found. **60**, 398–412 (2020)
26. Boukhari, Y., Boucherit, M.N., Zaabat, M., Amzert, S., Brahimi, K.: Optimization of learning algorithms in the prediction of pitting corrosion. J. Eng. Sci. Technol. **13**, 1153–1164 (2018)

Local Directional Strength Pattern for Effective Offline Handwritten Signature Verification

Naouel Arab[1(✉)], Hassiba Nemmour[2], and Yousef Chibani[1,2]

[1] Laboratoire d'Ingénierie des Systèmes Intelligents et Communicants (LISIC), University of Sciences and Technology Houari Boumediene (USTHB), Algiers, Algeria
{narab1,ychibani}@usthb.dz

[2] Faculty of Electronics and Computer Science, University of Sciences and Technology Houari Boumediene (USTHB), Algiers, Algeria
hnemmour@usthb.dz

Abstract. Offline handwritten signature verification is one the oldest and widespread biometric identification tools. In various daily life applications, the handwritten signature is used for documents approval and identity verification. Systems that automatically achieve this task are mainly composed of feature generation and verification modules. The more features are robust the best the verification score is. Presently, we introduce the use of the Local Directional Strength Patterns (LDSP) for handwritten signature characterization. This descriptor is associated with SVM classifier to perform the signature verification. Experiments conducted on two public datasets reveal the effectiveness of the proposed descriptor which outperforms several state-of-the-art techniques.

Keywords: Signature verification · Local directional strength pattern · Support vector machines · Local directional pattern

1 Introduction

Biometric systems based on offline handwritten signatures analysis are widely required in applications using document papers such as official contracts, transactions, and bank checks. In addition to being well accepted by society as a biometric identification tool, another reason behind the wide use of signatures is the cheap cost of developing Signature Verification Systems (SVS), which benefit from a long experience in forensics domain. Roughly, the offline handwritten verification system can be developed according to the writer dependent approach or the writer-independent approach. In the first case, an SVS is developed for each writer, whereas the second approach consists of developing one SVS for all writers. This makes the writer-dependent verification more effective since learning the signatures of a specific writer is much simpler than learning the signatures of a set of writers. For both approaches, the SVS is composed of two modules which perform respectively, features generation and verification. The state of the art reports a huge variety of methods proposed for the two stages. Regarding the verification task, machine learning techniques such as convolutional neural networks and support vector machines

B. Lejdel et al. (Eds.): AIAP 2021, LNNS 413, pp. 244–252, 2022.
https://doi.org/10.1007/978-3-030-96311-8_22

which are the most qualified to detect the forged signatures from the original ones [1–8]. Furthermore, various kinds of descriptors were utilized to generate features over the past years. Among them one can note mathematical transforms such as Wavelets, Contourlets, Curvelets, and Ridgelets which provide static information depicting the signature shape [4–6, 9]. Experiments conducted on various datasets revealed the medium performance of these transforms. That is what led researchers to focus on gradient and texture descriptors to extract pseudo dynamic features. In this respect, Histogram of oriented gradients, Scale invariant features transforms Local Binary Patterns (LBP) and then, Local Directional Patterns showed interesting results on several offline signature datasets [7, 8, 12–14]. Besides, some shape descriptors such as the Histogram of templates and Run length features showed very attractive scores [7, 15, 16]. On the other hand, some research works employed Convolutional Neural Networks (CNN) as writer-independent features generators [1–3]. However, a large amount of labeled data is required to derive robust features, whereas handwritten signature datasets contain few samples for each writer. Therefore, data augmentation techniques are commonly used to satisfy this training constraint. On the other hand, huge deep models such as VGG-Net and Res-Net induce a high computation cost, which doesn't automatically lead to higher performance than handcrafted features.

Roughly speaking, results collected on public datasets are sub-optimal and current SVS have not yet achieved the desirable performances. Therefore, researchers focus on combining multiple SVS or on developing more robust features, which can ensure the best tradeoff between intra-writer variability and inter-writer variability. In this respect, some new features were recently proposed such as the Local Difference Features [17–19]. Presently, we propose the Local Directional Strength Pattern (LDSP) as a new handcrafted descriptor for offline handwritten signature characterization. LDSP is an improved implementation of LDP that reduces the histogram size to only six bits highlighting the dominant orientation of the shape edges in the pixel vicinity [20, 21]. To achieve the verification task, LDSP is associated with the Support Vector Machines (SVM classifier. Experiments are performed on two public datasets that are CEDAR, and MCYT-75.

The remaining of this paper is organized as follows: Sect. 2 presents the proposed SVS based on LDSP features. Section 3 details the experimental evaluation of the proposed system, and the last section gives the main conclusions.

2 Methodology

The proposed SVS is developed by associating LDSP features with SVM classifier. The training process is conducted according to the writer dependent protocol, where each writer enrolled in the system has its own SVM. So, the verification task is a binary classification in which, genuine signatures of the considered writer compose the first training class, while random forgeries (genuine signatures of other writers) compose the second training class. The signature verification stage is carried out by using all remaining genuine signatures and imitated signatures (That are called skilled forgeries) of the considered writer.

Recall that LDSP is an improved implementation of the LDP, which considers relative edge responses in eight directions [14].

2.1 Local Directional Strength Patterns

The key element in Local Directional Patterns (LDP) computation is the use of kirsch detector to highlight image edges. The, LDP calculation begins by applying the Kirsch edge detector to extract edge images in eight directions. Then, the maximal edge values for each pixel are considered to derive a single maxima edge image. Each pixel neighborhood undergoes a local binary coding by attributing 1 to the K highest values while putting the others to 0 (K is a user-defined parameter). Derived codes composed of height digits which are converted into decimal values to elaborate the LDP code.

In order to reduce the histogram size and allow a more consistent characterization, the Local Directional Strength Pattern has been introduced [20, 21]. Specifically, LDSP modifies the coding of the maximal edge values image. First, a three-bit code is assigned for each edge orientation in the pixel surrounding. Then, codes of both maximal and minimal edge values are concatenated to compose a six-component sequence descriptor. More precisely, the LDSP calculation for negative image goes through the following steps:

1. Apply kirsch detector and generate the Maxima edge image.
2. Considering pixel vicinity, assign a three-bit binary code going from 000 to 111for each neighbor as shown in Fig. 1.c.
3. LDSP code corresponds to the concatenation of binary codes of both maximal and minimal values.
4. LDSP descriptor of the full image corresponds to the histogram of all LDSP codes.

Figure 2 illustrates the LDSP computed for a signature image.

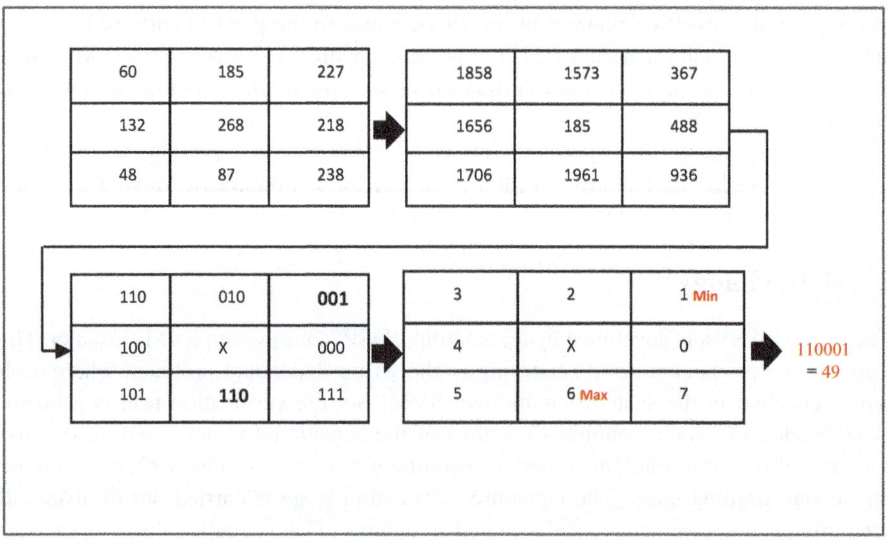

Fig. 1. LDSP computation for a central pixel

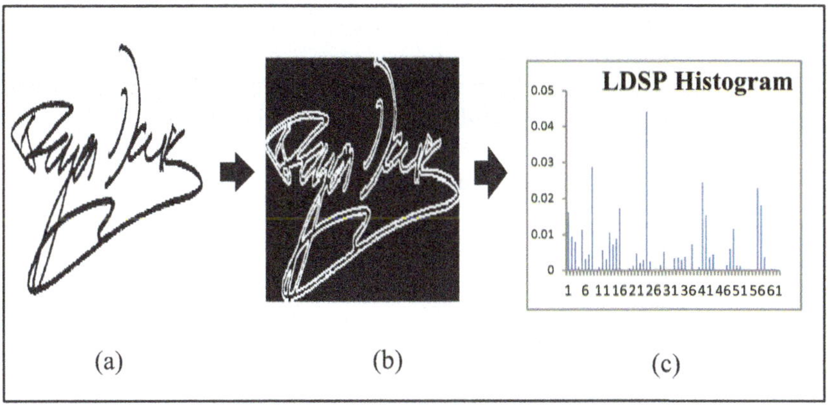

Fig. 2. LDSP histogram for a signature image: (a) Signature image, (b) Maxima edge image, (c) LDSP histogram.

2.2 Support Vector Machine

To achieve the verification task, which is a binary classification problem, Support Vector Machines classifier is utilized. This is a statistical learning theory method that is originally designed to solve binary classifications [22, 23]. The training process aims to find the optimal linear separation between to classes. Let consider a learning set $\{(x_i, y_i)\}$, where: $I - 1, \ldots, n$. x_i represent the training samples, while $y_i \in \{-1, +1\}$ are the class labels.

The decision function is given by the following equation:

$$F(x) = sign(\sum_{l=1}^{Sv} \alpha_i K(x_i, x) + b) \quad (1)$$

α_i: Lagrange multiplier of x_i.
S_v: Number of support vectors that are x_i having non-zero α_i.
b: Bias of the decision function.
K: Kernel function.

In this work, the Radial Basis Function kernel is utilized. This kernel is depicted in Eq. (2).

$$RBF(x_i, x) = exp(\|x_i - x\|2/\delta 2) \quad (2)$$

γ: User-defined parameter.

3 Experimental Results

To evaluate the performance of the LDSP-SVM verifier, two offline handwritten signature datasets were used in experiments:

- CEDAR corpus: it is containing 55 signers each of which has 24 genuine signatures and 24 skilled forgeries signatures.
- MCYT-75 corpus: it is composed of signatures belonging to 75 individuals. There are 15 genuine signatures and 15 skilled forgeries for each individual.

Figure 3 depicts some samples from adopted datasets.

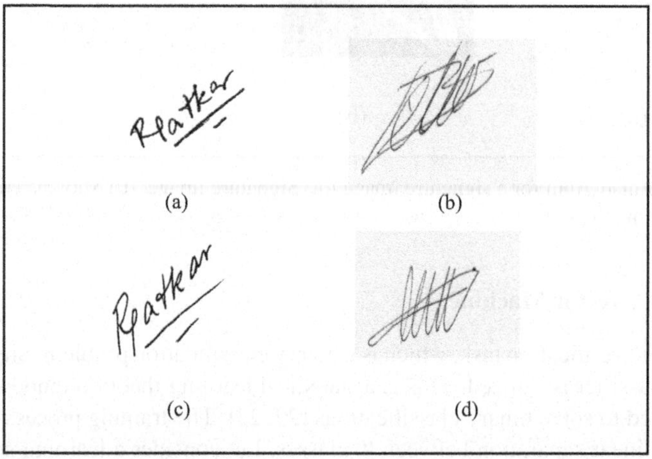

Fig. 3. Signature samples from adopted datasets. Genuine samples: (a) CEDAR, (b) MCYT 75. Forgeries: (c) CEDAR, (d) MCYT-75.

The performance assessment is based on three error types. The False Acceptance Rate (FAR) which is the percentage of skilled forgeries accepted as genuine by the system, the False Rejection Rate (FRR), which are genuine signatures considered as forgeries by the system, and the Average Error Rate (AER). Additionally, DET curves are utilized. For each writer, experiments were carried out by using two training sets. The first set contains 10 genuine signatures to train the SVM, while in the second test, a more challenging protocol is adopted since the number of training signatures is limited to 5 samples. In both experiments, all remaining genuine signatures as well as all skilled forgeries are used in the verification test. Error rates collected in these experiments are reported in Table 1.

Table 1. Error rates (%) obtained using two training sets.

Dataset	10 genuine signatures			5 genuine signatures		
	FRR	FAR	AER	FRR	FAR	AER
CEDAR	10.25	05.07	07.66	10.33	10.15	10.24
MCYT-75	13.60	03.82	8.71	13.73	8.62	11.17

In overall, the number of training signatures has a substantial influence on the FAR score, which jumps to a difference of 5% when the number of training signatures is reduced to 5. This outcome reveals that using more training samples, leads to more robust modeling of the signer's traits, and helps the system to make a better detection of imitated signatures. On the contrary, quite similar FRR scores are collected for the two sets. So, as expected, better AER scores are obtained for the largest training set. The improvement varies between 2.46%, and 2.58%, which indicates similar behavior for the two datasets.

Furthermore, DET curves highlight the effectiveness of the proposed SVS, since satisfactory Equal Error Rates (EER) calculated. These findings are proven when comparing the proposed system with the state-of-the-art results. As reported in Table 2, the LDSP provides the best tradeoff between accuracy and data size (Fig. 4).

Table 2. State of the art results of adopted datasets

Dataset	Reference	Features	Classifier	Training signatures	AER/EER%
CEDAR	[24]	Surroundedness	SVM	24	11.59
	[26]	Zernike moments	Harmonic distance	16	16.4
	[25]	Quad-tree HOT	AIRSV	10	08.76
	[24]	Chain code histogram	SVM	12	8.6
	Proposed system	**LDSP**	**SVM**	**10**	**7.66**
MCYT	[27]	Geometriccentroids	Degree of authenticity	9	21.61
	[28]	Slantmeasure	Variability mesure	10	22.13
	[29]	LDP	LS-SVM	10	11.54
	[30]	HOT	AIRS-SVM	10	11.07
	[25]	HOT	SVM	10	11.47
	Proposed system	**LDSP**	**SVM**	**10**	**8.71**

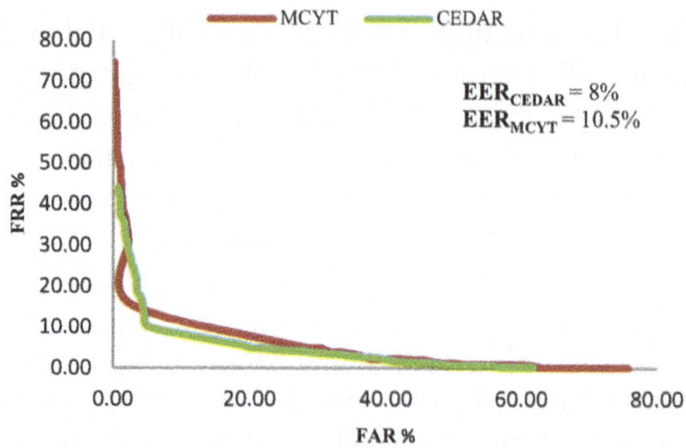

Fig. 4. DET curve obtained using 10 genuine signatures.

4 Conclusion

This work introduced the LDSP as a new descriptor edge descriptor for handwritten signature characterization. This descriptor was associated with SVM to develop the verification system according to the writer-dependent approach. The performance assessment was carried out on two public datasets, namely, CEDAR and MCYT. Compared to various state of the art features, the proposed LDSP can give similar and commonly better performance when it is associated with the same classification technique that is SVM. The AER is reduced by a gain of 0.94%, and 2.36% for CEDAR and MCYT, respectively. As future work, we are looking forward combining LDSP with other kinds of features to reinforce signature shape characterization.

References

1. Yapici, M.M., Tekerek, A., Topaloglu, N.: Convolutional neural network based offline signature verification application. In: 2018 International Congress on Big Data, Deep Learning and Fighting Cyber Terrorism (IBIGDELFT), pp. 57–61. IEEE (2018)
2. Yapıcı, M.M., Tekerek, A., Topaloğlu, N.: Deep learning-based data augmentation method and signature verification system for offline handwritten signature. Pattern Anal. Appl. **24**(1), 165–179 (2020). https://doi.org/10.1007/s10044-020-00912-6
3. Ruiz, V., Linares, I., Sanchez, A., Velez, J.F.: Off-line handwritten signature verification using compositional synthetic generation of signatures and Siamese Neural Networks. Neurocomputing **374**, 30–41 (2020)
4. Hamadene, A., Chibani, Y.: One-class writer-independent offline signature verification using feature dissimilarity thresholding. IEEE Trans. Inf. Forensics Secur. **11**, 1226–1238 (2016)
5. Guerbai, Y., Chibani, Y., Hadjadji, B.: The effective use of the one- class SVM classifier for handwritten signature verification based on writer independent parameters. Pattern Recogn. **48**, 103–113 (2015)
6. Nemmour, H., Chibani, Y.: Off-line signature verification using artificial immune recognition system. In: 10th International Conference on Electronics Computer and Computation, ICECCO 2013, pp. 164–167 (2013)
7. Serdouk, Y., Nemmour, H., Chibani, Y.: Combination of OC-LBP and longest run features for off-line signature verification. In: 10th International Conference on Signal-Image Technology and Internet-Based Systems, SITIS (2014)

8. Serdouk, Y., Nemmour, H., Chibani, Y.: An improved artificial immune recognition system for off-line handwritten signature verification. In: 13th International Conference on Document Analysis and Recognition (ICDAR), pp. 196–200 (2015)
9. Deng, P.S., Mark Liao, H.-Y., Ho, C.W., Tyan, H-R.: Wavelet-based off-line handwritten signature verification. Comput. Vis. Image Underst. **76**, 173–190 (1999)
10. Serdouk, Y., Nemmour, H., Chibani, Y.: Combination of OC-LBP and longest run features for off-line signature verification. In: 10th International Conference on Signal-Image Technology and Internet-Based Systems, SITIS (2014)
11. Serdouk, Y., Nemmour, H., Chibani, Y.: An improved artificial immune recognition system for off-line handwritten signature verification. In: 13th International Conference on Document Analysis and Recognition (ICDAR), pp. 196–200 (2015)
12. Yilmaz, M.B., Yanikoglu, B., Tirkaz, C., Kholmatov, A., Uekae, T.: Offline signature verification using classifier combination of HOG and LBP features. In: 2011 International Joint Conference on Biometrics (IJCB), pp. 1–7. IEEE (2011)
13. Yilmaz, M.B., Yanikoğlu, B.: Score level fusion of classifiers in off-line signature verification. Inf. Fusion **32**, 109–119 (2016)
14. Jabid, T., Kabir, M.H., Chae, O.: Local directional pattern (LDP); a robust image descriptor for object recognition. In: 2010 7th IEEE International Conference on Advanced Video and Signal Based Surveillance, pp. 482–487. IEEE (2010)
15. Tang, S., Goto, S.: Histogram of template for human detection. In: International Conference on Acoustics Speech and Signal Processing (ICASSP), pp. 2186–2189 (2010)
16. Serdouk, Y., Nemmour, H., Chibani, Y.: New histogram-based descriptor for off-line handwritten signature verification. In: 2018 IEEE/ACS 15th International Conference on Computer Systems and Applications (AICCSA), pp. 1–5. IEEE (2018)
17. Arab, N., Nemmour, H., Chibani, Y.: New local difference feature for off-Line handwritten signature verification. In: International Conference on Advanced Electrical Engineering (ICAEE) (2019)
18. Arab, N., Nemmour, H., Chibani, Y.: Improved multi-scale local difference features for off-line handwritten signature verification. In: 2020 1st International Conference on Communications, Control Systems and Signal Processing CCSSP, pp. 266–270 (2020)
19. Zhang, J., Deng, Y., Guo, Z., Chen, Y.: Face recognition using part-based dense sampling local features. Neurocomputing **184**, 176–187 (2016)
20. Uddin, M.Z., Khaksar, W., Torresen, J.: Facial expression recognition using salient features and convolutional neural network. IEEE Access **5**, 26146–26161 (2017)
21. Rokkones, A.S., Uddin, M.Z., Torresen, J.: Facial expression recognition using robust local directional strength pattern features and recurrent neural network. In: 2019 IEEE 9th International Conference on Consumer Electronics, ICCE-Berlin, pp. 283–288 (2019)
22. Vapnik, V.: The Nature of Statistical Learning Theory, p. 314. Springer, Heidelberg (1995)
23. Burges, C.J.A.: Tutorial on Support Vector Machines for pattern recognition. Data Min. Knowl. Disc. **2**, 121–167 (1998)
24. Kumar, R., Sharma, J.D., Chanda, B.: Writer-independent off-line signature verification using surroundedness feature. J. Pattern Recogn. Lett. **33**, 301–308 (2012)
25. Serdouk, Y., Nemmour, H., Chibani, Y.: Handwritten signature verification using the quad-tree histogram of templates and a Support Vector based artificial immune classification. Image Vis. Comput. J. **66**, 26–35 (2017)
26. Chen, S., Srihari, S.: A new off-line signature verification method based on graph matching. In: International Conference on Pattern Recognition, ICPR 2006, pp. 869–872 (2006)
27. Prakash, H.N., Guru, D.S.: Geometric centroids and their relative distances for offline signature verification. In: International Conference on Document Analysis and Recognition (ICDAR), pp. 121–125 (2009)

28. Alonso-Fernandez, F., Fairhurst, M.C., Fierrez, J., Ortega-Garcia, J.: Automatic measures for predicting performance in off-line signature. In: IEEE International Conference on Image Processing, pp. 369–372 (2007)
29. Grupo de Procesado Digital de Senales. http://www.gpds.ulpgc.es/download/index.html
30. Serdouk, Y., Nemmour, H., Chibani, Y.: A new handwritten signature verification system based on the histogram of templates feature and the joint use of the artificial immune system with SVM. In: Amine, A., Mouhoub, M., Mohamed, O.A., Djebbar, B. (eds.) Computational Intelligence and Its Applications, CIIA 2018, pp. 119–127. Springer, Cham (2018). https://doi.org/10.1007/978-3-319-89743-1_11

Ball Bearing Monitoring Using Decision-Tree and Adaptive Neuro-Fuzzy Inference System

Riadh Euldji[1](✉), Mouloud Boumahdi[2], Mourad Bachene[2], and Rafik Euldji[2]

[1] Laboratory of Automation and Industrial Diagnostic, University of Djelfa, Djelfa, Algeria
r.euldji@univ-djelfa.dz

[2] Department Mechanical Engineering, University of Medea, Médéa, Algeria

Abstract. This study aims to provide a methodology that relies on the combination of the following approaches: the decision tree, the neural network, and the fuzzy logic to monitor the evolution of bearing degradation. Data collected from the vibratory signals generated from the tests carried out on ball bearings mounted in an experimental fatigue platform, are used. The decision tree method is applied to select the most relevant monitoring indicator, which will be used to develop an Adaptive Neuro-Fuzzy Inference System (ANFIS). The training and test data required for model development have been classified according to the following states: normal, abnormal, and dangerous. These were defined from two thresholds: alert threshold and danger threshold. Then, the ANFIS model is trained from the indicators selected by the decision tree to predict the behaviour of the bearing in operation. The results confirm the effectiveness of the proposed approach for monitoring the health of ball bearing.

Keywords: Condition monitoring · Decision tree · ANFIS

1 Introduction

For several decades and until today, vibration analysis continues to be the primary tool for analysing the behaviours of rotating machines. This approach aims to assess the state of health of a machine in real time and to transform a set of raw data collected on the monitored machine, using a data mining approach, into health indicators whose extrapolation over time makes it possible to offer support for the decision-making.

In the monitoring of a rotating machine, several problems can be encountered, such as: (1) choice and configuration of degradation state indicators; (2) estimation of the remaining operating time before the total degradation of the rotating element; (4) predict their future behaviour; (5) extraction of decision rules; (6) maintenance decision-making; (7) unavailability of experts in the expertise of rotating machines.

This work will contribute to overcoming the difficulties encountered when monitoring the condition of a rotating machine. It focuses on monitoring the behaviours of a ball bearings.

In the state of the art, several techniques have been proposed in order to predict the future behavior of ball bearings, e.g., Artificial neural network networks [1–4], support vector machines [5, 6], decision tree [7–9], ANFIS [10].

B. Lejdel et al. (Eds.): AIAP 2021, LNNS 413, pp. 253–261, 2022.
https://doi.org/10.1007/978-3-030-96311-8_23

This paper aims to propose a methodology to model the prediction of the behaviour of ball bearings in operation. This methodology relies on the application of the decision tree, and ANFIS on a set of real data.

The rest of this article is organized as follows. In Sect. 2, we present a methodology based on a decision tree and ANFIS approaches. Section 3 focuses on the data collection. In Sect. 4, we present the results obtained from applying the proposed approaches to a dataset. Conclusions are enclose in Sect. 5.

2 Methodology

2.1 Decision Tree

A decision tree is a classification method. It aims to extract information contained in data by using classification algorithms. The construction of this tree requires the definition of the features and the classes which form the dataset. The classification algorithm allows to choose the most important feature by using the criteria entropy and Gain ratio. These criteria are defined as follows:

Entropy used to select input variable

$$Info(T) = -\sum_{i=1}^{k} |C_j|(|T|)^{-1} log_2(C_j(T)^{-1}) \tag{1}$$

$$Gain(X\_i, T) = Info(T)\text{-}Info(X\_i, T) \tag{2}$$

Gain ratio to select splitting attributes

$$GR(X_i, T) = Gain(X_i, T)(Split\ info(X_i, T))^{-1} \tag{3}$$

Where $X = \{X1, X2, \ldots, Xi, \ldots, Xn\}$ is the features set, n is the number of features, $C = \{C1, C2, \ldots, Cj, \ldots, Ck\}$ is the classes set, k is the number of classes, $|Cj|, j = 1, 2, \ldots, k$ is the number of examples belonging to the class, T is the set of training examples, and |T| is the total number of examples.

The feature chosen is the one that has the great gain ratio compared to the other features.

In order to build the desired decision tree, an algorithm must be used to classify the instances. Among the algorithms available, there are ID3 [11] and C4.5 [12]. The latter is the most widely used decision tree induction algorithm developed by Ross Quinlan. In this study, the J48 classification algorithm, a more developed version of C4.5, implemented in the WEKA software is used.

2.2 Adaptive Neuro-Fuzzy Inference System (ANFIS)

ANFIS is a hybrid learning algorithm that incorporates fuzzy logic and artificial neural networks (ANN) in order to give enhanced prediction. The fuzzy system represents the reasoning part, while the neural networks represents the learning part. ANFIS uses the fuzzy if-then rules involving premise and consequent parts of Sugeno-type fuzzy inference.

To describe this system, we assume that the examined fuzzy inference system has two inputs x and y and one output *f*. For a first-order Sugeno fuzzy model, a common rule set with two fuzzy if–then rules is defined as:

$$1.\ \mathit{If}\ x\ is\ A1\ and\ y\ is\ B1,\ then\ f1 = p1x + q1y + r1 \tag{4}$$

$$2.\ \mathit{If}\ x\ is\ A2\ and\ y\ is\ B2,\ then\ f2 = p2x + q2y + r2 \tag{5}$$

Where *p1, p2, q1, q2, r1* and *r2* are linear parameters in the consequent part and *A1, A2, B1* and *B2* are nonlinear parameters. The corresponding equivalent ANFIS architecture for two-input first-order Sugeno fuzzy model with two rules is shown in Fig. 1. The architecture of the ANFIS system consists of five layers, namely the fuzzy layer, product layer, normalized layer, de-fuzzy layer and total output layer.

Different layers of ANFIS have different nodes. Each node in a layer is either fixed or adaptive.

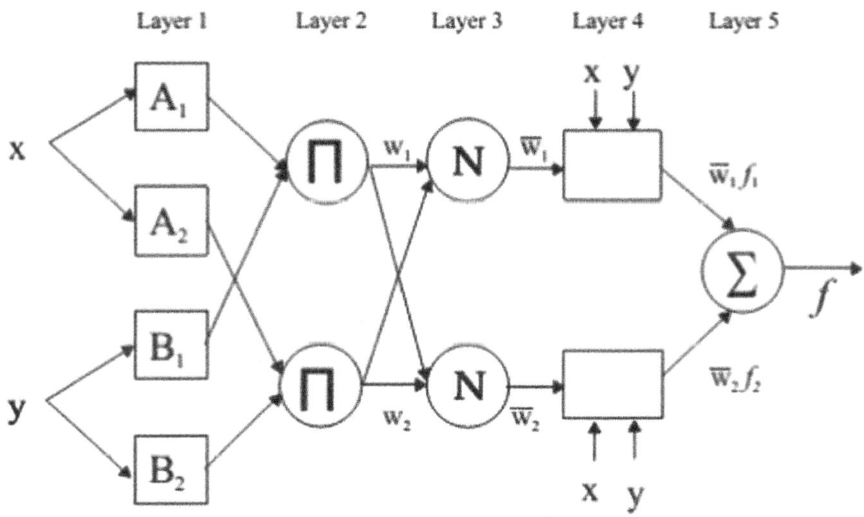

Fig. 1. ANFIS general structure.

3 Data Collection

To apply the proposed methodology, a data set was collected from the vibratory signals generated from the tests carried out on ball bearings mounted in an experimental fatigue platform, namely PRONOSTIA. This platform dedicated to test and validate bearings fault detection, diagnostic and prognostic approaches. The main objective of this platform is to provide real experimental data that characterize the degradation of ball bearings along their operational life [13].

The Nature of the data is vibration data measured by the vibration sensors during the time. Its features are namely: root mean square, kurtosis, Skewness, Peak, k factor, and the crest factor. Accounting, 2803 experimental data point samples were used to train the model. The mathematical description of each variable is presented in Table 1.

4 Results and Discussion

4.1 Feature Selection

There are three types of vibration analysis: time-domain analysis, frequency-domain analysis, and time–frequency analysis. The time domain analysis is a statistical analysis directly related to the time signal itself. In this analysis, several features are used in the vibratory follow-up of the bearing. We can quote to that end: root mean square, kurtosis, Skewness, Peak, k factor, and the crest factor (see Table 1). These features assess the state of global functioning of the machine but did not locate the defect. The presence of a defect can be detected if a feature exceeds a predetermined threshold.

In order to select the relevant feature, a decision tree algorithm, namely J48 is applied to the data set. This set contains information of three classes: normal, abnormal, and dangerous with six features (See Table 1). The classes were defined from two thresholds: alert threshold, and danger threshold, as follows: The alert threshold is the multiplication of the first measurement by two, which indicates the start of degradation, and the danger threshold is a multiplication of the first measurement by ten, which indicates the level of danger.

Table 1. Features used in the vibratory follow-up of the bearing.

Features	Description
Peak	$x_p = \max\lvert x(n)\rvert$
Root mean square	$x_{rms} = \sqrt{\frac{\sum_{n=1}^{N}(x(n))^2}{N}}$
Kurtosis	$x_K = \frac{\sum_{n=1}^{N}(x(n)-p_1)^4}{(N-1).STD^4}$
Skewness	$x_{sks} = \frac{\sum_{n=1}^{N}(x(n)-p_1)^3}{(N-1).STD^3}$
Crest Factor (CF)	$x_e = \frac{x_p}{x_{rms}}$
K Factor(KF)	$KF = x_p * x_{rms}$

Figure 2 shows a decision tree constructed from a training data set. According to this figure, it can be seen that the root mean square, as a root node, appears to be more reliable than the other features. This indicates that the root mean square is the most relevant feature for making a decision in the anomaly detection process. After getting the most relevant feature, we will use them to build the ANFIS model.

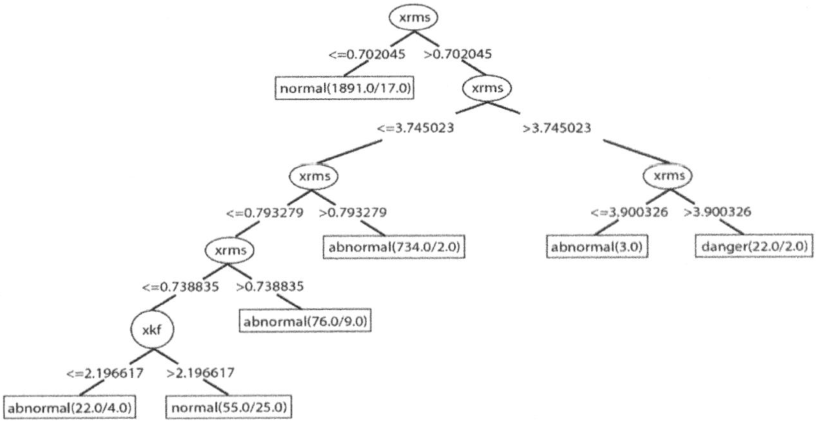

Fig. 2. Decision tree.

Table 2 shows the ranges of classification rate 0.97–0.98 and kappa statistics 0.95–0.89 which a classification rate of 1 means a perfect modelling and a kappa statistic of 0.7 or higher indicates a good statistics correlation.

Table 2. Performance of decision tree.

	Training data	Testing data
Classification rate	97.8951%	98.1132%
Kappa statistic	0.9514	0.8924
Mean absolute error	0.0216	0.0339
Number of instances	2803	1802

4.2 Prediction Using ANFIS

From the previous results, the ANFIS model is trained using only the root mean square as a feature to predict the behavior of the ball bearings in operation.

The ANFIS training was performed with the time series of xrms. A xrms (t + 6) prediction is performed by using xrms(t), xrms(t – 6), xrms (t – 12), xrms (t – 18) data as input, which corresponds to past values.

Figure 3 and Fig. 4 show that the feature root mean square presents a significant trend with regard to the evolution of the vibration level. The RMSE described in Table 3 is used to measure the prediction performance of the ANFIS.

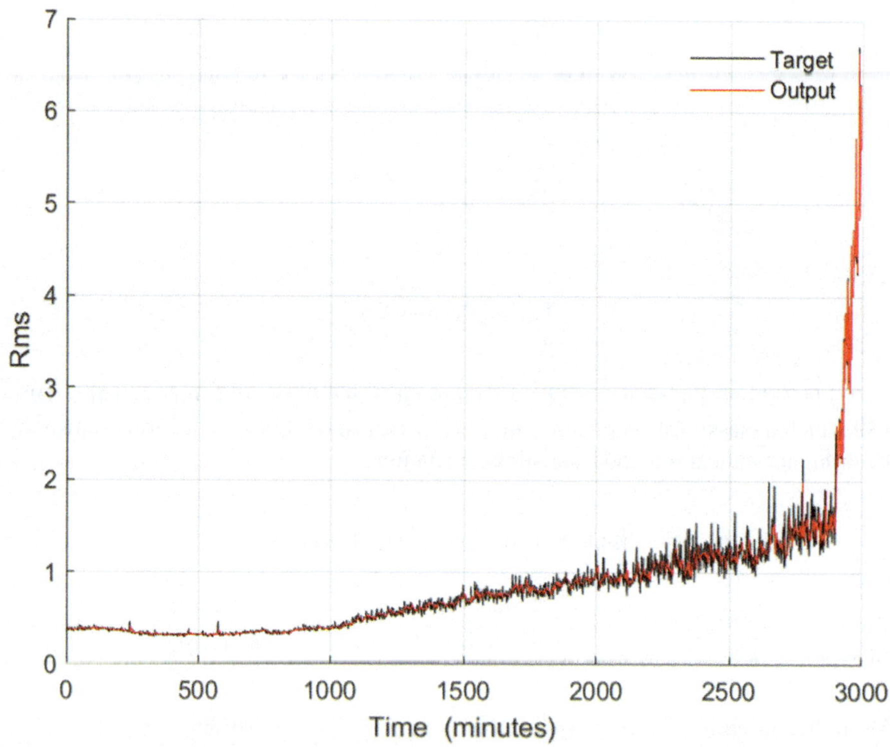

Fig. 3. Training data, experimental and predicted values of root mean square.

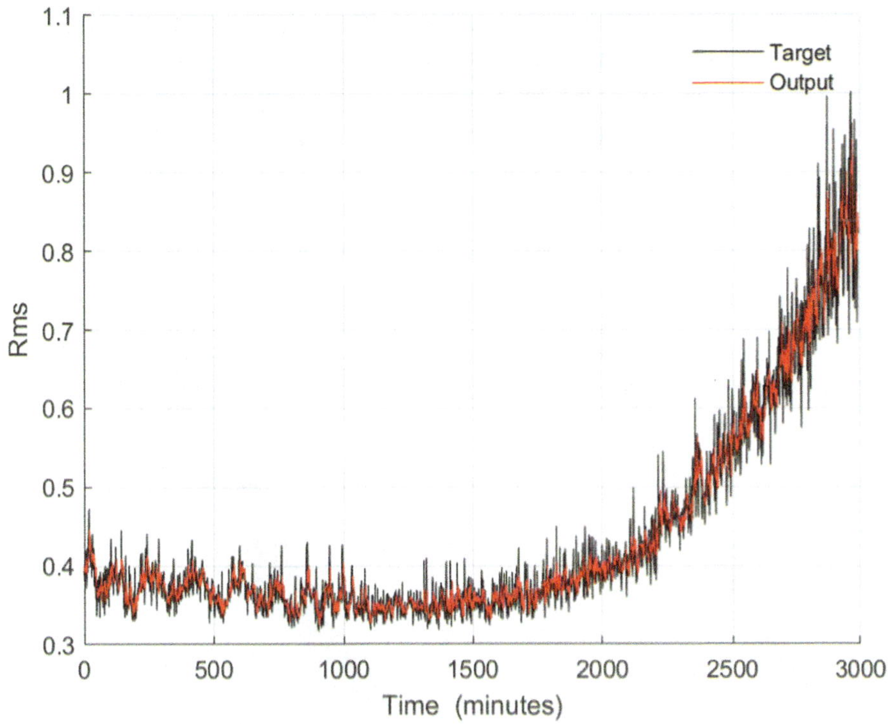

Fig. 4. Testing data, experimental and predicted values of root mean square.

Table 3. Performance of Anfis.

	Description	Training data	Testing data
MSE	$MSE = \frac{1}{n}\sum_{i=1}^{n}\left(y_i - \hat{y}_i\right)^2$	0.0072	0.0010
RMSE	$RMSE = \sqrt{\frac{1}{n}\sum_{i=1}^{n}\left(y_i - \hat{y}_i\right)^2}$	0.0850	0.0425

Figure 5 shows a good correlation among the ANFIS predictions and the experimental data ($R^2 = 0.99149$ for the training data and $R^2 = 0.96752$ for the testing data).

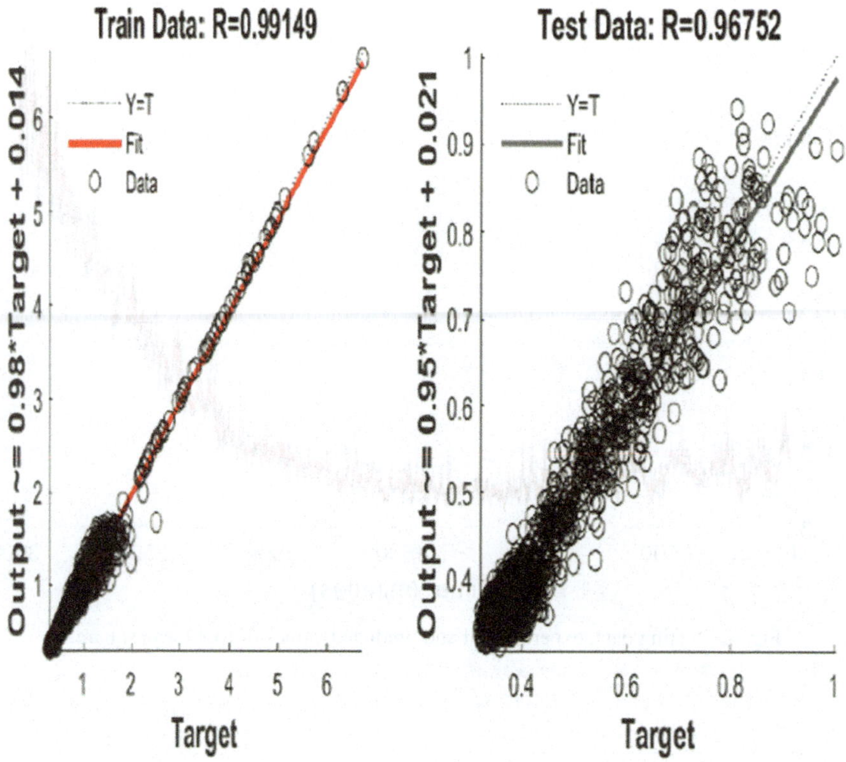

Fig. 5. Experimental and predicted values of root mean square.

5 Conclusion

In this study, we proposed a methodology to predict the behavior of ball bearings in operation. It relies on the vibration analysis and the application of the decision tree, and ANFIS on a set of real data collected from PRONOSTIA platform.

Two data set were used. The first dataset includes 2803 samples used for training, and the second includes 1804 samples used for testing. The data for each set were classified into three state: normal, abnormal, and danger. The application of the decision tree algorithm on theses set allowed to classify the states of ball bearing perfectly, the ranges of classification rate 0.97–0.98 and kappa statistics 0.95–0.89. These results indicate that the feature RMS is the relevant feature to detect de degradation of the ball bearing. Then, the ANFIS model is trained using only the root mean square as a feature to predict the behavior of the ball bearings in operation. The value of R^2 indicate a good correlation among the ANFIS predictions and the experimental data.

References

1. Samanta, B., Al-Balushi, K.R.: Artificial neural network based fault diagnostics of rolling element bearings using time-domain. Mech. Syst. Signal Process. **17**, 317–328 (2003)
2. Ali, J.B., Fnaiech, N., Saidi, L., Chebel-Morello, B., Fnaiech, F.: Application of empirical mode decomposition and artificial neural network for automatic bearing fault diagnosis based on vibration signals. Appl. Acoust. **89**, 16–27 (2015)
3. Unal, M., Onat, M., Demetgul, M., Kucuk, H.: Fault diagnosis of rolling bearings using a genetic algorithm optimized neural network. Measurement **58**, 187–196 (2014)
4. Chen, C., Vachtsevanos, G.: Bearing condition prediction considering uncertainty: an interval type-2 fuzzy neural network approach. Rob. Comput. Integr. Manuf. **28**(4), 509–516 (2012)
5. Rojas, A., Nandi, A.K.: Practical scheme for fast detection and classification of rolling-element bearing faults using support vector machines. Mech. Syst. Sig. Process. **20**(7), 1523–1536 (2006)
6. Sugumaran, V., Muralidharan, V., Ramachandran, K.I.: Feature selection using decision tree and classification through proximal support vector machine for fault diagnostics of rolling bearing. Mech. Syst. Signal Process. **21**, 930–942 (2007)
7. Yang, B.S., Lim, D.S., Tan, A.C.C.: VIBEX: an expert system for vibration fault diagnosis of rotating machinery using decision tree and decision table. Expert Syst. Appl. **28**, 735–742 (2005)
8. Karabadji, N., Seridi, H., Khelif, I., Azizi, N., Boulkroune, R.: Improved decision tree construction based on attribute selection and data sampling for fault diagnosis in rotating machines. Eng. Appl. Artif. Intell. **35**, 71–83 (2014)
9. Sugumaran, V., Ramachandran, K.I.: Automatic rule learning using decision tree for fuzzy classifier in fault diagnosis of roller bearing. Mech. Syst. Signal Process. **21**, 2237–2247 (2007)
10. Ertunc, H.M., Ocak, H., Aliustaoglu, C.: ANN-and ANFIS-based multi-staged decision algorithm for the detection and diagnosis of bearing faults. Neural Comput. Appl. **22**(1), 435–446 (2013)
11. Quinlan, J.R.: Induction of decision trees. Mach. Learn. **1**, 81–106 (1986)
12. Quinlan, J.R.: C4.5: Programs for Machine Learning. Morgan Kaufmann Publishers (1993)
13. Nectoux, P., et al.: PRONOSTIA: an experimental platform for bearings accelerated degradation tests. In: IEEE International Conference on Prognostics and Health Management, PHM 2012. IEEE Catalog Number: CPF12PHM-CDR, pp. 1–8 (2012)

Artificial Intelligent in Upstream Oil and Gas Industry: A Review of Applications, Challenges and Perspectives

Kenioua Abdelhamid[1](✉), Touat Brahim Ammar[2], and Kenioua Laid[3]

[1] Departement of Mechanical Engineering, University of Boumerdes, Boumerdes, Algeria
a.kenioua@univ-boumerdes.dz
[2] Schlumberger, Houston, USA
[3] Department of Information and Technology, University of El-Oued, El Oued, Algeria

Abstract. In the last two decades, oil and gas (O&G) industries are facing several challenges and issues in different levels; from the decrease in commodity prices to the dynamic and unexpected environment. There has been a constant urge to maximize benefits and attain values from limited resources. Traditional empirical and numerical simulation techniques have failed to provide comprehensive optimized solutions in little time due to the Immense amount of data generated on daily basis with various formats, techniques and process. The proper technical analysis of this "explosion of data" is to be carried out to improve performance of O&G industries.

Artificial intelligence (AI) has found extensive usage in simplifying complex decision-making procedures in practically every competitive market field, and O&G industry is not an exception. This paper provides a comprehensive state-of-art review in the field of machine learning and artificial intelligence to solve O&G industry problems. We focus on the upstream segment as the most capital-intensive part of oil and gas and the segment of enormous uncertainties to tackle. Based on a summary of various researchers work on machine learning and AI applications, we outline the most recent trends in developing AI-based tools and identify their effects on accelerating the process in the industry. This paper discusses also the main challenges related to non-technical factors that prevent the intensive application of AI in the upstream O&G industry.

Keywords: Artificial intelligence · Oil and gas industry · Upstream

1 Introduction

Digital transformation has a tremendous influence on business and society. With time it has been regarded as the "fourth industrial revolution", characterized by the convergence of technologies that blur the boundaries between the physical, digital and biological realms, such as artificial intelligence, robotics and autonomous vehicles. Artificial Intelligence (AI) technology is gaining considerable attention and becomes the most important general-purpose of today [1]. Because of its rapid response speed and robust

B. Lejdel et al. (Eds.): AIAP 2021, LNNS 413, pp. 262–271, 2022.
https://doi.org/10.1007/978-3-030-96311-8_24

capacity for generalization, it is quickly entering industries and creating potential of innovations and growth. AI triggered substantial changes and transformed the competition rules in media, transportation, finance and healthcare. Instead of relying on traditional and human-centered business processes, companies from these industries create value using AI solutions [2]. Advanced algorithms trained on large and useful datasets, and continuously supplied with new data drive the value creation process.

However, not only companies from digital-savvy industries are profiting from AI. Oil and gas companies are the latecomers to digitalization [3, 4], but they are also getting more and more dependent on AI solutions. Although the first applications of AI in the oil and gas industry were considered in the 1970s [5], the industry has started to look more proactively for AI application opportunities several years ago [6, 7]. It coincides with the exponential growth of AI capabilities and the industry's movement towards the Oil and Gas 4.0 concept, whose core goal is to achieve higher value utilizing advanced digital technologies [8].

As oil and gas companies are much quicker to adopt new technologies than to experiment with and change their business models [12], their AI's primary target (and other digitalization) efforts are to improve efficiency. In practice, that typically means to accelerate processes and reduce risks [5, 8].

The application of AI technology in the petroleum field, in order to organize regular relevant discussions. Based on the search result from the Onepetro platform, the number of articles on AI has increased significantly the last decade, whose main algorithms include the artificial neural network (ANN), fuzzy logic, support vector machine (SVM), hybrid intelligent system (HIS), genetic algorithm (GA), particle swarm optimization (PSO), etc. This suggests an increasing interest of the researchers in the application of artificial intelligence in the oil industry, and among all the algorithms, the ANN is the most studied one (Fig. 1).

Oil and gas deposits are often located thousands of feet below the earth's surface. The process of extracting oil and gas from these depths to the surface, and then converting it into a usable source of energy involves a large variety of operations. Figure 1 shows different sectors in the oil and gas industry operations. Broadly, the entire process of producing oil and gas is divided into the three industry sectors, upstream sector, midstream sector and downstream sector. The upstream summarizes the subsurface (mining) part of the industry, operations in the upstream industry are focused on identifying locations below the earth's surface, which have the potential of producing oil and gas. Following the identification of a potential reserve, detailed planning of exploration, drilling oil wells, and producing oil and gas also comes under the upstream industry. Midstream stands for transportation of oil and gas, and downstream is for refinery i.e., production of fuels, lubricants, plastics, and other products.

In this paper we focus our research and discussion on the upstream sector by explaining in detail many of the upstream activities, we discuss points where AI solutions are already applied and their results. We also highlight where we expect AI to be used and what results can come out of its application.

The upstream is of particular interest as it is the most capital-intensive and important of the three segments in the oil and gas business [9]. The saying "one rock, two geologists, three opinions" tells a lot about the high uncertainties and risks oil and gas companies have to deal with. Manual handling of theses enormous uncertainties and the rely to expert knowledge instead of the actual data can be very risky specially when making multibillion decisions on where and how to invest in the coming 5–20 years. However, despite the complex and uncertain nature of management problems in the sector, the single-criterion approaches have historically dominated decision-making [10]. To use existing field data to account for uncertainties associated with practitioners' subjective perception and decision-making based on experience, the first steps in using artificial intelligence and machine learning in the upstream are made, becoming increasingly popular [9]. The paper summarizes the different research works using AI to solve the problems and limitations in upstream sector. We covered the AI based researches in the whole spectrum of the upstream activities, geological assessment of the reservoirs, drilling optimization, reservoir engineering, field development, and production optimization.

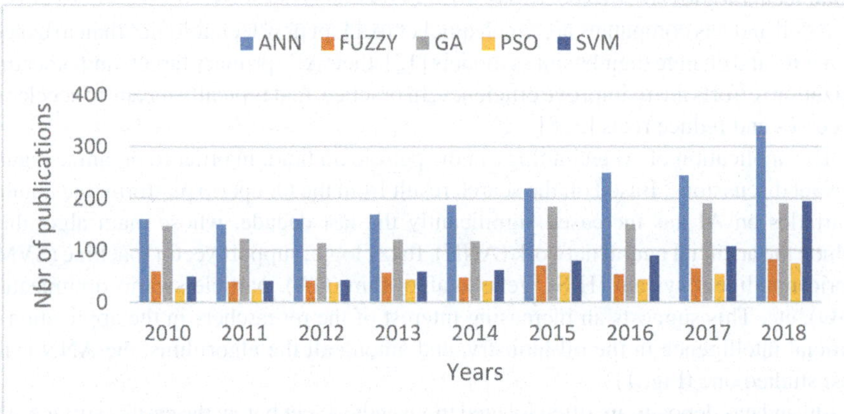

Fig. 1. AI the statistic of increasing number of articles on AI algorithms in Onepetro.

The paper is organized as follow, in Sect. 2, a brief overview of the AI approaches and algorithms is outlined with a description of the most used algorithms in the sector. In Sect. 3, we discuss the use of AI based tools in the different sections of upstream sector and identifying their effect on the accelerating process and enhancing performances. We discuss in Sect. 4 the main challenges face the AI intensive application in the industry. Finally, we conclude with perspective of how AI can change the upstream sector and probably the whole O&G industry in the nearest future.

2 Machine Learning Algorithm

Machine Learning is a subset of Artificial Intelligence. ML involves teaching the machines to handle large datasets for recognizing the patterns and extracting the relevant information with an enhanced efficiency. The computer science and technology focusing on building machines when combined with the statistical tools and inferences provides remarkable outcome as ML. As depicted in Fig. 2, ML has broadly been classified into: Supervised learning, Unsupervised learning, Semi-supervised learning and Reinforcement learning. In supervised learning, the data-driven model is built by processing a known labeled dataset that includes desired inputs (features) and outputs (labels/responses). Supervised learning finds the model that generates the outputs based on the inputs. A physics-driven model is a theory-based mathematical mapping that relates the input and output, whereas supervised learning identifies patterns in the available dataset, learns from observations, and makes the necessary predictions based on statistical mapping of inputs and outputs. During the process of building the supervised learning model, the predictions are compared to the output and the model is improved based on a loss function. This process continues until the data-driven model achieves a high level of accuracy and performance so the loss function can be minimized. Unsupervised learning processes datasets to identify patterns, relations, and commonalities without using examples, labels, and human instruction. This ML technique organizes the data in a certain way that describes the structure, variance, density, distribution, etc. of the dataset. This might mean grouping the data into clusters or arranging it in a way that looks more organized or easy to visualize. Few examples of unsupervised learning: dimensionality reduction, data compression, manifold learning, and feature extraction. An inherent issue with unsupervised learning is to assess the efficacy and reliability of the model. A training set encompassing unlabeled and labeled datasets characterizes semi-supervised learning algorithms. This approach can yield better results in some cases compared to supervised learning techniques. Finally, using trial and error, the algorithm repeatedly self-trains in reinforcement learning. Feedback is provided to the algorithm on the causes of error but no instruction for error rectification is provided (Fig. 3).

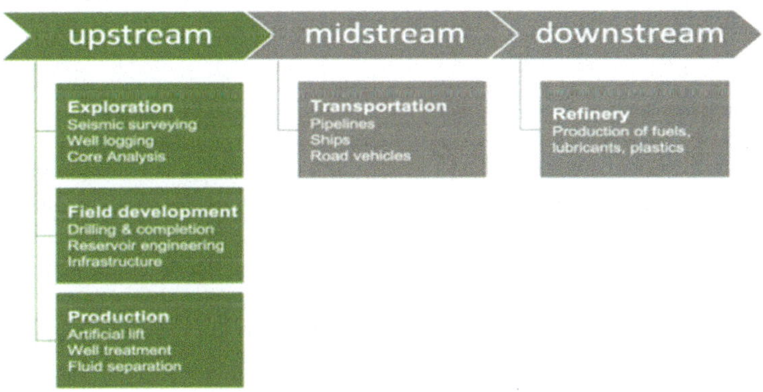

Fig. 2. Sectors of Oil and Gas industry

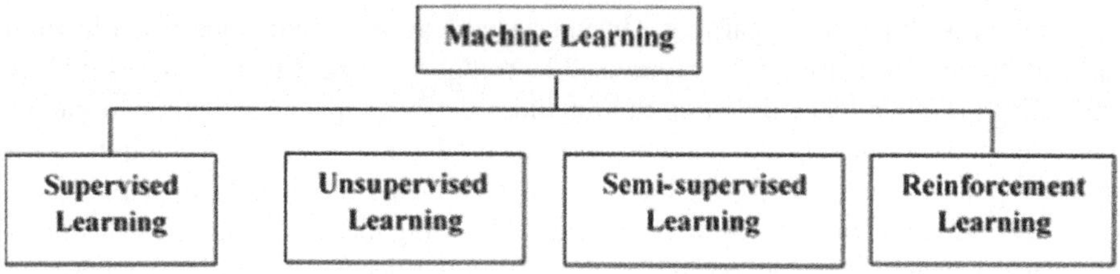

Fig. 3. Classification of ML approaches

3 Applications of ML in Upstream Oil and Gas

This section of the paper focuses on the researches that have been conducted to implement the ML tools and techniques in various sectors of upstream oil and gas industry which have been mainly categorized into Exploration & Drilling Operations, Reservoir Engineering, and Petroleum Production System. We have reviewed over 173 works compared machine learning techniques with traditional models. Most of them showed that the use of learning algorithms provided a more-accurate prediction than the use of traditional models. The reason for that is the capability of those models to capture non-linear relationship among the variables. Figure 2 shows the ML methods used in the studies as well as their dispatching per discipline in upstream sector. As shown in Fig. 2. a), ANN were employed 59 times followed by homogeneous ensembles (e.g. RF, GBM) which used 17 times then followed by the other techniques. In the scope of studies reviewed in this paper we can notify that ML used 78 times in drilling optimization followed by the production discipline with 41 studies then the exploration with 35 studies over the 173 works reviewed in this paper.

3.1 Exploration

Traditional methods of geological and geophysical modeling involve generating computerized representations of sub-surface observations obtained from various geological and geophysical surveys for evaluating the structural and stratigraphic description along with estimation of reservoir rock and fluid characteristics. These tasks typically consist of reservoir-scale seismic surveying, well logging, and lab core analysis.

Mahdi Rastegarnia et al. [11] utilized AI systems to extract electrofacies volumes and 3D Flow Zone Index (FZI) from large volume of 3D seismic data. Multi Resolution Graph based Clustering (MRGC) was employed to optimize the electrofacies volumes and multi attribute analysis was employed to create 3D FZI model. The electrofacies models are further improved using the Probabilistic Neural Network (PNN), and 3D FZI model was improved using Radial Basis Function (RBF) Network, Multilayer Feed Forward Network (MLFFN) and PNN. Modern pattern recognition techniques based on deep learning have started to dive into this seismic-related operation, accelerating the interpretation by a factor of 10–1000 [12]. Keran Qian et al. [13] proposed SVM-based algorithm in combination with multi-scale and multi-resource information from geology, seismic, drilling and logging for sweet spot prediction along and characterization of shale reservoirs with high accuracy. There is a low probability that the AI techniques will optimize the physical part (i.e., amount, cost, and placement layout of sensors) of the first seismic surveying at an asset. Still, they add value in the optimization of the secondary surveys at the same asset. The mathematics of recommender systems and interpolation capabilities of machine learning algorithms will enable proper recommendations on making the secondary surveys cheaper with a minor loss in the value of acquired information [14]. The petrophysical interpretation is a rather time-consuming process, and the result of the interpretation depends strongly on the interpreter. AI-aided technologies are the obvious way to accelerate and, maybe even more critical, to exclude the subjective part of the interpretation process. Wood DA [15] predicted the porosity, permeability and water saturation using optimized nearest-neighbor associated with data mining techniques. Meshalkin Y et al. [16] developed Robotized petrophysics workflow using machine learning and thermal profiling for automated mapping of lithotypes in unconventional. A numerical simulation model coupled with ANN and SVM classifier was developed by Kyung Jae Lee [17] to determine and classify the kerogen characteristics. Baraboshkin EE, et al. [18] used the deep learning to generate an automated well logs interpretation process for estimation rock types (Fig. 4).

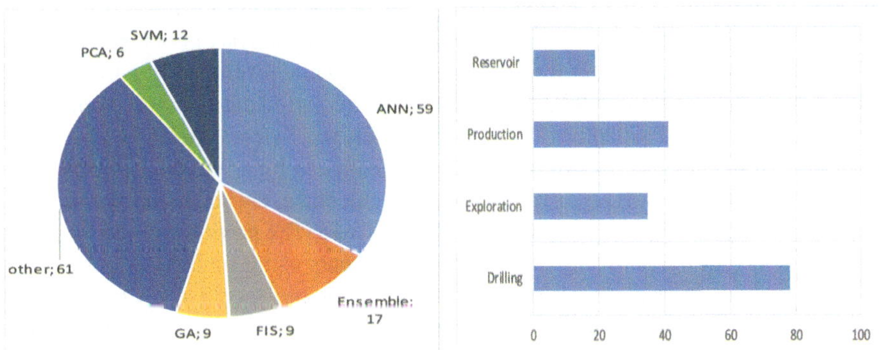

(a) (b)

Fig. 4. a) ML methods employed in upstream studies. b) Studies used ML in upstream by sector.

3.2 Reservoir Engineering

Once the initial geological model is built, it goes to reservoir engineers. They perform upscaling and build a reservoir model from the geological model using reservoir modeling software [19, 20]. This model can estimate reservoir flows at various field development schemes which contains the plan for well drilling and well operation. The result of each of the reservoir modeling runs is a forecast of oil/gas production for forthcoming years for a particular field development scheme. Performing many runs, the reservoir engineers select the optimal field development scheme and field development plan for both green-fields and brown-fields.

Deep neural networks technique is used for acceleration of reservoir modeling. Modern surrogate reservoir models with a new computation engine based on deep neural networks has been used in [21]; this technique can compress the mathematical problem dimensionality and approximate the time derivatives promise 100–1000 times the conventional models' speedup while keeping similar functionality. Simonov et al. [22] used different machine learning techniques to speed up the 3D modeling for the well in development. The upscaling process has also benefited from the machine learning algorithms to increase objectiveness and speed of the process by using deep learning algorithm trained on multiple cases of manual upscaling [23–25].

3.3 Drilling Engineering

Drilling a well is a challenging task as it involves no or minimal prior information of the sub-surface characteristics and the complexity increases with increasing depth or deviation of the well from its vertical path. Moreover, occurrence of drilling problems such as lost circulation, pipe sticking, shale sloughing, dogleg severity and others add up to make the job highly crucial. Most of well drilling cost is not product cost dependent, but time dependent [26]. Therefore, one of the main goals of drilling optimization is to reduce the total time, maintaining the risks as low as possible. One way to achieve it is through selecting of optimum drilling variables prior a run (e.g. selecting a suitable drill-bit and drilling fluid type). Another approach relies on real-time analysis in order to optimize operational parameters (e.g. bit weight, rotary speed) while drilling [27]. Farough Agin et al. [28] conducted research for estimation of lost circulation during drilling operations with the application of Adaptive Neuro-Fuzzy Inference System (ANFIS). The model was trained, tested and checked to predict lost circulation volume. Ekaterina Gurina et al. [29] have demonstrated ML algorithm to detect accidental events in directional wells by comparing the real – time MWD data with the past data. The proposed model performs anomaly detection by analyzing the similarity of events using time-series comparison and gradient boosting classification. Optimization of the pre-specified key performance indicators including ROP (ft/hr) and mechanical specific energy (psi) was carried out in the research [30] using coupled end-to-end model for drilling optimization by taking into account all the parameters of interest including Weight on Bit (WOB), rotary speed, flow rate and rock strength. The researchers utilized the Random Forest (RF) algorithm for developing models for each individual parameter under consideration that were coupled using ML algorithm.

3.4 Production Engineering

Producing reservoirs are attractive for AI-aided tools as well as the green fields. There are obvious machine learning applications for various pumps to implement predictive maintenance and select the optimal operation regimes concerning operational costs vs. production. Many of the pumps, including electric submersible pumps, pumps for injection wells, hydraulic fracturing, and other well treatment pumps, are equipped with a high number of sensors measuring pressures, temperatures, vibrations, flow rates, etc. There are many examples when an entirely data-driven or a hybrid model containing physics-driven and data-driven math helps optimize the regimes, prevent unexpected failures, and save on maintenance-on-schedule [31, 32]. Li et al. [33] used a Neural

Decision Tree (NDT) model for prediction of oil production by considering interconnectedness among the input variables. N. Chithra Chakra et al. [34] applied Higher Order Neural Network (HONN) to predict cumulative oil production (m3) from a conventional oil field with limited training data. It was found to be a satisfactory tool to forecast the cumulative oil production for short – term as well as long-term planning. Using electrical and frequency data as input features, Guo et al. in [32], developed an SVR workflow to forecast failures in Electrical Submersible Pump (ESP). Gupta et al., 2016, deployed a hybrid mathematical method consisting of an intelligent predictive monitoring KPI that automatically identifies imminent glitches, diagnose root causes, and prescribe corrective actions to abnormal ESP operational situations in real time raising alarms through predictive, diagnostic, and prescriptive analytics. There is an excellent opportunity to reduce the investment risks by accumulating the data from already produced well treatment jobs. Pioneering efforts on predicting the efficiency of hydraulic fracturing jobs in [35] and ML-based analysis of injectivity issues have already been made in [36].

4 Challenges and Perspectives

The success of artificial intelligence critically depends on human intelligence. AI solutions have to be customized to the business context and database of a company. Thus, to actively use AI in processes and products, companies must grow in-house teams composed of data and AI specialists. These teams should be able to support development of AI infrastructure (algorithms and datasets) and, at least to customize tools that companies will later utilize in their operations. That means that oil and gas companies will become (partially) data-driven companies and, that AI specialists will become irreplaceable in supporting almost all innovation efforts in oil and gas companies in the next 10 years. AI tools need the good quality data of a suitable volume to be trained and then to work properly in the operational mode. While using smarter algorithms may help in getting better results from datasets of limited size, no manipulation can help with bad data. Thus, access to big and quality data is a crucial enabler and barrier for AI applications' successful development. Artificial intelligence is born in open and collaborative environment as a consequence of academia being a leading force in AI research for decades, almost without any business influences. This created culture of free sharing and open publishing which companies across industries (and across the globe) had to embrace as a standard to succeed in the era of AI once they joined the race oil and gas companies should re-think strategies for collaborating and interacting with universities.

5 Conclusion

The aim of this paper is to present a review concerning the myriad of different applications and benefits of Artificial Intelligence and Machine Learning techniques in the upstream sector of oil and gas industry which cover disciplines of exploration, reservoir, drilling and production. The paper compiles the major workflows and achievements of the industry on a higher-level overview with a focus on its leverage over other traditional modelling techniques. The literature review of oil and gas industry is well-poised to take benefits of machine learning regarding their abilities of processing big data and fast computational speed. Machine learning has the potential of unequivocally changing the numerous critical actions made every day by administrators and engineers in the oil and gas sector. The future advantages of information can be achieved if appropriate techniques are used to implement different data types or structures and convert it into useful information that contributes to intelligent judgements. AI and machine learning will change the face of oil and gas industry and lead to speed up and de-risk many business processes associated to this critical and crucial industry.

References

1. Brynjolfsson, E., Mitchell, T.: What can machine learning do? Workforce implications. Science **358**(6370), 1530–1534 (2017)
2. Iansiti, M., Lakhani, K.R.: Competing in the age of AI. Harv. Bus. Rev. (2020). https://hbr.org/2020/01/competing-in-the-age-of-ai
3. Kohli, R., Johnson, S.: Digital transformation in latecomer. **10**(4), 141–156 (2011)
4. Kane, G.C., Palmer, D., Phillips, A.N., Kiron, D., Buckley, N.: Strategy, not technology, drives digital transformation. MIT Sloan Manag. Rev. (2015)
5. Li, H., Yu, H., Cao, N., Tian, H., Cheng, S.: Applications of artificial intelligence in oil and gas development. Arch. Comput. Methods Eng. (2020)
6. BCG Homepage: Going digital is hard for oil and gas companies—but the payoff is worth it. https://www.bcg.com/ru-ru/publications/2019/digital-value-oil-gas.aspx. Accessed 24 Aug 2021
7. BCG Homepage: Big oil, big data, big value. https://www.bcg.com/publications/2019/big-oil-data-value.aspx. Accessed 24 Aug 2021
8. Lu, H., Guo, L., Azimi, M., Huang, K.: Oil and gas 4.0 era: a systematic review and outlook. Comput. Ind. **111**(6), 68–90 (2019)
9. Shafiee, M., Animah, I., Alkali, B., Baglee, D.: Decision support methods and applications in the upstream oil and gas sector. J. Pet. Sci. Eng. **173**, 1173–1186 (2019)
10. Strantzali, E., Aravossis, K.: Decision making in renewable energy investments: a review. Renew. Sustain. Energy Rev. **55**, 885–898 (2016)
11. Rastegarnia, M., Sanati, A., Javani, D.: A comparative study of 3D FZI and electrofacies modeling using seismic attribute analysis and neural network technique: a case study of Cheshmeh-Khosh Oil field in Iran. Petroleum **2**(3), 225–235 (2016)
12. Cunha, A., Pochet, A., Lopes, H., Gattass, M.: Seismic fault detection in real data using transfer learning from a convolutional neural network pre-trained with synthetic seismic data. Comput. Geosci. 104344 (2019)
13. Qian, K.R., He, Z.L., Liu, X.W., Chen, Y.Q.: Intelligent prediction and integral analysis of shale oil and gas sweet spots. Pet. Sci. **15**(4), 744–755 (2018)
14. Portugal, I., Alencar, P., Cowan, D.: The use of machine learning algorithms in recommender systems: a systematic review. Expert Syst. Appl. **97**, 205–227 (2018)
15. Wood, D.A.: Predicting porosity, permeability and water saturation applying an optimized nearest-neighbour, machine-learning and data-mining network of well-log data. J. Pet. Sci. Eng. (2019)
16. Meshalkin, Y., Koroteev, D., Popov, E., Chekhonin, E., Popov, Y.: Robotized petrophysics: machine learning and thermal profiling for automated mapping of lithotypes in unconventionals. J. Pet. Sci. Eng. (2018)
17. Lee, K.J.: Characterization of kerogen content and activation energy of decomposition using machine learning technologies in combination with numerical simulations of formation heating. J. Pet. Sci. Eng. **188** (2020)

18. Baraboshkin, E.E., et al.: Deep convolutions for in-depth automated rock typing. Comput. Geosci. (2019). https://doi.org/10.1016/j.cageo.2019.104330
19. Gasda, S.E., Celia, M.A.: Upscaling relative permeabilities in a structured porous medium. Adv. Water Resour. **28**(5), 493–506 (2005)
20. Fanchi, J.R., Christiansen, R.L.: Introduction to Petroleum Engineering. Wiley, Hoboken (2016)
21. Temirchev, P., et al.: Deep neural networks predicting oil movement in a development unit. J. Pet. Sci. Eng. (2020)
22. Simonov, M., et al.: Application of machine learning technologies for rapid 3D modelling of inflow to the well in the development system (2018). https://doi.org/10.2118/191593-18rptc-ru
23. Barker, J.W., Thibeau, S.: A critical review of the use of pseudorelative permeabilities for upscaling. SPE Reserv. Eng. **12**(2), 138–143 (1997)
24. Farmer, C.L.: Upscaling: a review. Int. J. Numer. Methods Fluids **40**(1–2), 63–78 (2002)
25. Pickup, G.E., Stephen, K.D., Ma, J., Zhang, P., Clark, J.D.: Multi-stage upscaling: selection of suitable methods. In: Das, D.B., Hassanizadeh, S.M. (eds.) Upscaling Multiphase Flow in Porous Media, pp. 191–216. Springer, Heidelberg (2005). https://doi.org/10.1007/1-4020-3604-3_10
26. Lyons, W.C., Plisga, G.J.: Standard Handbook of Petroleum and Natural Gas Engineering, 2nd edn. Gulf Professional Publishing (2004)
27. Payette, G.S., et al.: Real-time well-site based surveillance and optimization platform for drilling: technology, basic workflows and field results. In: SPE/IADC Drilling Conference and Exhibition. Society of Petroleum Engineers, Hague, The Netherlands (2017)
28. Agin, F., Khosravanian, R., Karimifard, M., Jahanshahi, A.: Application of adaptive neuro-fuzzy inference system and data mining approach to predict lost circulation using DOE technique (case study: Maroon oilfield). Southwest Petroleum University (2019)
29. Gurina, E., et al.: Application of machine learning to accidents detection at directional drilling. J. Pet. Sci. Eng. **184** (2020)
30. Hegde, C., Pyrcz, M., Millwater, H., Daigle, H., Gray, K.: Fully coupled end-to-end drilling optimization model using machine learning. J. Pet. Sci. Eng. **186** (2020). https://doi.org/10.1016/j.petrol.2019.106681
31. Sneed, J.: Predicting ESP lifespan with machine learning (2017). https://doi.org/10.15530/urtec-2017-2669988
32. Guo, D., Raghavendra, C.S., Yao, K.T., Harding, M., Anvar, A., Patel, A.: Data driven approach to failure prediction for electrical submersible pump systems (2015). https://doi.org/10.2118/174062-ms
33. Li, X., Chan, C.W., Nguyen, H.H.: Application of the Neural Decision Tree approach for prediction of petroleum production. J. Pet. Sci. Eng. **104**, 11–16 (2013). https://doi.org/10.1016/j.petrol.2013.03.018
34. Chithra Chakra, N., Song, K.Y., Gupta, M.M., Saraf, D.N.: An innovative neural forecast of cumulative oil production from a petroleum reservoir employing higher-order neural networks (HONNs). J. Pet. Sci. Eng. **106**, 18–33 (2013). https://doi.org/10.1016/j.petrol.2013.03.004
35. Makhotin, I., Koroteev, D., Burnaev, E.: Gradient boosting to boost the efficiency of hydraulic fracturing. J. Pet. Explor. Prod. Technol. **9**(3), 1919–1925 (2019). https://doi.org/10.1007/s13202-019-0636-7
36. Orlov, D., Koroteev, D.: Advanced analytics of self-colmatation in terrigenous oil reservoirs. J. Pet. Sci. Eng. (2019). https://doi.org/10.1016/j.petrol.2019.106306

A Comparative Study of Road Traffic Forecasting Models

Redouane Benabdallah Benarmas(✉) and Kadda Beghdad Bey

Ecole Militaire Polytechnique - Chahid Abderrahmane Taleb (EMP),
PO Box 17, Bordj El Bahri, Algiers, Algeria

Abstract. In the context of Intelligent Transport Systems (ITS), the behaviour of road traffic has been the subject of many theoretical and experimental researches. In the last decade, road prediction is placed as the first line of research in this field. The problem has been solved with a variety of models to assist the traffic control, this includes, improving the efficiency of transport, guidance in the road, and smart coordination signals. This paper tries to synthesize the carried out, on three main approaches, namely based on statistical methods, time series and deep learning. A comparatives synthesis in terms quantitative and qualitative index of is provided in order to evaluate the performance and potential of the three forecasting approaches.

1 Introduction

Recent years have been marked by an exponential growth in the volume of road traffic. Intelligent Traffic Management systems (ITS) have been deployed in the face of social, economic and ecological challenges, for this purpose, an industrial and scientific partnership have been set up with the aim of improving the fluidity of road traffic. Furthermore, data collection and access to information techniques has been considerably expended, many researches aim to develop decision support tools in different contexts for this domain, including intelligent guidance, intelligent traffic control, or the construction of new roads as shown in Fig. 1.

The accuracy of prediction has been improved recently by the development of data sciences and machine learning (ML). Research in this field is moving towards the development of models capable of predicting road traffic under normal and abnormal conditions. The approaches adopted depend on the need for the prediction, the exact context and the field of application. Considerable effort, had begun research with classical models, such as the historical average, the Kalman filter proposed by [13] or K-NN. The development of modern statistical theory and machine learning methods have accelerated the pace of research on a variety of approaches, which usually revolved around classification and regression methods. The main goal of the present paper is to synthesize the traffic forecasting models on three main approaches, namely based on statistical methods, time series and deep learning. A quantitative and qualitative comparison is provided in order to evaluate the performance and potential of the three forecasting

B. Lejdel et al. (Eds.): AIAP 2021, LNNS 413, pp. 272–280, 2022.
https://doi.org/10.1007/978-3-030-96311-8_25

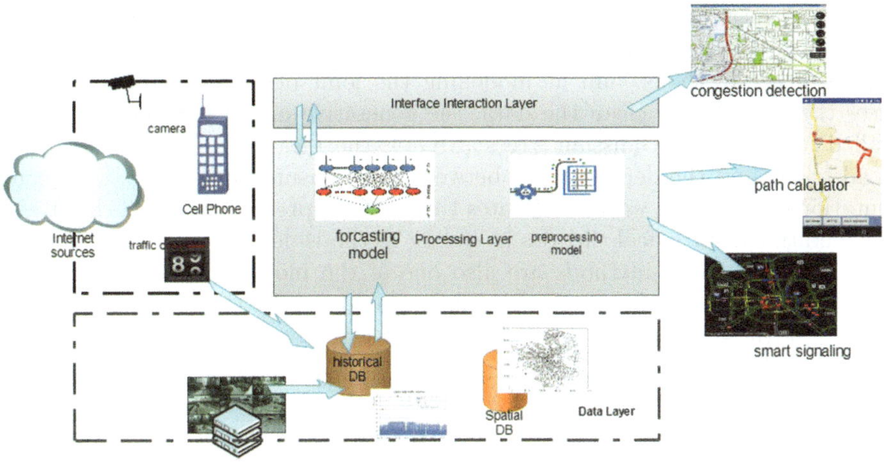

Fig. 1. Scope of road traffic prediction

approaches. This paper is organized into three sections, the first is an introduction, the second contains a review of the literature of approaches adopted in road prediction and the third concludes techniques giving satisfactory results in this domain.

2 Approaches Adopted in the Prediction of Road Traffic

Traffic forecasting methods in this context are based on modelling of road traffic data, for prediction methods based on machine learning, early works were boosted in the 2000s by statistical approaches, and continue to evolve using deep learning methods. Formally, the objective is to develop a method that can predict the average speed of flow at a point on the road network in the future, by using a data history. Many techniques have been proposed, and they would not be possible to make an exhaustive state of the art from previous works. An overview of the literature, allowed to synthesize the methods in three categories, Statistical approaches, Time Series approaches and Deep Learning approaches.

2.1 Statistical Approaches

Statistical methods are beginning to appear as powerful methods in the field of Machine Learning, starting with classical linear regression techniques such as K-NN, or kernel-based methods (SVM). In the statistics, Machine Learning is a problem of functional approximation or regression, as it proposed by [9], the model tries to find a relation between the future speed and the combination of the speeds observed before the moment of the forecast, this work is based on the unsupervised classification of daily traffic trends and gives inaccurate results in the very short term forecasting, these daily trends are modelled from the calendar measures assumed to be exact, the application of this model does not support missing or incorrect data.

An other methods, aim at modelling the joint density of probability and uses the inference to clean the data, the generative model proposed by [14] is called Multi-Varied Gaussian Trees (MGT), the idea is to characterizing the road traffic and the dependencies between the measured quantities, This computationally complex work, illustrates the power of prediction approaches based on conditional statistical theories including methods derived from Bayesian networks. Kernel-based methods are also one of the most promising approaches in predicting road traffic, the work was initiated by [23] then [12]. The performances of these techniques were also demonstrated by [16] exploiting a dataset of Chennei.

SVMs are categorized in the paradigm of supervised learning and implemented frequently in classification and regression analyses, in the case the flow of traffic is affected by many non-linear elements such as weather conditions, accidents or vacation days (see Fig. 2).

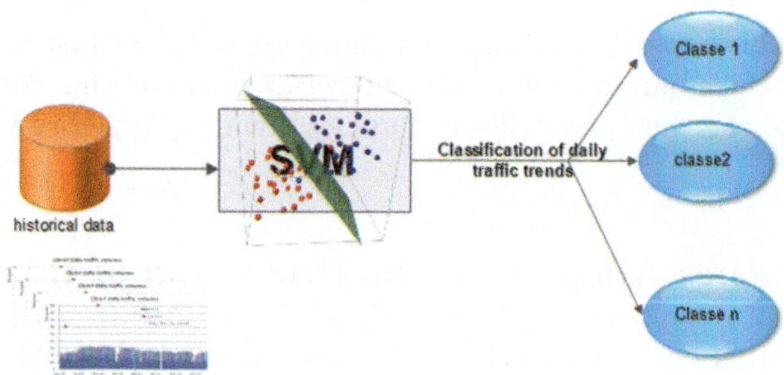

Fig. 2. SVM classification

Another variant LSSVMS (Least Square SVM), proposed by [20], in order to reduce the calculation compared to the use of basic SVMs, by transforming the quadratic programming problem into a problem solving a linear equation.

2.2 Time Series Based Approaches

Time Series are an important part of the data produced and available on the internet or in specific deployments namely, traffic control centers. ARIMA (Auto Regressive Means Average) or Jinkins-BOX methods have been reported on several studies in road prediction. The work shows that this method is the most adopted in the short term prediction.

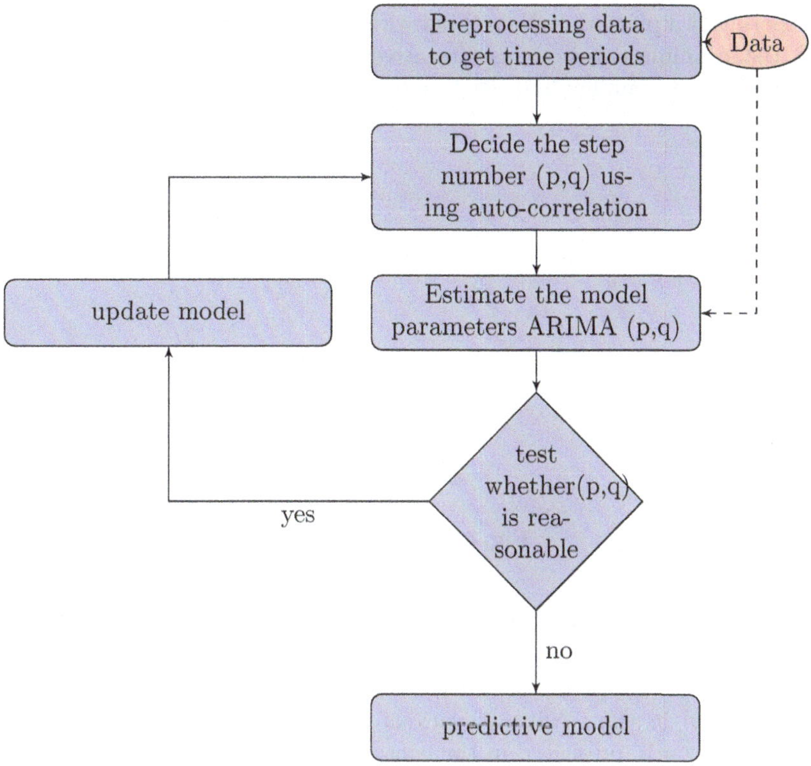

Fig. 3. ARIMA Process

The first results [10,15] show that cloud data availability is a major problem in the application of these methods. Current efforts are trying to define alternatives to out-discard this problem, the work of [6] with the SARIMA model (Seasonal ARIMA) gives conclusive results with a limited input data set, [24] also proposes the Switching ARIMA method to remedy the lack of data. These methods are a bit attractive because of their simplicity and efficiency, nevertheless, the no linearity must be insignificant for autoregressive models to show acceptable performance. The lack of data must be treated also in the time series, the work of [2] illustrate the use auto regressive in a time series completion tasks, by adding special constraints making it possible to model the relation between several series through a graph.

2.3 Deep Learning Approaches

In the past few years, there has been a resurgence in the use of ANN (Artificial Neural Network) for the treatment of predictive tasks which allowed the rebooting of the works in the different contexts, particularly in the field of road traffic management. The return was motivated by the availability of data and the experimental platforms, Neural networks have also evolved, in terms of modelling, the techniques are no longer iterative algorithms as the case of Perceptron, but with a more complex and deep architecture as the case of CNN and RNN (Fig. 4).

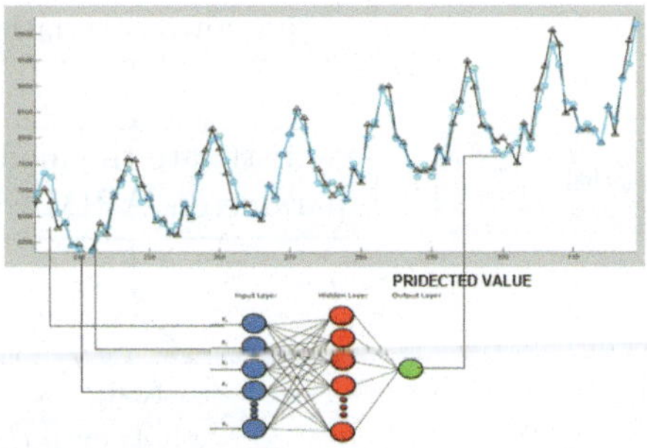

Fig. 4. ANN regression

ANN show their power following the work of [3,7] and [8], a FNN (Fuzzy Neural Network) is proposed by [4] for prediction in the urban road network using a self-adaptive predictor. Neural Network with deep architectures have been successful in the solving of various learning tasks, including image and speech and recently presented in the context of prediction as the most adopted techniques.

Yisheng et al. proposes a neural network with a SAE (Stacked Auto Encoder) architecture used for learning the generic characteristics of the traffic [21]. This model was trained, with data from a highways control center in California, by using an unsupervised learning algorithm, the results were compared with other classical NN architectures namely the DBNN (Deep Belief Neural Network) proposed by [18] and RBFN (Radial Basis Function Network) and SVM methods as an anther approach. The same set of data is applied in [22] by an architecture of an LSTM (Long Short Term Memory) network, then [19] with the proposal of a model called DeepTrend inspired by the classification of daily traffic trends, in this work, the architecture of the model is deeper and is based on two layers one for extracting daily traffic trends, and one based on an LSTM network for prediction.

Haiyang et al. develops a hybrid model called (SRCN) Spatial Recurve Convolutional Network [11], which combines two architectures, DCNN (Deep Convolutional Neural Network) and LSTM for the prediction of traffic speed in a road network. In this architecture, the DCNN layer has been used to capturing special dependencies of traffic in road network.

The combination of approaches is becoming more adopted in traffic modelling, especially for arterial roads, Spatio-temporal dependencies are modelled efficiency and in a finer way by using hierarchical representations of deep architectures [17].

3 Comparative Synthesis for the Three Approaches

In order to fully evaluate the performance and potential for field implementation of the three forecasting approaches, a performance index system is established in terms of two parts: quantitative and qualitative indexes. The quantitative index consists of Root Mean Square Error (RMSE). The qualitative index refers to the ease of implementing a forecasting model, which is evaluated based on time and effort required for model development, the required expertise, adaptability to changing temporal or spatial behaviour.

3.1 The Qualitative Index

The study of three modelling approaches allows to provide a brief comparison shown in the following table:

Qualitative index	Statistical approaches	Time series approaches	Deep learning approaches
Time and effort required for model development	The methods derived from baysien networks require design of a complex graph. SVM are the most statistical methods used for traffic prediction, the model is described by quadratic programming problem witch is complex to resolve	As shown in Fig. 3, the modelling process is simple but required a completion. For prediction traffic of many points in road network the relation between several series must be modelled in complex graph.	Artificial Neural network in traffic prediction are designed with a complex architecture. Recent works in this context use a combination of several model of Neural Network
Required skills	Require an experience in statistical theories of Machine Learning and the understand of the kernel functions	Require experience in statistical theories of time series.	Require experience in ANN (Artificial Neural Network) and Combination of several variants
Adaptative to change temporal and spatial behaviour	Statistic studies are developed for specific phenomena. The methods based on SVM are less adaptative when the flow of traffic is affected by many other no linear elements such as weather conditions, accidents or vacation days	Variants of ARIMA proposed in [5] and [23] are designed for specific context. The variables(exogenous inputs)added depend to a particular spatial context	The data-driven methods are better suited to develop adaptative model, this is the strong point compared to model-driven methods

The first models described in the first and the second approach, have proved a capacity and simplicity and contributed to the perfection of ITS, however they are considered classic and are developed for a simple contexts by using a limited datasets. It has been observed from complex parametric models that the accuracy of the prediction depends on the characteristics embedded in the spatial-temporal data collected. At this stage, the use of non-parametric methods, based on Neural Networks, are more promising and allow having a good prediction when the architecture is deeper.

3.2 Quantitative Index

In order to compare the approaches cited above, experiments were carried out on open dataset. The performance of the model has been evaluated by the Root Mean Square Error:

$$RMSE = \sqrt{(\frac{1}{N})\sum\nolimits_{i=1}^{n}(t_i - p_i)^2}$$

Where pi = predicted traffic flow; ti = actual traffic flow; N = the number of predictions.

In this experiment, traffic flow data was obtained from Baidu research Open Access dataset [1], A traffic dataset was provided for the 6th ring road, the speed of 15,073 road segments is recorded per minute, then averaged with a 15-minute time. This data was formatted as line delimited, in total there are 88,267,488 rows in the specific time period. We limit our study only for the highway. (See the studied segments explored by QGis tool in Fig. 5).

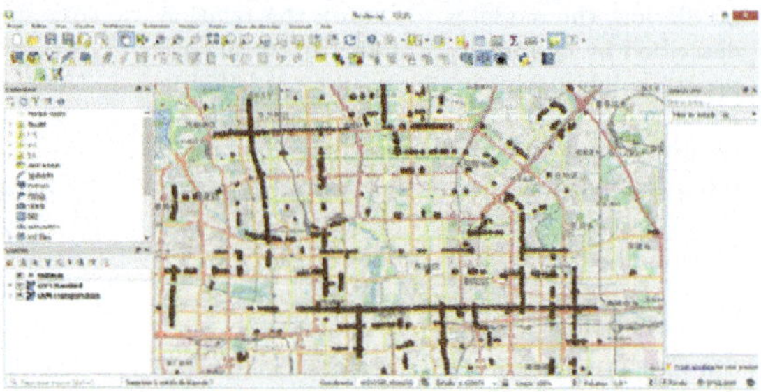

Fig. 5. The spatial distribution of studied segments

The result data is formatted as large matrix, then stored in CSV file. The data can be easily manipulated by Python libraries such as Pandas and explored by the elegant library matplotlib.

For all models, the experiments were performed under PC (Windows 8 Professional, CPU: Intel Xeon(R) E51650 3.50 GHZ, Memory: 8G), which installed TensorFlow TM (Version 1.0.0 CPU) and Python (Version 2.5.3). The Neural Networks model are fit using the efficient Adam version of stochastic gradient descent and optimized using the mean squared error, or 'MSE' loss function with the following parameter settings: Epoch = 50, Batch size = 72, result are obtained as follow:

Models	RMSE
ARIMA	95.140
SVR-LINEAR	102.847
SVR-POLY	101.144
SVR-RBF	101.843
SAE	62.19
LSTM	39.105
SE LSTM	35.458
CNN LSTM	31.585

The results are more conclusive for the architectures using deep learning models, especially for the hybrids architecture(CNN LSTM combination) which justifies the frequent adoption of these techniques in the most of current works.

4 Conclusion

This paper has describing and categorizing the different methods adopted in road traffic forecasting on three approaches; the statistical approaches based on statistical machine learning methods (KNN, SVM), time-series based approaches (ARIMA or box-Jinkins) and deep learning approaches (SAE, LSTM, CNN). A comparative study has been provided in terms of two parts: quantitative and qualitative index. It has been shown that the deep learning methods outperform the tow other approaches and improved by adopting an hybrid models.

The performance and accuracy of these methods were justified in the results and experiments compared to other approaches on the same datasets. It has been found from review, that recent researches are adopting models based on Deep Learning, more precisely, most popular works are currently focusing on the combination of the different architectures such as LSTM, CNN.

References

1. Baidu Research Open-Access Dataset. http://ai.baidu.com. Accessed 30 Apr 2021
2. Ziat, A.: Representation learning for time series classification and prediction (2017)
3. Chan, K.Y., Dillon, T.S.: On-road sensor configuration design for traffic flow prediction using fuzzy neural networks and Taguchi method (2013)
4. Bucur, L., Florea, A., Petrescu, B.: An adaptive fuzzy neural network for traffic prediction. Control and Automation (MED) (2010)
5. Williams, B.M., Hoel, L.A.: Modeling and Forecasting Vehicular Traffic Flow as a Seasonal ARIMA Process: Theoretical Basis and Empirical Results (2003)
6. Kumar, D.V., Vanajakshi, L.: Short-term traffic flow prediction using seasonal ARIMA model with limited input data (2013)
7. Kumar, K., Parida, M., Katiyar, V.: Short term traffic flow prediction for non urban highway using artificial neural network (2013)

8. Kumar, K., Parida, M., Katiyar, V.K.: Short term traffic flow prediction in heterogeneous condition using artificial neural network (2015)
9. Allain, G.: Prévision et analyse du trafic routier par des méthodes statistiques (2008)
10. Dong, H., Jia, L., Sun, X., Li, C., Qin, Y.: Road Traffic Flow Prediction with a Time-Oriented ARIMA Model (2009)
11. Yu, H., Wu, Z.: Spatiotemporal Recurrent Convolutional Networks for Traffic Prediction in Transportation Networks (2017)
12. Lingli, et al.: Traffic Prediction Based on SVM Training Sample Divided by Time (2013)
13. Okutani, I., Stephanides, Y.I.: Dynamic Prediction of Traffic Volume through Kalman Filtering Theory (1984)
14. Singliair, T.: Machine learning solutions for transportation network (2005)
15. Ahmed, M.S.: Analysis of Freeway Traffic Time-Series Data by Using Box-Jenkins Technique (1979)
16. Deshpand, M., Bajaj, P.: Performance Improvement of Traffic Flow Prediction Model Using Combination of Support Vector Machine and Rough Set (2017)
17. Wu, Y., Tan, H.: Short-term traffic flow forecasting with spatial-temporal correlation in hybrid deep learning framework (2016)
18. Huang, W.: Deep Architecture for Traffic Flow Prediction: Deep Belief Networks With Multitask Learning (2014)
19. Dai, X., Fu, R., Lin, Y., Wang, F.-Y.: DeepTrend: A Deep Hierarchical Neural Network For traffic Flow Prediction (2017)
20. Zhang, Y., Hou, Z.: Short Term Traffic Flow Prediction Based on Improved Support Vector Machine (2018)
21. Lv, Y., et al.: Traffic Flow Prediction With Big Data :Deep Learning Approach (2014)
22. Tian, Y., Pan, L.: Predicting Short-Term Traffic Flow by Long Short-Term Memory Recurrent Neural Network (2015)
23. Zhang, Z., Cao, C.: Road Traffic Freight Volume Forecast Using Support Vector Machine Combining Forecasting (2011)
24. Zhang, Y., Haghani, A.: A hybrid short-term traffic flow forecasting method based on spectral analysis and statistical volatility model (2014)

Machnine Learning for Sentiment Analysis Using Algerian Dialect

Nedioui Med Abdelhamid(✉) and Brahim Lejdel

University of EL-Oued, El-Oued, Algeria

Abstract. Text analysis has a great importance, especially in areas such as politics, productions, and services... etc. Currently, the social networks full of the texts in which, the Internet-users express in different subjects, the interest of their opinions is considerable, where the comprehension of the content conveyed by these texts is an essential element. We can say that the good manager who listens well to the opinions of citizens. In this sense, the Sentiment Analysis is very important to meet the needs of citizens. In this work, we will implement predefined algorithms that analyze and classify a set of publications derived from social networks. The classes that we have defined are: positive, negative or neutral. Our work is among the first one that uses and compares several comment classification algorithms on Facebook, using Algerian dialect.

Keywords: Sentiment analysis · Social network · Annotated corpus · Lexicon of sentiment

1 Introduction

The Internet is used frequently as a medium for the exchange of information. People can easily disseminate information including their personal subjective opinions on any topic on the internet. The user generates content in the Web in natural languages with unstructured-free-texts form. Today, a huge amount of information is available online, where we can find different types of Web documents such as Web pages, images, audio and video files, and a vast collection of different types of files. Also, we can find newsgroups, forums, blogs, and social network postings. Opinions' people express towards a subject are among the myriad types of information available online. In this context, we will study the use of dialects in social media. The dialect has to be considered in Arabic since the identification of the Arabic dialect helps to determine the context. The Arabic language has a standard version that is well-understood across the Arab world. It is known as Modern Standard Arabic (MSA). It is used alongside Arabic vernaculars in online content. Most of OSN users tend to use Dialectal Arabic (DA). The problems arising from using DA are way beyond those with MSA due to the lack of standardization of DA and the scarcity of tools for processing DA (Harrat, Meftouh, & Smaïli, 2017) [1]. Syiam et al. (2006) [2] declared that the opinion as a subjective belief, or the result of emotion or interpretation of facts, opinion

B. Lejdel et al. (Eds.): AIAP 2021, LNNS 413, pp. 281–290, 2022.
https://doi.org/10.1007/978-3-030-96311-8_26

is the result of a person's perspective, understanding, particular feelings, beliefs, and desires. It refers to unsubstantiated information different from knowledge and fact-based beliefs. Dialectal Arabic (DA) is significantly different from the Arabic language taught in schools and used in written communication and formal speech (broadcast news, religion, politics, etc.). Many existing approaches are proposed for analyzing Arabic language Sentiment Analysis (SA); however, they are generally restricted to Modern Standard Arabic (MSA) or some dialects of economic or political interest.

In this paper, we will explore the different works related to analyzing the sentiment in social networks using the different Arabic dialects (Sect. 2). In Sect. 3, we will present our contribution to analyzing the sentiment in social networks using the Algerian dialect. Then, we discuss the results of the experimentation (Sect. 4). Finally, we present a conclusion and some future works.

2 State of the Art

In this section, we will present main works that use principally the Arabic dialect (AD) for analyzing the sentiments in the reviews, comments or tweets.

Shoukry and Rafea (2012) [3] show an application on Arabic sentiment analysis by implementing a sentiment classification for Arabic tweets. The retrieved tweets are analyzed to provide their sentiments polarity (positive, or negative). Since this data is collected from the social network Twitter; it has its importance for the Middle East region, which mostly speaks Arabic. They collected 1000 tweets divided equally into 500 positives and 500 negatives. After filtering the tweets to remove non-Arabic words, HTML tags, pictures, etc., they used standard n-gram features and experimented with several classifiers (SVM and NB) through the Weka toolkit. In other work, they proposed a simple way to combine the corpus-based approach with the lexicon-based one. They focused on the Egyptian dialect and experimented on their dataset of 4800 tweets (split evenly across the positive, negative and neutral classes).

Al-Subaihin, Al-Khalifa, and Al-Salman (2011) [4] and Al-Subaihin and Al-Khalifa (2014) [5] proposed a novel lexicon-based technique to deal with dialectal Arabic. The novelty in their approach lied in the use of an online game to create a sentiment lexicon through what is called "human computation." In another work by the same group Albraheem and Al-Khalifa (2012) [6] discussed in detail the issues/challenges faced by lexicon-based approaches for dialectal Arabic SA. They collected 1000 tweets divided equally into 500 positive and 500 negatives. After filtering the tweets to remove non-Arabic words, HTML tags, pictures, etc., they used standard n-gram features and experimented with several classifiers (SVM and NB) through the Weka toolkit.

Nawaf et al. (2013) [7] address both approaches to Sentiment Analysis for the Arabic language. Since there is a limited number of publically available Arabic dataset and Arabic lexicons for SA, their work starts by building a manually annotated dataset and then takes the reader through the detailed steps of building the lexicon. Experiments are conducted throughout the different stages of this process to observe the improvements gained on the accuracy of the system and compare them to corpus-based approach.

Al-Kabi et al. (2013) [8] attempts to an opinion analysis study in Arabic language social media. Such social network content is mixed with Modern Standard Arabic (MSA) used in media and literature in addition to vernacular text. Opinion polarities are selected and each opinion or comment is assigned in class. Different domain classes are experimented and evaluated to see which selection can produce the best results in terms of accuracy in comparison with manual judgments. They developed an opinion analysis and classification tool dedicated to the Arabic language. A dictionary of positive, negative and neutral words in Arabic is assembled based on surveying a large number of documents and posts. Based on this polarity dictionary, they collected a large set of opinions or posts from social networks. Words in those posts are examined for polarity against the dictionary and a polarity class (e.g. positive strong, positive medium, positive weak, etc.) is given for each post based on the number of positive, negative or neutral words.

Mataoui et al. (2016) [9] propose a new lexicon-based sentiment analysis approach to address the specific aspects of the vernacular Algerian Arabic fully utilized in social networks. A manually annotated dataset and three Algerian Arabic lexicons have been created to explore the different phases of their approach. these approaches composed of four modules: common phrases similarity computation module; pre-processing module; language detection & stemming module; and polarity computation module. Our built lexicon is composed of three parts: keywords lexicon; negation words lexicon; intensification words lexicon. These three lexicons are enriched by a dictionary of emoticons and another dictionary of common phrases. Finally, they have built a test corpus for experimental purposes. This corpus was filtered and annotated to facilitate the evaluation process of our proposal. Experimental results show that their system obtains a performance with 79.13 % of accuracy.

Medhaffar et al. (2017) [10] focus on Sentiment Analysis of the Tunisian dialect. They use Machine Learning techniques to determine the polarity of comments written in the Tunisian dialect. First, they evaluate the Sentiment Analysis systems performances with models trained using freely available MSA and Multi-dialectal data sets. They collect and annotate a Tunisian dialect corpus of 17.000 comments from Facebook. This corpus shows a significant improvement compared to the best model trained on other Arabic dialects or MSA data.

Baly et al. (2017) [11] create the first Multi-Dialect Arabic Sentiment Twitter Dataset (MD-ArSenTD) that is composed of tweets collected from 12 Arab countries, annotated for sentiment and dialect. They use this dataset to analyze tweets collected from Egypt and the United Arab Emirates (UAE), intending to discover distinctive features that may facilitate sentiment analysis. They also perform a comparative evaluation of different sentiment models on Egyptian and UAE tweets. Results indicate superior performance of deep learning models, the importance of morphological features in Arabic NLP, and that handling dialectal Arabic leads to different outcomes depending on the country from which the tweets are collected.

Alomari et al. (2017) [12] introduce an Arabic Jordanian twitter corpus where Tweets are annotated as either positive or negative. They investigate different supervised machine learning sentiment analysis approaches when applied to Arabic user's social media of general subjects that are found in either Modern Standard Arabic (MSA) or Jordanian dialect. Experiments are conducted to evaluate the use of different weight schemes, stemming and N-grams terms techniques and scenarios. The experimental results provide the best scenario for each classifier and indicate that the SVM classifier using the term frequency-inverse document frequency (TF-IDF) weighting scheme with stemming through Bigrams feature outperforms the Naïve Bayesian classifier best scenario performance results.

Al-Twairesh (2018) [13] presented three hybrid sentiment analysis classifiers for Arabic tweets. The classifiers work on different levels of classification: two-way classification (positive and negative), three-way classification (positive, negative, and neutral) and four-way classification (positive, negative, neutral, and mixed). The approach was to incorporate the knowledge extracted from the lexicon-based method as features into the corpus-based method to develop the hybrid method. A set of features were extracted from the data then a backward selection algorithm was proposed to perform feature selection to reach the best classification performance.

Elouardighi et al. (2018) [14] use an approach based on machine learning. They proposed an approach that analyzes sentiment from the comments Facebook, real, shared especially in Moroccan dialect. The analysis of sentiments an a process during which the polarity (positive, negative or neutral) of a given text is determined and defined [18]. This process begins with collecting comments and annotating them using crowdsourcing. Then, they preprocess the text to extract the Arabic words reduced to their root. These words will be used for the construction of input variables using several combinations of extraction and ponderation process.

Many works create a huge dataset for testing their models as AAl-Obaidi and Samawi (2016) [15], which created the Opinion Mining Corpus for Colloquial Variety of Arabic language (OMCCA). The dataset consisted of 28,576 reviews annotated as positive, negative or neutral. The dialects of interest were Jordanian and Saudi. OMCCA is made publicly available. The authors reported experiments on OMCCA using different features and classification techniques. In Al-Suwaidi et al. (2016) [16], the size of the dataset is 1000 and the dialect of interest was the Emirati dialect (used in UAE), whereas, in Alomari et al. (2017) [12], the dataset is called the Arabic Jordanian General Tweets (AJGT) dataset and it consists of 1800 tweets. In addition, Assiri et al. (2016) annotated data set of 4700 for Saudi dialect sentiment analysis with (K = 0.807) [17].

Table 1. Number of comments per polarities

Polarity	Positive	Negative	Neural	Total
Number of comments	975	525	1391	2891

3 Contribution

Our contribution consists of four main points:

- we use the Algerian dialect with four classifiers like Support Vector Machines (SVM), Decision Tree (DT), Random Forest (RF) and Naïve Bayes (NB).
- We manually annotated the dataset of the Algerian dialect with 2891 comments.
- We create a dictionary of 1328 annotated words in the Algerian dialect.

3.1 Annotation

The process of determining the class of each comment by annotating it as positive, negative, or neutral. Two raters were used to determine manually the sentiment of the comments. They had a high degree of agreement in their classification of the comments, and for those comments that they disagreed about their sentiment; a third rater was used to determine its final sentiment. They use three polarities 1, −1, 0 which correspond respectively positive, negative and neutral sentiment. Table 1 shows the different polarities of the annotated comments of our dataset.

In Fig. 1, we some examples which describe some comments with their polarities.

Comments	Polarity
هذي جياحة نريسكي بعمري على حاجة تستاهل ماشي علاجال بوزوج رجلين يجري مور البالون	Negative
شفت المواطن مش واعي خلاص داير المربلة حذاه يكل منا او يرمي منا	Negative
في بسكرة خويا لعزيز طريق سيدي عقبة ؟	Neutral
هادو روشيات تاع رخام ماشي ثلج	Neutral
ربي يكثر من امثالك	Positive
درس في حسن ضيفة كل إحترام وتقدير ناس المدية	Positive

Fig. 1. Some comments with their polarities.

Table 2. The division of the words of our dictionary according the polarity.

Polarity	Positive	Negative	Total
Number of comments	565	763	1328

3.2 Dictionary

We have not found a specific lexicon of Algerian dialect words, In literature. Thus, we create our dictionary. A dictionary of positive, negative and neutral words in Arabic is assembled based on surveying a large number of documents. Based on this polarity dictionary, we collected a large set of comments from social networks. We take about 1 month of hard work for gathering 1328 words in the Algerian dialect from social media. Also, we asked the help of the friends of the north of Algeria, the south, the east, and the west to create a dictionary of Algerian dialect which covers as much as possible of the regions of our country. Table 2 represents the division of the words of our dictionary according to the polarity.

Figure 2 represents an example of positive words of the dictionary.

Fig. 2. Positive words of dictionary.

Figure 3 represents an example of the negative words of the dictionary.

4 Experimentation and Results

In this section, we use Python as the programming language in the experimentation, because of the many good (and fast) libraries to deal with natural language text, and cover all the tasks needed to build our tool, like CSV, Gensim, and Scikit-learn. Also, we use four algorithms, SVM, DT, RF, and NB. The experiments were run on a cluster of 2 machines with Intel Core2 Quad processors Q9400 (4 cores per processor) at 2.66 GHz and 4 GB memory.

عنف
فاشل
ردئ
مهزلة
خامجين
دمار
وسخ
تبلعيط
بلعاط
كارثة
النحس
رهج

Fig. 3. Negative words of dictionary.

Table 3. Feature used in our work.

Feature	Abbreviation	Signification
Has Positive Word	HPW	0 ou 1
Has Negative Word	HNW	0 ou 1
Positive Word Count	PWC	≥ 0
Negative Word Count	NWC	≥ 0
CommentLength	CL	Numeric > 0
SentimentLevel	SL	$-1 \leq V \leq 1$

4.1 Features

In our work, we use six main features. Table 3 represents the different features used in our works.

In our work, we use the supervised learning method and the lexicon-based approach. Thus, we have to divide our corpus into two parts, 80% for training and 20% for the test. We perform several tests, the Accuracy results are shown in Table 4.

We observe that we obtain the best results when we use all features with the classifier Random Forest (RF).

4.2 Comparison of Results

In this section, we will compare our work with other works that use other dialects as Tunisian, Moroccan, Egyptian, Saudi, and Jordanian. We can conclude that our classifiers have achieved good results. Table 5 summarizes the comparison that we make to validate our work.

Table 4. Features used in our work.

Test	Feature	Classifier			
		SVM	DT	RF	NB
1	All Features	85.14	83.07	85.31	84.28
2	HPW, HNW, PWC, NWC	84.11	84.45	84.28	82.38
3	PWC, NWC, CL, SL	84.62	83.24	85.31	83.76
4	PWC, NWC	84.11	84.28	84.45	67.87
5	HPW, HNW, CL, SL	84.11	84.11	84.11	48.35
6	HPW, HNW	84.11	84.11	84.11	72.30
7	CL, SL	47.49	47.66	48.18	48.18

Table 5. Comparison of results.

Dialects	Classifier	Results	
		Accuracy	F-measure
Our Work	RF	85.31 %	84.93 %
Jordanian [12]	SVM	87.2 %	/
Algerian [9]	/	79.13%	/
Tunisian [10]	MLP	/	78%
Moroccan [14]	SVM	/	78%
Egyptian [3]	SVM	/	72.5%
Saudi [13]	SVM	/	69%

5 Conclusion and Future Works

In this paper, we analyzed the sentiments of a corpus containing 2891 comments in the Algerian dialect, divided into three polarities: 975 positive comments, 525 negative comments, and 1391 neutral ones. We exploited four machine learning classifiers that are Support Vector Machine (SVM), Decision tree (DT), Random Forest (RF), Bayesian naive (NB) where the evaluation of these classifiers is done by 20% of the corpus. We used six features that are HasPositiveWord, HasNegativeWord, PositiveWordCount, NegativeWordCount, CommentLength, and SentimentLevel. We perform seven different tests, the first was done using all the features and the other tests were done by the substitution between these features. We compared the test results for the four classifiers. The results obtained are very encouraging, where we achieved a better Accuracy that of 85.31% when using the RF classifier.

We can cite some future works as:

1. The enrichment of our dictionary by more words of Algerian dialect covering more widely other areas because Algeria is very large and contains ten dialects,

2. The enrichment of Dataset by other comments in Algerian dialect to obtain precise results.
3. The application of other classifiers and other features.
4. Analysis by the use of the Mixed class in addition to the positive, negative and neutral classes.
5. The use of other configurations such as bigram, trigram and mixed.

References

1. Harrat, S., Meftouh, K., Smaïli, K.: Machine translation for Arabic dialects (survey). Inf. Process. Manag. **56**(2), 262–273 (2017)
2. Syiam, M.M., Fayed, Z.T., Habib, M.B.: An intelligent system for Arabic text categorization. Inf. Sci. **6**(1), 1–19 (2006)
3. Shoukry, A., Rafea, A.: Sentence-level Arabic sentiment analysis. In: SoMNet 2012, pp. 2–5 (2012)
4. Al-Subaihin, A.A., Al-Khalifa, H.S., Al-Salman, A.S.: A proposed sentiment analysis tool for modern Arabic using human-based computing. In: Proceedings of the 13th International Conference on Information Integration and Web-Based Applications and Services, pp. 543–546 (2011)
5. Al-Subaihin, A.S., Al-Khalifa, H.S.: A system for sentiment analysis of colloquial Arabic using human computation. Sci. World J. **2014**, 1–8 (2014)
6. Albraheem, L., Al-Khalifa, H.S.: Exploring the problems of sentiment analysis in informal Arabic. In: Proceedings of the 14th International Conference on Information Integration and Web-Based Applications & Services, pp. 415–418 (2012)
7. Abdulla, N.A., Ahmed, N.A., Shehab, M.A., Al-Ayyoub, M.: Arabic sentiment analysis: lexicon-based and corpus-based. In: AEECT 2013, pp. 1–6 (2013)
8. Al-Kabi, M., Gigieh, A., Alsmadi, I., Wahsheh, H., Haidar, M.: An opinion analysis tool for colloquial and standard Arabic. In: Proceedings of the 4th International Conference on Information and Communication Systems (ICICS) (2013)
9. Mataoui, M., Zelmati, O., Boumechache, M.: A proposed lexicon-based sentiment analysis approach for the vernacular Algerian Arabic. Res. Comput. Sci. **2016**, 55–68 (2016)
10. Medhaffar, S., Bougares, F., Esteve, Y., Hadrich-Belguith, L.: Sentiment analysis of Tunisian dialects: linguistic resources and experiments. In: Proceedings of the Third Arabic Natural Language Processing Workshop, pp. 55–61 (2017)
11. Baly, R., El-Khoury, G., Moukalled, R., Aoun, R., Hajj, H., Shaban, K.B.: Comparative evaluation of sentiment analysis methods across Arabic dialects. Procedia Comput. Sci. **117**, 266–273 (2017)
12. Alomari, K.M., ElSherif, H.M., Shaalan, K.: Arabic tweets sentimental analysis using machine learning. In: Proceedings of the International Conference on Industrial, Engineering and Other Applications of Applied Intelligent Systems, pp. 602–610 (2017)
13. Al-Twairesh, N., Al-Khalifa, H., AlSalman, A., Al-Ohali, Y.: Sentiment analysis of Arabic tweets: feature engineering and a hybrid approach. Computation and Language (cs.CL) (2018)
14. Elouardighi, A., Maghfour, M., Hammia, H., Aazi, F.Z.: Analyse des sentiments à partir des commentaires Facebook publiés en Arabe standard ou dialectal marocain par une approche d'apprentissage, Conférence Internationale sur l'Extraction et la Gestion des Connaissances, Paris, France, pp. 329–334 (2018)

15. Al-Obaidi, A., Samawi, V.: Opinion mining: analysis of comments written in Arabic colloquial. In: Proceedings of the World Congress on Engineering and Computer Science 2016 (WCECS 2016) (2016)
16. Al Suwaidi, H., Soomro, T.R., Shaalan, K.: Sentiment analysis for Emirati dialects on twitter. Sindh Univ. Res. J. **48**(4), 707–710 (2016)
17. Assiri, A., Emam, A., Al-Dossari, H.: Saudi twitter corpus for sentiment analysis. World Acad. Sci. Eng. Technol. Int. J. Comput. Electr. Autom. Control Inf. Eng. **10**(2), 272–275 (2016)
18. Al-Harbi, W.A., Emam, A.: Effect of Saudi dialect preprocessing on Arabic sentiment analysis. Int. J. Adv. Comput. Technol. **4**(6), 91–99 (2015)

Road Segments Traffic Dependencies Study Using Cross-Correlation

Benabdallah Benarmas Redouane(✉) and Kadda Beghdad Bey

Ecole Militaire Polytechnique - Chahid Abderrahmane Taleb (EMP), PO Box 17, Bordj El Bahri, Algiers, Algeria

Abstract. Traffic Prediction on a urban road network become more complex face to exponential growth in the volume of traffic, it is necessary to study the relationship between road segments before the prediction calculation. The spatial correlation theory has been well developed to interpret the dependency for understanding how time series are related in multivariate model. In large scale road network modeled by Multivariate Time Series, the Spatial-temporal dependencies detection can limit the use of only data collected from points related to a target point to be predicted. This paper present a Cross-Correlation as method to dependency analysis between traffic road segments, Scatterplot of Cross-Correlation is proposed to reveal the dependency, we provide a comparative analysis between a three correlation coefficients sush as Spearman, Kendal and Person to conclude the best one. To illustrate our study, the methodology is applied to the 6th road ring as the most crowded area of Beijing.

1 Introduction

Traffic flow prediction has become one of the important research fields in Intelligent Transportation System (ITS). The prediction of traffic flow information is important for control and guidance and makes the transport users better informed. The special characteristics of road network, such as their high scale and complex dependencies between segments, make the problem of prediction very challenging. The development of modern's statistical theory and machine learning methods has accelerated the pace of research on a variety of approaches, which usually revolved around classification and regression methods. The earliest traffic prediction methods mainly include Auto Regressive Integrated Moving Averaging (ARIMA) [11], Kalman Filter [12,13], Support Vector Machine (SVM) [14], Markov chain model [15] and Artificial Neural Network [16–18]. The first's solutions were provided on simpler contexts, which aim to predict the road flow at a given location, the model is qualified as uni-variate. The classical regression techniques were sufficient to solve the problem. Recently and for industrial needs, the problems are exposed in more complex and varied contexts, such as the prediction of several values in different places by using a data collected from different sources (See Fig. 1).

B. Lejdel et al. (Eds.): AIAP 2021, LNNS 413, pp. 291–300, 2022.
https://doi.org/10.1007/978-3-030-96311-8_27

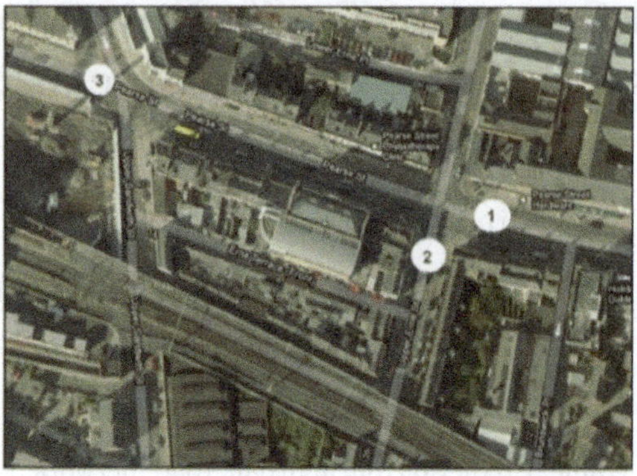

Fig. 1. Road network point positions

For this problem, the simple generalization of uni-varied models was insufficient, because on a road network, the flow is a stochastic phenomenon which evolves over time and which has an impact on others flows in neighboring points or located in other places, therefore, the modeling of spatial-temporal dependencies becomes necessary. Furthermore, the reliability and accuracy of prediction method is not based only on the used model but also on the choice and determination of the historical used data. The determination of relevant values used in the calculation aims to characterize in an efficient manner the spatial and temporal dependencies between a given point and different points in the road network. In road traffic prediction, the complexity is assessed in relation to the calculation time, the expertise and the effort required to providing a solution. Flexibility must also be achieved, so as to have a model less sensitive to the sizing of used data. For large scale road network, the detection of dependency between flows evolution captured from different road segments reduces significantly the data used for prediction calculation. The main goal in our work, is to demonstrate that a cross-correlation can interpret the dependency between road traffic segments, and reduce consequentially the data used for prediction calculation for a target point, furthermore, we provide a comparative study relatively to the coefficient used in cross-correlation calculation at second stage.

This paper is organized as follow. Section 2 present a brief review of related work. Section 3 consists in the definition of the working context which permit the problem formulation. Section 4 is devoted to the description of our dependency detection method. Experiment is performed and results are discussed in Sect. 5, finally we conclude some comments and future work directions.

2 Brief Review of Related Work

Traffic forecasting in large scale road network need an interpretation of the spatial and temporal traffic patterns in each segment and the relationship between each one. Spacial correlation is used to build a GIS system based method [20] to extract the real-time traffic information. These models were insufficient, because on a road network, the flow is a stochastic phenomenon which evolves over time and which has an impact on other flows in neighboring points or located in other places, for this purpose, spatial temporal modeling is necessary where the probability density must be defined in a joint way. Pan et al. Introduced the spatial-temporal correlation to the short-term traffic flow prediction by using the random region transmission framework [19]. Time auto-correlation analysis is carried out by [21] using journey time data collected on London's road network, the analysis was applied for uni-variate model.

Recently, the spatial-temporal correlation theory has been well developed to interpret the dependency for understanding how time series are related in multivariate model. At this stage, traffic data in large scale road network in most of studies are represented by multivariate time series, furthermore, a Cross correlation has been widely used in special analysis and in several contexts such as economics and environment [5], and present a potential for road traffic analysis. Many methods consider the spatial-temporal correlation as basic technique in the research on road traffic. A cross-correlation between network-aggregated density was proposed in [14] as a natural indicator of traffic phases for road networks [6]. Suggest that the method can be used to investigates the relationship between traffic flow series and the spatial distance of the road network sections [9]. Propose a de-trended cross-correlation analysis (DCCA) to measure the relationship between air pollution and traffic congestion in the urban area. The previous works merely consider the spatial-temporal correlation as technique to understand the interactions between different segments on road network, for traffic prediction, [8] use auto-correlation and cross correlation measure to find a seasonal patterns and provides a theoretical assumption for traffic forecasting. Recently [7] propose a new approach to identify the most influential locations, in this work, the captured correlation network between different locations might facilitate future studies on controlling the traffic flows.

3 Context Study and Problem Formulation

The problem formulation in the field of road prediction is based on the definition of a precise context on which the study will be carried out because an approach or a method's choice depends closely to the implementation. In literature, the theory of road traffic prediction is described on three aspects, the precision of the variables used in modeling, the dimension of the quantity measured and the prediction horizon. For the first, a choice between a macroscopic or microscopic modeling must be made to define the precision of the predicted variables, in a macroscopic model, the dynamic behavior of objects in the road is described as a homogeneous flow with a density or a speed. Our work study will be with a macroscopic modeling using a history of measured quantities and coming from different data sources, the prediction will be calculated from a history made up of several variables.

3.1 Data Collection and Modelling

Whatever the approach and methods adopted for traffic prediction, multi-varied modelling is based on the data definition and the relationships between them, making it possible to capture the dependencies between these data, then an adequate prediction method will be applied to arrive at prediction. There are different ways to get a road traffic data, it can be collected by a network of sensors, namely the detection loops and the traffic counters, another method is to use the GPS on board vehicles or installed in the phones. In any case, data is regularly reported in varying amounts to what is generally called a Traffic Management Center (TMC), using a data transmission network. At this level, traffic is aggregated in observation intervals into three main quantities: flow, speed and volume (density). Big data [4] was heavily used for predicting road traffic and allowed motivation for the adoption of the data-driven models.

3.2 Problem Formulation as a Multi-varied Time Series

A time series $T = t^1, t^2, ..., t^n$ is a sequence of real numbers obtained through repeated measurements over time. We represent the evolution of the variable collected over time and at given point in the road network, as a time series denoted as, $X = (x1, x2, \ldots, xT)$, For each t between 1 and T, x is a vector of dimension D, D represents the dimension of the vector x or the component is valued by the quantity collected in a place X^t belongs to R^D is the observation at time t. The problem is to predict the value $X^h, h = t + t^h$, h denotes the prediction horizon. For large scale road network, the global traffic state agregated in n time intervals can be expressed by a high-order matrix given a large-scale road network with m segments, this traffic state is denoted as Xmn.

$$\mathrm{X}^{mn} \begin{Bmatrix} x^1 1 & \ldots\ldots\ldots & x^1 n \\ x^2 1 & \ldots\ldots\ldots & x^2 n \\ .. & & .. \\ .. & & .. \\ x^m 1 & \ldots\ldots\ldots & x^m n \end{Bmatrix}$$

4 Dependency Detection Model

In the present work, detecting cross-correlations between traffic recorded in two points is the most usual way to diagnose and understand a complex large - scale road network. The simplest method is the traditional Pearson Cross-Correlation Analysis. In statistics, Pearson Correlation Coefficient (PCC) is one of the most popular statistical measures, and is frequently discussed in almost all fields, such as climatology [3], economics [5], and signal processing. As explained in previous sections, global traffic state in n time interval of large - scale road network can be represented in each point by m time series, then a cross - correlation is calculated between each time series x and y and stored in cross - correlation matrix denoted Xcc.

$$Xcc = \frac{\sum_{i=1}^{n}(x_i - \bar{x})(y_i - \bar{y})}{\sqrt{\sum_{i=1}^{n}(x_i - \bar{x})^2}\sqrt{\sum_{i=1}^{n}(y_i - \bar{y})^2}}$$

This matrix represents mutual traffic segments dependency on the road network, at this stage, the prediction is calculates by using only the data in point which the dependency is strong by means of cross-correlation, precisely given a parameter θ we consider j, when $Xcc(i, j) > \theta$ (Fig. 2).

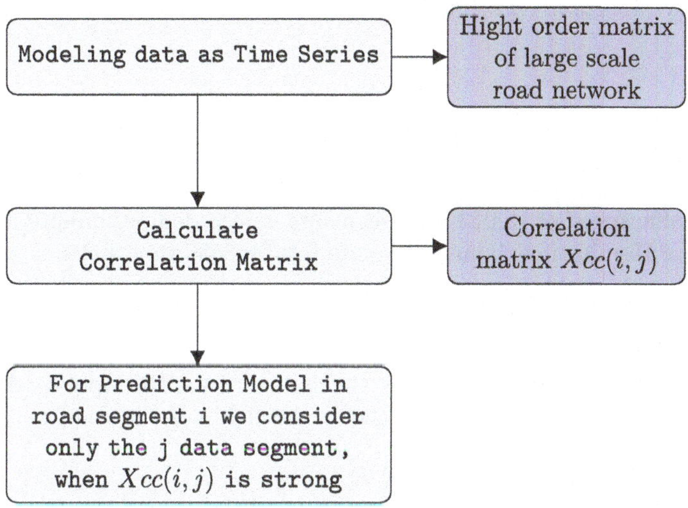

Fig. 2. Dependency detection process

5 Experiments

In this experiment, traffic data was obtained from Baidu research Open Access dataset [1]. Baidu is widely used by many researchers in experiments [2]. A large-scale traffic prediction dataset was provided for the 6th ring road (bounded by the lon, lat box of 116.10, 39.69, 116.71, 40.18), which is the most crowded area of Beijing. Figure 3 shows the spatial distribution of these road segments.

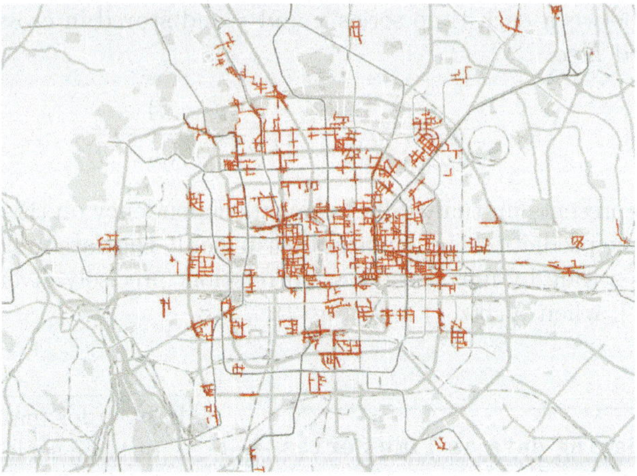

Fig. 3. Spatial representation of road segments

The traffic speed of 15,073 road segments is recorded per minute, then averaged with a 15-min time. Thus, there are totally 5856 time steps.

5.1 Experiment Setup

The data set used consists of two part, the first consist of the topology description of the road network. Table below shows the fields of the road network sub-dataset.

Field	Type	Description
link id	Char(13)	road segment id
width	Char(3)	width, 15:<3.0 m; (30:3.0 m, 5.0 m); 55:(5.5 m, 13 m); 130:<13 m
direction	Char(1)	direction, 0: unknown, default two-way; 1: two-way; 2: single-way, from start node to end node; 3: single-way, from end node to start node
snodegps	Char(30)	gps coordinate (lon, lat) of start node
enodegps	Char(30)	gps coordinate (lon, lat) of end

The second part is a data dump of speed represented as follow:

15257588940, 0, 42.1175
..., ..., ...
15257588940, 5855, 33.6599
1525758913, 0, 41.2719
..., ..., ...

The data is formatted as line delimited, in total there was 88,267,488 rows in the specific time period, this is 2.5 GB of zipped and approximately 8.5 GB unzipped. We limit our study only for the highway (length > 13 m) so we read only the line which the segment matched the identifier of the Highway (width field) stored in road network sub-dataset file, at this stage we use a fetch python program to extract the desired data from large dataset, then the result is formatted as large matrix and stored in CSV file, rows represent time stamp and column segment ID, figure below shows the variation of traffic for three segments (Fig. 4).

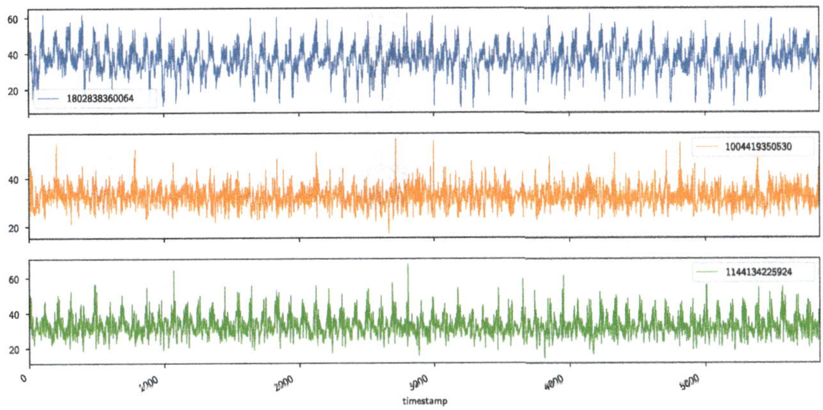

Fig. 4. Traffic averaged in three road points

The reason for formatting a data is to allow the reading in pandas data frame then calculate the cross- correlation with numpy LIB. Finally, Scatterplots of spatial cross-correlation was been used to visually reveal the causality behind road traffic segments.

Scatterplots of spatial cross-correlation can be used to reveal the causality between two variables visually (See Fig. 5). Based on the global cross-correlation coefficient, we can determine the data traffic segments used to predict the traffic in target points. A positive association means that both the variables are moving in same direction. If the coefficient is equal to 0, it does not necessarily mean that there is no relation between the two variables. It means that there is a no linear relationship, but there might be another type of functional relationship, for example, quadratic or exponential. If correlation is ±0.8 and above, high degree of correlation or the association between the dependent variables are strong. Correlation between ±0.5 to ±0.8, sufficient degree of correlation and less than ±0.5, weak correlation.

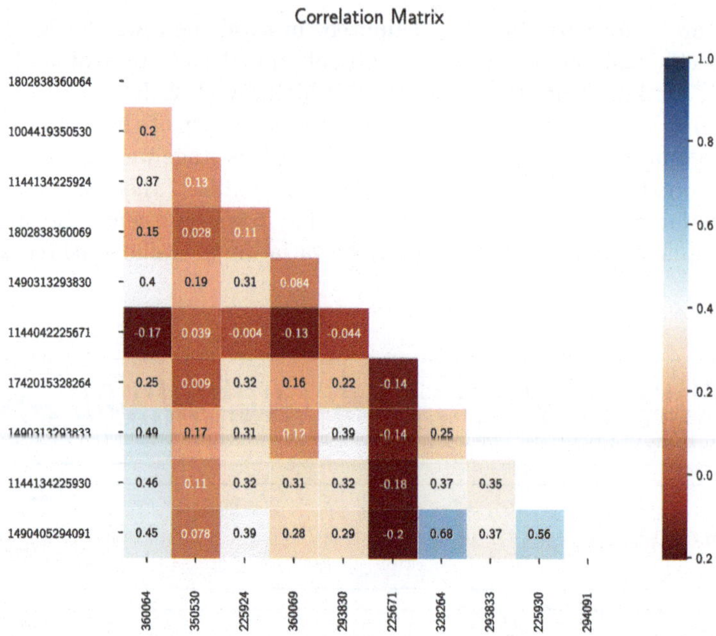

Fig. 5. Correlation matrix of 8 observation points

To conclude the best coefficient we make a comparison study between the three coefficients cited above [10], each coefficient extract the data used for prediction which is calculated by using use a legacy prediction method such ARIMA the following four case:

case 1: by using Person coefficient.

case 3: by using kendal coefficient.

case 2: by using Spearman coefficient.

Results are evaluated by means of root error.

$$RMSE = \sqrt{(\frac{1}{N})\sum_{i=1}^{n}(t_i - p_i)^2}$$

Where pi = predicted traffic flow; ti = actual traffic flow; N = the number of predictions. Table shows the prediction errors for different models.

5.2 Interpretation

For the three cases, the dependent segments set used in prediction calculation is not same, We consider the segments when $0.2 < \theta < 0.4$. The results are listed in the following table:

Correlation coefficient	RMSE
Person	70.144
Spearman	80.140
Kendal	89.847

It can be seen that the prediction is more accurate when using the dependents road data segments by means of Person cross-correlation. It was also observed that the prediction depends on the choice of used data segments. As shown in Fig. 6, the results are more conclusive if we limit data by considering a strong correlation.

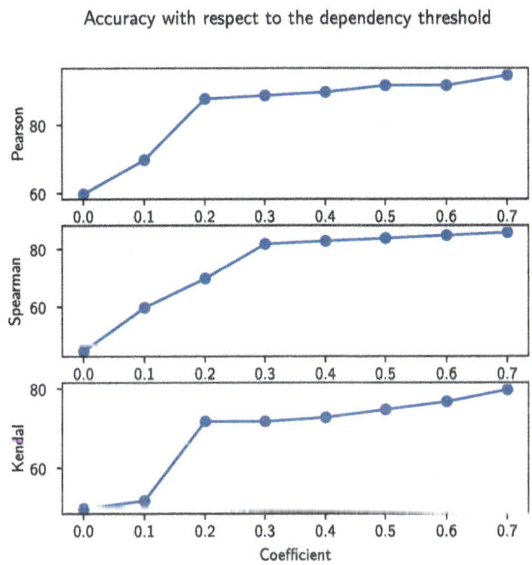

Fig. 6. Accuracy with respect to dependency threshold

6 Conclusion

This paper is devoted to laying the foundation for development of spatial cross-correlation theory in Road Traffic Forecasting. The basic measurements and analytical methods are put forward and applied to an urban study of China. Pearson's correlation coefficient and other coefficients can well reflect the relationship between data traffic denoted as dependency Road segments. Finally, on the basis of experimentation results and empirical analyses, we can conclude that statistical analysis for traffic forecasting can complement other approach such as machine learning methods and reduce data and time processing for the prediction calculation.

References

1. Baidu Research Open-Access Dataset. www.ai.baidu.com
2. Liao, B., et al.: Deep Sequence Learning with Auxiliary Information for Traffic Prediction (2018)
3. Yuan, N., Xoplaki, E., Zhu, C., Luterbacher, J.: A novel way to detect correlations on multi-time scales, with temporal evolution and for multi-variables (2016)
4. Chong, K., Sung, H.: Prediction of Road Safety Using Road/Traffic Big Data (2015)
5. Chen, Y.: A New Methodology of Spatial Cross-Correlation Analysis (2015)
6. Daxue, Q., Bao, X., Zhao, T., Zhang, Y., Zhou, Y., Feng, S.: Spatial cross correlations of traffic flows on urban road networks (2011)
7. Guo, S., et al.: Identifying the most influential roads based on traffic correlation networks (2019)
8. Su, F., Dong, H., Jia, L., Tian, Z., Sun, X.: Space-time correlation analysis of traffic flow on road network (2017)
9. Shi, K., Di, B., Zhang, K., Feng, C., Svirchev, L.: Detrended cross-correlation analysis of urban traffic congestion and NO2 concentrations in Chengdu (2018)
10. Hauke, J., Kossowski, T.: Comparison of values of person's and spearman's correlation coefficients on the same sets of data (2011)
11. William, B.M., Durvasula, P.K., Brown, D.E.: Urban freeway travel prediction: application of seasonal ARIMA and exponential smoothing models (1998)
12. Okutani, I., Stephanedes, Y.J.: Dynamic prediction of traffic volume through Kalman filtering theory (1984)
13. Xie, Y., Zhang, Y., Ye, Z.: Short-Term Traffic Volume Forecasting Using Kalman Filter with Discrete Wavelet Decomposition (2007)
14. Zhang, Y., Xie, Y.: Forecasting of Short-Term Freeway Volume with v-Support Vector Machines (2007)
15. Yu, G., Hu, J., Zhang, C., Song, G.: Short-term traffic flow forecasting based on Markov chain model (2003)
16. Wei, W., Wu, H., Ma, H.: An AutoEncoder and LSTM-Based Traffic Flow Prediction Method (2019)
17. Dai, X., Fu, R., Lin, Y., Li, L., Wang, F.Y.: DeepTrend: A Deep Hierarchical Neural Network for Traffic Flow Prediction (2017)
18. Chan, K.Y., Dillon, T.S.: On-road sensor configuration design for traffic flow prediction using fuzzy neural networks and Taguchi method (2013)
19. Pan, T.L., Sumalee, A., Zhong, R.X., Payoong, N.I.: Short-Term Traffic State Prediction Based on Temporal-Spatial Correlation (2013)
20. Baofeng, D.I., Kai, S., Kaishan, Z., Laurance, S., Xiaoxi, H.: Long-Term Correlations and Multifractality of Traffic Flow Measured By GIS for Congested and free-Flow Roads (2016)
21. Cheng, T., James, H.: Spatio-Temporal Autocorrelation of Road Network Data (2011)

On the Use of the Convolutional Autoencoder for Arabic Writer Identification Using Handwritten Text Fragments

Amina Briber(✉) and Youcef Chibani

Laboratoire d'Ingénierie des Systèmes Intelligents et Communicants (LISIC), Faculty of Electrical Engineering, University of Science and Technology Houari Boumediene (USTHB), 32, El Alia, Bab Ezzouar, 16111 Algiers, Algeria
{abriber,ychibani}@usthb.dz

Abstract. Convolutional autoencoders (CAE) are designed to reconstruct the input image to the output in a near-perfect way via a compact data namely encoded data containing relevant features. The encoded data can be used in various applications as for compressing or classifying the image. The present paper tries to investigate the use of the CAE for writer identification using handwritten text fragments. Hence, the CAE is used for generating features, which is fed to the distance-based classifier. Experimental evaluation is performed on the well-known IFN/ENIT dataset containing 411 writers. During training, a subset is selected from the 411 writers containing only 11 writers allowing to produce a lite CAE. Experimental results show an identification rate of 92.70% using the whole dataset when the feature vector is appropriately normalized.

Keywords: Writer identification · Handwritten · Text fragment · Convolutional autoencoders

1 Introduction

In recent decades, the writer recognition has been one of the most challenging and fascinating research areas in the field of individual recognition. The hypothesis that writing is an individualistic act has been proven by psychologists and psychoanalysts [1]. The writer recognition is divided in two categories: writer identification and writer verification. This study focuses on the writer identification problem, for which several studies have been reported using the whole document [2], paragraphs [3], lines of text [4], words [5], characters [6] and recently text fragments [7, 10]. In certain applications as for instance in forensics, few data are often available and therefore, the design of the writer identification system based on text fragments is considered as an interesting alternative way. Lastly, the writer identification from the text fragment has reached very interesting reliability levels as claimed by several authors [7, 8]. The writer identification on small text fragments is also effective in absence of the whole document.

Usually, a writer identification system is composed of various modules, which are preprocessing, feature generation, classification and decision. The feature generation

B. Lejdel et al. (Eds.): AIAP 2021, LNNS 413, pp. 301–311, 2022.
https://doi.org/10.1007/978-3-030-96311-8_28

is the cornerstone of the system. Indeed, this module aims to represent a handwritten document by a set of features to describe the writing style of the writer [7]. Hence, the feature generation can be performed via two approaches: handcrafted and feature learning methods. The handcrafted approach consists of manually developing targeted descriptors to extract only the desired information. Several descriptors have been developed and used in this context, as for instance, the texture namely LBP, LTP and LPQ [7]. In contrast, the feature learning approach uses deep learning algorithms to extract all the information. These algorithms are basically based on neural networks such as recurrent neural networks (RNN), convolutional neural networks (CNN) [11] or convolutional autoencoders (CAE) [12].

The present paper aims to explore for the first time the use of the convolutional auto-encoder (CAE) to extract features using handwritten text fragments. In contrast to CNN, the CAE does not require many data for training, which is considered as an advantage in real applications where, in certain circumstances, few data are available. For the classification module, the distance-based classifier is the most used to address the writer identification studies [8], since it offers an open system and quick execution time [8]. In this context, various distance kinds are used for classification when performed on handwritten text fragments [7–10]. In this paper, the simple Euclidean distance is used for writer identification using handwritten text fragments.

This paper is then organized as follows: Sect. 2 presents a brief review of convolutional autoencoders. Section 3 describes the proposed system. Section 4 devotes the experimental results and comparative analysis from the state of the art on text fragments for the Arabic writer identification. Finally, Sect. 5 presents out the conclusion and suggestions for future works.

2 Brief Review of the Convolutional Autoencoder

The standard autoencoder is an unsupervised neural network [13] designed to reconstruct the input data to the output in a near-perfect way via a compact data namely encoded data [14]. Formally, the autoencoder involves two parts, which are an encoding part and a decoding part as shown in Fig. 1.

The encoder allows producing an encoding data containing relevant features that are used next for reconstructing the image. Several kinds of autoencoders are proposed in the literature including sparse autoencoders [15], denoising autoencoders [15], convolutional autoencoders, variational autoencoders and so on [16].

Convolutional autoencoders (CAE) are well adapted for image processing for extracting all features from the input image [17]. CAE are based on a standard autoencoder architecture having convolutional encoding and decoding layers [18]. In convolutional autoencoders, the encoding process involves to performing a convolution operation on the input layer image in order to extract local features, thus obtaining a hidden layer input. The decoding process performs a deconvolution operation on the hidden layer data, so that the output image is reconstructed to the size of the input image [19].

This paper tries to investigate the use of the convolutional autoencoder as a feature generator associated to distance-based classifier for writer identification based on handwritten text fragments.

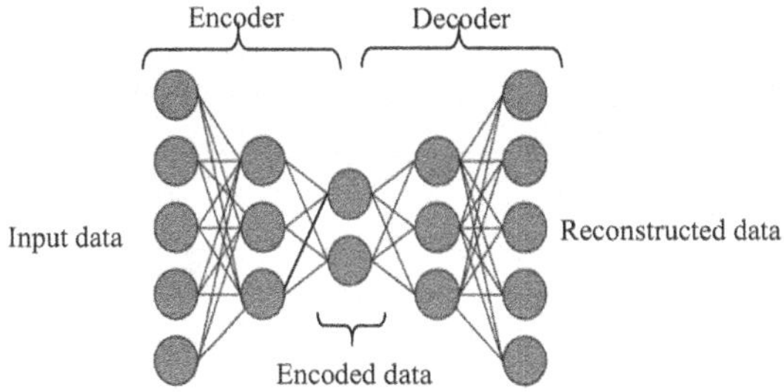

Fig. 1. Autoencoder architecture.

3 Description of the Proposed System

The proposed paper aims to propose an open system for writer identification using text fragments. Therefore, each writer is represented by a set of text fragments. As shown in Fig. 2, the proposed system contains two main modules: the feature generation module and the classification module.

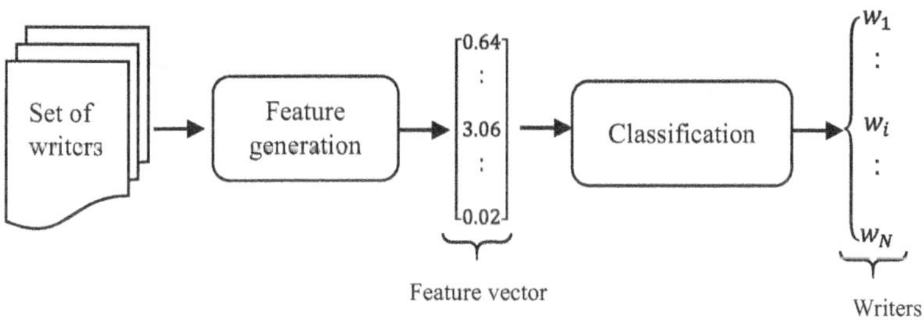

Fig. 2. Proposed writer identification system.

The feature generation is performed *via* the convolutional autoencoder, which is used as a feature generator for producing a feature vector. The resulting one is retrieved and then used as an input for the classification module using the Euclidean distance for identifying a writer. For better clarity, a more deeply description of each module is reported in the following sections.

3.1 Feature Generation

The feature generation is considered as a crucial step in the writer identification system. Its role is to extract features contained into the text fragments in order to form a feature vector that will be used as input for the classification module. As shown in Fig. 3, the design of the feature generator is performed into two steps recovering the feature vector *via* the CAE and then followed by a normalization way.

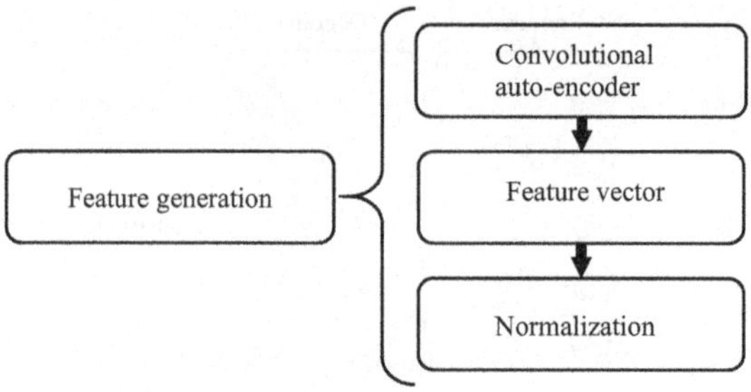

Fig. 3. Feature generation steps.

The CAE is used for representing the text fragment in a compact form, which defines the feature vector as shown in Fig. 4. The feature vector is considered representative when the output fragment is similar to the input fragment according to a measure criterion. The representative feature vector is found during the design step.

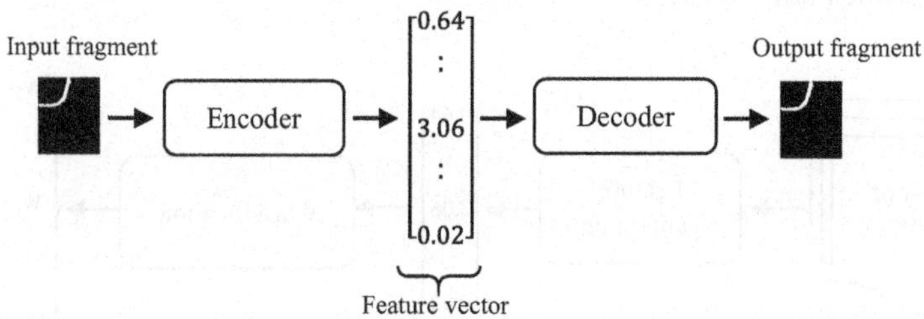

Fig. 4. Convolutional autoencoder for feature extraction.

The encoder and decoder of the proposed CAE are composed of convolution layers, an activation function (ReLU) and a pooling layer. Table 1 shows the architecture of the used CAE.

To ensure the homogeneity of the values contained into the feature vector, a normalization is performed for redistributing values in the range between zero and one. The mathematical formula for normalization is defined as follows [20]:

$$y_n = \frac{x_n^2}{\sum_{n=1}^{P} x_n^2} \tag{1}$$

P is the size of the feature vector, while x_n and y_n represent the no normalized and normalized feature vectors, respectively.

Table 1. Architecture of the proposed CAE.

	Layer type	Number of filters	Kernel size	Padding	Size of the output
Encode	Input image	–	–	–	100 × 100 × 1
	Convolution	5	3 × 3	Same	100 × 100 × 5
	ReLU	–	–	–	100 × 100 × 5
	MaxPooling	–	2 × 2	Same	50 × 50 × 5
	Convolution	10	3 × 3	Same	50 × 50 × 10
	ReLU	–	–	–	50 × 50 × 10
	MaxPooling	–	2 × 2	Same	25 × 25 × 10
	Convolution	20	3 × 3	Same	25 × 25 × 20
	ReLU	–	–	–	25 × 25 × 20
	MaxPooling	–	2 × 2	Same	13 × 13 × 20
Code	Flatten	–	–	–	3380
Decode	Convolution	20	3 × 3	Same	13 × 13 × 20
	ReLU	–	–	–	13 × 13 × 20
	UpSampling	–	2 × 2	–	26 × 26 × 20
	Convolution	10	3 × 3	Same	26 × 26 × 10
	ReLU	–	–	–	26 × 26 × 10
	UpSampling	–	2 × 2	–	52 × 52 × 10
	Convolution	5	3 × 3	Valid	50 × 50 × 5
	ReLU	–	–	–	50 × 50 × 5
	UpSampling	–	2 × 2	–	100 × 100 × 5
	Convolution	1	3 × 3	Same	100 × 100 × 1
	Output image	–	–	–	100 × 100 × 1

3.2 Classification

Inspired from [7], the proposed study also uses a dissimilarity measure between fragments of reference writer storage in the database and fragments of the query writer.

Let $r_j, j = 1, .., card(R)$ as the feature vector of the fragment belonging to the reference writer R and $card(R)$ is the number of fragments, and let $q_i, i = 1, .., card(Q)$ as the feature vector of the fragment belonging to the query writer Q and $card(Q)$ is the number of fragments. The dissimilarity measure namely Γ is calculated in order to compare between two writers, which is defined as follows:

$$\Gamma(Q, R) = \frac{1}{card(Q)} \sum_{i=1}^{card(Q)} \min_{r_j \in R}\left(d\left(q_i, r_j\right)\right) \tag{2}$$

Where $d(q_i, r_j)$ is the distance between two fragments (q_i, r_j), which can be computed *via* the Euclidean distance defined as follows:

$$d(q_i, r_j) = \sqrt{\sum_{n=1}^{P} (q_{ni} - r_{nj})^2} \quad (3)$$

P defines the size of the feature vector.

For identifying a writer, the dissimilarity measure is performed against all writers stored in the database. A decision rule is then performed for selecting the minimal dissimilarity measure among all calculated dissimilarity measures. Formally, the index of the writer namely W_{index} is obtained through the following equation:

$$W_{index}(Q) = \arg\min_{k=1,\ldots,K}(\Gamma(Q, R_k)) \quad (4)$$

Where K is the number of writers stored in the database.

4 Experimental Results

4.1 Dataset Description

For evaluating the proposed writer identification system, the well-known IFN/ENIT dataset is used. It includes 2,200 documents with more than 26,000 names of Tunisian cities and town villages written in Arabic collected from 411 different writers. In this paper, the text fragments are produced for training and testing according to Hannad et al. [7] for a fair comparison. Figure 5 shows some samples of the text fragments from the same writer [21].

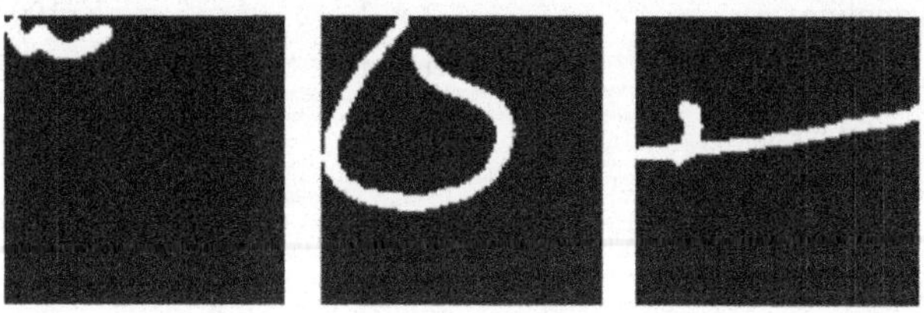

Fig. 5. Samples of the IFN/ENIT text fragments of the same writer.

4.2 Evaluation Criterion

The performance of the system is evaluated using the identification rate (IR), which is equivalently to Accuracy. It is defined as the ratio of the number of well identified writers to the total number of writers as follows:

$$IR(\%) = \frac{\text{Number of writers well identified}}{\text{Total number of writers}} \times 100 \quad (5)$$

This measure is used for a fair comparison against the state-of-art.

4.3 Experimental Design

For finding the optimal architecture of the CAE model offering the best identification rate, the first eleven (11) writers are selected for designing the proposed system while the remaining 400 writers are used for evaluation. The selected training fragments belonging to 11 writers are divided into two subsets. The first subset containing 80% of training fragments is used for training the CAE while the second subset containing 20% of training fragments is used for validation. Testing fragments of the eleven (11) writers are used for evaluating the robustness of the CAE. No data augmentation is performed during the training of the CAE. The number of writer's fragments varies from writer to another. Once the system is validated, the evaluation is performed on the whole dataset. For the training, the number of epochs is fixed at 350 and the binary cross entropy is used as a loss function. Figure 6 shows the accuracy (Identification Rate) and the loss of the model for training and validation samples.

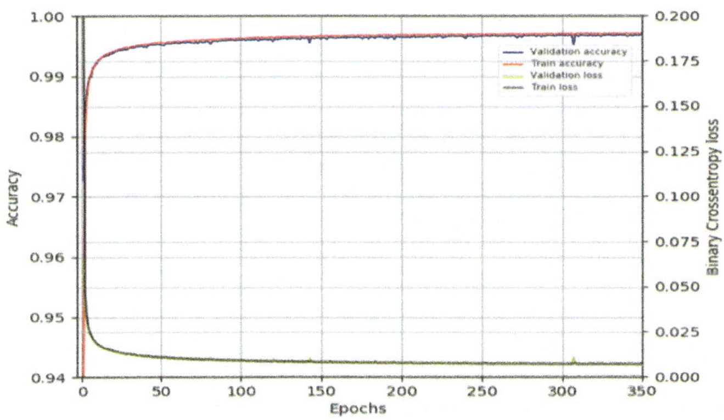

Fig. 6. Accuracy and the loss for training and validation.

As it can be seen, the model is well trained without overfitting and with a high accuracy, which means that the input image is reproduced on the output of the CAE as almost identical. Therefore, the feature vector retrieved from the CAE can be considered sufficiently relevant for using in the writer identification.

4.4 Experimental Evaluation

The proposed system is evaluated on the whole IFN/ENIT dataset containing 411 writers. The model is evaluated firstly on a subset of the IFN/ENIT dataset containing 11 writers and the remaining writers are added progressively in order to keep an open system. At first, the evaluation is performed on the feature vector directly without normalization. Afterward, a normalization is performed on the feature vector as described in Eq. 2 in order to show its influence. Figure 8 depicts the identification rate using normalized and no normalized feature vector, respectively.

As expected, the identification rate decreases when the number of writers increases. However, the normalized feature vector is better since it keeps its stability performance compared to the no normalized feature vector. It can be seen also that the normalized feature vector gives an encouraging identification rate of 92.70% with an increase of 6.57% from 86.13% using the feature vector without normalization. This experiment clearly shows that the normalization allows a considerable influence for improving the identification rate. Furthermore, the evaluation shows the stability performance of the proposed system against the number of writers. Indeed, when a new writer is added to the system, the identification rate remains almost stable.

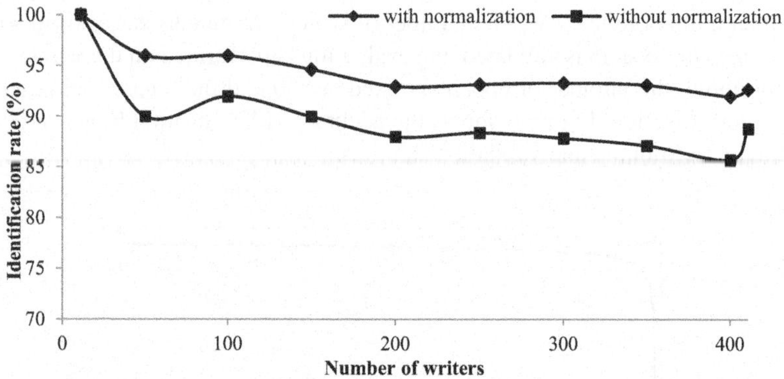

Fig. 8. Identification rate versus number of writers.

4.5 Comparative Analysis

In order to situate the proposed work among the works carried out from the state-of-art, Table 2 shows only selected single systems (no combined systems) for a fair comparison when using the fragmented IFN/ENIT dataset. As can be seen, all existing single systems use handcrafted features whereas the proposed system use a feature learning method using the convolutional autoencoder (CAE).

Hannad et al. [7, 9] used different descriptors associated to the Hamming distance classifier. The LPQ descriptor showed the best performance with an IR of 94.89%. While Hadjadji et al. [8] evaluated several descriptors with different classifiers. The LPQ descriptor associated to the k-means classifier has giving an IR of 93.67%.

Through this comparison, it can be deduced that the IR obtained by the proposed system is ranked in 4th position, with an IR of 92.70%, compared to the rates of the existing systems.

It is worth noting that systems designed by Hannad et al. [7] and Hadjadji et al. [8] deleted a component from the LPQ descriptor to improve the identification rate. According to [7, 8], the text fragments contain no-pertinent fragments that can influence the result. In the proposed system, all feature vector components are retained.

Among the existing systems, the proposed system is an open system. For instance, when a new writer is added to the system, the re-training of the CAE is not required. Furthermore, a small amount of data is required for training the CAE with only 11 writers without data augmentation.

Table 2. IR (%) obtained by the offered system against the state-of-the-art methods using text fragments for the IFN/ENIT dataset.

Method	Reference	Feature extraction	Classifier	IR (%)
Handcrafted	Hannad et al. [7]	LBP	Hamming distance	73.48
		LTP	Hamming distance	87.12
		LPQ	Hamming distance	94.89
	Hannad et al. [9]	HOG	Hamming distance	86.62
	Hadjadji et al. [8]	Run Length	k-Means	93.18
			k-Centers	88.07
			Hamming distance	92.45
	Hadjadji et al. [8]	oBIF	k-Means	82.23
			k-Centers	78.34
			Hamming distance	90.51
	Hadjadji et al. [8]	LPQ	k-Means	93.67
			k-Centers	88.32
Feature learning	**Proposed**	**CAE**	**Euclidian Distance**	**92.70**

The presented system is promising and offers an encouraging identification rate while it remains a very simple and lite system offering fast and effective results.

5 Conclusion

This paper proposed to investigate a new Arabic writer identification open system using convolutional autoencoder (CAE) and distance-based classifier from text fragments. The used approach is based on the CAE model for generating features performed on a subset of writers, each one is represented by all his fragments. In the sequel, the same model is used for identifying the query writer from all writers contained into the IFN/ENIT dataset using the distance-based classifier without retraining the model. The identification rate achieved 92.70% with a lite CAE model.

The use of CAE such as feature extractor for writer identification task-based text fragments shows an encouraging performance for capturing the relevant features of the writer style, despite the lite model used and the small amount of data used in the training step. For further works, an investigation is planned for reducing the size of the feature vector by trying to improve the performance.

Acknowledgement. This work was supported by the Direction Générale de la Recherche Scientifique et du Développement Technologique (DGRSDT) grant, attached to the Ministère de l'Enseignement Supérieur et de la Recherche Scientifique, Algeria.

References

1. Saks, M.J., Commentary on: Srihari, S.N., Cha, S.H., Arora, H., Lee, S.: Individuality of handwriting. J. Forensic Sci. **47**(4), 856–872 (2002). J. Forensic Sci. **48**(4), 916–920 (2003)
2. Chawki, D., Labiba, S.-M.: A texture based approach for Arabic writer identification and verification. In: 2010 International Conference on Machine and Web Intelligence (ICMWI), pp. 115–120. IEEE (2010)
3. Bulacu, M., Schomaker, L., Brink, A.: Text-independent writer identification and verification on offline Arabic handwriting. In: Ninth International Conference on Document Analysis and Recognition (ICDAR 2007), vol. 2, pp. 769–773. IEEE, September 2007
4. Abdi, M.N., Khemakhem, M.: A model-based approach to offline text-independent Arabic writer identification and verification. Pattern Recogn. **48**(5), 1890–1903 (2015)
5. Abdi, M., Khemakhem, M., Ben-Abdallah, H.: A novel approach for off-line Arabic writer identification based on stroke feature combination. In: 24th International (2009)
6. Idicula, S.M.: A survey on writer identification schemes. Int. J. Comput. Appl. **26**(2), 23–33 (2011)
7. Hannad, Y., Siddiqi, I., El Kettani, M.E.Y.: Writer identification using texture descriptors of handwritten fragments. Expert Syst. Appl. **47**, 14–22 (2016)
8. Hadjadji, B., Chibani, Y.: Two combination stages of clustered one-class classifiers for writer identification from text fragments. Pattern Recogn. **82**, 147–162 (2018)
9. Hannad, Y., Siddiqi, I., El Merabet, Y., El Youssfi El Kettani, M.: Arabic writer identification system using the histogram of oriented gradients (HOG) of handwritten fragments. In: Proceedings of the Mediterranean Conference on Pattern Recognition and Artificial Intelligence, pp. 98–102, November 2016
10. Tang, Y., Wu, X., Bu, W.: Offline text-independent writer identification using stroke fragment and contour based features. In: 2013 International Conference on Biometrics (ICB), pp. 1–6. IEEE, June 2013
11. He, S., Schomaker, L.: Fragnet: Writer identification using deep fragment networks. IEEE Trans. Inf. Forensics Secur. **15**, 3013–3022 (2020)
12. Zhu, Y., Wang, Y.: An offline text-independent writer identification system with SAE feature extraction. In: 2016 International Conference on Progress in Informatics and Computing (PIC), pp. 432–436. IEEE, December 2016
13. Dong, C., Xue, T., Wang, C.: The feature representation ability of variational autoencoder. In: 2018 IEEE Third International Conference on Data Science in Cyberspace (DSC), pp. 680–684. IEEE, June 2018
14. Ng, A.: Sparse autoencoder. CS294A Lecture Notes **72**(2011), 1–19 (2011)
15. Bank, D., Koenigstein, N., Giryes, R.: Autoencoders. arXiv preprint arXiv:2003.05991 (2020)
16. Guo, X., Liu, X., Zhu, E., Yin, J.: Deep clustering with convolutional autoencoders. In: Liu, D., Xie, S., Li, Y., Zhao, D., El-Alfy, E.S. (eds.) Neural Information Processing, ICONIP 2017, vol. 10635, pp. 373–382. Springer, Cham (2017). https://doi.org/10.1007/978-3-319-70096-0_39
17. Gondara, L.: Medical image denoising using convolutional denoising autoencoders. In: 2016 IEEE 16th International Conference on Data Mining Workshops (ICDMW), pp. 241–246. IEEE, December 2016
18. Masci, J., Meier, U., Cireşan, D., Schmidhuber, J.: Stacked convolutional auto-encoders for hierarchical feature extraction. In: Honkela, T., Duch, W., Girolami, M., Kaski, S. (eds.) Artificial Neural Networks and Machine Learning – ICANN 2011. LNCS, vol. 6791, pp. 52–59. Springer, Heidelberg (2011). https://doi.org/10.1007/978-3-642-21735-7_7

19. Kundur, D., Hatzinakos, D.: Blind image deconvolution. IEEE Signal Process. Mag. **13**(3), 43–64 (1996)
20. Christlein, V.: Handwriting analysis with focus on writer identification and writer retrieval (2019)
21. Awaida, S.M., Mahmoud, S.A.: State of the art in off-line writer identification of handwritten text and survey of writer identification of Arabic text. Educ. Res. Rev. **7**(20), 445–463 (2012)

Security Issues in Self-organized Ad-Hoc Networks (MANET, VANET, and FANET): A Survey

Sihem Goumiri[1](✉), Mohamed Amine Riahla[2], and M'hamed Hamadouche[2]

[1] Ingénierie des Systèmes et Télécommunications Laboratory, University of Boumerdes, 35000 Boumerdes, Algeria
s.goumiri@univ-boumerdes.dz

[2] LIMOSE Laboratory, University of Boumerdes, South Campus, 35000 Boumerdes, Algeria
ma.riahla@univ-boumerdes.dz

Abstract. Self-organized AdHoc networks have become one of the most interested and studied domains, especially with the rapid development of communication technologies and electronic devices. These networks regroup wireless and self-configuring nodes that communicate independently without a fixed infrastructure. Many applications operate with the AdHoc network due to its rapid deployment and low costs. Security in AdHoc networks is a crucial aspect that protects the exchanges between users and improves network performances. In this paper, a presentation of three AdHoc networks: MANET (Mobile AdHoc Network), VANET (Vehicle AdHoc Network), and FANET (Flaying AdHoc Network) is performed with the focus on their security issues. The paper blends the security requirements and the different attacks faced to the three reviewed networks.

Keywords: Security · Self-organized networks · AdHoc networks · MANET · VANET · FANET

1 Introduction

A self-organized AdHoc network is a dynamic, autonomous, and wireless system composed of a group of mobile devices able to communicate independently in the network area [1]. Each mobile device is an autonomous and self-configuring node that acts in the network without needing any central administration [1, 2]. The number of nodes and links is varied over time, which frequently changes the network topology. Recently, these networks are present in many fields and applications because they offer great benefits like rapid deployment and low costs. A self-organized AdHoc network is a large definition that gathers very diverse network technologies like those introduced in this survey: MANET (Mobile Ad hoc network), VANET (Vehicular Ad hoc Network), and FANET (Flaying AdHoc network). These networks encountered many serious problems and challenges about maintaining the normal function of the network or improving its performances. Indeed, ensuring security in such an environment is the greatest challenge

B. Lejdel et al. (Eds.): AIAP 2021, LNNS 413, pp. 312–324, 2022.
https://doi.org/10.1007/978-3-030-96311-8_29

for searchers. The network must be resilient to different risks and provide many alternative solutions faced with attacks. In addition, nodes and data must be present in a safe environment to accomplish a predefined mission or ensure the purpose of the network deployment.

This paper reviews the security issues and requirements in the self-organized AdHoc networks (MANETs, VANETs, and FANETs). The remaining parts of this survey are planned as follows:

The first section presents background about three existing and popular AdHoc networks (MANETs, VANETs, and FANETs). The second section describes the security issues and services in AdHoc networks. The third section lists the potential attacks and threats on MANET, VANET, and FANET. The last section concludes the paper and designates our future direction of searches.

2 Self-organized AdHoc Networks

2.1 Background

In the last years, the AdHoc network interested industry and academia due to its intended fields of application. By adjusting dimensional parameters and using new technologies of devices, new AdHoc network subcategories emerged like MANET (Mobile Ad hoc network), VANET (Vehicles Ad hoc network), and FANET (Flying Ad hoc network).

MANET (Mobile Ad hoc network) [3] is a wireless mobile system composed of nodes that communicate with wireless links Fig. 1. Its main characteristics are the absence of any fixed infrastructure and the self-configuring nodes, which makes them able to establish communications, exchange information, and ensure network functionalities. The network size of MANET is frequently changed over time due to nodes newly joined the networks and those dynamically leaved (roaming). Today with the popularity of mobile devices (smartphones, sensors, pc, etc.) MANET is present in many military and civil fields like room class conferences, emergency rescue operations, and military control.

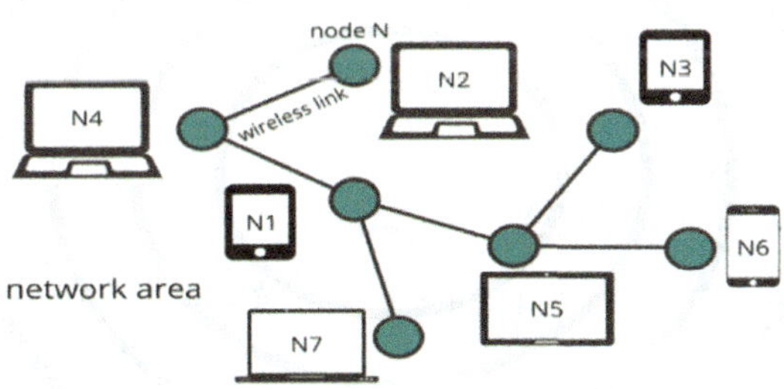

Fig. 1. Mobile AdHoc network (MANET)

VANET (Vehicular AdHoc Network) is a technology for managing road traffic and providing a safe driving environment [4]. The network is composed of a set of vehicles present in the road Fig. 2. Vehicles communicate and exchange information with each other by using two communication modes. The first one is direct Vehicle-to-Vehicle communication (V2V) that allows establishing immediate communication between vehicles in the same network. The second mode is a vehicle to interface (V2I) that requires the connection to a fixed infrastructure unit called RSU (roadside unit). This interface allows communication between vehicles, monitors them, and provides them access to the Internet cloud.

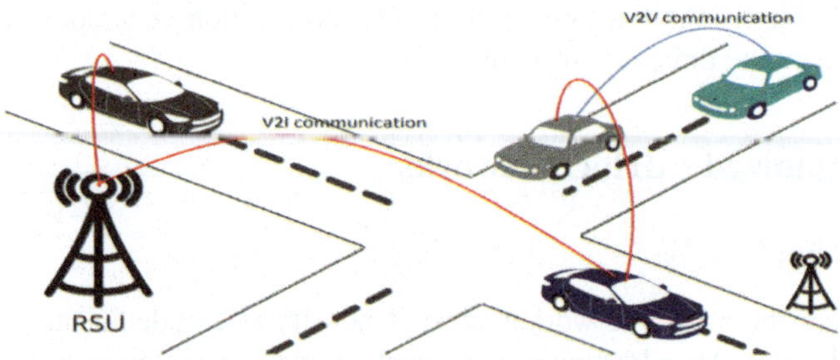

Fig. 2. Vehicle AdHoc network (VANET)

FANET (Flying AdHoc Networks) is a subset of MANETs that uses AdHoc communication in a three-dimensional plane. The network is composed of a collection of UAVs (Unmanned Aerial Vehicles) [5] able to execute a predefined mission Fig. 3. UAVs are small aerial vehicles equipped with sensors and advanced computing devices. FANET inherits the same features of MANET except that nodes can fly autonomously in the network producing higher mobility degrees. Two communication modes are distinguished: Air-to-air wireless communications (A2A) using the AdHoc mode and air-to-ground wireless communications (A2I) using infrastructures like ground stations or satellites. Overall, this portion of networks is called to ensure dangerous tasks related to disasters, target detection for security services or rescue operations, monitoring, etc.

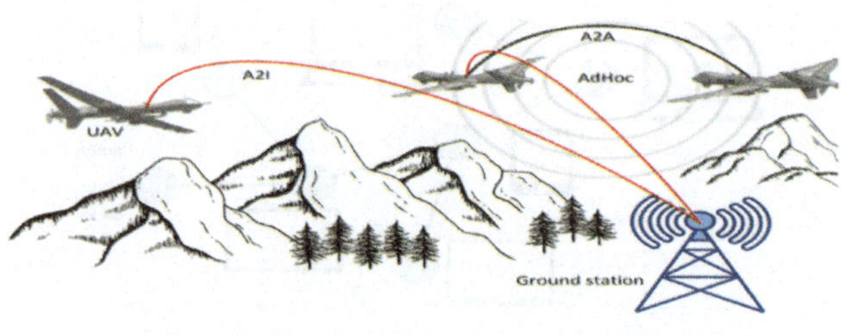

Fig. 3. Flaying AdHoc network (FANET)

2.2 Main Features

Self-organized AdHoc networks introduced in this paper have some features in common. The main ones are:

- The self-organized nodes,
- The use of existing nodes for managing the network traffic (i.e. node acts as hosts and routers at the same time),
- The multi-hop rooting to transfer information,
- The distributed mobile system in which the topology is randomly changed.
- The limited physical security due to the movement of nodes.

Other specific characteristics that make the difference between MANETs, VANETs, and FANETs are listed as follow:

- The nodes used in the network: heterogeneous or homogeneous in type, nodes interacted in dynamic networks are numerous and depend on the purpose of designing the network. Vehicles, sensor devices, or UAVs are the famous types of equipment that could be present with the existing networks.
- The environment dimension: this indicates the movements of nodes in the coverage area of the network. In some technologies, nodes move close to the ground, and in others, nodes can fly in free space.
- The speed of nodes: random, height, or slow movements of nodes characterize networks. This metric identifies the mobility level that changes the network topology.
- The energy of nodes: This feature differs for each dynamic network. Depending on the mission of nodes, some technologies require devices with a High capacity of energy, but others tolerate equipment with a low capacity of energy.

Table 1. Differences between the network features

Networks	Nodes	Environment	Node speed	Node energy	Node density
MANET	PC, sensor, smart phone,	Random 2-D	Lower	Low	Low
VANET	Vehicles	linear trajectory 2-D	Higher	High	High
FANET	UAVs systems	Random 3-D	Most higher	High	Low

3 Security Issues and Services

Security is an important factor and a veritable challenge in dynamic networks [6]. The main objective here is to preserve the security of nodes, data, and services. Indeed, the network must protect core services [7] that ensure some policies and requirements. It includes authentication, privacy, and non-repudiation of nodes on the one hand and integrity and confidentiality of the exchanged data on the other hand. Moreover, the network must be resilient faced with different attacks. These objectives are challenged in the AdHoc mode due to its specific feature previously detailed. The services of self-organized AdHoc networks (MANET- VANET- FANET) to be protected are highlighted and defined bellows:

- Availability refers to ensure the operation of the service provided by the network [6]. The network must ensure the role of all nodes during their life cycle (even those attacked). Before deploying any dynamic networks, it is essential to implement alternate solutions that always ensure communications between nodes in case of attacks.
- Authentication provides trustable communications between the network nodes. This service ensures the real identity of nodes by using methods like certification [8]. Researches in this area are numerous and handled many challenges because of the limited features of dynamic networks.
- Confidentially is the way to define permissions that allow nodes accessing to a specified data and services. This service ensures transiting information securely between nodes [8]. The main application to ensure confidentiality employs encryption methods. However, improving this service in dynamic networks is challengeable, making researches always opened.
- Integrity means not manipulate the message circulated in the network. Therefore, attacks against integrity attempt to modify or delete the content of packets transiting between nodes [6].
- Non-repudiation service associates delivered data and behavior to the correct node that sent any packet in the network [8]. Such a service is essential to have traceability and prevent erasing information related to an attack.

4 Attacks on AdHoc Networks

Various attacks against AdHoc networks were examined [9–11] in this paper. All these attacks share similar effects that paralyze the network services and degrade performances. Depending on the status, the behavior, or the purpose of the attacker, authors in [12] define three attacks classifications. In the first one, the main parameter is whether the attacker belongs to the network or not, which specifies the intern and the extern attacks. The intern attacks are the most challenged because existing solutions proposed for the extern attacks are not applicable here. The second classification includes active and passive attacks. Here, the active attacks attempt only to monitor and analyze the network traffic and make in danger the confidentiality service. The passive attacks execute actions like modification, injection, or damage of messages. They disturb the correct operation of the network and target availability, integrity, authentication, or non-repudiation services. Specific classes of attackers existed in the literature for VANET. Authors in [13] define Malicious vs. Rational attackers, Local vs. Extended attackers, and Independent vs. Colluding attackers. The former category determines the benefit of the adversary while executing an attack in the network. Rational to seek personal benefit, else the attacker is malicious. The second category is based on the area focused by the attacker: local for a limited area and extended for a large scoop. The last category includes the cooperation between adversaries to execute an attack: One attacker and no cooperation define independent attackers in the different cases the attackers are colluding.

MANET, VANET, and FANET are vulnerable to many attacks. The medium used in such networks helps attackers to penetrate the networks, thus causes problems like seize control, sabotage the mission on the network or just capture sensitive data. In the literature, many studies have classified the Ad-hoc security attacks into categories [14–16] as explained in the following:

4.1 Sniffing Attacks

A sniffing attack is also known as an eavesdropping or snooping attack. The wireless AdHoc networks are prone to such attacks due to their specific features like the shared wireless medium [14]. Some studies in the literature have classified the sniffing attack as a passive or an active attack [15, 16]. The main objective of a malicious node or an attacker is to target personal or sensitive data transmitted between nodes to grab confidential information. Reaching these objectives with a passive attack means just listening to the transited messages during the wireless transmissions. In an active attack, the attacker sends some queries to the target nodes to build a friendly relationship and exchange information.

4.2 Modification Attacks

Modification attacks affect the integrity of the messages in the network. Some of these attacks are based on changing the routing packets by using different methods as needed. The self-organized nodes in AdHoc networks make the routing messages in danger face to attackers. Indeed, these malicious nodes apply packets misrouting by changing some routing information like the sequence number, the count hops, or the source node. A different scenario of modification attacks is called a spoofing attack [17]. Here, the attacker falsifies its real identity to create some weakness in the networks. Executing such an attack is based on different methods like changing the IP or the MAC addresses.

4.3 Fabrication Attacks

Similar to the modification attacks, the fabrication attacks aim at the routing messages [18]. However, in this case, attackers do not modify the existing packets; they create their ones and damage the network services. This category includes attacks that generate false route error messages. Thus, destruct valid routes and cause a denial of service or sleep deprivation. Other attacks under this category exist in state-of-the-art like Rushing Attacks, Wormhole Attacks, and Gray Hole Attack, which will be described further with more details.

4.4 Selfish Behavior Attacks

Nodes in AdHoc mode need to cooperate to ensure the operation of the network [19]. Some nodes in (MANET, VANET, and FANET) are uncooperative due to selfish reasons. Consequently, some important tasks like rooting are not correctly performed. The selfish behavior stops or slows the traffics at the malicious node, which can interrupt the operation of the whole network.

4.5 Routing Attacks

Routing is a service to find routes for exchanging data between sources and destinations. This process is ensured using routing protocols that have been designed depending on the network characteristics and constraints. Given the importance of this service, many routing protocols have suffered from various attacks, and searches have explored different challenges in this area.

5 Potential Attacks and Threats on MANET VANET and FANET

5.1 Attacks on MANET

Mobile AdHoc Networks (MANETs) are vulnerable to numerous attacks [9, 20]. The following table briefly describes some exiting attacks by focusing on their types, the affected layer of the OSI model, and the targeted security services.

5.2 Attacks on VANET

Vehicles AdHoc Networks (VANETs) security has received several attentions from researchers and industry [10]. VANETs security aims to provide safety applications that manage road traffic and avoid the loss of human life. VANETs are a subset of AdHoc networks; thus, common attacks described in Table 1 also exist with this portion of networks. Table 2 lists some of the most popular attacks encountered in VANET and those examined in [13, 21, 22]:

5.3 Attacks on FANET

Flaying AdHoc Networks (FANETs) are a subclass of AdHoc networks, where the need for security is highlighted as a crucial aspect. Indeed, FANETs hold all the classical security issues previously discussed and designs new problems. Table 3 lists the potential attacks targeting the stability of the network services in FANETs [11, 23, 24] (Table 4).

Table 2. Different type of attacks on MANETs

Attacks	Concepts	Types	Layers	Services
- Eavesdropping	- Capture and extract sensitive and personal information (data and routing packets)	- Passive or - Active	- Network	Authentication
- Jamming attack	- Injection of noisy signals at the physical layer and affect communication - Leads to a denial of service	- Active - External	- Physical	- Availability Service integrity
- Collision attack	- Injection of false control messages in parallel times with other transmissions to make a collision and drop messages - Leads to a denial of service	- Active - External	- Link	Availability Data integrity
- Wormhole- - Blackhole - Greyhole	- Modifying routes and pretend node to have the optimal route - Redirecting the network traffic through particular links - Collecting, modifying, or deleting data - Leads to a denial of service	- Active	- Network	Availability Data integrity Authentication
- Routing (Table Overflow, Table Poisoning, Replication, Rushing…etc.)	- Attacks against the routing protocols implemented in the networks - Creating false routes, sending pretended updates, replicating expired packets, etc. - Prevent creating routes, isolating some parts of the network, height consummation in bandwidth, and power energy	- Active	- Network	Availability Data integrity

(*continued*)

Table 2. (*continued*)

Attacks	Concepts	Types	Layers	Services
- Sybil	- Duplicating the malicious node with multiple identities of nonexistent nodes - Affecting the cooperation between nodes - Leads to a denial of service	- Active	- Network	Availability Service integrity
- IPspoofing	- Impersonation of identity using falsified IP addresses - Denying services and avoid locating the source of the attack - Making a trust relationship between two nodes to take control of one of them	- Active	- Network	Availability Service integrity Non-repudiation
- State pollution	- Falsifying parameters in replay - Allocating an occupied IP address to new nodes - Flooding the network by a broadcast of duplication address detection messages	- Active	- Network	Availability Data integrity
- SYN Flooding	- Sending many SYN requests for establishing connections with a victim node	- Active	- Transport	Availability

Table 3. Different type of attacks on VANETs

Attacks	Concepts	Attacker	Services
- Denial of service - DoS I distributed DoS	- Many artificial massages in successive transmission by one (DoS) or a group of attackers (Distributed DoS)for flooding and jamming the network	- DoS: Outsider, Active, Local, Independent - Distributed DoS: Insider, Active, Colluding	Availability
- Blackhole	- Area with no participating node in communication - Loss of messages and data packets	- Passive, Outsider	Availability
- GPS Spoofing	- Giving neighbor nodes false information of locations to fake positions	- Active, Insider, Independent	Authentication
- Sybil attack	- Multiple identities for one attacker - Illusion nodes by multiple messages transmitted from different sources	- Insider, Active, Local	Authentication
- Illusion Attack	- An application to fabricate message using a voluntarily placed sensor - Transmitting false traffic information	- Insider, Malicious	Data integrity Data trust
- Greedy drivers	- Using the network resources for the own benefit of the attackers - Altering traffic information and jamming the network by false massages to deviate another node from the road	- Insider, Rational, Active	Data integrity
- Timing	- Delaying messages and do not transmit critical information to neighbors in real-time leading to fatal consequences	- Insider, Rational	Data integrity

(continued)

Table 3. (*continued*)

Attacks	Concepts	Attacker	Services
- Wormhole	- Two attackers are positioned in the most vital place in the network and create a band wormhole (tunnel) - Intercepting communications and tunneling packets from one location to other vehicles in different locations	- Insider, Extended, Passive, Colluding	Authentication, Confidentiality

Table 4. Different type of attacks on FANETs

Attacks	Concepts	Types	Layers
- Denial of service	- Making the network resources unavailable, so paralyzing the network services	- Internal or external - Active	- Availability
- GPS Jamming	- Harm on the navigation to prevent the correct supervision of location, routes, altitude, and direction - Losing the fly control	- External	- Availability
- Worm hole	- Recording packets from a vital place in the network - Tunneling the gathered packets to another place to retransmit them	- Internal	- Availability
- Sinkhole	- The attacker broadcast false routing information to attract all the network traffic	- Internal	- Availability
- Power-consuming	- Initiating a big volume of data in the network to make reacting all the network nodes	- External	- Availability
- Man in the middle	- The attacker places itself between the UAV and its Ground Control Station to capturing data from their communication link	- External	- Integrity

(*continued*)

Table 4. (*continued*)

Attacks	Concepts	Types	Layers
- Rooting loop	- Modifying the routing information to prevent nodes from finding the valid routes	- External	- Integrity - Availability
- Malware Software	- Injecting malware like a backdoor in the internal navigation system of UAVs and the control station - Automatically accept commands by the attacker to take control or pull sensitive data	- Internal - External	- Integrity - Availability - Privacy

6 Conclusion

The self-organized AdHoc system is the new network generation that offers very significant applications for users. Their specific features like the dynamic topology, the self-configuring nodes, and the wireless communications make the end users or the whole network prone to different attacks. In this paper, we interest in MANET, VANET, and FANET as important real-life examples of the AdHoc network. The paper introduces the concept of each network and presents the security requirements with the existing attacks. By the end of this review, searchers must invest in ad-hoc network security as a hot topic of search to provide new safety applications and enhance numerous fields. Through the various researches carried out in this area of research, we define our future interests. Our attention will be centered on finding solutions for specific attacks by combining or using methods newly introduced in the literature.

References

1. Ganesan, S., Loganathan, B.: A survey of ad-hoc network: a survey. Int. J. Comput. Trends Technol. (IJCTT) **4** (2013)
2. Student, V.R.P., Dhir, R.: A study of ad-hoc network: a review. Int. J. **3**(3) (2013)
3. Basagni, S., Conti, M., Giordano, S., Stojmenovic, I. (eds.): Mobile Ad Hoc Networking. Wiley, Hoboken (2004)
4. Zeadally, S., Hunt, R., Chen, Y.S., Irwin, A., Hassan, A.: Vehicular ad hoc networks (VANETS): status, results, and challenges. Telecommun. Syst. **50**, 217–241 (2012)
5. Bekmezci, I., Sahingoz, O.K., Temel, Ş: Flying ad-hoc networks (FANETs): a survey. Ad Hoc Netw. **11**(3), 1254–1270 (2013)
6. Zhou, L., Haas, Z.J.: Securing ad hoc networks. IEEE Netw. **13**(6), 24–30 (1999)
7. Loo, J., Mauri, J.L., Ortiz, J.H. (eds.): Mobile Ad Hoc Networks: Current Status and Future Trends. CRC Press (2016)
8. Liu, G., Yan, Z., Pedrycz, W.: Data collection for attack detection and security measurement in mobile ad hoc networks: a survey. J. Netw. Comput. Appl. **105**, 105–122 (2018)
9. Abdel-Fattah, F., Farhan, K.A., Al-Tarawneh, F.H., AlTamimi, F.: Security challenges and attacks in dynamic mobile ad hoc networks MANETs. In: 2019 IEEE Jordan International Joint Conference on Electrical Engineering and Information Technology (JEEIT), pp. 28–33. IEEE, April 2019

10. Kumar, A., Bansal, M.: A review on VANET security attacks and their countermeasure. In: 2017 4th International Conference on Signal Processing, Computing and Control (ISPCC), pp. 580–585. IEEE, September 2017
11. Singh, K., Verma, A.K., Aggarwal, P.: Analysis of various trust computation methods: a step toward secure FANETs. In: Computer and Cyber Security: Principles, Algorithm, Applications, and Perspectives, pp. 171–194. CRC Press (2018)
12. Di Pietro, R., Guarino, S., Verde, N.V., Domingo-Ferrer, J.: Security in wireless ad-hoc networks–a survey. Comput. Commun. **51**, 1–20 (2014)
13. Malhi, A.K., Batra, S., Pannu, H.S.: Security of vehicular ad-hoc networks: a comprehensive survey. Comput. Secur. **89**, 101664 (2020)
14. La Polla, M., Martinelli, F., Sgandurra, D.: A survey on security for mobile devices. IEEE Commun. Surv. Tutor. **15**(1), 446–471 (2013)
15. Anand, M., Ivesy, Z.G., Leez, I.: Quantifying eavesdropping vulnerability in sensor networks. In: Proceedings of the 2nd International VLDB Workshop on Data Management for Sensor Networks (2005)
16. Wang, Q., Dai, H., Zhao, Q.: Eavesdropping security in wireless ad hoc networks with directional antennas. In: 2013 22nd Wireless and Optical Communication Conference, Chongqing, China, pp. 687–692 (2013). https://doi.org/10.1109/WOCC.2013.6676462
17. Al-shareeda, M.A., Anbar, M., Manickam, S., Hasbullah, I.H.: Review of prevention schemes for modification attack in vehicular ad hoc networks. Int. J. Eng. Manag. Res. **10** (2020)
18. Sbai, O., Elboukhari, M.: Classification of mobile ad hoc networks attacks. In: 2018 IEEE 5th International Congress on Information Science and Technology (CiSt), pp. 618–624. IEEE, October 2018
19. Rajesh, M.: A review on excellence analysis of relationship spur advance in wireless ad hoc networks. Int. J. Pure Appl. Math. **118**(9), 407–412 (2018)
20. Meddeb, R., Triki, B., Jemili, F., Korbaa, O.: A survey of attacks in mobile ad hoc networks. In: 2017 International Conference on Engineering & MIS (ICEMIS), pp. 1–7. IEEE, May 2017
21. Hezam, M.A., et al.: Classification of security attacks in VANET: a review of requirements and perspectives (2018)
22. Saggi, M.K., Sandhu, R.K.: A survey of vehicular ad hoc network on attacks and security threats in VANETs. In: International Conference on Research and Innovations in Engineering and Technology (ICRIET 2014), pp. 19–20, December 2014
23. Bekmezci, İ, Şentürk, E., Türker, T.: Security issues in flying ad-hoc networks (FANETS). J. Aeronaut. Space Technol. **9**(2), 13–21 (2016)
24. Sumra, I., Sellappan, P., Abdullah, A., Ali, A.: Security issues and challenges in MANET-VANET-FANET: a survey. EAI Endorsed Trans. Energy Web **5**(17) (2018)

A Comprehensive Study of Multicast Routing Protocols in the Internet of Things

Issam Eddine Lakhlef(✉), Badis Djamaa, and Mustapha Reda Senouci

Distributed and Complex Systems Lab, Ecole Militaire Polytechnique, Algiers, Algeria
issameddinelakhlef@gmail.com

Abstract. IP multicast is a desired communication feature in the Internet of Things (IoT) as it provides noticeable resource savings, especially for Low-power and Lossy Networks (LLNs). Indeed, multicast allows cost-, energy-, and time-efficient networking for a multitude of LLN applications ranging from over-the-air programming and information sharing to device configuration and resource discovery. In this context, several multicast routing protocols have been recently proposed for LLNs including Stateless multicast RPL Forwarding (SMRF), Enhanced SMRF (ESMRF), Bi-Directional multicast Forwarding Algorithm (BMFA), and multicast Protocol for LLNs (MPL). Nevertheless, each protocol has been evaluated under different conditions, topologies, and traffic flow, which prevents making comprehensive comparisons of their characteristics and performance. In this paper, we provide an overview of recent LLN multicast protocols followed by a multidimensional performance evaluation of the most popular ones to extract their advantages and drawbacks under different traffic conditions, routing scenarios, and network topologies. Obtained results from extensive realistic simulations using Cooja show that, although each protocol is dominant under specific conditions, MPL remains the best in terms of packet delivery ratio in all scenarios at the expense of extra energy consumption, which requires new resources-aware multicast solutions for the IoT.

Keywords: Internet of Things · IP multicast · MPL · SMRF · ESMRF · BMFA

1 Introduction

The Internet has now connected the whole world starting from mainframes and servers to personal devices and objects. This becomes possible thanks to the massive explosion of low-cost smart objects in our lives along with their inevitable connection to the global Internet to offer unprecedented opportunities and services in a multitude of fields, including smart healthcare, smart agriculture, and smart surveillance. Such opportunities have given rise to the so-called Internet of Things (IoT). The things in the IoT are made up of sensors and/or actuators

B. Lejdel et al. (Eds.): AIAP 2021, LNNS 413, pp. 325–335, 2022.
https://doi.org/10.1007/978-3-030-96311-8_30

that perform a specific function and they are part of an infrastructure allowing the transport, storage, processing, and access to gathered data [4]. Such objects, however, usually operate under limited resources in terms of energy, computation, storage, and bandwidth.

These constraints impose strict challenges on all the layers of the TCP/IP networking stack, especially at the network layer, which needs to provide efficient routing protocols adapted to the constraints of this environment. Depending on the need, we have two main types of routing, the first is unicast and the second is multicast. Unicast is the most used mode for data exchange and it is fulfilled by the recently standardized Routing Protocol for Low-power and Lossy Networks (RPL) [12]. Nevertheless, other real-world IoT functionalities and applications, such as network configuration, resource discovery, and security management would be better served by efficient multicast routing protocols like SMRF [9], ESMRF [1], and MPL [7]. This mode of IoT routing is still challenging, under active research, and is the main focus of this paper.

In order to guide future work and optimize the use of object resources within multicast routing protocols a comprehensive study of their performance should be conducted. Currently, and to the best of our knowledge, no quantitative comparison is available in the IoT literature. Therefore, we conducted this work to provide the research community with such insights.

The remainder of this paper is organized as follows. Section 2 gives a background on representative multicast routing protocols in the IoT, details their operations, and discusses their features. This is followed by the design of our comprehensive study in Sect. 3, and the discussion of the obtained results under different network settings in Sect. 4, respectively. The paper ends in Sect. 5 with a conclusion and ideas for future research.

2 Background on Multicast Routing in the IoT

IP multicast is a form of diffusion of IP datagrams from a transmitter (single source) to a group of interested receivers in a single transmission [3,6]. Each group is identified by a single multicast address that is used as a destination address by each host that is a member of the group, as shown in Fig. 1. This

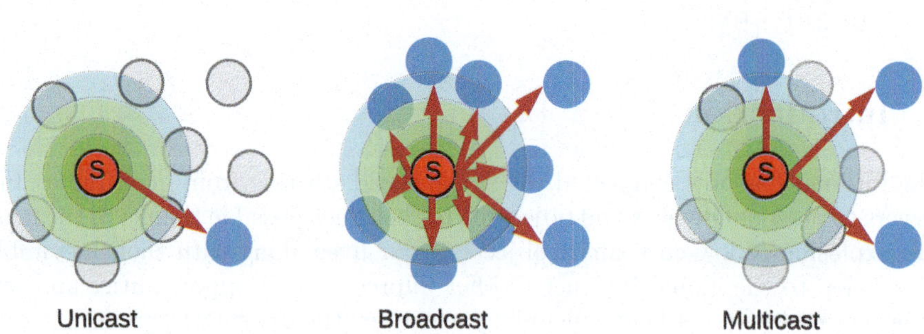

Fig. 1. IP packet routing primitives

technique allows the efficient use of bandwidth and energy and prevents duplicate transmissions [2]. Multicast routing is very different from unicast and hence more challenging. First, the source sends the traffic to a group of dynamic receivers. To reach all the members, the multicast delivery path must create multiple branches across the network to build a distribution tree. Second, the source address plays an important role in the creation of the distribution tree, hence multicast routing paths are generally shaped by the source, instead of the destination. Third, multicast routing, generally, relies on a unicast routing protocol to optimize generated overhead. These challenges are intensified taking into account the limitations of IoT devices.

There are several multicast routing protocols proposed in the IoT, which usually follow one of the two basic modes: dense or sparse, with the dense mode as the most deployed one. Indeed, all the prominent IoT multicast solutions discussed below fall under this class. Nevertheless, they differ w.r.t. the dependence on the unicast routing protocol (RPL), as shown in the taxonomy of Fig. 2.

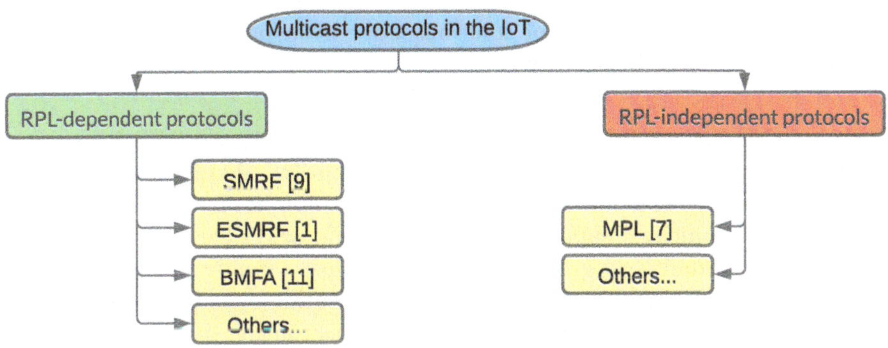

Fig. 2. Taxonomy of multicast protocols in IoT.

2.1 RPL-Independent Multicast

There is only one IoT multicast routing protocol falling under this category, namely the Multicast Protocol for Low-Power and Lossy Networks (MPL), which was standardized in 2016 by RFC 7731 [7]. MPL avoids the need to build or maintain a multicast topology, by broadcasting messages using the Trickle algorithm [8] to all MPL Forwarders in an MPL Domain. The protocol presents a proactive and a reactive mode of operations along with blind flooding. It should be noted that reactive and proactive modes can be enabled simultaneously [7].

In the proactive mode, if an MPL Seed (source node) wants to transmit a multicast message in an MPL Domain, it generates an MPL Data Message. If the destination address is different from the MPL Domain Address, IP-in-IP tunneling is used to encapsulate the multicast message in an MPL Data Message, preserving the original IPv6 destination address. Upon receipt of an MPL Data Message, the MPL Forwarder extracts the MPL Seed ID and message sequence number and determines whether the message has been received previously based on the MPL Seed set and the Buffer Messages set for a given MPL Domain. If the sequence number is less than a lower bound kept in the MPL Seed set or if a message with the same sequence number exists in the Buffer Message set, the MPL Forwarder marks the MPL Data Message as old. If not, the message is marked as new and it updates the MPL Seed set, adds the MPL Data Message to the Buffer Message set, and performs its processing and multicast of the message.

In the reactive mode, an MPL Forwarder periodically broadcasts information contained in the MPL Seed set and the Buffer Message set of an MPL Domain to the local neighbors using MPL Control Messages. MPL Forwarder determines whether or not there are new MPL Data Messages that have not yet been received by the control message source, and multicasts these MPL Data Messages.

2.2 RPL-Dependent Multicast

In this category, a multitude of solutions have been proposed. The following subsections detail the most deployed ones.

Stateless Multicast RPL Forwarding: Stateless multicast RPL Forwarding (SMRF) is a lightweight stateless multicast forwarding algorithm presented as an alternative to MPL for RPL networks [9]. It uses RPL's storage mode of operation with multicast support (MOP 3) [10], and works based on RPL parent information and multicast group membership. Thus, parent nodes are supposed to know the multicast group addresses of their children and make entries for those advertised multicast group addresses in their routing tables.

With SMRF, multicast traffic can only flow downward in the Destination Oriented Directed Acyclic Graph (DODAG). The root of the RPL network is the only node capable of generating a multicast message. The other nodes only accept multicast packets received from their preferred parent in the DODAG tree. Subsequently, the node checks if it is a member of the multicast group indicated in the received packet, if so, it processes the multicast packet in its network stack. In the end, instead of blindly forwarding all datagrams to all nodes, and based on the information provided by the RPL, the node checks if there is at least one child node interested in this traffic to forward the packet, if not it drops it. Multicast datagrams in SMRF will only reach those parts of the network that have expressed interest in the flow by joining a multicast group.

Enhanced SMRF: Qorany and al. [1] proposed an improvement of the SMRF protocol in their Enhanced SMRF (ESMRF) protocol that supports bidirectional multicast traffic flows downward and upward in the DODAG. The main idea of ESMRF is that sources of multicast traffic encapsulate their packet in an ICMPv6 delegation packet and send it to the root of the RPL tree, which transmits the multicast packet on behalf of the original source.

Bi-directional Multicast Forwarding: Georgios and al. [11] proposed Bi-directional Multicast Forwarding Algorithm (BMFA) for RPL based networks. In order to support bidirectional traffic and avoid routing loops, BMFA uses the 20-bit flow label of the IPv6 header. A node will accept an incoming message if the Link-Local Address (LLA) of this message is the LLA of the preferred RPL parent or the LLA of one of its children, and if and only if the value of the flow label in the IPv6 header does not contain its own LLA. BMFA also uses the information provided by the RPL group membership system.

In BMFA, a source can use flow label to tag packets for which the source requests a special treatment by IPv6 nodes. For example, a source may require a real-time or default quality of service. The disadvantage of BMFA is that it is impossible to use the flow label in a non-standard way and hence might prevent interoperability. Table 1 summarizes the main objectives, strengths, and weaknesses of the studied protocols.

Table 1. Summary table of the studied multicast routing protocols.

Protocol	Objectives	Strengths	Weaknesses	Year
MPL [7]	- Ensure multicast data transmission in LLNs without maintaining a routing table	- Ensures retransmission of packets in case of loss and provides a high delivery rate thanks to the buffering principle - Maintains network consistency and ensures reliability - Fast update and adaptation to network mobility and density	- Can cause communication overhead due to retransmission of lost packets also control packets consumes resources - Low device storage capacity can limit buffer size, resulting in reduced performance - High End-to-End latency	2016
SMRF [9]	- Lightweight algorithm for multicast data transmission based on RPL	- Reduced resource consumption - Easy and simple to implement and understand - Transfer messages in the right order	- Only allows multicast transmission downwards - No retransmission mechanism in case of loss	2013
ESMRF [1]	- SMRF enhancement to enable upstream and downstream multicast data transmission	- Resolves the SMRF gap by allowing upward multicast traffic	- Monitors communication by using the root node to route all multicast traffic - May result in high End-to-End latency - Additional control messages	2015
BMFA [11]	- An alternative solution to improve the SMRF to ensure upward and downward packet delivery	- Bidirectional - Reduced consumption of resources - No additional control messages	- No retransmission mechanism - Misuse of the IPv6 Flow field	2017

3 Experimental Design and Tools

To achieve the objectives of this study we evaluate the performance of the studied multicast routing protocols using the Contiki operating system. Contiki is a lightweight open-source operating system designed for the IoT, which includes the kernel, libraries, program loader, and a set of processes, developed by the Swedish SICS research team designed for the resource-constrained device [5]. We perform the simulations using the Cooja simulator.

3.1 Experimental Protocol

In order to compare the performance of the studied protocols, it is necessary to measure them under the same conditions in different scenarios. We studied the performance of the protocols with the variation of the parameters shown in the Table 2. The reception probability is a configurable simulation parameter that defines the probability of receiving a transmitted message from a neighbor.

For the simulations where we study the effect of varying the reception probability, the number of sources, and the number of members, we used a topology with 100 nodes of type Z1 and a density of 9. Concerning the variation of the number of nodes, we used 4 topologies: 25, 50, 100 and 400 nodes with a density of 3, 6, 9, and 16 respectively, the simulation setup and approximate current consumption of the Z1 mote are shown in Table 2.

Table 2. Simulation setup.

Parameter		Value(s)
Duration		10 min
Number of repetitions (RandomSeed)		5 times
Transmission rate of the multicast source		1 pkt/40 s
Network layer protocol (unicast)		RPL - MOP3
RDC - MAC - PHY		CX-MAC - CSMA - 802.15.4
CSMA - MAX RETRIES		5
Reception probability		[0.2, 1.0]
Number of nodes		{25, 50, 100, 400}
Number of senders		{1, 2, 4, 8, 16}
BMFA, SMRF, ESMRF and MPL		Default parameters in Contiki
CC2420	TX @ 0 dBm/RX	17.4 mA/18.8 mA
MSP430f2617	Active @ 8 MHz/LPM	4 mA/0.5 uA

3.2 Comparison Criteria

- Energy consumption: the average energy consumption (CPU and radio activity) per node, we use the "energest.h" library and extract the energy consumption in (mW) by the following formula:

$$\frac{ENERGEST\_TYPE \times Current \times Voltage}{RTIMER\_SECOND \times simulation\ time} \quad (1)$$

- Packet delivery ratio (PDR): N is the number of unique multicast datagrams sent by traffic sources to the group's IPv6 address, M is the number of multicast group members and R is the total number of unique multicast datagrams received by the members.

$$\frac{R}{N \times M} \tag{2}$$

- End-to-End delay: the average time it takes for a message to be received by all members.

4 Performance Evaluation Results

According to the theoretical study, SMRF only works in RPL networks where the DODAG root is the multicast data source, and in this case, SMRF and ESMRF work in the same way. Our study focuses on real cases where the data source is not necessarily the root node. Therefore, we have studied only the ESMRF protocol.

4.1 Variation of the Reception Probability

In this experiment, We varied reception probability in the interval [0.2, 1.0]. Figure 3 present the obtained results.

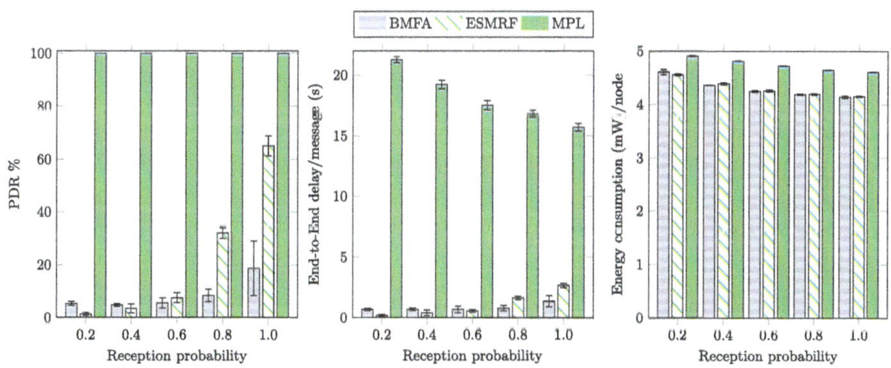

Fig. 3. Variation of message reception probability.

As can be seen in this figure, when the reception probability is equal to 0.2, all protocols except MPL show a decrease in PDR until it reaches about 0. This is because MPL, unlike the others, tries to maintain consistency between nodes in terms of data through a retransmission mechanism implemented using the trickle algorithm. To ensure this functionality, MPL needs a higher number of multicast messages sent than other protocols, and with the decrease of packets reception probability more packets are lost, so the number of these messages increases and the End-to-End delay also increases. On the other hand, ESMRF and BMFA do not have a retransmission mechanism for lost packets. Thus, when the value of the reception probability decreases, fewer multicast packets are received by the network nodes, which justifies the decrease in energy consumption and the End-to-End delay will not be significant because the packets are not received by most members of the group.

4.2 Variation of the Density and the Number of Nodes

This time we tested the behavior of each protocol in four different topologies: 25 nodes density equal to 3, 50 nodes density equal to 6, 100 nodes density equal to 9, and 400 nodes density equal to 16. Results are depicted in Fig. 4.

In the case of MPL, we note that as the number of nodes increases, the PDR remains equal to 100%, and the End-to-End delay increases, since delivering a packet to 23 nodes takes less time than to 398 nodes. We also notice that as the size of the network, and thus the density, increases, the average energy consumption per node decreases. This can be explained by the fact that each node in the network will have more neighbors and thus more sources of the multicast stream, the probability of receiving a multicast packet multiple times increases, and since the constant K is equal to one, the probability of canceling the sending of a packet by triggering a consistency increases which reduces the energy consumption per node. The total number of multicast messages flowing through the network increases with the size of the network, but on a per-node basis, it decreases.

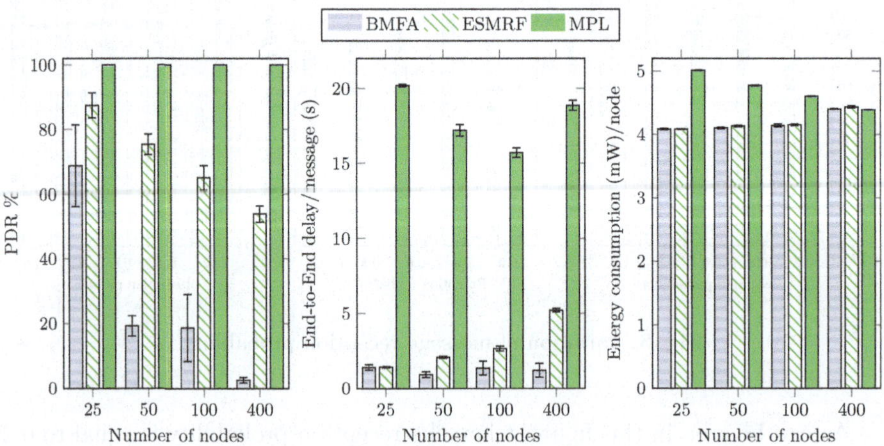

Fig. 4. Variation of the density and the number of nodes.

In the other case, we notice that the other protocols have lost their performance in terms of packet delivery ratio with the increase in the number of nodes, and this is mainly due to the increase in the number of messages sent which generates more collisions and a large part of the messages sent are lost. Noting that BMFA is the weakest protocol, it has the lowest PDR, especially in large topologies.

4.3 Variation of the Number of Multicast Sources

The goal is to study the performance of multicast routing protocols by varying the number of multicast traffic sources from 1 to 16 sources.

With the hardware characteristics of the Z1 mote, we could not increase the number of MPL multicast sources by more than 4 and the messages buffer is equal to 6 messages.

As can be seen in the Fig. 5, MPL with a saddle or multiple sources remains to provide a PDR equal to 100% contrary to the other protocols that they presented a decrease of the latter.

Regarding the number of packets sent in the network, and with only 4 sources and 10 multicast messages sent from each source, MPL achieved over 2185 multicast packets sent per node, which consumes bandwidth and energy of the nodes that are critical constraints in this type of network. The resource consumption, in this case, is a major drawback of this protocol.

4.4 RAM and ROM Consumption

MPL uses three data structures: Buffered Messages set to store the latest MPL Data Messages from an MPL seed in an MPL Domain, Seed set to store information about the MPL Data sources in an MPL Domain and Domain set to store the addresses of the MPL Domains where the node registers. The Table 3 shows the consumption of each MPL data structure. SMRF, ESMRF, and BMFA do not need an additional data structure to operate but only use the data provided by the RPL unicast routing protocol.

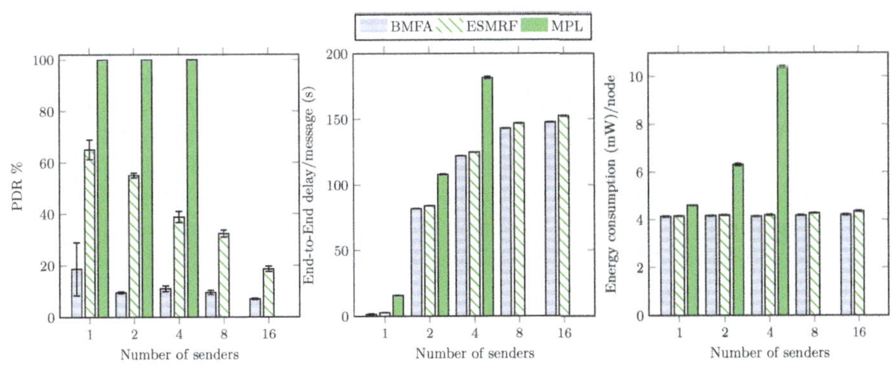

Fig. 5. Variation of the number of senders.

Table 3. The size of MPL data structures.

Data structures	Unit size	Total table size (Octet)
Buffered Messages set	212	MPL_BUFFERED_MESSAGE_SET_SIZE*212
MPL Domain	82	MPL_DOMAIN_SET_SIZE*82
Seed set	28	MPL_SEED_SET_SIZE*28

In our topologies we have three types of nodes: source, sink and root. The Table 4 shows the code size of each node according to the multicast routing protocol enabled. The code size of nodes with MPL is the largest, followed by ESMRF and BMFA, this is mainly due to the complexity of the protocol, but overall with all three protocols the ROM consumption does not exceed 50 kilobytes.

Table 4. ROM consumption.

Multicast protocol	Code size of the nodes (Octet)		
	Source	Sink	Root
MPL	49521	49161	49481
ESMRF	47795	47527	47709
BMFA	44441	44213	44401

5 Conclusion and Future Work

In this paper, we analyzed the behavior of the most studied multicast protocols in the literature with respect to several factors that influence their performance. The results obtained showed that the MPL protocol was able to maintain a PDR equal to 100% in all scenarios, which shows the strength of this protocol in terms of information dissemination, the huge consumption of resources, especially energy and RAM remains the main problem of this protocol. ESMRF and BMFA require RPL as a unicast routing protocol, unlike MPL they support sparse mode by relying on the group member management information provided by the RPL protocol in MOP3. With the increase in the number of nodes or the decrease in the probability of reception, both protocols and especially BMFA show a considerable decrease in PDR which shows the weakness of these protocols under these conditions.

Our future work will focus on the design of a generic multicast routing protocol independent of the unicast protocol used in the network and guaranteeing a good PDR with reasonable resource consumption, this protocol should support sparse mode.

References

1. Abdel Fadeel, K.Q., El Sayed, K.: ESMRF: enhanced stateless multicast RPL forwarding for IPv6-based low-power and lossy networks. In: Proceedings of the 2015 Workshop on IoT challenges in Mobile and Industrial Systems, pp. 19–24 (2015)
2. Carzaniga, A., Khazaei, K., Kuhn, F.: Oblivious low-congestion multicast routing in wireless networks. In: Proceedings of the Thirteenth ACM International Symposium on Mobile Ad Hoc Networking and Computing, pp. 155–164 (2012)
3. Deering, S.E.: RFC1112: host extensions for IP multicasting (1989)
4. Dorsemaine, B., Gaulier, J.P., Wary, J.P., Kheir, N., Urien, P.: Internet of things: a definition & taxonomy. In: 2015 9th International Conference on Next Generation Mobile Applications, Services and Technologies, pp. 72–77. IEEE (2015)
5. Dunkels, A., Gronvall, B., Voigt, T.: Contiki-a lightweight and flexible operating system for tiny networked sensors. In: 29th Annual IEEE International Conference on Local Computer Networks, pp. 455–462. IEEE (2004)
6. Gould, K.: Methods and apparatus for efficient IP multicasting in a content-based network, US Patent 7,693,171, 6 April 2010
7. Hui, J., Kelsey, R.: Multicast protocol for low-power and lossy networks (MPL). https://doi.org/10.17487/RFC7731
8. Levis, P., Clausen, T., Hui, J., Gnawali, O., Ko, J.: The trickle algorithm. Internet Engineering Task Force, RFC6206 (2011)
9. Oikonomou, G., Phillips, I.: Stateless multicast forwarding with RPL in 6LowPAN sensor networks. In: 2012 IEEE International Conference on Pervasive Computing and Communications Workshops, pp. 272–277. IEEE (2012)
10. Oikonomou, G., Phillips, I., Tryfonas, T.: IPv6 multicast forwarding in RPL-based wireless sensor networks. Wirel. Pers. Commun. **73**(3), 1089–1116 (2013)
11. Papadopoulos, G.Z., Georgallides, A., Tryfonas, T., Oikonomou, G.: BMFA: bi-directional multicast forwarding algorithm for RPL-based 6LoWPANs. In: Mitton, N., Chaouchi, H., Noel, T., Watteyne, T., Gabillon, A., Capolsini, P. (eds.) InterIoT/SaSeIoT -2016. LNICST, vol. 190, pp. 18–25. Springer, Cham (2017). https://doi.org/10.1007/978-3-319-52727-7_3
12. Winter, T., Thubert, P., Brandt, A., Hui, J., Kelsey, R.: RFC 6550: RPL: IPv6 routing protocol for low-power and lossy networks (2012). https://tools.ietf.org/html/rfc6550

Efficient Auto Scaling and Cost-Effective Architecture in Apache Hadoop

Warda Ismahene Nemouchi(✉), Souheila Boudouda, and Nacer Eddine Zarour

A. Mehri - Constantine 2 University, Constantine, Algeria
{warda.ismahene,souheila.boudouda,
nasro.zarour}@univ-constantine2.dz

Abstract. In the age of Big Data Analytics, Cloud Computing has been regarded as a feasible and applicable technology to address Big Data Challenges, from storage capacities to distributed processing computations. One of the keys of its success is its high scalability which refers to the ability of the system to increase its performance, resources and functionalities according to the workload. This flexibility has been seen as an appropriate way to decrease datacenters' energy consumption and thus assures cost-saving and efficiency without effecting performance of the system. In order to handle Big Data operations, Cloud Computing has implemented various platforms and tools such as Apache Hadoop and provides distributed processing of very large data sets across multiple clusters. This paper proposes an auto scaling architecture based on the framework of Hadoop; it adjusts automatically the computation resources depending on the workload. In order to validate the effectiveness of the proposed architecture, a case study about Twitter data analysis in a cloud simulated environment has been implemented to improve the cost-effectiveness and the efficiency of the system.

Keywords: Big data · Cloud computing · Apache hadoop · Auto scaling

1 Introduction

Big Data refers to the increasing amount of data generated every second from e business applications, smartphones and more and more connected objects. This explosion in the digital world has led to the evolution in different technologies in order to store, process, and analyze this important volume of data [1]. For that, multiple solutions have been proposed namely Apache Hadoop framework which allows distributed processing for large scale data sets across multiple computers in a cluster. Hadoop offers both massive storage and huge processing capacities in master slave architecture [2]. Hadoop Distributed file system (HDFS) is responsible for storing data in form of blocs of 64 Mb or 120 Mb. Each bloc is stored on a slave node, the information related to all data blocs are stored in the master node. To assure data availability, HDFS follows a replication policy where each bloc is replicated n times (n = 3).

Despite the important value that Big Data has added to big companies through its delivered insights and knowledge, its requirements of having the appropriate infrastructure to deal with such growing volume and variety of data seem highly expensive.

B. Lejdel et al. (Eds.): AIAP 2021, LNNS 413, pp. 336–345, 2022.
https://doi.org/10.1007/978-3-030-96311-8_31

Hence the need for another technology capable of meeting both the needs of Big Data at a reduced cost and a fully managed infrastructure [3]. It's Cloud Computing (CC), a paradigm for managing and delivering services over the internet with its characteristics of elasticity and scalability to millions of instances in a completely transparent way to the final user [4].

Big Data and CC present the perfect combination to process huge amounts of data on a platform that is scalable and has the resources to analyze massive data [5]. However, there exist some challenges to overcome in particular the energy waste of data centers that are kept running without actually being used [6].

For this, the ability to automate the scaling process so it can provision or reduce resources based on workload automatically is needed in order to reduce the energy consumption of the cluster and thus reduce related costs. In a Big Data context, controlling Hadoop' resources automatically are more challenging due to its replication policy as powering off nodes is a time-consuming operation, data blocs need to be transferred at first before shutting down the node [7]. In this context, an auto-scaling approach has been investigated to overcome the energy waste and related costs problem. This paper is organized as follows: Sect. 2 discusses some research works related to Big Date auto-scaling applications, Sect. 3 presents the proposed approach and details its different components whereas Sect. 4 shows the implementation of the approach. Finally, some conclusions and research lines are presented in Sect. 5.

2 Related Works

Multiple methods and models have been proposed in the context of implementing a dynamic scaling architecture in a cloud based Big Data applications. Some of these works are presented in this section.

To manage resources efficiently, authors in [8] have proposed a Distributed Dynamic and Customized Load Balancing (ddclb) algorithm in Amazon Elastic Compute Cloud (EC2). The proposed algorithm takes in consideration the CPU, RAM utilization of the cluster along with response time of each instance and assign requests to instances with the lowest metric. Although, the work proposes a dynamic scaling approach, it does not support Hadoop for processing large volume of data [10]. An efficient processing framework of large geospatial datasets has been proposed in [9]. The proposed framework is applied to Hadoop where a node separation to data/compute has been adopted to make unused nodes removing easier and faster. Also, a predictive algorithm has been implemented to calculate the number of resources needed for the workload. The experiments of the framework showed an 80% reduction of the resource's utilization. [10] proposed an auto-scaling framework for analyzing Big Data in the Cloud, it controls the cluster' metrics (CPU, Map/Reduce Tasks and job's state) through the use of Amazon Cloud Watch to perform scaling actions (adding or removing nodes). Authors in [11] have identified container allocation as a key factor that affects Hadoop performance. To ease the overhead occurring they have come up with three methods of data redistribution. The experiments of this work show that adding resources to the cluster dynamically without redistributing data has improvements in term of time response. In [12], researchers have proposed a dynamic energy efficient data placement and cluster reconfiguration algorithm for MapReduce framework. The algorithms turn off and on nodes running MapReduce jobs based on current workload, so when the cluster utilization rises above or fall under thresholds predefined by the administrator, scaling up or down action are performed. The results show energy reduction of 33% under average workloads and up to 54% under low workloads.

To perform scaling down of operational clusters, Leverich and Kozyrakis [13] have proposed to store at least one replica in a covering subset that is not taken in consideration while scaling down operations. This proposition has led to better cluster performance (CPU use, energy efficiency). [14] have proposed a Berkeley Energy Efficient MapReduce system called BEEMR where a cluster separation into batch ad interactive jobs is implemented. The use of BEEMR system has achieved 40% of energy saving [15]. In the same context of covering subsets, Kaushik et al. [16] have separated the cluster into hot and cold zones, the hot zone contains files frequently accessed and new created ones whereas the cold zone contains the remaining ones. The scaling down operation is performed only on nodes situated in the cold zone. The work was able to meet all the scaling down mandates and has achieved up to 24% energy reduction.

Other works like [17, 18] have opted for balancing the workload and while in [19] a scheduling policy is another way to improve the cluster utilization.

In order to analyze and evaluate the research works presented above in a meaningful way, Table 1 presents a comparison between the approaches which are similar to our research work. We have specified the following criteria: Cloud platform, Controlled metrics and datasets used in each work.*

In the same context, the main goal of this work is to improve the resources utilization of a Hadoop cluster by scaling dynamically the cluster based on multiple metrics. The next section presents the main focus of this paper, the proposed approach, the motivations and challenges of this work.

Table 1. Comparative study between the related research works.

Related work	Platform used	Controlled metrics	Datasets	Objective
[8]	Amazon Web Services (AWS)	CPU, RAM	Amazon EC2' users requests	- Manage dynamically requests of Amazon EC2 users - Scale EC2 instances up or down
[9]	Public or Private Cloud	CPU, RAM	Geospatial applications	- Scaling dynamically Hadoop cluster by provisioning the right number of resources based on workload
[10]	Amazon Web Services (AWS)	Workload	Social network Twitter	- Reduce resources utilization costs - Scaling Hadoop cluster resources and improving data processing and analyzing in rcal time
[11]	Cloud (SAAS)	CPU, MapReduce jobs	Social network Twitter	- Predicting when adding dynamically MapReduce nodes
[12]	/	Workload, MapReduce jobs	/	- Reduce datacenters' energy consumption by reconfiguring the cluster

3 Auto Scaling Approach

The aim of this paper is to propose a dynamic scaling approach in Apache Hadoop to process efficiently large-scale data on a cloud environment. From what's have been established before, we notice that controlled metrics are the key value to decide exactly when adding or removing instances is required.

In this contribution, we have chosen to monitor the CPU, RAM along with pending tasks, the later refers to the number of CPU cores needed for their execution. Each node has a maximum number of tasks that it can execute simultaneously according to the number of its CPU cores. We also have used Nagios as a cluster monitor, it collects data about the system, the network and the infrastructure. Figure 1 presents the architecture of the proposed approach with its different modules.

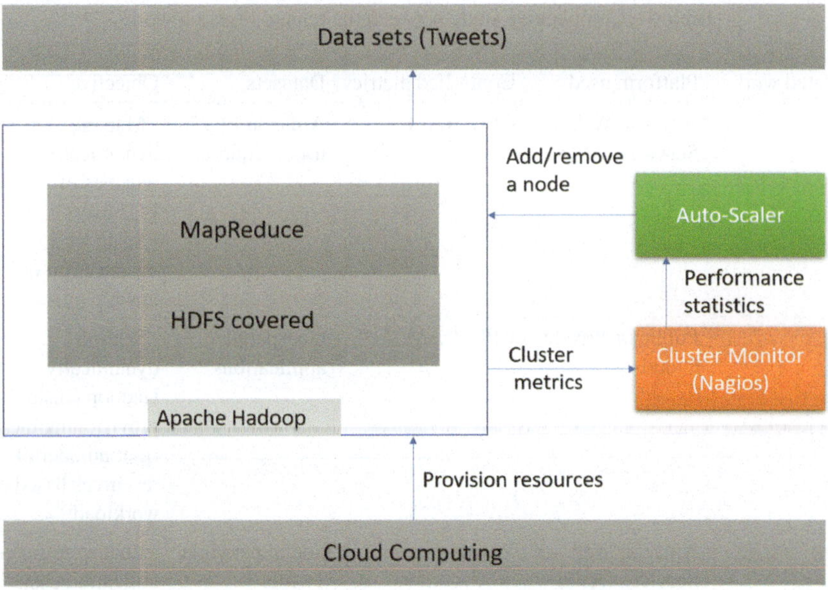

Fig. 1. Architecture of the proposed approach

The approach is based on CC as a working environment that englobes the architecture, each component plays its role in the system and assures the good proceeding of each phase (storage, processing, auto-scaling). Two components have been added to the Hadoop framework in order to control its resources, Nagios, the cluster monitor, to control all the system's components and the Auto-Scaler to manage and scale the cluster's resources based on Nagios' information.

3.1 Cluster Monitor Nagios

In order to guarantee the proper functioning of the system, it is fundamental to supervise the cluster. For that, we have chosen Nagios as a monitoring tool to control resources, application status and system operation. Nagios has a built-in alert-based notification system. It is a supervision component; it alerts us in the event of failures encountered on the various monitored components. It can be used to approach different perspectives of surveillance:

- Obtain instant information about the Hadoop infrastructure.
- Generate and receive alerts in case of system failures.
- Analyze, produce reports and graphs on usage and make decisions on future material acquisition.
- Monitor queue depletion and search for available nodes to run jobs.

For HDFS, Nagios checks space usage, files replication and slave nodes balancing. It also allows to monitor the master node (Name Node) which is considered as a sensitive point of the system (its failure leads to a complete shutdown of the system).

For MapReduce, Nagios checks the status of JobTracker and TaskTracker. It provides information related to the jobs being executed and the number of jobs in the queue, this information is almost important in deciding whether we need to add a new worker node or not.

In an autoscaling context, we use metrics provided by Nagios to trigger the resize operation, we can even set up an alarm that notifies us when a threshold condition is trigged.

3.2 Auto-Scaler

Auto-Scaling is a mean of automatically increasing or reducing the cluster, it adds compute nodes when the workload exceeds the capacity of the cluster (Scale-up), or it frees inactive ones when the load is low (Scale-down).

Depending on workload of the cluster; we can transparently increase the number of compute nodes without affecting the general functioning of the system. In case of unused capacity for a period of time, a decrease in node number is realized. However, the Scaling-Down operation causes some services to restart and therefore all running or pending tasks to fail. To avoid such a problem, nodes with running map or reduce tasks will not be considered (will not be released).

To have the auto-scaler perform resizing operations automatically, we define two scaling rules based on CPU usage. These rules tell the Auto-Scaler to add or remove nodes based on average CPU usage and the number of queues.

Once the metrics to be considered are trigged, we can proceed to an auto-scaling according to the defined rules. We specify the minimum and maximum threshold that the auto-scaler must reach to trigger an automatic resizing event.

Two scaling operations are defined, adding a node or Scaling-Up and removing a node or Scaling-Down. If average usage reaches target usage, auto-scale adjusts nodes in a Hadoop cluster either by increasing or decreasing as needed (Fig. 2). A cool-down time between auto-resize events allows the system to stabilize.

Scaling-Up

We have chosen, in our contribution, to monitor the CPU component and therefore take into consideration all the related information. We also set the threshold for maximum CPU usage (70%). As each Map/Reduce task is executed by a CPU core, we will refer to the number of queued tasks by the number of CPU cores needed for their executions.

Figure 2 shows an algorithm that describes the flow of the Automatic Scaling-Up operation. When average CPU usage hits a threshold, Auto Resize dynamically and conveniently allocates resources in real time. The maximum number of nodes in the cluster must be determined by the user through a configuration interface.

Scaling Down

During a scaling-down operation, it is necessary to define a minimum number of nodes that must remain active to maintain the performance of the cluster and manage the tasks that have occurred in a given time. It is also essential not to remove nodes with unfinished spots or whose reduction spots are in progress. Since sizing operations are applied just

```
Node_max = maximum number of worker node
CPU_max = maximum CPU use
Begin
   nodes = number_of_actif_nodes;
   While (nodes < Node_max) then
         CPU = current_CPU_use;
         If (CPU < CPU_max) then
            CPU_queue = number_of_pending_jobs;
            CPU_free =  number_of_free_CPU
            if (CPU_queue < CPU_free )then
               Execute_Tasks();
            Else
               Provision_node();
            End if
          Else
               Provision_node();
           End if
    End While
End
```

Fig. 2. Scaling up algorithm

on Worker type nodes, removing a node is applied instantly and does not affect cluster operation or data availability.

Figure 3 describes the flow of the Automatic Scaling-Down operation, when the average CPU usage reaches the minimum threshold, the auto-scaler decreases the number of active compute nodes which reduces costs and minimizes consumption. energy of the cluster.

```
Node_min = minimum_number_of_worker_node;
CPU_min = minimum_CPU_use;
Begin
   nodes = number_actif_nodes;
   while (nodes > Node_min) then
         CPU = utilisation courante du CPU
         if (CPU <= CPU_min) then
           state= verify_running_jobs;
            if (state == false ) then
                delete_node();
            end if
         End if
    end while
End
```

Fig. 3. Scaling down algorithm

4 Implementation and Results

In order to validate the proposed approach, we have installed Hadoop in a virtual machine environment to simulate the Cloud. We have chosen a Hadoop single node cluster to observe the results of the implemented algorithms. In addition, we have installed the cluster monitor Nagios and we have configured it with Hadoop so it can control its resources, Fig. 4 shows Nagios interface after configuration.

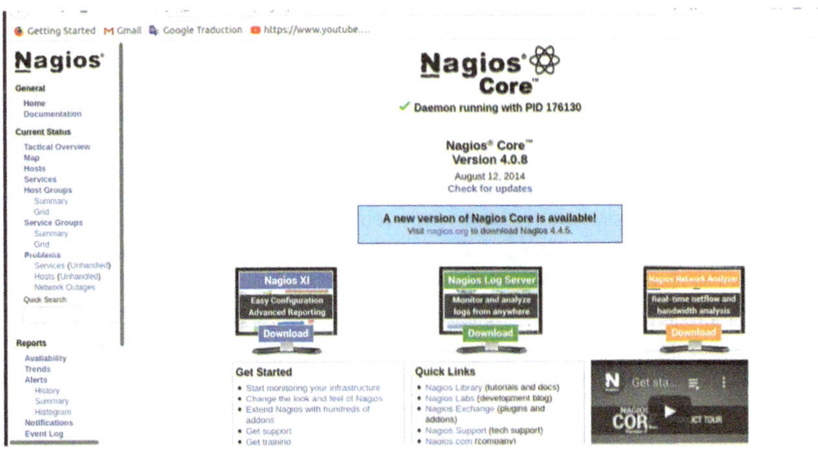

Fig. 4. Cluster monitor Nagios

As an execution scenario, we have chosen a file of more than 10000 tweets with multiple hashtags related to the corona virus. These tweets are processed with MapReduce to count the number of each hashtag and store it in HDFS under the supervision of Nagios, Fig. 5 shows the results of the execution in HDFS.

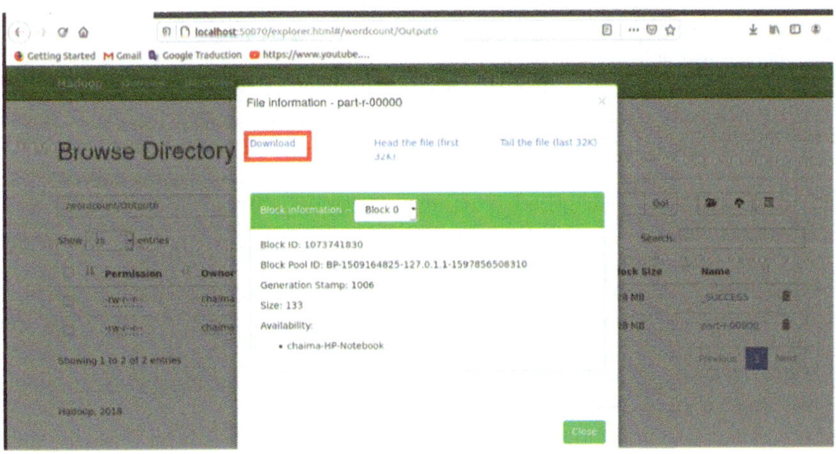

Fig. 5. File output of the execution

The objective of this implementation is to retrieve cluster metrics from Nagios and pass them to the Auto-Scaler. This last compare them to the predefined thresholds and trigger the appropriate action (Scaling Up or Down). Since the approach is implemented in single node architecture, we have managed to display notifications rather than actually adding or removing instance to just validate the efficiency of the Auto-scaling Algorithm. The results showed that in a single node cluster The Auto-scaler module has displayed notifications in the exact time according to the controlled metric and the specified thresholds.

5 Conclusion

The cooperation between Big Data and Cloud Computing has led to a significant waste of energy and high costs, from where comes the need to propose an auto scaling mechanism to control dynamically the cluster according to workload. Hence, we have proposed an approach based on Hadoop to add resources or remove nodes whenever needed according to CPU, RAM and pending jobs metrics extracted by Nagios and sent to. We have implemented the approach in a single node cluster and a virtual machine environment. We have also chosen to process more than 10000 tweets to count hashtags related to corona virus using MapReduce. The simulation showed a promising execution of the proposed algorithm displaying notification about the appropriate scaling action needed.

We aim in the near future to test our approach in a real cloud environment under a various workload (high and low). We also aim to take in consideration other metrics and even predict the right number of resources required to execute jobs and tasks.

References

1. Wamba, S.F., Gunasekaran, A., Akter, S., Ren, S.J.-F., Dubey, R., Childe, S.J.: Big data analytics and firm performance: effects of dynamic capabilities. J. Bus. Res. (2017)
2. Hashem, I.A.T., Anuar, N.B., Mokhtar, S.: The rise of "big data" on cloud computing: Review and open research issues. Inf. Syst. (2015)
3. Talia, D.: Clouds for scalable big data analytics. Computer (2013)
4. Mell, P., Grance, T.: The NIST definition of Cloud Computing. National Institute of Standards and Technology, special publication (2012)
5. Balachandran, B.M.; Prasad, S.: Challenges and benefits of deploying big data analytics in the cloud for business intelligence. Procedia Comput. Sci. (2017)
6. Barroso, L., Hölzle, U.: The datacenter as a computer: an introduction to the design of warehouse-scale machines. Synth. Lect. Comput. Archit. **4**(1), 1–108 (2009)
7. Maheshwari, N., Nanduri, R., Varma, V.: Dynamic energy efficient data placement and cluster reconfiguration algorithm for mapreduce framework. Future Gener. Comput. Syst. **28**(1), 119–127 (2012)
8. Shah, V., Trivedi, H., et al.: A distributed dynamic and customized load balancing algorithm for virtual instances (2015)
9. Li, Z., Yang, C., Liu, K., Hu, F., Jin, B.: Automatic scaling hadoop in the cloud for efficient process of big geospatial data (2016)
10. Jannapureddy, R., Vien, Q., Shah, P., Trestian, R.: An auto-scaling framework for analyzing big data in the cloud environment (2019)
11. Fu, Q., Timkovich, N., Riteau, P., Keahey, K.: A step towards hadoop dynamic scaling (2018)
12. Maheshwari, N., Nanduri, R., Varma, V., et al.: Dynamic energy efficient data placement and cluster reconfiguration algorithm for MapReduce framework (2011)

13. Leverich, J., Kozyrakis, C.: On the energy (in)efficiency of hadoop clusters. Oper. Syst. Rev. **44**(1), 61–65 (2010)
14. Chen, C.C., Hasio, Y.T., Lin, C.Y., Lu, S., Lu, H.T., Chou, J.: Using deep learning to predict and optimize hadoop data analytic service in a cloud platform, pp. 909–916 (2017)
15. Jam, M.R., Khanli, L.M., Akbari, M.K.: Survey on improved autoscaling in hadoop into cloud environment. In: 15th Conference on Information and Knowledge Technology (IKT) (2013)
16. Kalagiakos, P., Karampelas, P.: Cloud computing learning. In: The Proceedings of IEEE International Conference on Application of Information and Communication Technologies, Baku, pp. 1–4 (2011)
17. Domanal, G.S., Reddy, M.G.R.: Optimal load balancing in cloud computing by efficient utilization of virtual machines. In: proceedings of the Sixth International Conference on Communication Systems and Networking (COMSNETS), pp. 1–4 (2014)
18. Mahalle, H.M., Kaveri, P.R., Chavan, V.: Load balancing on cloud data centres. Int. J. Adv. Res. Comput. Sci. Softw. Eng. IJARCSSE, 1–4 (2013)
19. Wang, X., Lu, Z., Wu, J., Zhao, T., Hung, P.: InSTechAH: an autoscaling scheme for hadoop in the private cloud. In: IEEE International Conference on Services Computing (2015)

GA-Based Approaches for Optimization Energy and Coverage in Wireless Sensor Network: State of the Art

Khalil Benhaya(✉), Riadh Hocine, and Sonia Sabrina Bendib

LaSTIC Laboratory, Computer Science Department, University of Batna 2, 05000 Batna, Algeria

Abstract. Wireless sensor networks (WSNs) have become one of the leading research subjects in computer science over the last few years. WSNs are resource-constrained concerning available energy, bandwidth, processing power, and memory space. Thus, optimization is essential to get the best results of these constraining parameters. Due to the advantages of genetic algorithms, different GA methods have been implemented to optimize different objectives like energy, coverage, QoS, and many other metrics. This paper presents a survey on the current state of the art during the last four years in wireless sensor network optimization using genetic algorithms to optimize energy consumption and the coverage of WSNs to give an up-to-date background to researchers in this field. Also, a classification of the works, based on the used methods, is provided.

Keywords: Optimization · Genetic algorithm · Wireless sensor network · Energy · Coverage

1 Introduction

Sensor networks are tiny sensor nodes having elements created for specific operations like sensing the environment, processing data, and changing information with other nodes. When many sensor nodes are used to sense their material environmental conditions, they create a sensor network consisting of a sink node and sensor nodes which can be started from a few hundred to thousands in number [1, 2]. WSNs have attracted significant attention recently from different research groups, and through their use, several applications are developed in the current and coming system [3]. WSNs are designed for various domains like monitoring events, agriculture, health care, and surveillance, which are classified into military, commercial, and medical applications [4]. In consideration of the application scenarios, WSNs may depend on crucial performance metrics to be optimized, like the energy consumption and network lifetime, because sensor nodes are powered by batteries, whose changing them is usually challenging and impossible. Moreover, the network coverage, latency, and many other metrics are critical for the quality of WSNs efficiency [5, 6]. The aforementioned metrics often conflict with each other, thus balancing the trade-offs between them is very important in the matter of obtaining the optimal performance of real applications in WSNs. Consequently, multi-objective optimization (MOO) can be used for solving the previous problem [7, 35]. MOO has been an

B. Lejdel et al. (Eds.): AIAP 2021, LNNS 413, pp. 346–355, 2022.
https://doi.org/10.1007/978-3-030-96311-8_32

essential topic of interest to researchers for solving various multi-objective optimization problems, in which many objectives are treated concurrently subjected to a set of limitations [7]. Nevertheless, it is unattainable for multiple objectives to reach their particular optima simultaneously. Therefore, there may not be a unique optimal solution, which is the most desirable regarding all objectives. However, a set of Pareto-optimal, which is named Pareto front (PF), could be obtained. In other words, the PF is formed by a particular set of solutions, for which none of the several objectives can be developed without sacrificing the other objectives [8]. In order to find the PSs of multi-objective problems, Various approaches have been proposed, such as nature-inspired, metaheuristics, and mathematical programming-based scalarization methods. Multi-objective problems are usually solved by bio-inspired approaches, such as swarm intelligence algorithms [9] and evolutionary algorithms [10]. Due to the advantage of genetic algorithms, they have been the most broadly used approach in the family of multi-objective evolutionary algorithms [11]. This paper provides a study on the current state of the art during the last four years in wireless sensor network optimization. This article examines the papers with GA-based methods to optimize energy consumption and the coverage of WSNs to give an up-to-date background to researchers in this field. Also, we classify these papers based on the type of the used GA methods to present a smooth overview and clear arrangement of ideas to readers.

1.1 Classification of GA-Based Methods

Based on the methods used to optimize different objectives, Table 1 shows a classification of all the papers in this work upon the type of the method used (scheduling, energy-efficient, optimal path, clustering, mobility), and Table 2 shows the additional method used beside the GA and the optimization objectives (energy, coverage…) of each reference.

1.2 Scheduling

Wireless sensor network lifespan for large-scale monitoring systems is represented as the period that all targets can be covered. One method to prolong the lifetime is to separate the deployed sensors into disjoint subsets of sensor covers so that all targets can be covered by every sensor cover and operate by turns (scheduling). Therefore, the high number of sensor covers that can be reached, the more prolonged sensor network lifetime can be achieved [12]. Obtaining the highest number of sensors covers can be done via conversion to the Disjoint Set Covers (DSC) problem, which has been determined to be NP-complete. For this, the existing heuristic algorithms either get inadequate solutions or take exponential time complexity. Thus, the authors in [13] propose a genetic algorithm to resolve the DSC problem by using a new parameter called the Difference factor (DF). In [14], an unsupervised learning method for topology control is offered to increase the lifetime of ultra-dense WSNs. Further, it schedules some members in the cluster to sleep to save the node energy utilizing geographically adaptive fidelity. For the purpose of achieving continuous coverage in tracking and monitoring applications, the target needs to be covered by more than one sensor concurrently. Mohamed Elhoseny et al. used a GA-based K-coverage approach to find the optimum sensor covers for K-coverage

Table 1. Classification of GA-based methods.

Ref	Scheduling	Optimal path	Clustering	Energy-efficient	Mobility
12	X				
13	X				
14	X				
15	X				
16	X				
17	X			X	
18			X	X	
19				X	
20				X	
21				X	
22		X	X		
23		X			
24		X			
25		X			
26		X			
27		X			X
28			X		
29			X		
30			X		
31			X		
32		X	X		
33					X
34					X

environments. Then a covers control method that shifts between several covers to enhance the network lifetime is implemented [15].

Because of the large-scale WSNs, current set cover algorithms cannot afford adequate performance for WSNs scheduling. The authors in [16] have developed a Kuhn-Munkres parallel genetic algorithm for the set cover problem and used it for the lifespan maximization of large-scale WSNs. They used the divide-and-conquer procedure of dimensionality reducing. Firstly, the target field is separated into various subareas, and then individuals are evolved separately in every subarea until the state factor arrives at a predefined value. The developed algorithm is then used to splice the solutions achieved in each subarea to generate the whole problem's global optimal solution. Otherwise, to enhance the global performance, another sensor schedule strategy is improved.

Table 2. Classification based on optimization objectives.

Ref	Other used method	Optimization objectives
12	Maximum number of covers	Energy, Coverage
13	Maximum number of covers	Energy, Coverage
14	Unsupervised learning	Energy
15	K-coverage	Energy, Coverage
16	Kuhn-Munkres parallel GA	Energy, Coverage
17	3D protocols	Energy, Coverage, data delivery reliability
18		Energy
19		Energy
20	Probabilistic sensor detection	Energy, Coverage
21		Energy, Coverage
22		Energy, Coverage
23		Energy
24	Fuzzy Algorithm, agent node selection	Energy
25	Routing	Energy, data delivery reliability, Fault tolerance
26	Multiple sinks	Energy
27	Adaptive balance function	Energy
28		Energy
29	Artificial bee colony	Energy, Coverage
30	Fuzzy algorithm	Energy, received data packets
31	Relay nodes	Energy
32	Cluster division region	Energy
33		Energy, Coverage, optimal movement of mobile sensor
34		Energy

Even though the previous WSN approaches are created to be used on Two-Dimensional (2D) areas under models that rely on measuring the Euclidean distance between sensors, in reality, sensors are deployed in the 3D field in several applications. Thereby, Riham Elhabyan, Wei Shi, and Marc St-Hilaire [17] proposed a multi-objective method (NSGA-CCP-3D) to design an energy-efficient, scalable, reliable, and coverage-aware network configuration protocol for 3D WSNs. The principal purpose of the proposed approach is to find a simultaneous solution to conserve full connectivity and coverage in the 3D field by reaching the optimal status (cluster head, active, or inactive) for every sensor in the network.

1.3 Energy-Efficient Protocols

The number and the position of cluster-heads are strictly affecting the whole energy consumption. Therefore, A Zahmatkesh et al. [18] introduced a multi-objective Genetic Algorithm to create energy-efficient clusters for wireless sensor networks. So, as the first objective is to create an optimal number of cluster heads and cluster members, the distance between sensor nodes for data transmission is considered as the second objective. Thus, the approach minimizes the nodes' energy consumption and the cost of transmission in the network.

Another approach based on a GA is introduced in [19] to find the optimal number of sensors in each cover set which are covering critical targets for a fixed duration (working time) to maximize the network lifetime of WSN. The authors formed the target coverage problem as a maximum network lifetime problem (MLP) and represented it by applying linear programming.

Jie Jia et al. [20] have proposed a new energy-efficient coverage control algorithm (ECCA) in wireless sensor networks. The object of ECCA is to activate only the required number of sensor nodes in a densely deployed environment. Two constraints control the algorithm: one is the specific coverage rate, and the other is the number of the selected nodes from the complete network. Likewise, it can avoid partial optimized solutions due to exploring the entire state-space. Although an accurate probabilistic sensor detection model is carried for a realistic approach, the ECCA algorithm can achieve balanced performance on various detection sensor models while preserving an excellent coverage rate. The authors have also explained how the model can be utilized in the coverage control scheme.

In contrast to the methods mentioned above, the GA approach in [21] aims to optimize the number of potential positions to provide sensor m-connected and target K-covered. Therefore, the method can reach an excellent trade-off between target coverage and energy consumption.

The authors in [22] aim to present an energy-efficient clustering protocol called Hybrid Weight-based Coverage Enhancing Protocol (WCEP) for area monitoring to prolong lifetime. The WCEP helps choose suitable cluster heads and their corresponding cluster members using the weighted sum approach to minimize the energy consumption while maintaining complete coverage and find optimal routing path by using a GA.

1.4 Optimal Path

Due to the impact of direct transmission on enhancing energy consumption when the Cluster head is far from the base station (BS), there has been increased interest to address this problem. The work proposed in [23] concentrates on developing an optimal multi-hop path between a source (CH) and a destination (BS), thereby decreasing energy consumption, which enhances the network lifetime compared with the direct transmission process. A genetic algorithm is used to achieve an optimal path by introducing a new fitness function. Moreover, changes in the CHs selection are introduced to improve the performance of the GA concerning the execution duration and the quality of the chromosomes. Instead of the arbitrary selection of cluster head in the previous work, the used method in [23] selects the CHs efficiently (via three levels) and introduces a new mechanism that can attain an optimal multi-hop path in WSNs. Furthermore, the proposed mechanism decreases the length of the chromosomes of GA and thus the execution time is decreased, by contrast to conventional GA.

Authors in [24] choose to work with a fuzzy logic mechanism using the distance from the base station, the trust value of the node, and energy consumption as parameters to select a particular agent node that collects data and transmits it to the base station. Moreover, in the second phase, they implemented the transmission and receiver energy consumption with a new fitness function in GA to prolong the network lifetime by finding the optimal multipath route. Among the significant factors to the sufficient overall network and application operation are energy consumption and data delivery. Therefore, a multi-objective integer problem (MOIP) is presented in [25] to obtain fitting solutions regarding such trad-off in routing problems using a Non-dominated Sorting Genetic Algorithm in the network with only one sink. On the other hand, the authors in [26] proposed a GA-based optimization routing path in WSNs but with the deployment of multiple sinks, where the nodes forward the packets towards the nearest sink.

In contrast to the previous static wireless sensor network, Shanthi et al. [27] introduced new progress in the Genetic algorithm and named it a Dominant Genetic Algorithm to define the optimum energy-efficient route path connecting sensor nodes and also determine the optimal trajectory for the mobile node that gather data. Although the proposed method has been applied under two different scenarios, it proves that it has faster convergence and high reliability over the conventional GA in various experiments.

1.5 Clustering

Some of the significant factors that affect the network lifetime are the sink distance and the cluster distance. Nevertheless, the available work neglects the influence of the network's general consumption and network energy consumption balance on clustering. Hence, the authors of [28] created an extension model to LEACH protocol, the many-objective energy balance model of cluster head. They considered four objectives to determine the cluster head node: the sink node distance, cluster distance, the total energy consumption of the network, the network energy consumption balance. Meantime, A new approach is proposed (LEACH-ABF) based on adaptive balance function to resolve the model. ABF merges the diversity and convergence functions and uses genetic operations to produce a more desirable solution. Experiments prove that LEACH-ABF has a better balance of energy consumption and prolongs the lifetime of WSN compared to other existing approaches. Also, the approach in [29] used a genetic algorithm for selecting the cluster heads based on four parameters (residual energy, density, centrality, and distance). Additionally, the Artificial bee colony method is applied for selecting nodes in each cluster of the chosen CH.

Only two parameters are chosen [30] to develop a new protocol based on Fuzzy logic and genetic algorithm for WSNs. The fuzzy logic approach for selecting CHs relies on two principal factors: the distance between the node and the BS and node residual energy. The Genetic Algorithm is used as a fitness function for arranging the fuzzy rule table. As a result, cluster heads' selection becomes more effective, and the cluster forming gets more precise. Due to all the nodes nearly die simultaneously, the network lifetime of WSN is prolonged, and the number of data packets received in the sink is maximized. Another routing protocol [31] is an aggregate of Micro Genetic algorithm with LEACH protocol. The μGA-LEACH protocol gives strength to the cluster head (CH) selection and decreases the network's energy consumption using the relay nodes to make communication easier between CHs and BS at large distances.

In [32], the genetic algorithm-based clustering and Routing (LECR-GA) method was created to maximize the WSN lifetime and improve its quality of service by choosing the best of genetic algorithm operations, chromosome representation, and fitness function. Thus, the approach has reached higher system operability and enhance data rate with keeping satisfactory complexity and minimize the energy consumption of nodes by choosing the nearest CH with the appropriate parameters (more energy and less distance to achieve data).

1.6 Mobility

The mobile wireless sensor cannot cover the whole target moving path due to its inadequate number and short sensing range. Thus, in [33], a new approach is proposed to obtain complete coverage for the moving target on a preselected trajectory with a restricted number of mobile sensors. Mobile sensors must change their previous position to a new position in the path to affording complete coverage. The authors minimized the total moving distance of mobile sensors. The farthest movement distance is minimized using a Genetic algorithm-based approach to save energy and prolong the network lifetime.

In [34], an optimizing algorithm for Connected Dominating Set based on anchor nodes was proposed. It used arbitrary mobility for the anchor and the unknown nodes. The optimization method uses the genetic algorithm with elitism procedure so that the fittest solution can be maintained for fast convergence of the global solution. So as the anchor nodes execute the necessary and significant computations in the proposed algorithm, the network lifetime increases.

2 Synthesis and Open Issues

The previous papers showed good results with the experiments, but they are not sufficient in real-life applications. For instance, excepting one work that considered 3D, all the approaches they work on 2D area, Also the majority used the binary models, ignoring the probabilistic nature of WSN.

In addition, these approaches suffer from the high complexity of GA nature which can be reduced by the innovation of a new distributed GA approach. Also, the papers ignore the obstacles factor which can limit and prevent the connectivity and the communication between sensor nodes or degrade the QoS of the whole network.

3 Conclusion

In this paper, we have collected the most important articles concerned with addressing optimization in WSNs. Due to their advantage over the other optimization methods, we have chosen the GA-based approaches in this work. Also, we have limited the time range of the studied articles to the last four years to make the researchers understand the latest findings in this field. Additionally, we have provided a classification of the articles based on the utilized method type and we have afforded a table of each paper's Optimization Objectives. This work allows a better understanding of the proposed approaches during the last four years and is, consequently, a basis for future ideas and works.

References

1. De Gante, A., Aslan, M., Matrawy, A.: Smart wireless sensor network management based on software-defined networking. In: 2014 27th Biennial Symposium on Communications (QBSC), pp. 71–75 (2014)
2. Moon, Y., Lee, J., Park, S.: Sensor network node management and implementation. In: 2008 10th International Conference on Advanced Communication Technology, vol. 2, pp. 1321–1324 (2008)
3. Nagpurkar, A.W., Jaiswal, S.K.: An overview of WSN and RFID network integration. In: 2015 2nd International Conference on Electronics and Communication Systems (ICECS), pp. 497–502 (2015)
4. Yick, J., Mukherjee, B., Ghosal, D.: Wireless sensor network survey. Comput. Netw. **52**(12), 2292–2330 (2008)
5. Wang, F., Liu, J.: Networked wireless sensor data collection: issues, challenges, and approaches. IEEE Commun. Surv. Tutor. **13**(4), 673–687 (2010)
6. Cheng, L., Niu, J., Cao, J., Das, S.K., Gu, Y.: QoS aware geographic opportunistic routing in wireless sensor networks. IEEE Trans. Parallel Distrib. Syst. **25**(7), 1864–1875 (2013)
7. Konstantinidis, A., Yang, K., Zhang, Q., Zeinalipour-Yazti, D.: A multi-objective evolutionary algorithm for the deployment and power assignment problem in wireless sensor networks. Comput. Netw. **54**(6), 960–976 (2010)
8. Coello, C.A.C., Pulido, G.T., Lechuga, M.S.: Handling multiple objectives with particle swarm optimization. IEEE Trans. Evol. Comput. **8**(3), 256–279 (2004)
9. Zhang, Z., Long, K., Wang, J., Dressler, F.: On swarm intelligence inspired self-organized networking: its bionic mechanisms, designing principles and optimization approaches. IEEE Commun. Surv. Tutor. **16**(1), 513–537 (2013)
10. Ripon, K.S.N., Tsang, C.-H., Kwong, S.: Multi-objective evolutionary job-shop scheduling using jumping genes genetic algorithm. In: The 2006 IEEE International Joint Conference on Neural Network Proceedings, pp. 3100–3107 (2006)
11. Rajan, S.D.: Sizing, shape, and topology design optimization of trusses using genetic algorithm. J. Struct. Eng. **121**(10), 1480–1487 (1995)
12. El-Sherif, M., Fahmy, Y., Kamal, H.: Lifetime maximisation of disjoint wireless sensor networks using multiobjective genetic algorithm. IET Wirel. Sens. Syst. **8**(5), 200–207 (2018)
13. Lai, C.-C., Ting, C.-K., Ko, R.-S.: An effective genetic algorithm to improve wireless sensor network lifetime for large-scale surveillance applications. In: 2007 IEEE Congress on Evolutionary Computation, pp. 3531–3538 (2007)
14. Chang, Y., Yuan, X., Li, B., Niyato, D., Al-Dhahir, N.: A joint unsupervised learning and genetic algorithm approach for topology control in energy-efficient ultra-dense wireless sensor networks. IEEE Commun. Lett. **22**(11), 2370–2373 (2018)
15. Elhoseny, M., Tharwat, A., Farouk, A., Hassanien, A.E.: K-coverage model based on genetic algorithm to extend WSN lifetime. IEEE Sensors Lett. **1**(4), 1–4 (2017)

16. Zhang, X.-Y., Zhang, J., Gong, Y.-J., Zhan, Z.-H., Chen, W.-N., Li, Y.: Kuhn–Munkres parallel genetic algorithm for the set cover problem and its application to large-scale wireless sensor networks. IEEE Trans. Evol. Comput. **20**(5), 695–710 (2015)
17. Elhabyan, R., Shi, W., St-Hilaire, M.: A full area coverage guaranteed, energy efficient network configuration strategy for 3D wireless sensor networks. In: 2018 IEEE Canadian Conference on Electrical & Computer Engineering (CCECE), pp. 1–6 (2018)
18. Zahmatkesh, A., Yaghmaee, M.H.: A genetic algorithm-based approach for energy-efficient clustering of wireless sensor networks. Int. J. Inf. Electron. Eng. **2**(2), 165 (2012)
19. Chand, S., Kumar, B.: Genetic algorithm-based meta-heuristic for target coverage problem. IET Wirel. Sens. Syst. **8**(4), 170–175 (2018)
20. Jia, J., Chen, J., Chang, G., Tan, Z.: Energy efficient coverage control in wireless sensor networks based on multi-objective genetic algorithm. Comput. Math. with Appl. **57**(11–12), 1756–1766 (2009)
21. Chen, Y., Xu, X., Wang, Y.: Wireless sensor network energy efficient coverage method based on intelligent optimization algorithm. Discret. Contin. Dyn. Syst. **12**(4&5), 887 (2019)
22. Sohal, A.K., Sharma, A.K., Sood, N.: A hybrid approach to improve full coverage in wireless sensor networks: (full coverage improving hybrid approach). In: 2019 International Conference on Communication and Electronics Systems (ICCES), pp. 1924–1929 (2019)
23. Al-Shalabi, M., Anbar, M., Wan, T.-C., Alqattan, Z.: Energy efficient multi-hop path in wireless sensor networks using an enhanced genetic algorithm. Inf. Sci. (Ny) **500**, 259–273 (2019)
24. Ah, M.A.: Design of super-agent node and energy aware multipath routing using fuzzy logic and genetic algorithm for WSN. J. Gujarat Res. Soc. **21**(14), 499–516 (2019)
25. Jeske, M., Rosset, V., Nascimento, M.C.V.: Determining the trade-offs between data delivery and energy consumption in large-scale WSNs by multi-objective evolutionary optimization. Comput. Netw. **179**, 107347 (2020)
26. Panhwar, M.A., Deng, Z., Khuhro, S.A., Hakro, D.N.: Distance based energy optimization through improved fitness function of genetic algorithm in wireless sensor network. Stud. Inform. Control **27**(4), 461–468 (2018)
27. Shanthi, D.L.: Energy efficient intelligent routing in WSN using dominant genetic algorithm. Int. J. Electr. Comput. Eng. **10**(1), 500 (2020)
28. Wu, D., Geng, S., Cai, X., Zhang, G., Xue, F.: A many-objective optimization WSN energy balance model. KSII Trans. Internet Inf. Syst. **14**(2), 514–537 (2020)
29. Zangeneh, M.A., Ghazvini, M.: An energy-based clustering method for WSNs using artificial bee colony and genetic algorithm. In: 2017 2nd Conference on Swarm Intelligence and Evolutionary Computation (CSIEC), pp. 35–41 (2017)
30. Alwafi, A.A.W., Rahebi, J., Farzamnia, A.: A new approach in energy consumption based on genetic algorithm and fuzzy logic for WSN. In: Md Zain, Z., et al. (eds.) Proceedings of the 11th National Technical Seminar on Unmanned System Technology 2019. LNEE, vol. 666, pp. 1007–1019. Springer, Singapore (2021). https://doi.org/10.1007/978-981-15-5281-6_72
31. Radhika, M., Sivakumar, P.: Energy optimized micro genetic algorithm based LEACH protocol for WSN. Wirel. Netw. **27**(1), 27–40 (2020). https://doi.org/10.1007/s11276-020-02435-8
32. Hamidouche, R., Aliouat, Z., Gueroui, A.: Low energy-efficient clustering and routing based on genetic algorithm in WSNs. In: Renault, É., Boumerdassi, S., Bouzefrane, S. (eds.) Mobile, Secure, and Programmable Networking, pp. 143–156. Springer, Cham (2019). https://doi.org/10.1007/978-3-030-03101-5_14
33. Liang, C.-K., Lin, Y.-H.: A coverage optimization strategy for mobile wireless sensor networks based on genetic algorithm. In: 2018 IEEE International Conference on Applied System Invention (ICASI), pp. 1272–1275 (2018)

34. Kumar, G., Rai, M.K.: An energy efficient and optimized load balanced localization method using CDS with one-hop neighbourhood and genetic algorithm in WSNs. J. Netw. Comput. Appl. **78**, 73–82 (2017)
35. Benghelima, S.C., Ould-Khaoua, M., Benzerbadj, A., Baala, O.: Multi-objective optimisation of wireless sensor networks deployment: application to fire surveillance in smart car parks. In: 2021 International Wireless Communications and Mobile Computing (IWCMC), pp. 98–104 (2021)

The Internet of Things Security Challenges: Survey

Inès Beggar(✉) and Mohamed Amine Riahla

M'hamed Bougara University, Avenue de l'indépendance, 35000 Boumerdès, Algeria
{i.beggar,ma.riahla}@univ-boumerdes.dz

Abstract. The Internet of Things (IoT) and its security issues have been gaining interest in recent years. It is more than necessary to tackle these issues as soon as possible and to find specific solutions to the IoT which will allow it to attain its full maturity and enable us to take advantage of the facilitations it brings to our daily life. In order to do so, it is necessary to identify and to master the ins and outs of the problem which is developed in the present paper. This paper aims on the one hand to present the Internet of Things in a succession of point, and on the other hand to address the security challenges regarding each one of the said points. Moreover, we discuss the properties of the IoT in relation to traditional networks, the level of security required according to the area of application, and security from a point of view of the actors of the IoT ecosystem. Besides, the existing architectures are examined, a comparative table is drawn up, and the results are discussed in order to allow the positioning of future research and to better understand the security issue.

Keywords: Internet of Things · Security · Actors of IoT ecosystem

1 Introduction

The world has now become digital. Cell phones with different types of sensors and applications have become commonplace, as well as pets with collars, autonomous cars, industrial plants, heart sensors, cameras, and so on. It seems that everything is connected and in every field; the number of online devices that work together is constantly increasing. According to Huawei's estimation, around 100 billion devices will be connected by 2025 [1]. This type of connectivity goes by the name of "Internet of Things" (IoT), and it is becoming an integral part of our daily life. The Internet of Things can be described as the interconnection of physical objects via embedded computing devices such as sensors, software, and network connectivity that allows these objects to collect and exchange data [2].

Humans are endlessly interacting with these objects, which is inducing the rapid development and commercialization of new IoT devices. The number of devices is growing exponentially creating an increase in the number of security threats and invasions of privacy. This can negatively impact our lives as the damage caused by a cyber-attack

B. Lejdel et al. (Eds.): AIAP 2021, LNNS 413, pp. 356–366, 2022.
https://doi.org/10.1007/978-3-030-96311-8_33

in such a context has a far greater impact than what can be caused by an intrusion, data theft, or a denial of service as we experience them today.

The future of IoT can be jeopardized if the security aspect is not rapidly taken care of. So, the protection of devices becomes essential although it poses many challenges. The first is to be able to protect the elements of a very heterogeneous IoT environment. In fact, it can integrate entities of very variable origin as a multitude of platforms, protocols, and specifications must coexist [3]. The second challenge is that IoT is accepted as an extended version of some different technologies including wireless sensor networks [2], which already have various security flaws that make it vulnerable to wireless security attacks such as denial of service, wiretapping, message injection, identity theft, and jamming [4]. The third is that one cannot apply a common security solution to all IoT devices. A security solution suitable for one IoT device may not be suitable for another. There is also the issue of defining who is responsible for the security of an IoT device, knowing that they are designed, supplied and deployed by different companies. Finally, IoT devices are lightweight, limited in sources such as energy; storage capacity (memory), and computational power. Most traditional network security countermeasures are based on voracious algorithms and resource intensive protocols. Thus, it would be very difficult to implement these solutions on IoT devices [3]. To overcome this type of challenges; it is essential to already understand the IoT ecosystem with all its complexity and security requirements, to identify the domain and scope of application and its sensitivity, as well as the vulnerabilities of each party in order to propose a coherent and adapted security policy based on technical solutions; such as the ones that use low-cost protocols and less greedy computation algorithms that can provide strong authentication and encryption to IoT devices. This is the interest of drawing up a state of the art on security in IoT, which is the objective of this work.

The present paper is organized into two main parts. The first part defines the Internet of Things in Sect. 2, describes the properties of IoT in Sect. 3, and deals with the domains and scope of application in Sect. 4. Moreover, the actors of the IoT ecosystem are presented in Sect. 5 of this part with the analysis of the architecture and technology of IoT. The second part details the concepts of security related to IoT in Sect. 7 as follows: after the definition and presentation of the families of risks, the security will be approached only according to the points of view treated previously such as the properties of IoT compared to traditional networks, the level of security required according to the application domain, the security from the point of view of the IoT ecosystem actors, and the security in relation to the architecture and technologies. Section 8 mentions the comparison of the security levels according to the 4 points of view and discusses the results. Finally, Sect. 9 concludes the article.

2 Definition

The IoT for "Internet of Things" is a buzzword. It was first coined in 1999 by Kevin Ashton, executive director of the Auto-ID Center at the Massachusetts Institute of Technology. The term has been widely adopted but there is no unanimously accepted definition of it. However, the common point between all the definitions is that the first version of the Internet connected computers or data were created by people, while the IoT connects objects or data that can be created by objects (see Fig. 1). An object by definition is a physical or virtual machine, which has a capacity of calculation and memorization; it is; therefore, 'intelligent', 'autonomous', not requiring human intervention for a treatment and can be 'connected' with any object in a transparent and flexible way [5]. A smartphone, a smartwatch, a connected television or systems of detection of presence, and so on, constitute concrete examples of connected objects.

The CERP-IoT "Cluster of European Research Projects on the Internet of Things" defines the Internet of Things as: "... an integral part of the Internet of the future. It could be defined as a dynamic global network infrastructure with self-configuring capabilities based on interoperable communication standards and protocols, where physical and virtual "objects" have identities, physical attributes, virtual personalities and use intelligent interfaces, and are seamlessly integrated into the network" [6].

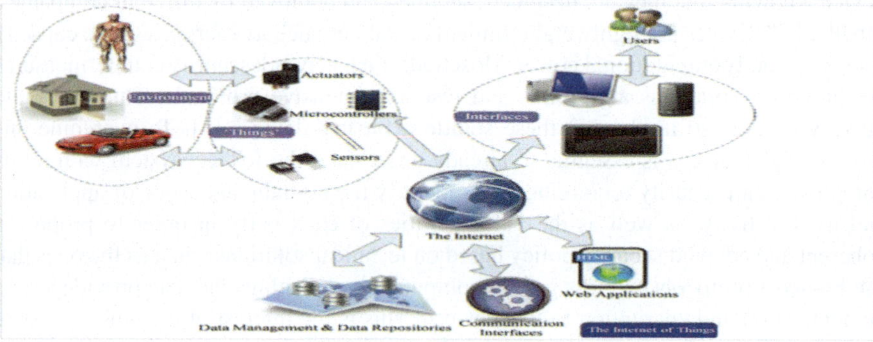

Fig. 1. Illustration of internet of things [7]

3 IoT Properties

For the IoT to be fully realized; a number of challenges must be addressed while considering the combination of IoT properties that make it unique. Vasilomanolakis et al. [8] identified four distinctive properties: the uncontrolled environment, heterogeneity, the need for scalability, as well as the limited resources used in IoT: 1) Limited resources in terms of energy (battery), computing capacity (micro sensors) and storage space (memory) to be taken into account for security mechanisms.2) The IoT is an uncontrolled environment mainly due to the mobility of objects, the extended possibilities to access them physically, and the lack of trust. 3) Heterogeneity: an IoT environment can integrate entities from very different origins (different platforms, communication protocols, suppliers...) to take into account the compatibility of versions and interoperability. 4) The scalability is related to the quantity of objects that can be interconnected. It requires highly scalable protocols and influences the security mechanisms.

4 Areas and Scope of Application

The rapid growth of IoT technology and the high potential it promises if it reaches full maturity will disrupt modern life as we know it and transform every aspect of our daily lives [9]. In other words; the IoT will eventually touch almost every area of our daily lives and cover a wide range of applications. Table 1 gives an overview of some of the main areas and sectors of application of IoT [10].

Table 1. Examples of IoT application areas.

IoT application domains	
Army; Energy; Telecommunications;	Pharmaceuticalfield; Automotive
City Management/Intelligent Buildings	Safety, Security, Privacy and Recycling
MedicalTechnology, Healthcare; Oil and Gas	Environment Monitoring
Logistics, Supply Chain Management and Retail	People and Goods Transportation
Manufacturing, Product Lifecycle Management	Agriculture and Breeding; Insurance

For implementing a proper IoT security solution, it is crucial to determine the scope of the system. Some IoT systems operate primarily on a local scale, for instance, smart homes that are largely autonomous. Other systems operate on a cosmopolitan scale, for example, a system of sensors deployed across continents and collecting environmental data could feed into devices to analyze climate change or phenomena [9]. However, the data collected at a (local) scale could be integrated into a larger (macroscopic) system.

In addition, IoT systems can also be integrated into systems of systems and sometimes span more than one domain. Data collected from one domain can be used in another domain and play a role in strategic decision making; e.g. in the management of the Coronavirus 2019 health crisis, data from the air transport system, originally collected as part of the management of passenger flows, were combined with data from health systems to track the spread of the disease from one region to another.

5 Actors of thc IoT Ecosystcm

The actors of the IoT ecosystem can be distinguished into two main categories: the manufacturers and the users whose priorities are different. The manufacturers are the economic actors from different sectors, including industry. Its main actors are: designers and manufacturers of connected objects, manufacturers of computer components for these objects, operators and managers of data flow transmission networks, managers of data collection and processing platforms, designers of software interfaces between objects and users, service providers who collect, analyze, and exploit user data provided by connected objects, and public regulators who ensure compliance with laws in terms of respect for life and private data, as well as security standards for connected objects.

The priorities of manufacturers are linked to several factors, including cost, preservation of the brand's image, the ability to scale up regardless of the number of users, and the identification of the objects, so that the data collected can be associated with them and value-added analysis can be performed. As for users, several profiles are identified for them: companies, local authorities, craftsmen, or "simple" individuals who use objects on the move or at home. The priorities of users, considering their profiles, converge on several dimensions, from the price, knowing that the IoT market is very competitive, to the respect of the confidentiality of information. Moreover users have gained in maturity and take security more and more into consideration during acquisitions [11, 12] as well as the regulatory context which is more and more restrictive. Reliability is also considered a priority with a level of sensitivity that depends on the user's profile, his sensitivity to security issues, and also on the application domain.

We suggest taking into consideration a third actor, "the authorities," in view of the important role this function plays in the future development and emergence of IoT, bringing together policy makers, public regulators, regulatory bodies and industry alliances developing standards and guidelines to secure IoT devices [11–15]. As an example, we cite the National Information Security Reference System 2020 'L06-Final version of the RNSI 2020' which applies to administrations and public sectors, as well as any infrastructure hosted on the Algerian national territory and dealing with sensitive information according to the laws and regulations in force, proposed by the Algerian Ministry of Post and Telecommunications (MPT), it provides, through a set of recommendations, an approach to securing information based on risk management with regard to the confidentiality, integrity and availability of information. The security measures related to the Internet of Things are defined in domain 12 of the standard.

6 Architectures and Technology

The Internet of Things, due to its complexity and peculiarities, is very broad and unlimited, which is a major problem in the implementation of its concept. There is no uniform architecture that can be applied to all domains. The development and proper functioning of IoT involves an assortment of several technologies such as RFID [16], wireless sensors and actuators, networks, protocols, machine-to-machine (M2M) communications [17], and computing, among others [18].

The four-layer architecture: we synthesize the works published in the literature [2, 15] and [19] presenting an architecture that can be extended to the actual development of applications and guide theoretical research (see Fig. 2). The perception (physical) layer: using its sensors, it interconnects and identifies unique devices and provides the discovery service [2, 11] collecting information from the physical world [20]. The network layer: is responsible for the communication and connectivity of all devices in the system and the transfer of information collected by them to an information processing system using several protocols. The support layer (middleware): it acquires data from the network layer, connects the system to the database or cloud to store the data, and also involves information processing systems that take the information in one form and transform it into another. It also meets the requirements of the application layer by providing APIs. And the last layer, the Application (Service) Layer: provides practical applications developed according to user requirements or industry specifications. it provides specific services to end users [12], hence the designation service layer.

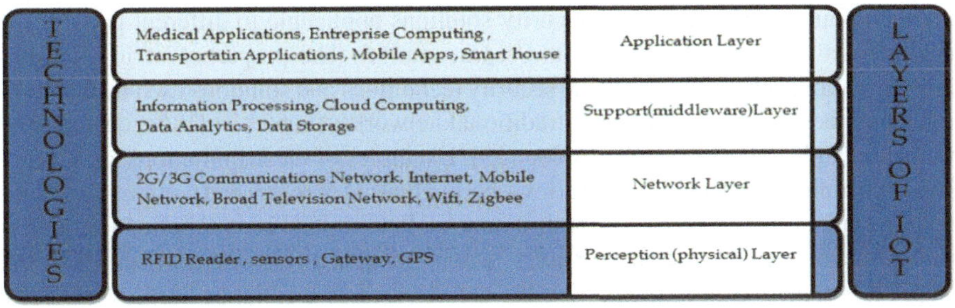

Fig. 2. IoT architecture

1) Sensor technology, intelligence embedded technology, nanotechnology and RFID are located in the perception/physical layer. 2) Fiber optic and 2G/3G communication networks, Wifi, Zigbee, large TV networks, fixed telephone networks and others are located in the network layer. 3) Databases and the cloud are located in the support/middleware layer. 4) Specific applications and system integration are located in the application/service layer, e.g. smart traffic, smart home, etc.

7 The Security

IoT security is defined in the work of Hammi in 2018 [5] as ensuring the proper functioning of a system and guaranteeing the expected results of its design. The set of policies and practices adopted to monitor and prevent misuse, unauthorized access and modification, or denial of a computer operation thus represent security. The threat of cyber-attacks makes IoT security one of the major issues which hinder the rapid deployment and evolution of this technology of technologies.

The impacts in case of an attack are varied [21], the impact differs according to the type, use and functionality of the objects. The families of risks are common to all of them: from denial of service, to loss of confidentiality and integrity of measurements made by sensors, to leakage of personal data, or even worse, to breach of personal safety [9]. There are three main categories: Privacy risks, Systemic risk and Other risks associated with poorly secured IoT devices.

In what follows; security will be discussed from different perspectives.

7.1 From a Point of View: Properties of IoT and the Security of Traditional Networks

IoT systems coexist with traditional systems in the same computer networks; they are faced with various cyber-attacks. In order to cope with the many security threats that affect computer networks, many security solutions applicable to different parts of the networks have emerged (firewalls and segmentation). The properties of IoT systems have limitations when faced with the security techniques and solutions used by the traditional methods for the protection of traditional networks; such as isolation, device-level protection and network-level protection [15]. Examples of these limitations are given. - Resource limitation of an IoT device (low energy; limited memory and computing power) makes it vulnerable to even the simplest attacks. Security solutions applied for device-level protection in traditional computer networks such as anti-virus or anti-malware cannot be adopted. -Interoperability is the cohabitation of disjoint devices, systems and mechanisms and the possibility to make them cooperate and interact in all flexibility. Its most basic form is the accessibility of IoT objects from traditional computer networks. But the coexistence of vulnerable and insecure IoT devices and non-IoT devices is unavoidable in some cases or bridges between the two initially isolated networks are built and eventually compromises the security of the entire enterprise network.- Heterogeneity: the heterogeneous nature of IoT systems and device types: each with its own behaviors and vulnerabilities; it is difficult for devices used for network-level protection such as a firewall or IDS appliance to distinguish between normal traffic and abnormal traffic that could be symptomatic of an attack. - Scalability: it is difficult to monitor each individual device using traditional techniques; this leads to increased maintenance costs. Also, centralized approaches, such as hierarchical public key infrastructures (PKI), and distributed approaches, such as pairwise symmetric key exchange systems, cannot scale with the IoT.

From what has been presented, it can be seen that the security problems in both networks can be similar, but different approaches and techniques are used to address each security problem depending on the network [22]. Therefore, it is essential to develop specific security solutions for objects with strong resource constraints having multiple wireless communication methods. For example, a solution would be to design a protocol based on robust algorithms, but at the same time light and flexible, adaptable to different types of objects, from the weakest to the most powerful without degrading the security performance.

7.2 From a Point of View: Area and Scope of Application

From what was presented in Sect. 4, it is clear that IoT infrastructures can almost touch all areas of our daily lives and cover a wide range of applications but also have different scopes, so it becomes difficult to impose a standard in all these areas, as the security requirements of a home network may be different from those of a critical infrastructure [15]. Furthermore, it would be more prudent to secure the most critical parts of the IoT, namely those in sensitive areas such as the military and critical infrastructure, rather than consumer goods [9].

In addition to the application domain; determining the scope of an IoT system can tell us about the complexity of its architecture; whether its operation is at a local scale or integrated into systems of systems the security solution to be implemented will depend on its functional architecture.

7.3 From a Point of View: Actors of the IoT Ecosystem

In order to reduce IoT-related cyber threats, security must be considered and assessed by all stakeholders (manufacturers, users, authorities and service providers) each as it relates to them. Manufacturers include constructers and service providers as well as public regulators. We have proposed to consider a third actor in its own right "authorities" and in what follows we will distinguish the service providers from the manufacturers. Manufacturers are under competitive pressure for cheaper products and shorter time-to-market; especially since there are no credible means (trustmarks, certifications,...) for consumers to distinguish the security level of one vendor from another and cyber security skills and security testing are scarce, however, failing to react to threats will tarnish the brand image. IoT service providers are required to support the security of IoT systems which is an expensive task without the ability to quantify the security assurance provided. Users are consumers of IoT technology (property or enterprise managers and network managers). Although these users follow strict procedures when purchasing and install in their networks only secure devices with strong encryption and proper maintenance according to the regulatory recommendations in force; they remain victims of limited skills for network monitoring; lack of sufficient knowledge; inadequate operational testing; and lack of automatic resource management.

Authorities: Standards and guidelines are developed by authorities; they pose other major challenges [15]: (a) limited attention to security: the system is potentially vulnerable because the standards are limited to a subset of the security aspect and create a partially trustworthy environment, (b) imprecision: the recommended guidelines are qualitative and subject to human interpretation, (c) lack of legacy support: devices that are already on the market are not regulated, and (d) lack of mandate: the difficulty of imposing a standard in all of these areas has been demonstrated.

7.4 From a Point of View: Architectures and Technology

Based on the variety, richness and specificities of the technologies located in the layers of the architecture models, the architecture of an IoT solution varies from one system to another based on multiple criteria: 1) the communication technology used; 2) the data processing in the cloud (computing power) or relying on local computing capabilities (computing speed) or relying on other smart devices in the vicinity; 3) the type of object used: physical object equipped with IoT element or digital object existing in the real world; the smart object communicates directly with the cloud or indirectly.

8 Comparative Table of the Different Points of View

Table 2. Comparison of security levels according to points of view.

A point of view	IoT security			
	Devices protection	Networks protection	Data protection	Modularity of the failure
Traditional networks	0	1	3	0
Area and scope of application	5	5	6	5
Actors of the IoT ecosystem	4	2	4	1
Architectures and technology	6	6	5	6

The existing conventional security architecture is limited and does not meet the properties of IoT. It would be more prudent to secure the most critical parts of the IoT, namely those in sensitive areas and focus on data. Security requirements are enhanced by the provisions of the standards; they ensure data protection, service continuity and device security; eventually, the bar for device acceptance will be raised prior to installation. Also, the challenges posed by the development of the aforementioned standards and guidelines do not support network protection and scalability in the event of failure (Table 2 and Fig. 3).

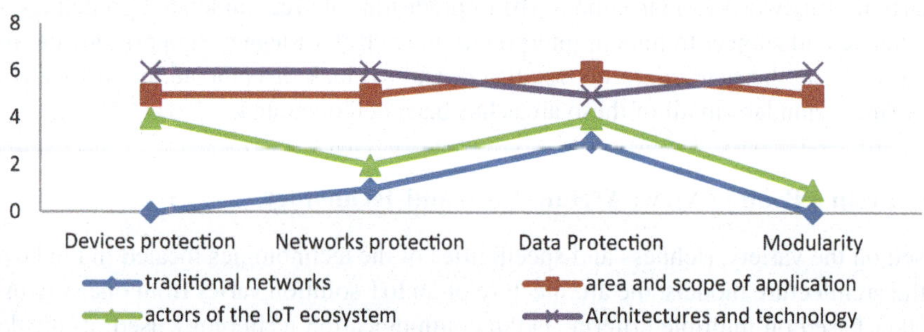

Fig. 3. The levels of security according to the points of view.

The technological portfolio and the flexibility in the IoT architecture seem to offer infinite possibilities of IoT system solutions.

IoT security will strongly depend on its architecture, on the technologies employed, and also on its scope and the sensitivity of its application domain for the choice of the crucial parts to secure. The analysis of the data (Table 2 and Fig. 3) leads us to propose the following steps for an efficient handling of the IoT security issue, namely: 1) Specify the domain and scope of application 2) Take into consideration the regulations, standards and norms in the field 3) Take into consideration the existing (traditional network) 4) Determine the functional architecture of the IoT system 5) Adopt a data-centric approach to security 6) Rely on the new IoT technologies, tools and solutions. In our study, we limit ourselves to these conclusions. For the implementation of security there are multiple solutions in the literature; we cite as an example, and according to the layers of the IoT architecture, the work of Leloglu [2] which classifies all types of security threats that can be critical in the development and implementation of IoT in different domains, and provides recent solutions to these threats. This could be the subject of a future research. Nevertheless, we return to the data-centric approach to security; its primary focus is the protection of (valuable) data since there will always be a way to penetrate systems, even the ones using the best cyber security tools. Understanding the infrastructure, flows, and risks related to data is essential but so is the classification of sensitive data, while monitoring and controlling its use.

9 Conclusion

Several researches have been carried out on IoT security in recent years; however, there are still many key issues that need more efforts in order to be resolved. The importance of security in the development of the Internet of Things, which does not only depend on the possibility to make cooperate intelligent and autonomous objects with connection means or to make this technology adapt to our lives in all areas of everyday life, is no longer to be proven. It is essential for reliable and secure infrastructures, harmonization, IoT security guidelines, and recommendations to simultaneously exist in order to stimulate its adoption. It is important that standardization processes remain aligned with the technology.

In this article, the first part was devoted to the Internet of Things, which was discussed in detail in the form of points: its definition, its properties, the domains and scope of application, the actors of the ecosystem, and its architecture were exposed. Different technologies were located in the layers of the architectural models presented to get on a functional architecture. In the second part, it is the security concepts of the IoT that were largely reviewed according to the same points exposed in the first part. The properties of IoT compared to traditional networks, the level of security required according to the application domain, the security from the point of view of the actors of the IoT ecosystem and the security according architectures and technology were compared and discussed.

References

1. Manyika, J., et al.: The Internet of Things: Mapping the value beyond the hype (2015)
2. Leloglu, E.: A review of security concerns in Internet of Things. J. Comput. Commun. **5**, 121–136 (2016)
3. Restuccia, F., D'Oro, S., Melodia, T.: Securing the internet of things in the age of machine learning and software-defined networking. IEEE Internet Things J. **5**, 4829–4842 (2018)
4. Chen, K., et al.: Internet-of-things security and vulnerabilities: taxonomy, challenges, and practice. J. Hardw. Syst. Secur. **2**, 97–110 (2018). https://doi.org/10.1007/s41635-017-0029-7

5. Hammi, M.T.: Sécurisation de l'Internet des objets, Paris Saclay (2018)
6. Sundmaeker, H., Guillemin, P., Friess, P., Woelfflé, S.: Vision and challenges for realising the Internet of Things. Cluster Eur. Res. Projects Internet Things Eur. Comm. **3**, 34–36 (2010)
7. Mahmoud, M.S., Mohamad, A.A.: A study of efficient power consumption wireless communication techniques/modules for internet of things (IoT) applications (2016)
8. Vasilomanolakis, E., Daubert, J., Luthra, M., Gazis, V., Wiesmaier, A., Kikiras, P.: On the security and privacy of internet of things architectures and systems. In: 2015 International Workshop on Secure Internet of Things (SIoT), pp. 49–57. IEEE (2015)
9. Tonin, M.: The internet of things: Promises and perils of a disruptive technology, Science & Technology Committee, NATO Parliamentary Assembly (2017)
10. Challal, Y.: Sécurité de l'Internet des Objets: vers une approche cognitive et systémique, Université de Technologie de Compiègne (2012)
11. Andrea, I., Chrysostomou, C., Hadjichristofi, G.: Internet of things: security vulnerabilities and challenges. In: 2015 IEEE Symposium on Computers and Communication (ISCC), pp. 180–187. IEEE (2015)
12. Dean, A., Agyeman, M.O.: A study of the advances in IoT security. In: Proceedings of the 2nd International Symposium on Computer Science and Intelligent Control, pp. 1–5 (2018)
13. Fagan, M., Megas, K., Scarfone, K., Smith, M.: Recommendations for IoT device manufacturers: foundational activities and core device cybersecurity capability baseline (2nd Draft). National Institute of Standards and Technology (2020)
14. I.A. Australia: Strategic Plan to Strengthen IoT Security in Australia (2017)
15. Hamza, A., Gharakheili, H.H., Sivaraman, V.: IoT network security: requirements, threats, and countermeasures, arXiv preprint arXiv:2008.09339 (2020)
16. Stockman, H.: Communication by means of reflected power. Proc. IRE **36**, 1196–1204 (1948)
17. V. ETSI: Machine-to-machine communications (M2M): Functional architecture, Int. Telecommun. Union, Geneva, Switzerland, Technical report TS, 102, 690 (2011)
18. Gigli, M., Koo, S.G.: Internet of things: services and applications categorization. Adv. Internet Things **1**, 27–31 (2011)
19. Alaba, F.A., Othman, M., Hashem, I.A.T., Alotaibi, F.: Internet of things security: a survey. J. Netw. Comput. Appl. **88**, 10–28 (2017)
20. Farooq, M.U., Waseem, M., Khairi, A., Mazhar, S.: A critical analysis on the security concerns of internet of things (IoT). Int. J. Comput. Appl. **111**, 1–6 (2015)
21. Hantouche, C.: Peut-on sécuriser l'Internet des Objets? Sécurité et stratégie **22**, 31–38 (2016)
22. Kasinathan, P., Costamagna, G., Khaleel, H., Pastrone, C., Spirito, M.A.: An IDS framework for internet of things empowered by 6LoWPAN. In: Proceedings of the 2013 ACM SIGSAC Conference on Computer & Communications Security, pp. 1337–1340 (2013)

Hybrid Approach to WebRTC Videoconferencing on Mobile Devices

Bakary Diallo(✉), Abdelaziz Ouamri, and Mokhtar Keche

University of Science and Technology of Oran - Mohamed BOUDIAF (USTO - MB), Oran, Algeria
{bakary.diallo,ouamri.abdelaziz,mokhtar.keche}@univ-usto.dz

Abstract. This paper provides an in-depth comparative study and an interoperability study between a WebRTC browser-based P2P videoconferencing solution and a hybrid mobile app based one, built with the React Native framework. The comparison is in terms of CPU load, RAM occupancy, and network data usage. To carry out our experiments, we designed and implemented a WebRTC P2P videoconferencing prototype, including a signaling server and two separate client applications based on the same algorithm written in JavaScript. The first application is a WebRTC web client (compatible desktop and mobile) and the second is a WebRTC React Native hybrid mobile application. According to the results obtained after several video calls performed over WLAN and LTE networks, our WebRTC hybrid app consume less CPU ($\sim -10\%$) compared to the web browser-based one. The two types of applications show comparable RAM occupancies. In short, our results showed that implementing WebRTC real-time video streaming in a hybrid mobile app can be a better alternative in the WebRTC videoconferencing on mobile devices, while most of the scientific researches carried out around WebRTC still focus on the web browser.

Keywords: WebRTC · Videoconferencing · Hybrid App · Web App · CPU consumption · RAM occupancy · Bandwidth occupancy · Quality of Experience (QoE)

1 Introduction

Nowadays, especially since the Covid-19 pandemic, the growing demand for real-time multimedia data transmission applications such as videoconferencing, distance education, IP television, video on demand, medical tele-operation, and remote video surveillance are increasingly challenging, on the one hand, the application developers in terms of technologies (API, framework and platform) and topologies, and on the other hand, the device manufacturers in terms of physical performances (CPU, RAM, and power) and mobile operators in terms of bandwidth and coverage. Moreover, most of the solutions available in the literature are paid, proprietary or require some external plugins. WebRTC is a free and open-source framework that does not require any plugin and integrates several powerful tools for encoding, decoding, and securing audio and video

B. Lejdel et al. (Eds.): AIAP 2021, LNNS 413, pp. 367–378, 2022.
https://doi.org/10.1007/978-3-030-96311-8_34

streams [1]. The technologies behind WebRTC are implemented as an open web standard and available as regular JavaScript APIs in all major web browsers. For native and hybrid clients, like Android and iOS applications, a library is available that provides the same functionality.

WebRTC is intended to work in web browsers, native and hybrid clients on different kind of devices (desktop, mobile and IoT devices). To the best of our knowledge, there is no research that rigorously studied the usage of WebRTC outside the web browsers. In this paper, we present an in-depth comparative study and an interoperability study between a WebRTC browser-based videoconferencing solution and a hybrid mobile app based one. The comparison is in terms of CPU load, RAM occupancy, and network occupancy on mobile devices, on different types of networks (WLAN, and LTE). To carry out our experiments, we designed and implemented a WebRTC videoconferencing solution containing a signaling server and two separate client applications based on the same algorithm written in JavaScript: The first one is a responsive web application, compatible with a mobile and a desktop. The second client is a WebRTC hybrid mobile application, developed with the React Native framework [2].

To push further our study, the "video calling functionality" of two mobile multimedia applications "Facebook Messenger" and "Facebook WhatsApp Messenger" were included in our experiments to find out whether a simple WebRTC React Native based hybrid application can be compared, on some technical performance aspects (CPU load, RAM occupancy and network data usage), to these high-end applications supported by the dynamic working groups of Facebook.

The rest of this paper is organized as follows. Section 2 presents the related works. Section 3 describes our proposed prototype. Experimental evaluation and results obtained are presented in Sect. 4. Section 5 provides a discussion on the obtained results. Finally, a conclusion is drawn and future works are suggested in Sect. 6.

2 Related Works

Since its introduction by Google in 2011, WebRTC attracted the curiosity of several developers and researchers around the world. Several studies and researches were carried out on the subject.

The authors of [3] describe some measurements collected from a WebRTC implementation operated from real mobile nodes within the pan European MONROE platform. They experimentally investigated and compared the performance of WebRTC for static and mobile cellular users of several networks across Europe. They observed that mobility is still an important challenge for WebRTC, since mobile broadband operators do not yet cope with full quality coverage for users on the move. Their studies were limited to Goggle Chrome web browser and the factors related to network (video frame rate, video bit rate, packet delay, and jitter delay).

Asif et al. [4] compared WebRTC video conferencing functionality against a Skype-based solution to determine whether an integrated approach could provide an experience as good as or better than the off the shelf solution on certain aspects. They achieved this by implementing WebRTC into an existing groupware web application, PowerMeeting, and compared this with PowerMeeting's existing Skype-based solution. They found that whilst users felt that WebRTC was capable of delivering a solution that could be used without any major issues, the quality and reliability of the Skype solution provided a more stable experience for groupware activities overall. Since the implementation of their solution, there was many progresses in WebRTC technology. Their solution was based on the SimpleWebRTC library, which was developed before the WebRTC first stable release [5]. A WebRTC hybrid application could be a better solution.

Kundan et al. [6] presented seven different cross platform apps built using Chrome App (for desktop) and Apache Cordova (for mobile) frameworks and tools. These apps use WebRTC for real-time audio and video streaming. They described some challenges and techniques (like media capture and playback, network connectivity, interoperability, and multi-way call) related to audio, video, network, power conservation, and security in such applications. The authors of this paper did not present any comparative study between web-based WebRTC solutions and hybrid-based ones in terms of technical performances related to hardware, such as CPU load and RAM occupancy, which can drastically impact the quality experienced by a user in a videoconferencing solution.

Authors of [7] created and implemented a WebRTC videoconferencing solution that can offer bi-directional communication over different networks, such as wired and Wi-Fi of LAN and WAN networks. A deep evaluation of the physical implementation was done regarding CPU performance, bandwidth consumption and QoE. They concluded that the bandwidth consumption of audio communication in WebRTC exhibited (53–54 Kbit/s) bandwidth rate over LAN and (48–50 Kbit/s) over WAN, while the CPU exhibited a range of 13% to 17% as an average needed rate, according to their testing environment. Their studies were only based on the Google Chrome web browser and they did not address the case of mobile devices.

While implementing a P2P videoconferencing system based on WebRTC, the authors of [8] tracked the CPU and memory states according to the number of users in the conference. Although brief and limited to web browsers, this study is a good introduction to technical performances (CPU time, and RAM occupancy) evaluation of a P2P videoconferencing solution based on WebRTC.

3 The Proposed Solution

A WebRTC solution comprises two parts: a client part and a server part. The client part can be a web or mobile or desktop application, coded in JavaScript/HTML/CSS or other native languages (Java, C++, Oject-C). The HTML/CSS part is used for HMI (Human-Machine Interface) and the JavaScript part for communications with the servers (signaling, STUN and TURN) and the exploitation of WebRTC functions (connection establishment, data exchange, etc.).

Our WebRTC videoconferencing prototype includes a signaling server and two separate client applications.

3.1 Signaling Server

WebRTC repose on three main APIs:

- MediaStream, which represents a media stream composed of audio/video tracks obtained via the mediaDevices.getUserMedia() method allowing access to user's multimedia resources (camera and microphone), or from other multimedia data files (.mp3, .mp4) already available on the user's device.
- RTCPeerConnection, which makes it possible to establish a P2P connection between two users via the ICE protocol to exchange audio /video streams.
- RCTDataChannel, which represent an arbitrary data channel between two end users.

Since WebRTC standard does not specify a signaling protocol between the clients and the signaling server, each developer has the possibility to implement its own signaling mechanism. In this work, we designed and implemented a signaling server based on the Node.js framework [9], the JavaScript Session Establishment Protocol (JSEP) [10], and the WebSocket API [11, 12]. Our signaling protocol includes five control messages: "connexion", "offer", "answer", "candidate", and "fin".

Figure 1 illustrates a scenario in which client 1 sends a message "offer" to client 2, via the signaling server, to initiate a call. Client 2 receives this message and sends a message of type "answer" to client 1, to answer its call. Then the two clients exchange "candidate" messages via the signaling server until a P2P connection (RTCPeerConnection) is fully established between them (for further information on WebRTC signaling, see [1] and [10]).

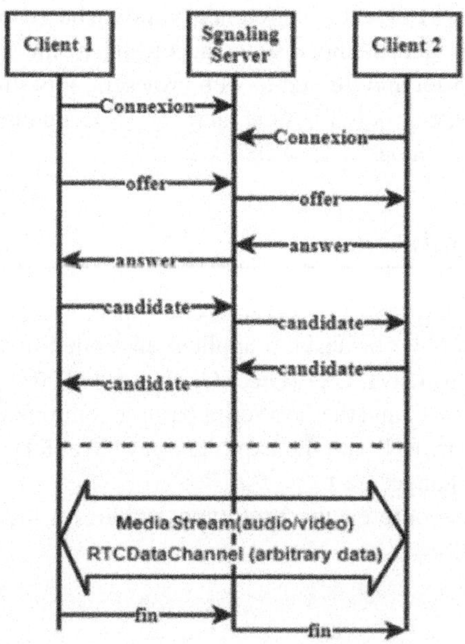

Fig. 1. Our proposed signaling protocol

3.2 Client Applications

We implemented two WebRTC client applications based on the same main algorithm written in JavaScript. The first one, is a responsive web application (for desktop and mobile), and the second one, is a hybrid mobile application developed with the React Native framework. Some screenshots of these applications are shown in Fig. 2.

Fig. 2. Our WebRTC web and hybrid clients

4 Evaluation and Results

4.1 Experimental Environments

We carried out experiments according to the two network topologies illustrated in Fig. 3. The first one is a WLAN (Wireless Local Area Network). The second network represents the topology that was used on the LTE network. It consists of the two smartphones (Smartphone 1 and Smartphone 2) equipped with a LTE USIM card from the mobile telephony operator Djezzy in Algeria.

Evaluated Parameters. The 4 main parameters evaluated in our studies are: the CPU consumption, the RAM occupancy, the Number of bytes sent per second, and the Number of bytes received per second. These parameters can significantly affect the user perceived quality of experience in a video streaming application.

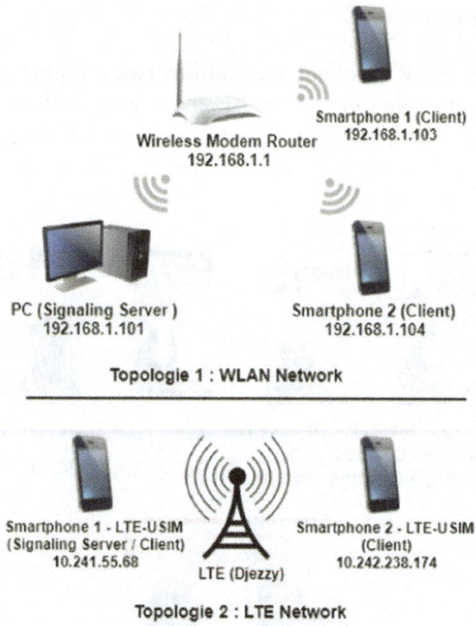

Fig. 3. Our test bed

To accomplish our studies, we performed 15 video calls over the WLAN and the LTE networks, each video call took 5 min. We performed 61 measurements, at a rate of one measurement every 5 s, of CPU consumption, RAM occupancy, number of bytes sent per second, and number of bytes received per second by using the Android mobile application "Simple System Monitor".

The purpose of calls performed over the WLAN network was to accomplish a comparative study, in terms of CPU consumption, RAM and bandwidth occupancies, and an interoperability study between the WebRTC web client and the WebRTC hybrid based one.

The purpose of video calls performed over the LTE network was to evaluate our WebRTC hybrid mobile client over the LTE and to establish a comparative study between it and the "video calling functionality" of two popular and major multimedia mobile applications "Facebook WhatsApp Messenger" and "Facebook Messenger", in terms of CPU consumption, RAM and bandwidth occupancies.

The tests were conducted with a video definition of 320 × 240 pixels and a frame rate of 25 fps, in each client application.

4.2 Evaluation Over WLAN

Comparative Study. Over WLAN, we performed 6 video calls between the two smartphones, which we divided into 2 scenarios:

- Scenario 1: In which we performed, between the two smartphones, a video call in Google Chrome Mobile browser, a video call in Mozilla Firefox Mobile browser and a video call in our WebRTC hybrid mobile app.
- Scenario 2: We made the same calls as in Scenario 1 by changing the sense of calls (the calling smartphone becomes the called one).

The average values over the 61 measurements performed from the CPU consumption, the RAM occupancy, the number of bytes sent per second, and the number of bytes received per second are displayed in Table 1 for Smartphone 1, and Table 2 for Smartphone 2, for both Scenario 1, and Scenario 2.

The numerical average values displayed in Table 1 show that the mobile web browsers Google Chrome and Mozilla Firefox consume more CPU (~100%) compared to our React Native hybrid mobile application (~85%) on the Smartphone 1, and that the difference is negligible in RAM occupancy (~55%). They also show that the Google Chrome Mobile browser consumes much more data (~65 KB/s) compared to Mozilla Firefox Mobile (~17 KB/s) and our hybrid application (~15 KB/s).

The results obtained on the Smartphone 2 (Table 2) confirm those of Smartphone 1. From this table, we can observe that the Google Chrome and Mozilla Firefox mobile browsers display an average CPU load of (~45%), while our hybrid app displays an average CPU load of (~36%). The difference in RAM occupancy is still negligible (~58%). We can observe again that the Google Chrome Mobile browser exceed the two other applications in terms of binaries data usage (~50 KB/s vs. ~15 KB/s) during a video call. Besides this, we can observe that the two scenarios of tests give approximately the same average values on both smartphones.

Table 1. Average values of CPU consumption, RAM occupancy, number of bytes sent/s and number bytes received/s on Smartphone 1 (in Chrome Mobile, Firefox Mobile, and React Native app) - Scenario 1 and Scenario 2.

Measurement	Before calls	Application					
		Chrome mobile		Firefox mobile		React native hybrid app	
		Sce[a]. 1	Sce. 2	Sce. 1	Sce. 2	Sce. 1	Sce. 2
CPU consumption (%)	50,21	100,00	100,00	100,00	100,00	85,29	84,17
RAM occupancy (%)	43,49	55,59	53,24	55,52	53,75	56,23	57,35
Bytes sent/s (KB/s)	1,28	54,14	60,84	11,00	23,16	17,38	14,75
Bytes received/s (KB/s)	0,07	67,92	81,97	13,17	23,76	16,55	11,58

Interoperability Study. To check the interoperability between our WebRTC brower-based solution and the hybrid-based one, we performed 3 video calls between the two smartphones over the WLAN, which we considered as Scenario 3 (on the WLAN). In this scenario, we performed a call between Google Chrome Mobile browser (Smartphone 1) and Mozilla Firefox Mobile (Smartphone 2), a call between Google Chrome Mobile (Smartphone 1) and our hybrid mobile application (Smartphone 2), finally, a video call between Mozilla Firefox Mobile (Smartphone 1) and the WebRTC hybrid mobile application (Smartphone 2).

Table 2. Average values of CPU consumption, RAM occupancy, number of bytes sent/s and number of bytes received/s on Smartphone 2 (in Chrome mobile, Firefox mobile, and React Native app) - Scenario 1 and Scenario 2.

Measurement	Before calls	Application					
		Chrome mobile		Firefox mobile		React native hybrid app	
		Sce. 1	Sce. 2	Sce. 1	Sce. 2	Sce. 1	Sce. 2
CPU consumption (%)	36,43	43,24	48,31	46,48	41,74	38,35	33,82
RAM occupancy (%)	51,03	63,13	59,72	58,66	58,38	59,94	58,40
Bytes sent/s (KB/s)	1,12	54,98	38,89	16,45	19,82	16,51	13,59
Bytes received/s (KB/s)	0,10	48,48	59,94	4,19	6,74	13,57	13,82

All mixed video calls were performed successfully, (see Table 3), where we can observe that the web browsers display average values that are similar to those obtained in the precedent study (Table 1 and Table 2). The same observation holds for the hybrid app also. On the basis of this study, we can affirm that the latest release of WebRTC (WebRTC 1.0) ensures the interoperability between the mobile web browsers Google Chrome and Mozilla Firefox on one hand, and on the other hand, between these browsers and a WebRTC hybrid mobile application developed with the React Native framework, on the Android operating system.

4.3 Evaluation over LTE

For a further evaluation of our approach (implementing WebRTC videoconferencing in a mobile hybrid application), we included in our studies the "video calling functionality" of two out of range multimedia applications "Facebook WhatsApp Messenger" and "Facebook Messenger". To figure out how our approach can be situated among the current state of art of mobile videoconferencing solutions, according to certain technical aspects, such as CPU consumption, RAM occupancy, and mobile data usage, which are some major factors that impact the quality experienced by a user.

LTE (Long Term Evolution) also known as 4G, represents a major evolution in the mobile networks field. It is the extension of GSM (2G) and UMTS/HSPA (3G/3G+). LTE has the advantage to be an all-IP network, any device connected to an LTE network is automatically assigned with an IP address, internal to this network [13]. This IP address can then be used to reach the device across the operator's mobile network. This is the

Table 3. Average values of CPU consumption, RAM occupancy, bytes sent/s and bytes received/s on Smartphone 1 and Smartphone 2 (in Chrome Mobile, Firefox Mobile, and React Native app) - Scenario 3.

Measurement	Call scheme					
	Chrome to Hybrid		Firefox to Hybrid		Chrome to Firefox	
	Chrome (Sma[b]. 1)	Hybrid (Sma. 2)	Firefox (Sma. 1)	Hybrid (Sma. 2)	Chrome (Sma. 1)	Firefox (Sma. 2)
CPU consumption (%)	99,51	34,19	100,00	34,11	100,00	40,17
RAM occupancy (%)	58,66	59,02	56,75	59,09	55,86	59,97
Bytes sent/s (KB/s)	4,70	14,11	2,32	1,75	6,42	11,35
Bytes received/s (KB/s)	20,58	2,69	2,49	2,35	13,41	5,02

[a]Sce. means Scenario in the Table 1 and Table 2.
[b]Sma. means Smartphone in Table 3.

functionality that we used in this study to make video calls over the LTE network, with our hybrid mobile application.

Over the Algerian Djezzy LTE mobile network, we performed 6 video calls which we divided into 2 scenarios:

- Scenario 1: We performed, between the two smartphones, a video call in "Facebook WhatsApp Messenger", a video call in "Facebook Messenger" and a video call in our WebRTC hybrid mobile app.
- Scenario 2: We repeated the same calls as in Scenario 1 by switching the sense of calls (the calling smartphone becomes the called one).

As on the WLAN, we performed 61 measurements, at a rate of one measurement every 5 s, of CPU consumption, RAM occupancy, number of bytes sent per second, and number of bytes received per second on each smartphone. The average values over the 61 measurements performed from the CPU usage, the RAM occupancy, the number of bytes sent per second, and the number of bytes received per second are displayed for the two scenarios in Table 4 for Smartphone 1, and Table 5 for Smartphone 2.

According to these values, our WebRTC hybrid mobile application shows a little better CPU performance (~67% on Smartphone 1, and ~33.34% on Smartphone 2), compared to the two other applications (~73% on Smartphone 1, and ~33.40% on Smartphone 2). The hybrid app shows comparable RAM occupancy with WhatsApp Messenger on both smartphones (~53% on Smartphone 1 and, ~57% on Smartphone 2), while Facebook Messenger displayed a little higher RAM occupancy (~57% on smartphone 1, and ~61% on Smartphone 2). In terms of data usage, the WhatsApp Messenger is much better optimized than our hybrid app and Facebook Messenger.

These results show that the WebRTC, running in a hybrid mobile app built with React Native framework, can be a better and up to date alternative to real-time video streaming on the mobile devices.

Table 4. Average values of CPU consumption, RAM occupancy, number of bytes sent/s and bytes number of received/s on Smartphone 1 (by Facebook Whatsapp Messenger, Facebook Messenger and our React native Hybrid app) over LTE - Scenario 1 and Scenario 2.

Measurement	Application					
	WhatsApp		Messenger		React native hybrid app	
	Sce. 1	Sce. 2	Sce. 1	Sce. 2	Sce. 1	Sce.2
CPU consumption (%)	77,11	89,92	73,43	73,31	65,92	68,24
RAM Occupancy (%)	54,06	54,53	57,86	53,79	52,07	53,43
Bytes Sent/s (KB/s)	0,08	0,15	17,33	9,76	20,82	19,23
Bytes Received/s (KB/s)	0,04	0,55	5,96	7,56	11,63	10,27

5 Discussion

In this paper, a WebRTC browser-based videoconferencing solution is compared to a hybrid-based one, in terms of CPU consumption, RAM occupancy and network data usage on mobile devices. A check of the interoperability between these solutions is also performed.

Table 5. Average values of CPU consumpancy, RAM occupation, number of bytes sent/s, and number of bytes received/s on Smartphone 2 (by Facebook Whatsapp Messenger, Facebook Messenger and the React Native hybrid app) on LTE - Scenario 1 and Scenario 2.

Measurement	Application					
	WhatsApp		Messenger		React native hybrid app	
	Sce. 1	Sce. 2	Sce. 1	Sce. 2	Sce. 1	Sce. 2
CPU consumption (%)	33,41	33,40	33,42	33,36	33,34	33,34
RAM occupancy (%)	57,15	57,24	61,09	62,63	58,67	57,23
Bytes sent/s (KB/s)	3,77	1,01	49,95	61,18	64,84	40,71
Bytes received/s (KB/s)	0,64	0,02	51,55	53,81	86,86	8,13

The results obtained after several video calls performed over WLAN and LTE networks showed that our WebRTC React Native hybrid app consumes less CPU (around

10%), compared to web browser-based one on mobile devices, while the difference in RAM occupancy is insignificant. It was also found that our hybrid mobile app and Mozilla Firefox Mobile consume less data than Google Chrome Mobile.

According to our experiments, a WebRTC hybrid app is more customizable, smoother and more efficient compared to a web browser-based one. For the end-user, there is no difference between a hybrid app and a native based one. However, a WebRTC browser-based solution is simpler and faster to develop, distribute and to update compared to a hybrid-based one. The WebRTC is not directly available in the React Native framework, it needs a free and open-source third-party module "react-native-webrtc" [14] that requires a lot of configuration to be supported in React Native.

Experiments carried out over the LTE network allowed us to evaluate our React Native hybrid app over LTE, and compare it against the "video calling functionality" of two popular multimedia applications "Facebook WhatsApp Messenger" and "Facebook Messenger", in terms of CPU consumption, RAM occupancy, number of bytes sent per second and number of bytes received per second during a video call.

We found out that our WebRTC hybrid app shows results that are comparable to those of these two major applications, in terms of CPU load and RAM occupancy. This demonstrates the efficiency of the WebRTC technology and the prodigality of our approach to implement the WebRTC video streaming in a hybrid mobile app.

Finally, our interoperability study revealed that the first stable release of WebRTC, WebRTC 1.0: Real-Time Communication between Browsers, supports the interoperability between a WebRTC hybrid app, built with React Native framework, and a browser-based one, which gives the possibility for a developer to provide web and hybrid versions of its application, according to necessities.

6 Conclusion and Future Works

In this paper, we demonstrated the feasibility of the implementation of WebRTC real-time videoconferencing in a React Native hybrid mobile application, and the interoperability between it and a browser-based one.

We designed and implemented a WebRTC videoconferencing solution containing a signaling server and two separate client applications based on the same algorithm written in JavaScript: a WebRTC browser-based client, and a WebRTC hybrid mobile app built with the React Native framework.

After several video calls performed over WLAN and LTE networks, we found out that our WebRTC React Native hybrid app consumes less CPU (~-10%) compared to the web browser-based one, while showing comparable RAM occupancy. We also found out that the interoperability is ensured between the two types of application.

Our prototype can find many applications in: E-Learning; E-commerce (video customer service); and Healthcare (doctor-patient communication system).

In future works, we plan to introduce the iOS operating system, and other hybrid mobile technologies, like Google Flutter framework, which is becoming an undisputable competitor to the React Native framework. We also plan to include some hybrid desktop technologies like Electron, to make a comparison on the desktop also.

References

1. Real-time communication for the web. https://webrtc.org. Accessed 25 June 2021
2. React Native. https://reactnative.dev. Accessed 25 June 2021
3. Moulay, M., Mancuso, V.: Experimental performance evaluation of WebRTC video services over mobile networks. In: IEEE INFOCOM 2018 - IEEE Conference on Computer Communications Workshops (INFOCOM WKSHPS), Honolulu, pp. 541–546 (2018). https://doi.org/10.1109/INFCOMW.2018.8407020
4. Hussain, A., Wang, W., Xu, D.-L.: Comparing WebRTC video conferencing with Skype in synchronous groupware applications. In: 2017 IEEE 21st International Conference on Computer Supported Cooperative Work in Design (CSCWD), Wellington, pp. 60–65 (2017). https://doi.org/10.1109/cscwd.2017.8066671
5. WebRTC 1.0: Real-Time Communication between Browsers. https://www.w3.org/TR/webrtc. Accessed 25 June 2021
6. Singh, K., Buford, J.: Developing WebRTC-based team apps with a cross-platform mobile framework. In: 2016 13th IEEE Annual Consumer Communications & Networking Conference (CCNC), Las Vegas, NV, pp. 236–242 (2016). https://doi.org/10.1109/ccnc.2016.7444762
7. Edan, N. M., Al-Sherbaz, A., Turner, S.: Design and evaluation of browser-to-browser video conferencing in WebRTC. In: 2017 Global Information Infrastructure and Networking Symposium (GIIS), St. Pierre, pp. 75–78 (2017). https://doi.org/10.1109/GIIS.2017.8169813
8. Apu, K.I.Z., Mahmud, N., Hasan, F., Sagar, S.H.: P2P video conferencing system based on WebRTC. In: 2017 International Conference on Electrical, Computer and Communication Engineering (ECCE), Cox's Bazar, pp. 557–561 (2017). https://doi.org/10.1109/ECACE.2017.7912968
9. Node.js. https://nodejs.org/en. Accessed 25 June 2021
10. JavaScript Session Establishment Protocol. https://tools.ietf.org/html/draft-ietf-rtcweb-jsep-26. Accessed 25 June 2021
11. The WebSocket API (WebSockets). https://developer.mozilla.org/en-US/docs/Web/API/WebSockets_API. Accessed 25 June 2021
12. The WebSocket Protocol. https://tools.ietf.org/html/rfc6455. Accessed 25 June 2021
13. Yannick, B., Éric, H., François-Xavier, W.: LTE et les 4G. Eyrolles, Paris (2012)
14. React Native WebRTC. https://github.com/react-native-webrtc. Accessed 25 June 2021

Modeling and Simulation of Urban Mobility in a Smart City

Saheb Faiza(✉) and Ahmed H. Habib

LISCO Laboratory, Computer Science Department, Technology Faculty,
Badji Mokhtar – Annaba University, BP 12, 23000 Annaba, Algeria

Abstract. Agent based urban modeling for smart cities face many challenges dealing with complexity of systems and requiring expertise in various domains. Simulation models could be used to offer decision support and a tool to evaluate different scenarios and modeling. In this paper, we have used agent based modeling to create a model of the mobility urban of smart city. We have simulate different scenarios of a case study.

Keywords: Smart city · Urban mobility · Simulation · Modeling · Agent based approach · CR

1 Introduction

According to recent studies the world's urban population is on the rise. Consequently, cities will face challenges concerning growth and urban concentration, competitiveness, and residents' livelihoods. The global urbanization trends and the increased number of residents living in urban areas generate additional mobility. Most cities in the world are confronting to establish a sustainable traffic system, which is essential for maintaining and improving the urban environment [12].

The potential risk of worldwide climate change is another argument for a strong need for urgent actions, to reconceive the way to consume and produce the energy that is need. The integration of renewable energy sources into urban energy networks and the increase in energy efficiency in cities are some of the core topics to be addressed in the near future. As urbanization is progressing worldwide and due to the fact that almost two thirds of our energy is consumed in urban environments, intelligent cities will play a significant role in the cities of the future and urban transformation. There is an urgent need to improve the understanding of cities and their metabolism, however, is pressed not only by the social relevance of urban environments, but also by the availability of new strategies for city-scale interventions that are enabled by emerging technologies. Leveraging advances in data analysis, sensor technologies, and urban experiments, Smart City approach can provide new insights into creating a data-driven approach to urban design and planning. To apply this approach is need for a scientific understanding of cities that considers the built environments and the people who inhabit them [10].

The term "smart city" was coined towards the end of the 20th century. It is rooted in the implementation of user-friendly information and communication technologies

B. Lejdel et al. (Eds.): AIAP 2021, LNNS 413, pp. 379–394, 2022.
https://doi.org/10.1007/978-3-030-96311-8_35

developed by major industries for urban spaces. Its meaning has since been expanded to relate to the future of cities and their development [1].

Smart cities emerge as the result of many smart solutions across all sectors of society: Smart mobility, smart safety, smart energy, water and waste, smart building and living, smart education, smart finance, smart tourism and leisure, smart retail and logistics, smart manufacturing and, construction, smart government.

The emergence of the smart cities permits to have goals as:

- Economic growth,
- quality of life,
- a good city to live in,
- ecological footprint

these goals can permit the challenges as:

- controlled transision of labor market to automation.
- winning the war on talent between metropolitan areas.
- Social cohesion, inclusiveness, solidarity.
- Secure digital environnement, privacy.
- Resilience.

Smart cities forward-looking, progressive and ressource-efficient while providing at the same time a high quality of life. They promote social and technological innovations and link existing infrastructures. They incorporate new energy, traffic and transport concepts that go easy on the environnement. Their focus is on new forms of governance and public participation. Intelligent decisions need to be taken at the strategic level if cities want to become smart. It takes more than individual projects but careful decisions on long-term implementation. considering cities as entire systems can help them achieve their ultimate goal of becoming smart.

Smart cities fully tack the current global challenges, such as climate change and scarcity of ressources. Their claim is also to secure their economie competitiveness and quality of life for urban populations continuously on the rise.

Main fields of action in this context are energy, mobility, the environment, the economy, society, politics, administration and quality of life. Some of the above are intertwined and increasingly networked with the support of IT. Technical, economic and social innovations provide the foundation for such activities. Smart cities build on sustainability but also on resilience in the sense that cities as systems are made more resistant and adaptable to influences from inside and out [1, 4, 5].

In the literature several studies have been carried out on environment sustainability, concerning every individual which is one of the most critical problems in world. The main objective in this area is to preserve scarce resources and reduce CO2 emissions in order to prevent environmental degradation. In recent years the potential of information systems (IS) as a driver for environmental sustainability has emerged under the term "Green IS" and other researches on the theme of urban development. Given that cities represent a huge share of environmental degradation due to factors such as mobility, energy and water consumption, and waste production, the municipal domain offers huge potentials in terms of sustainability. The advent of smart cites is an attempt to address this concern. In IS environmental sustainability research a crucial topic is presents and two different approaches are addressed: Green IT and Green IS. Whereas Green IT considers information technology (IT) to be a cause of environmental pollution, Green IS regards information systems and the inherent IT involved as a possible solution for reducing environmental degradation [11].

In these context our research on mobility is to participate in presenting the information system in order to find a solution to the problem by modeling and simulating of mobility in the city, in order to facilitate the leaders to take decisions. The decisions based on information given by simulation to improve mobility, reduce CO2 emissions, preserve sustainability and facilitate human life in cities.

Our work is about smart mobility which means innovative traffic and transport infrastructure that saves resources and builds on new technologies for maximum efficiency. Accessibility, affordability and safety of transport system, as well as compact urban development are essential factors in this context. New user-friendly will make it easier for pepole to switch to integrated transport systems focused on environ-mentally friendly transport modes. Joint utilization, i.e. "car sharing", instead of private owner ship is what counts these days when using motor véhicles [6], to study intelligent mobility we must use new technologies in order to model the system and then simulate it in order to make some decisions concerning the planning of intelligent mobility or to add other infrastructures that will allow us to make intelligent mobility.

Agent-based models (ABM) include mobile and interacting agents in a spatially large urban context. Agent-based models mainly differ from Cellular Automata(CA) in that the used agents are objects without a fixed location. Agents can interact with each other as well as the environement in a acting autonomously. Succinctly, an ABM has the following characteristics:

- Agents are explicitly designed to represent a particular mobile object (e.g., a person); there may be more than one agent type in single simulation thus the agents are implicitly distributed throughout the environment.
- Agents can sense and act within their environment in one of several ways: behavior can be reactive (e.g., they behave solely on their surroundings) or deliberative (e.g., they have their own agenda or goal which drives their behavior); clearly the design of agents sensing and acting is critical to a simulation. Altogether, agents exhibit a form of autonomous behavior and thus lend themselves to a variety of simulated behaviors including emergent patterns. Within urban simulation, many works have focused on examining cities as self-organizing complex systems, and solutions have been designed to explore the emergent properties of agents with relatively simple behavioral rules embedded by the modeler. However, relatively other research has used the smart city representation using the 3D geometric representation which emits the cities from static and dynamic view. This presentation is not as easy to use and requires experts in modeling because some real situations are difficult to present, more that, modeling cities and urban spaces in general is a daring task for computer graphics, computer vision and visualization. Little attention has been paid to issues of validating models using observed data or trends and as with (CA) models, traditional ABM urban simulation have behaviors influenced only by a localized context [2, 3, 6, 7].

We propose a modeling of intelligent urban mobility, an example of a simulation case based on agents of the proposed modeling executed on the Netlogo platform. Individual car agents and pedestrians have been placed on the road network. Experiments out to asses the mobility system in a smart city. The following section briefly presents the agent-based PASSI modeling approach used in the context of this work. Section 3 introduces into the Netlogo platform the simulation example at the level of which presenting us results and comments on these results. A conclusion finalizes this document.

2 The Design Approach

Multi-Agent Systems (MAS) differ from non-agent based systems because agents are intended to be autonomous units of intelligent functionality. As a consequence, Agent-Oriented Software Engineering (AOSE) methods must complement standard design activities and representations with models of the agent. Some methods coming from artificial intelligence community adress social knowledge and relationships but have high-level design abstractions as their end points. In our work we use PASSI [9], which is a method for developing multiagent software that integrates design models and philosophies from both object-oriented software engineering and MAS using UML notation. The method has evolved through several stages, it has been previously applied in the synthesis of embedded robotics software and they are currently exploring its applications to the design of various agent-based information systems. In PASSI an agent is a significant unit of software at both the abstract (low-fidelity) and concrete (high-fidelity) levels.

According to this view, an agent is an instance of an agent class that is the software implementation of an autonomous entity capable of pursuing an objective through its autonomous decisions, actions and social relationships. An agent may occupy several functional roles during interactions with other agents to achieve its goals. Where a role is the collection of taskes perrformed by the agent in pursuing a sub-goal. A task, in turn is defined as a purpose ful unit of individual or interactive behavior.

The methodology is composed of five models, in which the agents are present [9]. Now we will present the diagrams of our approach to develop our application. We will follow the steps outlined of the approach in [9].

2.1 System Requirements Model

An anthropomophic model of the system requirements in terms of agency and purpose. Developping this model, involves four steps:

- **Domain Description:** A functional description of the system using conventional use-case diagrams.
- **Agent Identification:** Separation of responsability concerns into agents, represented as steriotyped UML packages.
- **Role Identification:** Use of sequence diagrams to explore each agent's responsibilities through role specific scenarios.

- **Task Specification:** Specification through a use case diagram and auxiliary descriptions of the capabilities of each agent.

Domain Description: Case name "Car Agent", Fig. 1.
Agent Identification:
The main agents: Car (Cognitive agent)
This is the car driver he has knowledges of:

- Traffic lights status
- The traffic code
- Objects, people, animals, cars

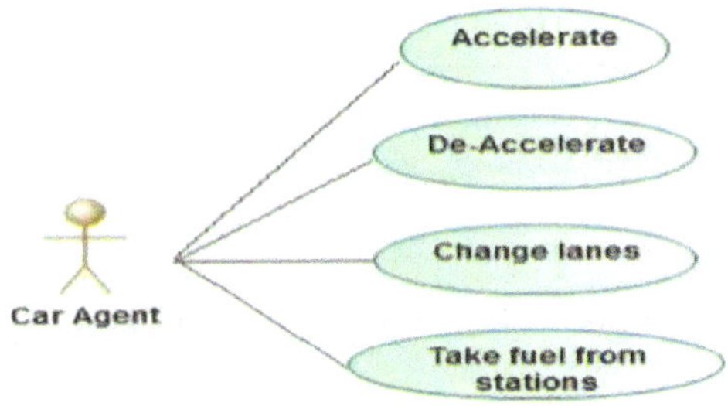

Fig. 1. Car agent use case diagram

Cars can make their own decisions, they can even turn at intersections.
The task specification of Car Agent is described by Fig. 2.

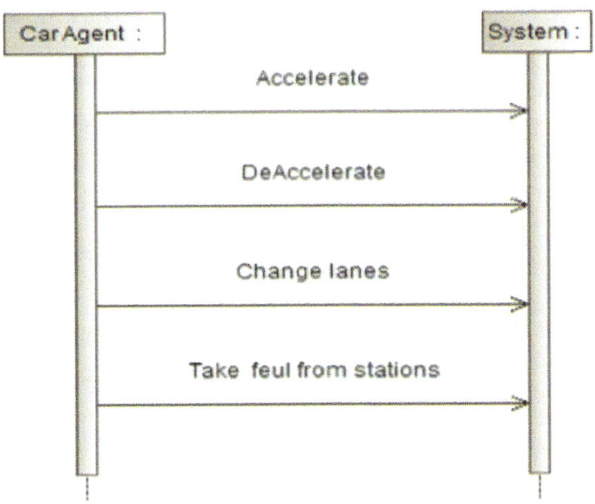

Fig. 2. Car agent sequence diagram

Domain Description:
Case name: "Agent pedestrian", Fig. 3.
Agent Identification:
The main agent: pedestrian (Cognitive agent).
He has knowledges of:

- Traffic lights status
- The traffic code
- Objects, people, animals, cars

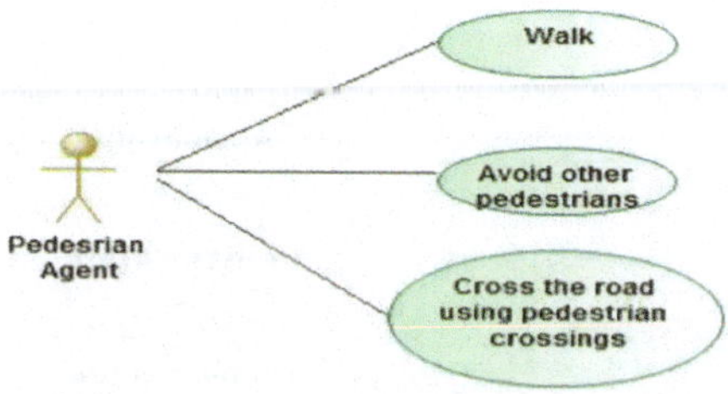

Fig. 3. Pedestrian agent use case diagram

The pedestrian agent who can walk, avoid other pedestrians cross the road using the "crossings".

The task specification of Pedestrian Agent is described by Fig. 4.

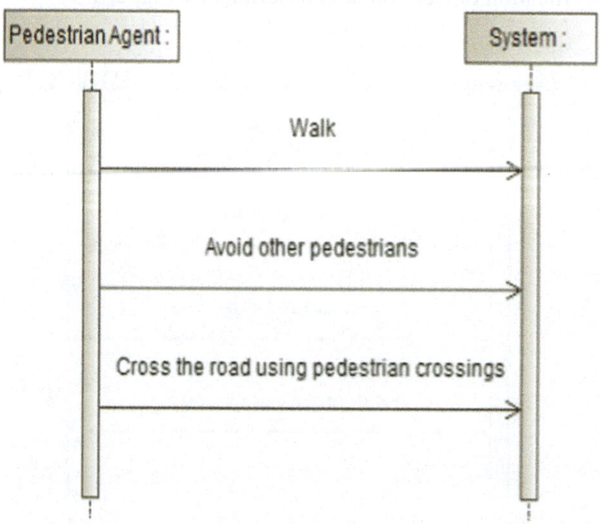

Fig. 4. Pedestrian agent communication protocol diagram

2.2 Agent Society Model

Amodel of the social interactions, Fig. 5 is presented by the sequence diagram which descibes the communication protocols:

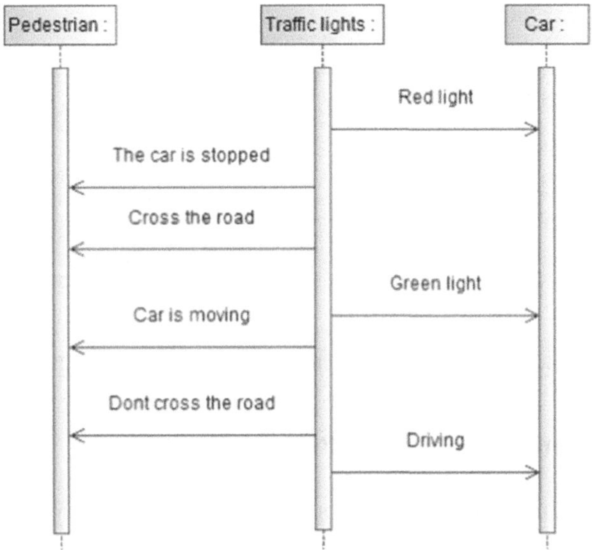

Fig. 5. Communication protocol diagram

The traffic light agent sends two types of signals to the car, if the signal is red the light informs the pedestrian that the car is stopped and that he can cross the road. If the opposite case the trafic light sends a light green towards the car to allow it to move and should not cross the road.

2.3 Description of the Domain

Used to describe the agent society from a domain point of view (class diagram called domain diagram) and the exchange of information between agents (class diagram called communication diagram).

The Fig. 6 shows the relationships between the road, car, pedestrian and traffic light classes and when can pedestrians cross the road by operating on the traffic light, if it is green or red and the cars that can move, respecting the signaling order given by the fire. The construction of domain diagram is to built the database of the system to be implemented.

In this section we have presented the design of our system using the PASSI approach that is proposed for the agent-based modeling of a city's transport mobility flow. The implementation of our system will be presented in the next section. In this section, we will detail the implémentation part of our application. First we present NETLOGO platform. Then, we expose parts of the source code of our system and finally we will finish with some screenshots of the application.

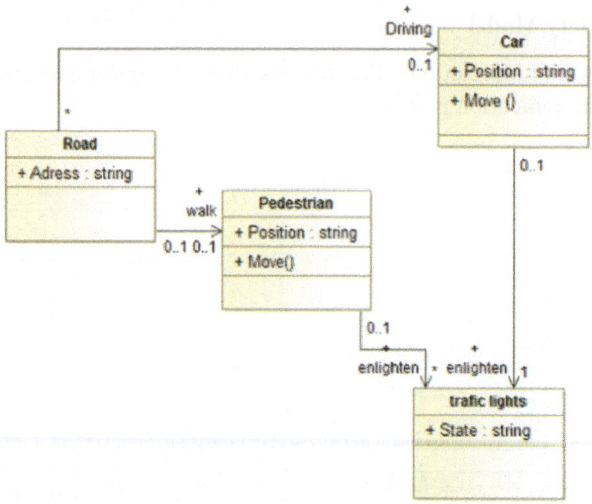

Fig. 6. Ontology class diagram of the city's transport mobility flow diagram

3 Development Tools Netlogo Plateform

Netlogo is a programming environment for the modeling /simulation of natural collective phenomena well suited to the modeling of complex systems composed of hundreds, thousands of agents acting in parralel. Possibility to « play» with many simulation in sociology, biology, medecine, phisics, chemistry, mathematics, computer science, economics and social psychology. Possibility of creating your own models [8].

3.1 Agent Concepts

The world of Netlogo is made up of agents, who can follow instructions. The activities of the different agents run simultaneously. There are three agent types: turtles, patches, and observer.

- Turtles are the agents that move around the world. They correspond to the agents seen in progress.
- The world is in 2D or 3D, divided according to a grid (toric or not) of patches.
- A patches is a portion of the ground on which turtles can sit and move. The patches correspond to the concept of environnement seen in progress.
- The Observer looks at the word of turtles and patches from the outside (is not located in the world)

3.2 Simulation of the Application

In the next part we present some main screenshots of our application.

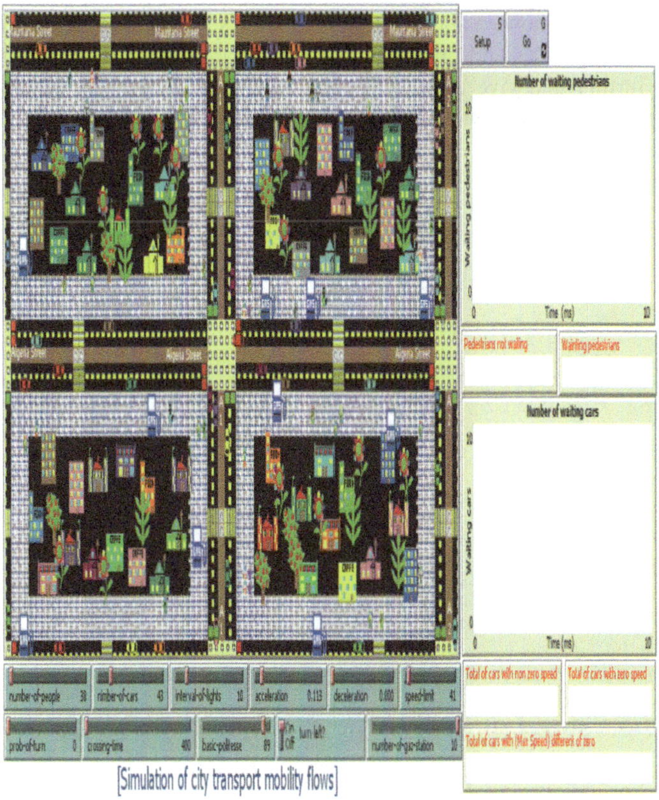

Fig. 7. Dashboard of the application

The Fig. 7 shows the dashboard of the application for example the simulation screen which shows the houses, the trees, the flowers, the cars and fuel stations, the roads and the people who live in the city and other tools, it offers the Netlogo platform such as sliders, switches, buttons. All the tools that exist in the simulation were not shown due to the screenshot witch is small.

```
breed[cars car]
;breed[trucks truck]
breed[houses house]
breed[trees tree]
breed[crossings crossing]
breed[lightsR lightR]
breed[lightsL lightL]
breed[lightsU lightU]
breed[lightsD lightD]
breed[persons person]
breed[stations station]
```

Fig. 8. Breeds definitions

The Fig. 8 above shows a slice of code and shows that Netlog allows you to define different breeds of turtles and breeds of links. Once these breeds are defined, you can go one step further and give these different breeds behaviors. For example, you can have called persons and sidewalks and then have the persons walk on the sidewalks and vehicles travel on the roads.

```
patches-own[
   meaning
   will-cross?
   used
   traffic
]
globals[
   redV
   greenV
   redH
   greenH
   speedLimit
]
persons-own[
   speed
   walk-time
   crossing-part
   waiting?
]
cars-own[
   speed
   maxSpeed
   will-turn?
   turnX
   turnY
   politeness
   will-stop?
]
```

Fig. 9. Definition of variables

The Fig. 9 shows keywords, like:

- **Patches-own:** it defines variables that all patches can use,
- **Globals:** it defines new global variables the "global" variables are accessible by all the agents and can be used anywhere in a model.
- **The turtles-own:** it defines the variables belonging to each turtle as persons-own and cars-own, and its keywords can only be used at the start of a program, before any function definition. It defines the variables that all patches can use. All patches will then have the given variables and can use them.

Example: At the cars-own level we have the speed of the car speed, then the maximum speed that the cars can reach **maxSpeed,** then **will-turn**? which represents whether the car is going to turn or not and the turn X/Y which represents left or right turning then we politeness of cars, that means how often they will stop and let people cross the road, then the will-stop? that means if the car will stop or not.

```
to setup
  ca
  set speedLimit speed-limit
  draw-sidewalk
  draw-roads
  draw-houses&trees
  draw-crossings

  place-cars
  ;place-trucks
  place-lights
  place-people
  place-stations


  reset-ticks
  tick

end

to go
  move-cars
  ;move-trucks
  control-traffic-lights
  move-people
  plot-waiting
  plot-waitings
  tick
end
```

Fig. 10. The setup and go procedures

In the Fig.10 we have the procedure to setup which starts to define a procedure named **«setup»**, **«ca»** is the abbreviation of clear-all. This command will clear the screen, initialize any variables you might have to 0, and remove all turtles. Basically it cleans up the slate for a new run of the project, **«set»** define the variable to set the value given here draw-roads, draw houses and trees, draw crossings this means to draw the roads then the houses and the trees and the crossings, the procedure togo lance the simulation for example to make move the cars and the people, to control the traffic lights.

```
to plot-waiting
  set-current-plot "Number of pedestrians waiting"
  set-current-plot-pen "Pedestrians waiting"
  plot (count persons with [waiting? = true])
end

to plot-waitings
  set-current-plot "Number of cars waiting"
  set-current-plot-pen "Cars waiting"
  plot (count cars with [speed = 0])
end
```

Fig. 11. Number of pedestrians and cars waiting

The previous Fig. 11 shows the code of the number's waiting pedestrians and the number's waiting cars.

The Fig. 12 below shows the results in the "**plots**" on the dashboard of the simulation number's waiting pedestrians and the number's cars waiting in milliseconds.

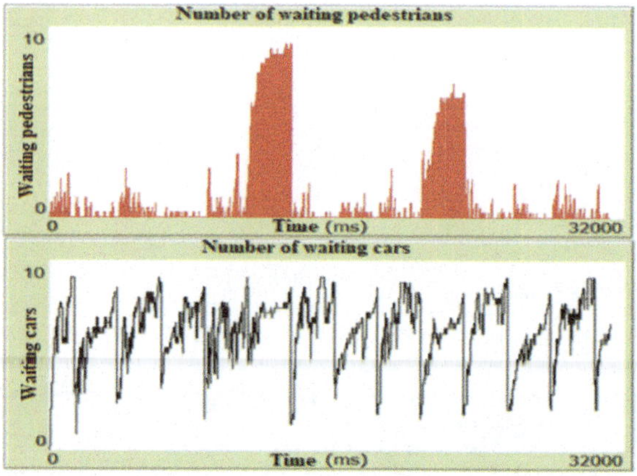

Fig. 12. The display of pedestrian "plots" and waiting cars

The Fig. 13 below shows the "sliders" and the "switches" number of people, cars and fuel stations has the role of increasing or to decrease the number of people and cars and fuel stations in the city. The acceleration and deceleration sliders have the role of increasing or decreasing the speed of the cars. The interval-of- lights "slider" defines the traffic light interval, the speed-limit slider to set the speed limit of cars in the city, the prb-to-turn slider sets the probability of turning, that is, indicates the probability of the cars turning on the right / left at the junction, the time-of-crossing slider says how long pedestrians have to walk before they start looking for a passage and decide to cross the road, the basic-politeness "slide" sets the value of input for calculating the courtesy of cars, finally, do we have the left-turn switch? Says whether cars will be able to turn left or not.

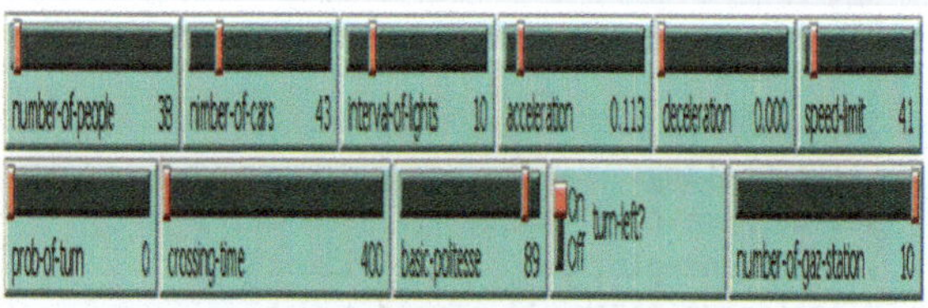

Fig. 13. Sliders and switches

This Fig. 14 shows the number of pedestrians waiting or not waiting in the city.

Fig. 14. The number of pedestrians waiting and not waiting in the city

The Fig. 15 shows the number of cars whose speed is different from zero, i.e. the cars which circulate in the city then the cars which do not move i.e. the cars whose speed is zero and the cars whose maximum speed is different from zero.

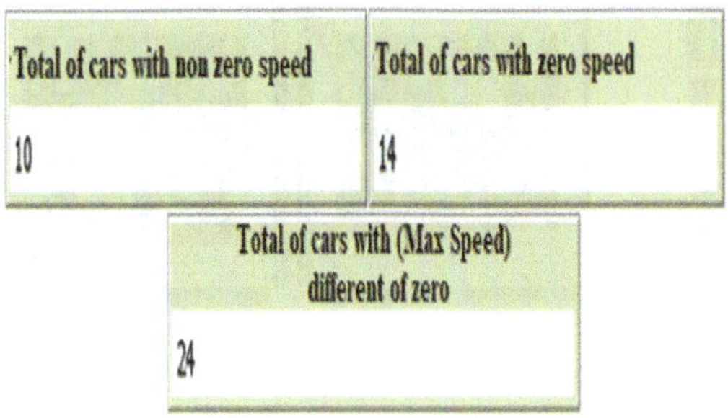

Fig. 15. The number of cars wich zero and maximum speed and different to zero

The following Fig. 16 shows the number of fuel stations in operation in the city and the number of cars and the number of pedestrians in the city.

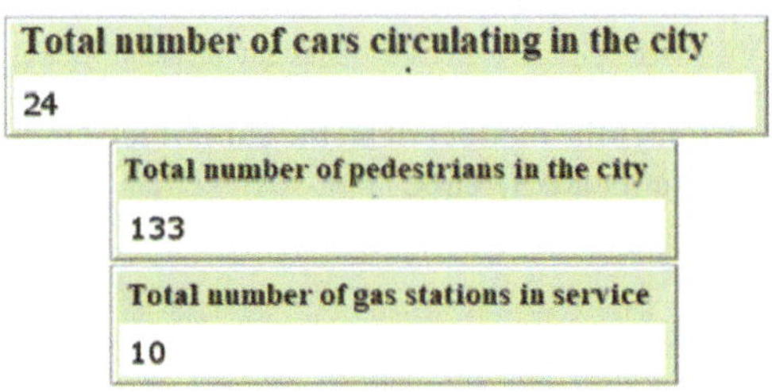

Fig. 16. Stations, cars and pedestrians in the city

The Fig. 17 shows the street names Algeria Street, and Mauritania Street and streets A1 and A2 which represent A1and A2.

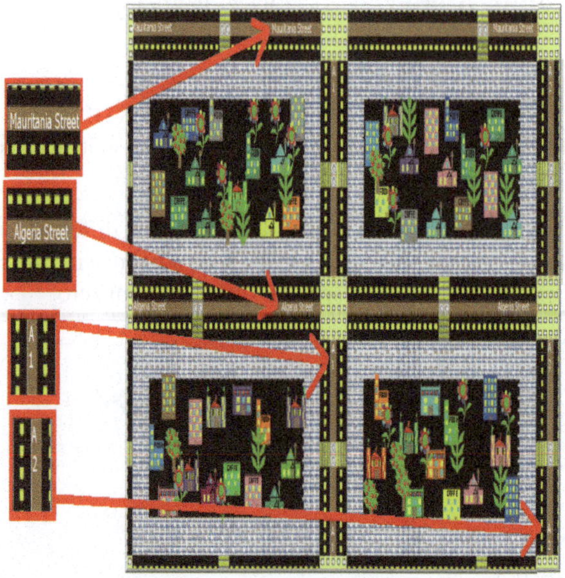

Fig. 17. The streets of the city

Total number of cars on the streets of (Mauritania) & (Algeria)

37

Fig. 18. Total number of cars on the streets Algria and Mauritania

The Fig. 18 above displays the number of all the cars which circulate on the horizontal streets Algeria and Mauritania are 37 cars.

4 Discussion

We have provided a challenge for smart urban mobility modeling and simulation that is not too detailed, but the simulation model can offer decision support and can provide a tool with which different scenarios and designs can be compared and evaluated in an iterative manner with a team of leaders and experts in urban mobility to address the socio-technical nature of cities.

There are key challenges to change or improve mobility to make it smart, such models require teamwork and collaboration between experts in order to give a reliable model and which also requires data collection in cities using information systems if they exist to avoid redoing because it requires hardware (such as cameras and software) that can provide us with the data necessary to study the simulation with real data and from that the experts according to the needs of the cities will propose solutions to the master plan.

Through the use of agent-based modeling the master plan can be iteratively evaluated in real time as part of the normal workflow, this is where the real power lies, as complex planning and design decisions are made can be tested and verified instantly by simulating the real impacts that a new urban proposal will have on a city.

In order to provide solutions for urban mobility it is necessary to rethink the way in which people can move through the new road traffic, the means of transport used, the car parks which help in the ease of parking and avoid congestion by certain means of transport.

We also have to think about how to make traffic flow smoothly during rush hour to reduce the increase in CO2 emissions from transport. All of this requires smart techniques and advanced intelligent information systems to make city mobility in cities, smart.

5 Conclusion

Transport was one of the sectors to integrate digital devices to better manage flows in the city. Globally, three dimensions are used to capture the transport and intelligent mobility of a city:

1) improving logistic flows to ensure greater efficiency of businesses through increased knowledge of the network;
2) providing real-time digital information to users on traffic conditions;
3) helping to develop collaborative or alternative modes of travel for people in the transition to mobility less dependent on the private car.

In the example of the simulation we have used two types of agents which are vehicles and pedestrians. Our objective is to simulate an intelligent digital system which reflects the real system and which contains all of these agents. Each agent works individually but in collaboration with the other agents is the better regulate traffic and to ensure adequate pedestrian passage and to avoid traffic jams on the roads.

References

1. Smart City. https://www.wien.gv.at/stattentwicklung/studien/pdf/b008403j.pdf
2. Aliaga, G.: 3D design and modeling of smart cities from a computer graphics perspective. International Scholarly Research Network, ISRN Computer Graphics, vol. 2012, Article ID 728913, 19 p. (2012)
3. Van Dam, K.H., Koering, D., Bustoos-Turul, G., Jones H.: Agent-based simulation as an urban design too, iterative evaluation of a smart city masterplan. Conference Paper (2014). https://www.reseachgate/publication274077771
4. Jovanovic, D., et al.: Building virtual 3D city model for smart cities application, a case study on campus area of the University of Novi Sad. ISPRS Int. J. Geo-Inf. (2020)

5. Trindale, E.P., Phebe Farias Hinnig, M., Moreira da Costa, E., Sabatini Marque, J., Cid Bastos, R., Yigitcanlar, T.: Sustainable development of smart cities, a systematic review of the literature. J. Open Innov. Technol. Market Complex. (2017)
6. Wegal, J., Glake, D.: Large scale traffic simulation for smart planning with Mars. SummerSim-MSaaS, 22–24 July, Berlin, Germany (2019)
7. Lpes, C.V., Lindstrom, C.: Virtual cities in urban planning: the Uppsala case study. J. Theoret. Appl. Electron. Commer. Res. **7**(3), 88–100 (2012)
8. Muelle, C., Klein, R.U., Hof, A.: An easy-to- use spacial simulation for urban planning in smaller municipalities. Comput. Environ. Urban Syst. J. (2018). www.elsevier.com
9. Cossentino, M., Potts, C.: Acase tool supported methodology for the design of multi-agent-systems. In: Engineering Research and Practice (SERP'02) (2002)
10. Aeleneia L., et al.: Smart City, a systematic approach towards a sustainable urban transformation. In: International Conference on Solar Heating and Cooling for Buildings and Industry, SHC 2015, Energy Procedia, vol. 91, pp. 970–979 (2016)
11. Brauer, B., Eisel, M., Kolbe, L.M.: The state of the art in smart city research – a literature analysis on green IS solutions to foster environmental sustainability. In: Pacific Asia Conference on Information Systems, PACIS 2015 Proceedings (2015)
12. Brcic, D., Slarvulj, M., Jurat, J.: The role of smart mobility in smart cities. In: 5th International conference on Road and Rail Infracture, CETRA 2018, Zadar, Croatia (2018)

OAIDS: An Ontology-Based Framework for Building an Intelligent Urban Road Traffic Automatic Incident Detection System

Samia Hireche[1(✉)], Abdeslem Dennai[1], and Boufeldja Kadri[2]

[1] Smart Grids and Renewable Energies (SGRE) Laboratory, Cloud Computing and Artificial Intelligence (CCAI) Team, TAHRI Mohamed University, 08000 Bechar, Algeria
hireche.samia@pg.univ-bechar.dz,
Abdeslem.dennai@labrier-univ-bechar.dz

[2] Smart Grids and Renewable Energies (SGRE) Laboratory, Electronics and Power Electronics Applications in Energy Conversion Systems Team, TAHRI Mohamed University, 08000 Bechar, Algeria

Abstract. Handling interoperability of data exchange among road traffic sensor devices, connected vehicles, infrastructure components and heterogeneous traffic management center applications has become an important and basic requirement nowadays. To meet this requirement, this paper proposes an ontology based framework to capture the knowledge domain about traffic automatic incident detection system (AIDS) based on Connected Vehicles (CVs) technology. This ontology addresses the semantic data interoperability needed between different heterogeneous entities constituent this AIDS. This contribution aims at modeling and capturing the semantic of the anomaly information used in the incident detection process and describing the AIDS components, their observations, measurements and communications messages features. First, to achieve this goal, NeOn methodology was adopted. Then, we defined the basic concepts and observations of a traffic sensor and CVs that has been extended to define concepts related to the data sensing and gathering layer of this framework based on ontology concepts. In addition, to ensure data interoperability and identify ontology's restrictions, we used the OWL (Web Ontology Language) language. Furthermore, to build this ontology, we used the OWL under Protégé tool. Finally, OAIDS consisted of 93 concepts and 33 object properties. OntoMetrics was used to confirm the effectiveness of this proposed ontology to carry out the interoperability of CV's sensor data in the urban road AIDS domain.

Keywords: Traffic incident · Automatic Incident Detection System (AIDS) · Interoperability · Ontology · Web Ontology Language (OWL) · NeOn

1 Introduction

In modern smart cities, an Automatic Incident Detection System (AIDS) is an indispensable component used to improve the performance of transportation systems and to provide suitable and reliable safety services based on the data generated from the movement of smart vehicles on different road infrastructures. Actually, the AIDS contain two

B. Lejdel et al. (Eds.): AIAP 2021, LNNS 413, pp. 395–405, 2022.
https://doi.org/10.1007/978-3-030-96311-8_36

major components, which are data sensing, and gathering component, and data processing and incident detection component. The first component ensures the effectiveness of the obtained traffic data quality, while the second component is the more intuitive one to understanding, analyzing, and making decisions about the real-time traffic incident within an efficient transportation system.

However, one of the major requirements of AIDS is the collection and the dissemination of real time data in order to process, monitor, and manage it with heterogeneous components. These requirements can help end users to understand the real-time road situations such as road traffic congestion, closure of lanes due to traffic incidents, travel speed limits and also sharing urgency alert notification information especially on urban areas. To these end, AIDS are based on the following traffic data collection technologies: intrusive and non intrusive sensors (e.g. Inductive-Loop, Seismic Sensors, Radar, Ultrasound, Wireless Vehicle Identification Sensors and Video Image Processing sensors), and sensors located into CVs. Therefore, the data generated from these technologies are heterogeneous in format, structure, semantics, organizations, and accessibility. That makes its implementation by the second AIDS's component without any delay or miscommunication problems a severe challenge. For that, the researchers used some communication standards under Vehicular Ad-Hoc Network (VANET) that is identified as one of the most promising technologies for managing future ITSs data exchange and communication. However, according to the last development research recommendations, this proposition cannot give a complete solution to handling the data interoperability, especially for sensor data.

Actually, the primary challenge for any AIDS based on CV data is how to deliver these data and then transforming it into useful visual representation in unique and standard form? To answer this question, we must first understand the relationships between these CV sensors and AIDS entities, including road traffic sensors, Road Side Units (RSUs), Traffic Management Center (TMC) and traffic information systems. Unfortunately, sharing and retrieval CVs data represent a challenge. Since this operation needs an interoperability among different AIDS entities involved in ITS.

One of the simplest ways of tackling this problem is to use "a formal explicit specification of shared conceptualization" [1] defined as ontology. As ontologies provide a common vocabulary in a given domain and allow defining, with different levels of formality, the meaning of terms and the relationships between them. In an intelligent transportation system (ITS), Ontologies can describe the live traffic situation of the movement of CVs and the interaction between ITS infrastructure components and different involved sensors. Such usage can be done across the design of a standardized methodology of conceptual schemes to allow communication and information exchange between different ITS monitoring system entities. Nevertheless, different contributions have been developed in this regard. The researchers have been trying to provide a set of ontologies to describe a specific scenario in Advanced Driving Assistance Systems, vehicle sensing system failures, modeling context information utilized pervasive transportation services, prediction of congestion and Traffic Signal Controller (TSCs), and management of traffic situations. However, there has been very few research efforts conducted into using ontology to give a unique semantic interpretation of information collected by the integrated sensors in CVs, RSUs and TSCs to describing the semantic data of anomaly information events in a signalized intersection. In particular, there is a gap to deal with these sensory data in the AIDS field.

In spite of this, a new approach based on the concept of ontology is therefore needed to tackle the problem of incident traffic data exchange on the whole transportation network. Taking all this into account, our paper aim is to focus on fill this gap by implementing a unified ontology under traffic incident cases. This is to achieve the mounting need to serve CV's Basic Safety Message Data (BSM) data in real-time and to provide a detailed description about the knowledge data in this type of system.

As a solution, the contributions of this work can be summarized as follows: First, describe the OAIDS ontology structure; Second, Adopt the NeOn Methodology to build OAIDS ontology. Finally, developed and modeled the anomaly traffic information and its specifications according to the OAIDS Ontology.

The paper is organized as follows: Sect. 2 discusses the state of the art of different proposed ITS ontologies based on sensor data. We introduce our proposed ontology and its empirical evaluation in Sect. 3. Finally, we conclude our paper in Sect. 4.

2 Related Works

In this section, we review ITS's ontologies proposed in these last years.

In 2016, the authors in [2] proposed a new ontology to improve the driving environment through a traffic sensor network. To describe the different concepts and the relationships between them, Semantic Sensor Network (SSN) ontology is applied. To validate this proposed ontology, Web Ontology Language- Resource Description Framework (OWL-RDF) language and the Protégé tool are used. The results obtained confirmed the greater decision-making capability after using this ontology. By using the same ontology that is SSN, authors in [3] tried to manage the sensor information to perform the automatic traffic light settings allowing to predict and avoid traffic accidents and the routing optimization. The authors used RDF concepts to mapping sensor data into SSN ontology. Other ontology to support road traffic assistance transportation management applications is also proposed in [4]. The concepts of this ontology are related to vehicles and the elements related to road infrastructure. In term implementation, they used OWL-RDF language using the protégé tool and SPARQL query language. Authors in [5] introduced a visualization oriented ontology that formalizes the knowledge about urban mobility events and visualization techniques. OWL is used to develop this ontology. Allow the semantic annotation of urban mobility events is the main advantage of this ontology. Adaptive I-parking system ontology is also discussed in [6]. First, authors identified ontology's concepts. Then, they used OWL to define restrictions between them. Moreover, the adaption rules are identified by using the SWRL. Finally, Protégé tool is used to develop this ontology.

In 2017, authors proposed two ontologies to describe the safe driving for Advanced Driver Assistance Systems [7]. The first is for autonomous vehicles and the second is for the knowledge base that integrated map, car and control ontologies. The RDF is used to convert sensor data and C-SPARQL Query Engine to observe it. Moreover, the SWRL rule is used as a reasoning method. Experimental results prove the capability of this proposition to enhance real-time decision making systems. New ontology for traffic event using autonomous vehicles is presented also in [8]. An annotation for predicting future traffic accidents is also developed. Their results confirmed that accident occurrence depending on varies traffic environment scenarios as presented using Protégé tool. In [9], the authors proposed the SSN ontology to manage the sensor information in an ITS. Their aim was to develop a semantic integrator module to map sensor data to SSN ontology in automatic way. For this, they used RDF concepts. For validation, the semantic integration system was applied. Using machine learning with rule based system is suggested to enhance the semantic annotation process.

In [10, 11], authors presented vehicle signal specification ontology (VSSO).To this end, they used the concepts of Sensor, Observation, Sample, and Actuator (SOSA) and SSN ontology in order to represent observation of car signals. Comparison results of integration of VSSO, SOSA/SSN and STEP to other ontology demonstrates it effectiveness in term sensor coverage, semantics, and trajectory enrichments metrics.

VANETs is another ITS domain where the researchers developed a specific ontology named Messaging Ontology for VAnets (MOVA) to enhance performance multi-hop message dissemination over VANETs as presented in [12]. OWL and Protégé tool are used to design and organize the information transmitted between vehicles.

In 2020, several ontologies in ITS domain are proposed. First, the authors in [13] developed a new ontology on the foundations of SOSA ontology. The aim of this contribution is to provide a suitable ontology for managing data coming connected vehicle's sensors. The authors used Semantic Web technologies (SWT) and RDF to support the operation of CAVs within urban roads. BSMs data are used to translate CVs data onto SOSA ontology. To validate this contribution, Apache Jenna Fuseki and SPARQL queries are used. The results obtained demonstrate their effectiveness in term query response time compared to SSN ontology. Secondly, authors in [14] proposed Visualization-oriented Urban Mobility Ontology (VUMO) for integration and visualization of urban mobility data. The objective of this ontology is to lay the semantic foundation for integrating urban mobility data from heterogeneous sources, and building knowledge-assisted visualization tools, annotating visualization techniques and expert knowledge. Thirdly, the work [15] presented ontology to understanding the common terms used in the transport system. For this, the authors used Multi-Agent systems coupled with semantic web in order to help making decisions. To implement this ontology, they used Protégé, OWL 2 Web Ontology Language and RDF. Finally, route suggestion system is another aspect to model it by ontology as presented in [16]. In this contribution, researchers developed an architecture called Ontology-based Route Suggestion by using the OWL with realistic data. To validate this ontology, SWRL is used to develop semantic rules and Protégé tool.

Table 1 presents comparative analysis between these reviewed ontologies. As it can be seen, different domain ITS ontologies are proposed to describe the knowledge and semantic relations between ITS components based on SSN ontology. Also, it can be found that different types of ontology languages are used, the most one are RDF and OWL. To this end, Protégé tool is chosen to implement knowledge, because this tool provides rich features compared to others. In sum, although there was a vast literature on ontology based ITS as described in Table 1, to our knowledge, there is no actual ontology for an urban AIDS that use the concept of CVs. Consequently, there is a dire need for finding an efficient solution to visualizing real time traffic incident detection data based on unified concepts and knowledge. For thus, our system focuses on the development of a novel ontology to assist end users to use a traffic data monitoring based on ontology principles and data semantic interoperability.

Table 1. Comparative analysis for ITS's ontologies.

Ref	Year	ITS subfield scope	Language	Tool
[2]	2016	Improving the driving environment through a traffic sensor network	OWL/RDF	Protégé
[3]	2016	Managing the sensor information for TSC	RDF	–
[4]	2016	Management of advanced driver assistance (ADA) in different road traffic situations	OWL/RDF	Protégé
[5]	2016	Characterizing urban mobility events and visualization knowledge	OWL	Protégé
[6]	2016	Adaptation service for intelligent parking	OWL	Protégé
[7]	2017	Decisions making for ADA based on AVs	RDF	Protégé
[8]	2017	Preventing traffic accidents	OWL	Protégé
[9]	2017	Managing the sensor information in an ITS	RDF	Protégé
[10]	2018	Describing trajectories from car signals	RDF	–
[11]	2018	Describing the car signals and sensors data	RDF	–
[12]	2018	Addressing the multi-hop message dissemination scope over VANETs	OWL	Protégé
[13]	2020	Understanding the state of CAVs data	RDF	Jenna
[14]	2020	Integration and visualization of urban mobility data from ITS	OWL2RL	–
[15]	2020	Management of Tramway and Bus data	OWL2RDF	Protégé
[16]	2020	Developing routes suggestion system	OWL2	Protégé

3 Proposed Ontology Framework

In our solution, we propose creating domain ontology for an AIDS based on the CV technology. The ontology is called "Ontology for Automatic Incident Detection System" (OAIDS). More details about the conception of this ontology are given below.

3.1 AIDS Structure

Since an AIDS consists of a number of heterogeneous components, it is important to define a shared representation between them. For that, OAIDS would provide a structure that is flexible, and that naturally organizes the information in multidimensional ways. We divide our functional architecture in four layers as shown in Fig. 1. Each of these interconnected layers has its own data control and data distribution layer. The data control layer is responsible for the management of data and its operations, and the data distribution layer is the responsible for the data circulation between the layers. In our approach, these four layers are interconnected with each other to provide a real-time incident data for road traffic management efficiency and safety. In the following subsections we provide further details of these layers. The TDC layer consists of intrusive and non intrusive sensors with CVs in a road infrastructure. The TDPS consists of a data acquisition and data storage services. TDA consists of a set of services, which have the goal of detection of data anomaly and then the detection of incident. The last layer, which is IDAV is contained the distribution of incident alarm notification to end users and the second the prevention of route suggestion services.

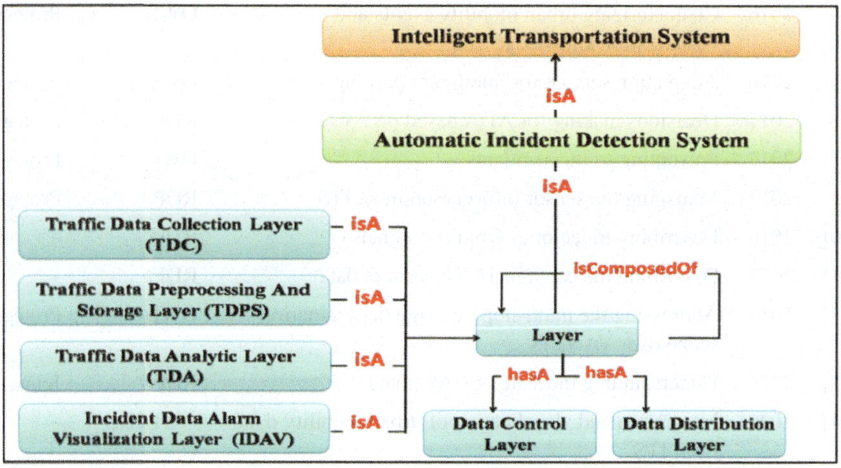

Fig. 1. The AIDS skeleton structure.

3.2 OAIDS Development

In this subsection, we describe our proposed ontology structure in detail using the methodological guidelines proposed in NeOn methodology [17].

The NeOn Methodology. The NeOn is one of the proven ontology methods. "It is a scenario-based methodology that provides accurate details about key aspects of the ontology engineering process, paying special attention to the reuse and reengineering of ontological and non-ontological resources. This framework is founded on four pillars: (1) a glossary of processes and activities; (2) a set of nine scenarios for building ontologies and ontology networks; (3) two modes of organizing ontology developments, called ontology life-cycle models; and (4) a set of precise methodological guidelines for performing specific processes and activities" (p. 110) [18, 19].

The Proposed AIDS Ontology. The proposed method for building OAIDS is illustrated in Fig. 2. A detailed description of each component is presented next.

Ontology Requirements. In this phase, we used ontology requirements specification documents to describe the purpose of building the ontology, the scope covered by this ontology, the intended uses, the end users who may use this ontology, and then, the requirements that the ontology should perform. And also, the glossary of terms that will be formally represented in the ontology by means of concepts, attributes and relations. The output of these tasks is the OAIDS ontology (see Table 2).

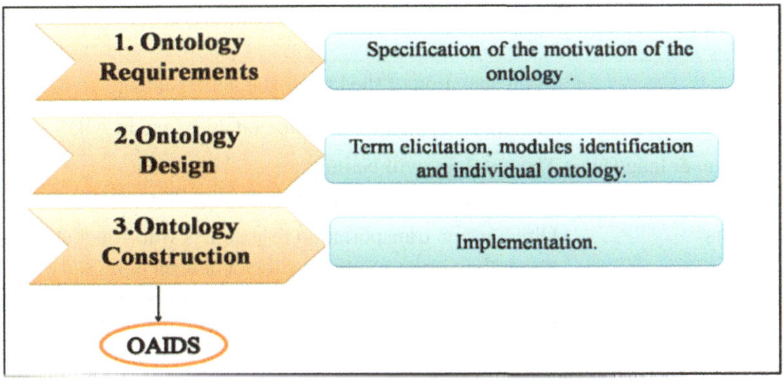

Fig. 2. Phases and procedures used for the proposed methodology of OAIDS.

Ontology Design. This phase organize AIDS domain knowledge in a shared and conceptual model. First, we define the concepts classification trees. Then, we identify class attributes and restrictions. Finally, we identify the instances.

- Identification of Concepts: To cope with the multi-objective aspect of an AIDS, we present the knowledge in our ontology as 8 groups of concepts (See Fig. 3), each covering an objective aspect, namely: (1) infrastructure, (2) vehicle, (3) detector (4) network, (5) information, (6) incident, (7) service and (8) Person. Those concepts are the key for conceiving real time incident detection.
- Identification of restrictions, object properties and definition of individuals. Figure 4 illustrates the main OAIDS data property, object property and instances.

Ontology Construction. We have designed and implemented our proposed OAIDS ontology using Protégé 5.2.0 tool. The designed ontology is extracted in OWL 2 format. For purpose of sharing and reusing, the ontology open source is available at: https://github.com/SAMIAHCC/OAIDS/tree/V1.

OntoMetric web-based tool [20] was used to display statistics about ontology metrics as shown in Table 3. The results of this validation present that OAIDS is composed of 931 axioms, 93 classes, 33 object properties, 73 data properties, and 47 individuals. The level of Data Logic (DL) expressivity is ALEHQ (D). After checking this ontology, we observed that these obtained results demonstrate that our OAIDS ontology give clarity to recognize all concepts, relationships and their correspondences. Moreover, regarding to the size of the ontology, OAIDS has the largest number of classes, individuals, and properties compared to the reviewed ontologies. That makes it very usable to understand a detailed description about the knowledge data in this type ITS's system. To sum, this evaluation shows acceptable measures to validate the effectiveness of this OAIDS. However, inconsistency, incompleteness and redundancy measures are required to give more accurate evaluation in the future.

Table 2. Summary of the ontology requirements specification document for OAIDS.

Purpose	Representation of the knowledge in the domain of AIDS
Scope	The ontology will on information exchange in AIDS
Implementation & language	The ontology will be implemented in the ontology representation language OWL 2 using Protégé tool
Intended Users	Direct users: transportation researchers who conduct academic research in order to improve incident detection. Software engineering for developing ontology driven information systems End users: drivers and TMC authorities who are searching incident detection notification. Public or private transportation service which offers services to gather, to analyze and to prepare some data and statistics about the situation of traffic safety
Intended use case	Use case 1: to share a common understanding of traffic data knowledge between different AIDS components such as type of message, format of data, structure of packets
Ontology requirements	What are the major devices used in AIDS? What is the traffic sensor behaviors sent from CVs? What are the main activities for each AIDS component? Which type of VANET messages is used? What types of information are circled in AIDS? What are the symptoms used to judge the incident detection?
Pre-glossary of terms	Traffic, CV, detectors, signal, message, detection, incident, anomaly, VANET, collecting, distribution, intersection, road
Identified questions	What is meant by an incident? Which information is used to define a traffic anomaly data? What are the main factors used to define an incident alarm? Who are the people associated with incident detection? Which mechanisms are used to distribute traffic data?

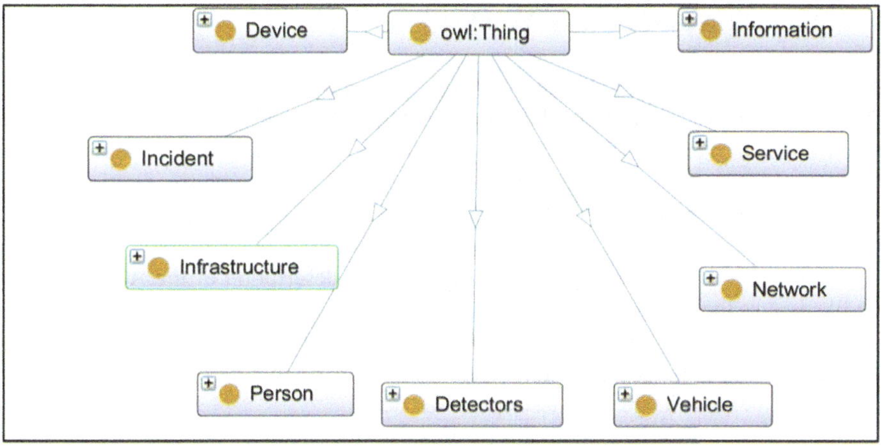

Fig. 3. Main OAIDS concepts.

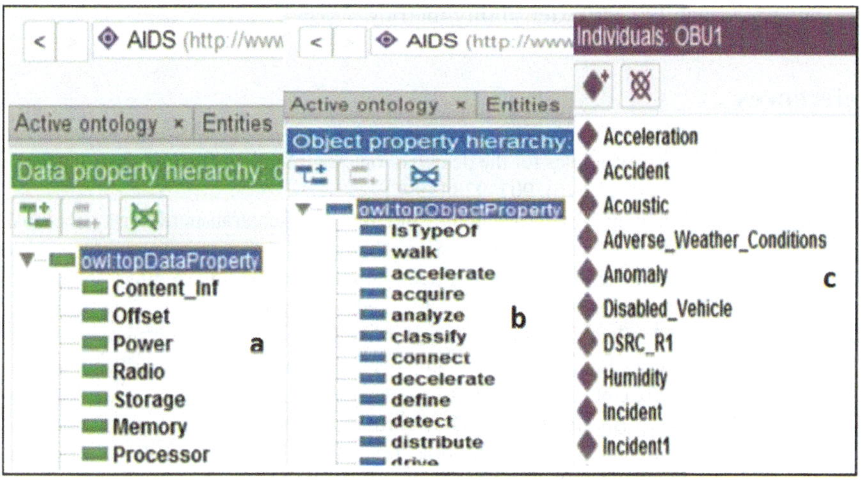

Fig. 4. (a) OAIDS data property. (b) Object property. (c) Instances.

Table 3. General OAIDS ontology metrics.

Base metrics		Graph, Schema & Knowledgebase Metrics	
Axiom	931	Absolute sibling cardinality	93
Logical axiom count	675	Absolute depth	244
Class count	93	Absolute breadth	93
Object property count	33	Total number of paths	93
Data property count	33	Attribute richness	0.784946
Individual count	74	Class/relation ration	0.794872
Annotation Property count	1	Average population	0.505376

4 Conclusion and Future Works

Several propositions exist in literature which constructs a domain ontology for different ITS domains. However, none of them explains the AIDS ontology engineering process which might be very helpful for ontology developers, domain experts, and end-users to improve ITS's safety and efficiency. In this work, we deal with the development of OAIDS, an ontology representing the domain knowledge specific of AIDS, particularly for the urban areas. It was built by adapting NeOn ontology methodology. Through this paper, the authors give the comprehensive semantic description of the AIDS domain. It will serve as the semantic knowledge base to facilitate and simplify the development of urban AIDS as mentioned in results. Consequently, the size of OAIDS and richness of attributes and population demonstrated the semantic data interoperability details to understand the knowledge data description needed.

In the future, we will extend and re-organize this ontology for representing other AIDS management knowledge and then prove compliance of the world model to the world modeled formally of OAIDS by adding ontology evaluation phase to evaluate performance, scalability and extensibility metrics.

References

1. Gruber, T.R.: Toward principles for the design of ontologies used for knowledge sharing. Int. J. Hum. Comput. Stud. **43**(5–6), 907–928 (1995)
2. Fernandez, S., et al.: Ontology-based architecture for intelligent transportation systems using a traffic sensor network. Sensors **16**(8), 1287 (2016)
3. Fernandez, S., Ito, T.: Using SSN ontology for automatic traffic light settings on intelligent transportation systems. In: 2016 IEEE ICA Proceedings. IEEE (2016)
4. Fernandez, S., Ito, T., Hadfi, R.: Architecture for intelligent transportation system based in a general traffic ontology. In: Lee, R. (ed.) Computer and Information Science 2015, pp. 43–55. Springer, Cham (2016). https://doi.org/10.1007/978-3-319-23467-0_4
5. Sobral, T., Galvão, T., Borges, J.: VUMO: towards an ontology of urban mobility events for supporting semi-automatic visualization tools. In: 2016 IEEE 19th International Conference on Intelligent Transportation Systems Proceedings. IEEE (2016)
6. Ghannem, A., Makram, S., Hany, A.: An adaptive I-parking application: an ontology-based approach. In: Future Technologies Conference Proceedings (2016)
7. Zhao, L., et al.: Ontology-based driving decision making: a feasibility study at uncontrolled intersections. IEICE Trans. Inf. Syst. **E100.D**(7), 1425–1439 (2017)
8. Akagi, Y.: Ontology based collection and analysis of traffic event data for developing intelligent vehicles. In: 6th GCCE Proceedings. IEEE (2017)
9. Fernandez, S., Ito, T.: Semantic integration of sensor data with ssn ontology in a multi-agent architecture for intelligent transportation systems. IEICE Trans. Inf. Syst. **100**(12), 2915–2922 (2017)
10. Klotz, B., et al.: Generating semantic trajectories using a car signal ontology. In: Companion Proceedings of the Web Conference 2018 (2018)
11. Klotz, B., et al.: VSSo: the vehicle signal and attribute ontology. SSN ISWC. 2018, pp. 56–63 (2018)
12. Bibi, A., Rehman, O., Ahmed, S.: An ontology based approach for messages dissemination in vehicular ad hoc networks. EAI Endors. Trans. Scalable Inf. Syst. **5**(16) (2018)

13. Viktorović, M., Yang, D., de Vries, B.: Connected traffic data ontology (CTDO) for intelligent urban traffic systems focused on connected (semi) autonomous vehicles. Sensors **20**(10), 2961 (2020)
14. Sobral, T., Galvão, T., Borges, J.: An ontology-based approach to knowledge-assisted integration and visualization of urban mobility data. Expert Syst. Appl. **150**, 113260 (2020)
15. Larioui, J., El Byed, A.: Towards a semantic layer design for an advanced intelligent multimodal transportation system. Int. J. **9**(2) (2020)
16. Çintaş, E., Özyer, B., Hanay, S.: Ontology-based instantaneous route suggestion of enemy warplanes with unknown mission profile. Sakarya Üniversitesi Fen Bilimleri Enstitüsü Dergisi **24**(5), 803–818 (2020)
17. Suárez-Figueroa, M.C.: Neon methodology for building ontology networks: specification, scheduling and reuse. Thèse de doctorat. Informatica (2010)
18. Suárez-Figueroa, M.C., Gómez-Pérez, A., Fernández-López, M.: The NeOn methodology for ontology engineering. In: Suárez-Figueroa, M., Gómez-Pérez, A., Motta, E., Gangemi, A. (eds.) Ontology Engineering in a Networked World, pp. 9–34. Springer, Heidelberg (2012). https://doi.org/10.1007/978-3-642-24794-1_2
19. Suárez-Figueroa, M., Gómez-Pérez, A., Fernández-López, M.: The NeOn methodology framework: a scenario-based methodology for ontology development. Appl. Ontol. **10**(2), 107–145 (2015)
20. OntoMetric (2021). https://ontometrics.informatik.uni-rostock.de/ontologymetrics/

A Study of Wireless Sensor Networks Based Adaptive Traffic Lights Control

Sofiane Benzid(✉) and Ahmed Belhani

University of Constantine 1, Constantine, Algeria
{sofiane.benzid,ahmed.belhani}@umc.edu.dz

Abstract. With the rising impact of congestion in cities, implementing an adaptive traffic light system as part of an Intelligent transportation system (ITS) is more than necessary to control traffic lights at intersections. These adaptive systems have the ability to sense traffic in real time and adjust the lights accordingly, contrary to the existing preprogrammed control systems which they fail to adapt to traffic variation causing more traffic jams. The existing industrial adaptive control systems use expensive equipment impeding their spread worldwide, but thanks to the advances in embedded systems, Wireless Sensor Networks (WSN) are emerging as a potential solution for the expensiveness of the existing adaptive schemes. In this paper, relevant works of designing an adaptive traffic lights control system using WSN are reviewed with a focus on the different sensing technologies, architectures and the algorithms used.

Keywords: Congestion · Adaptive traffic systems · ITS · Average waiting time · WSN

1 Introduction

Cities and especially highly populated ones are facing a big challenge of handling congestions on their roads. Congestion arises when the transport infrastructures and the number of vehicles are not growing at the same pace leaving more and more vehicles hungry for road space [1]. According to the Urban Mobility Report published on August 2019 by the Texas A&M Transportation institute [2], in 2017, congestion made urban Americans waste 8.8 billion hours on extra travel time and consume 11.3 billion liters of extra fuel rising the cost of congestion to $179 billion. In 2013 congestion costed France €17 billion and would rise to €22 billion in 2030, €4 123 would be spent by a Parisian due to congestion compared to €2 883 in 2013 [3]. Locally in Algeria and according to a study conducted by the National Polytechnic School in 2018 covering seven highly populated cities [4], congestion costs the country €100 million per year. Traffic congestion also causes emissions contributing to air pollution and impairing human health, researches from Harvard School of Public Health (HPLS) estimates more than 2200 premature deaths and health cost of $18 billion annually in the 83 largest urban areas in USA due to congestion emissions [5].

B. Lejdel et al. (Eds.): AIAP 2021, LNNS 413, pp. 406–417, 2022.
https://doi.org/10.1007/978-3-030-96311-8_37

One of the primary sources of congestion in cities are linked to the inability of existing traffic lights systems to adequately manage the flow of vehicles at intersections. Most of the existing Traffic Monitoring Controllers (TMCs) use a fixed time control to set the light plan. In the fixed time control, the sequence of the phases and the green time duration for each phase are both fixed, a phase is a combination of movements allowed to occur simultaneously without conflict and the sequence of phases where every movement is at least selected once is called a cycle [6]. The problem of this type of control is that it doesn't always favor the movement with the highest number of vehicles and it doesn't detect accidents or give priority to emergency vehicles which increases traffic congestion. As opposite to the fixed time control, the adaptive traffic control continuously detects traffic at intersections and dynamically adjusts the order of the phases and the green time duration [6]. The objective of the adaptive traffic light control is to alleviate traffic jams through reducing the average waiting time (AWT) and the average queue length (AQL) of vehicles at intersections and also give priority to emergency vehicles.

The key of establishing an intelligent transportation system (ITS) is the use of sensors to collect information about traffic condition. Commonly, ITS systems rely on expensive sensors which are wired to a central entity limiting their deployment and make them more prone to faults, as a malfunctioning of the central entity leads to the malfunctioning of the entire system and cause several TLCs to be out of service or working in a predetermined manner, regardless of traffic variation. As a solution to the aforementioned problems, wireless sensor networks (WSNs) are gaining more and more attention in the literature as they are made of tiny non expensive entities called sensor nodes. These sensor nodes are easily deployed and communicate wirelessly allowing them to cover larger areas. A typical sensor node is composed of one or more sensors, a processing unit, a storage unit and a communication unit. In the adaptive traffic light systems, WSNs are used to collect information about traffic and take decisions about the traffic plan i.e. order of the phases and the green time duration. Sensors are placed in different parts of the roads or in vehicles to continuously count the number and speed of vehicles on each lane in an intersection and also to detect emergency vehicles or accidents, this information are then passed wirelessly in a hierarchical manner up to a TMC where the traffic light algorithm is executed and hence updating the traffic plan, Fig. 1 shows a typical adaptive traffic light control system using road side sensors to collect traffic information.

In the present state of the art, we review some of the relevant works regarding the implementation of adaptive traffic light controllers using WSNs focusing on the architecture and the algorithms used to reduce the AWT of vehicles at intersections and help avoiding congestions. In the second section we present some of the sensing technologies being used to detect traffic, Sect. 3 reviews relevant works and the different strategies used to manage intersections adaptively using WSNs, Sect. 4 summarizes the different approaches by highlighting the pros and cons of each technique and discusses some aspects towards designing a complete adaptive traffic controller. Section 5 concludes the paper.

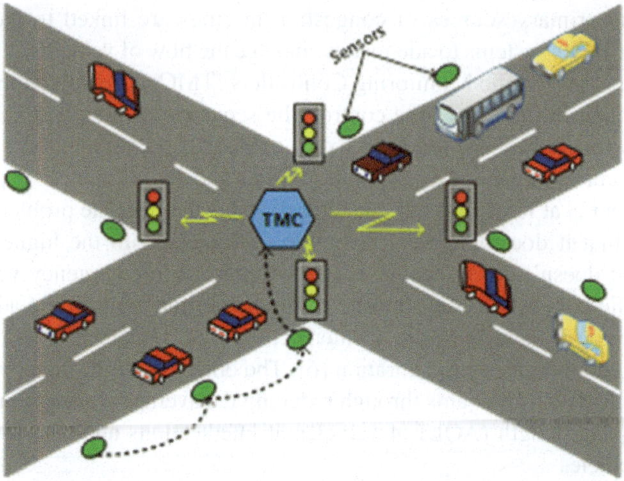

Fig. 1. WSN-based adaptive traffic light control system

2 Sensing Technologies

Sensors are being used in ITS to detect traffic and also sometimes weather conditions affecting traffic flow. They are being used to count the number of vehicles on lanes, to measure the speed of passing vehicles and detect special events like accidents or the presence of an emergency vehicle.

2.1 Traffic Sensors Classification

Traffic sensors fall into two main categories depending on their placement configuration:

Intrusive Sensors. This type of sensors often require pavement cut or sensors being embedded in potholes on roads. The advantage of these sensors is their higher capacity of detecting vehicles however they cause traffic disturbance during both phases of installation and maintenance raising their cost of use [7]. Inductive loops, pneumatic road tubes and magnetic sensors (when being installed under the pavement) are the most famous intrusive sensors.

Non-intrusive Sensors. Or sometimes called above the ground sensors have emerged as a solution to the drawbacks of intrusive sensors thanks to their ease of installation and maintenance, making them non-disruptive to traffic flow as they are often installed on lanes, road sides or on poles like cameras and magnetic sensors, other popular above the ground sensors are [7]: RFID (Radio Frequency Identification) sensors, Acoustic sensors, Ultrasonic sensors and Infrared sensors.

2.2 Anisotropic Magnetoresistance Sensors

Among all the above-mentioned traffic sensors, Anisotropic Magnetoresistance sensors (AMR) are the most used sensors in the literature for traffic detection. Given that the earth's magnetic field is uniform, theses sensors sense the field disturbance caused by a passing ferrous object like vehicles. AMR sensors outperforms the other sensors for numerous reasons [8]: they are solid and immune to vibration, they are highly sensitive and stable when it comes to recording the magnetic signature in different climatic conditions and last but not least, they are cheap and small sized allowing their deployment on a larger scale.

Liu et al. [9] employed a single 3 axis AMR sensor on a lane line to detect and measure the vehicles flow on three positions: left of the line, on the line and right of the line. The AMR used is the BMM 150. The authors developed two double window algorithms: 1) Vehicle mixed algorithms (VMR) to detect two vehicles passing at the same time, one on the left lane and the other on the right lane, with an accuracy of 98%. 2) Vehicle Motion-State Discrimination Algorithm (VMSDA) which can identify the position of the passing vehicle: left lane, right lane or on the line with an accuracy of 96,4%. The study focused on low speed vehicles without classification limiting its use only for vehicles count.

In [8] the authors designed a system for vehicle detection and classification to be used in ITS systems unlike the work in [9] where only vehicle detection was considered. The authors focused on the types of vehicles specific to the Australian roads namely: sedan, van, truck and bus. For detection purposes, the system relies on a single AMR sensor installed on the road-side 60 cm away from passing vehicles, this configuration doesn't disrupt traffic during installation or maintenance. To identify the type of the vehicle, the signal feature extraction and vector quantization techniques were applied and for system training, Dynamic Time Wrapping (DTW) algorithm is used to select the most suitable representation among the samples recorded from each passing vehicle. The effect of the distance between the AMR sensor and the passing vehicle was not investigated and hence the choice of 60 cm distance is not justified. Furthermore, the system can only detect one vehicle on a single lane hindering its use for multiple lane roads.

Santoso et al. [10] used two 3-axis AMR sensors of type HMC 5883L installed on a roadside and spaced by a distance of 1m to measure the flow and speed of passing vehicles. The authors used a moving average filter to enhance the raw signal obtained by the AMR sensors and for speed measurement they took the derivative of the magnitude of both sensors to detect the steepest ascent in each signal and hence compute the time a vehicle takes to travel the 1m distance, having the time and the distance, computing the speed becomes straightforward. The results were validated using 250 fps camera. The authors didn't make mention of the distance between the sensors and vehicles and also the possibility of using the system in multiple lane roads was not evoked.

Apart from AMR sensors, other sensors are used for traffic detection: In [11] the authors used dual loop inductive detector for vehicle classification, RFID technology was used in [12] to detect emergency vehicles and give them priority at intersections and the authors in [13] surveyed many works and challenges related to traffic monitoring at intersections using cameras.

3 Relevant Works

Among the existing commercial adaptive traffic light control systems, SCOOT (Split Cycle Offset Optimization Technique) and SCATS (Sydney Coordinated Adaptive Traffic Systems) are the two most widely used ones. SCOOT works on minimizing delays and stops by predicting traffic and adjusting traffic plan through optimizing Splits, Cycles and Offsets [14]. Traffic sensors in SCOOT are placed 90 to 120 m before an intersection so it can update traffic lights before a queue is formed [15]. The implementation of SCOOT system by Siemens [14] in a corridor made of 33 intersections in the city of Seattle led to reducing travel times by 21% during rush hours, magnetometers and video detection cameras were used for traffic detection in the project. SCATS is distributed among three levels of control: local, regional and central for the purpose of coordinating multiple intersections and cover large areas [16], it relies on inductive loops for vehicle detection and push buttons mounted on traffic light poles for pedestrian detection [17]. SCATS is employed in more than 55 000 intersections in 28 countries worldwide and in which it reduced journey times by 28%, fuel consumption by 12% and emissions by 15% [17].

The abovementioned techniques and other similar techniques are expensive to implement impeding their spread in the world and especially in the developing countries, for instance on average SCOOT costs $49 000 per intersection and SCATS $60 000 per intersection without taking into account detection costs of $20 000 per intersection [18]. Designing adaptive traffic systems using WSN can be a suitable substitute for the expensive existing techniques as they rely on non-expensive and off the shelf components.

To achieve the goal of designing an adaptive traffic control system using WSN, different techniques and strategies are found in the literature:

3.1 Queuing Theory

Yousef et al. [19] developed their adaptive technique based on queuing theory, each movement at the intersection is modeled as M/M/1 queue with its own Arrival rate λ and departure rate μ, average queue length Q and average waiting time W. Using little's law $W = Q/\lambda$ and the queue length Q_j of the jth cycle is given by the following formula:

$$Q_j = Q_{j-1} + \lambda T_G - \mu T_G + \lambda T_R \tag{1}$$

Q_{j-1} is the remaining vehicles from the previous cycle, T_G is the green light time and T_R is the red light time. The phases are formed using a conflict matrix and sorted to give priority to the movements with the highest queue lengths, the phase order determination is cycle based i.e. updated at the end of each cycle and the green time of each phase is set in accordance to discharge the movement that has the longest queue. For traffic detection, sensor nodes are placed in protected holes on the roads with two sensors for each lane (before and after the traffic light) to count vehicle arrivals and departures. All sensors communicate with a Base Station using TDMA protocol, the base station aggregates the received information and pass them to a Control Box where the TSTMA (Traffic Signal Time Manipulation Algorithm) is executed to set the appropriate order and timing of the different phases, the algorithm developed by the authors has an objective of reducing the AQL and the AWT. The simulation results show the advantage of using the adaptive traffic algorithm to dynamically control the traffic light in comparison with a fixed time control. The authors in their work assumed that all vehicles have the same speed and all sensor nodes were placed in the coverage range of the BS limiting their deployment and hence the efficiency of the system, also the distance between sensor nodes on each lane was not mentioned.

3.2 Score Function

The authors in [20] considered a single intersection of four directions (N, E, S, W) and two lanes for each direction. Two sensor nodes arc placed on each lane, one at the junction and the other one at a distance $D = T_G^{Max}.V$ where T_G^{Max} is the maximum allowed green time and V is the speed of vehicles. The authors adopted 12 phases and each phase has its own traffic light as shown in Fig. 2. The proposed algorithm starts by detecting the departure and arrival rates, and the type of vehicles using the different sensor nodes. The next step of the algorithm uses the traffic information obtained from the previous step to set the sequence of the phases relying on a score function that reflects the degree of demand of a green light for each phase based on five weighted factors as follow:

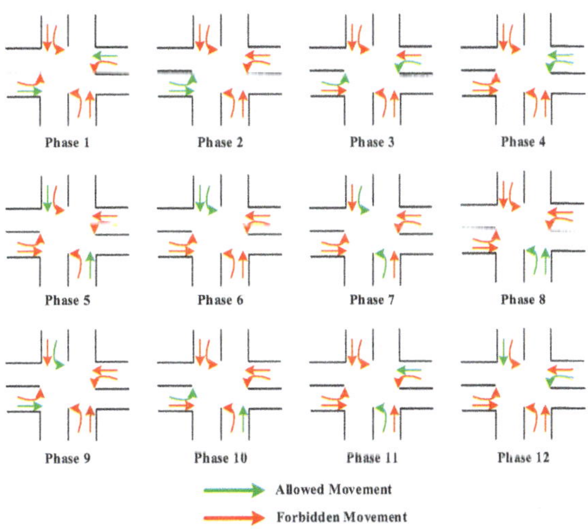

Fig. 2. All possible 12 phases

$$SF = a_1TV + a_2W + a_3HL + a_4BC + a_5SC \tag{2}$$

Where TV the traffic volume reflects the number of vehicles, W is the average waiting time, HL is the Hunger Level that reflects how many or how few a given phase has been attributed a green light before, the Hunger Level is used here to prevent famine i.e. a situation where a given phase is not selected for a long period of time. The Blank count (*BC*) reflects segments of a lane where no vehicle is present and how far these segments are from the intersection, the higher the *BC* the lower is the priority. Special Circumstances (*SC*) factor reflects special events like the presence of an emergency vehicle or accident where higher priority is given for an emergency vehicle and lower priority for an accident. The coefficients $a_1 \ldots a_5$ are set to give priorities for the previously mentioned factors from highest to lowest in a descending order as follow: *SC*, *BC*, *HL*, *TV* and finally waiting time *W*. The phase which has the highest value of *SF* is selected next to have a green light and the green time duration is set in a way that all the vehicles of the selected phase pass the intersection and is bounded by T_G^{Max}. Through simulation the authors demonstrated the superiority of their algorithm in comparison with an actuated and fixed time control systems in terms of reducing the average waiting time AWT and increasing throughput i.e. the rate of vehicle departures at the intersection. In the proposed algorithm, unrealistic assumptions are made as all vehicles are assumed to run at the same speed and also the turning right movements are not considered. In [21] the authors extended their work to cover multiple intersections, the green time of a phase is recalculated adding an offset to account for coming vehicles from neighboring intersections and hence creating green waves.

Like most of the literature Faye et al. [22] use two sensor nodes per lane as shown in Fig. 3 separated by a distance $D = N.L$ where L is the average vehicle length and N is the maximum number of vehicles allowed to pass when $T_G = T_G^{Max}$, $N = (T_G^{Max} - T_s)/T_h$ where T_s is the starting time when the light switches from red to green and T_h is the average time separating two discharging vehicles. All sensor nodes are assumed of the same type and have a magnetometer as a sensing element.

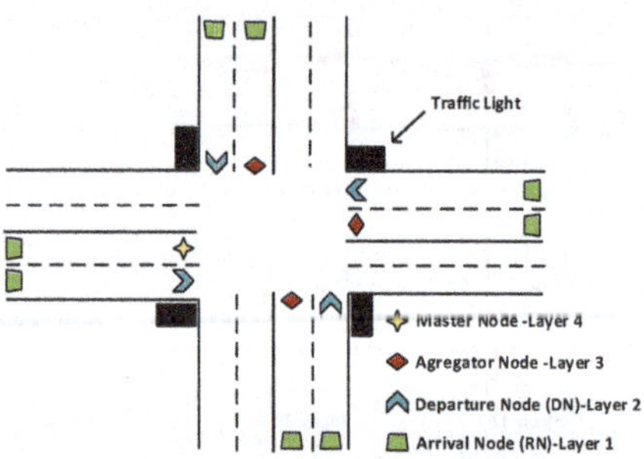

Fig. 3. Intersection architecture

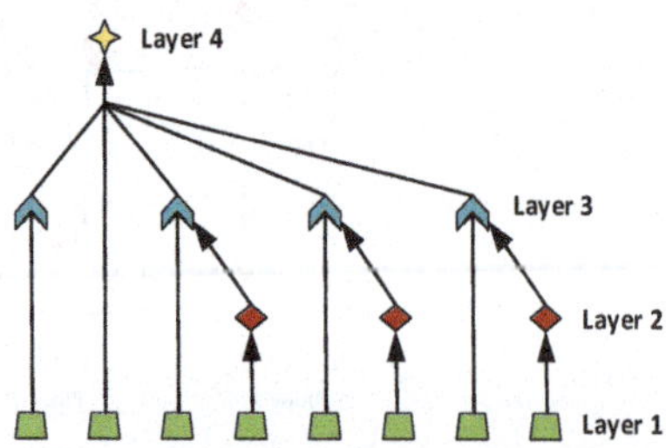

Fig. 4. Hierarchical architecture

The architecture adopted by the authors follows a hierarchical scheme, Fig. 4, where the sensors are distributed among four layers. Layer 1 is composed of Arrival Nodes (RN) counting the number of vehicles approaching the intersection on each lane and pass them to layer 2 nodes, Departure Nodes (DN) in layer 2 count the number of vehicles leaving the intersection and with the information received form layer 1, they keep track of the number of vehicles occupying each lane. Layer 3 nodes are elected among departure nodes to count the number of vehicles for each possible movement and also count the time since the movement was last selected. Similar to the idea in [21], the master node in layer 4 uses the information of layer 3 and assigns a score for each movement k based on the queue length and hunger level as follow:

$$SF_k = \alpha \cdot \left(\frac{N^k}{\sum_{p=1}^{M} N^p} \right) + \beta \cdot \left(\frac{T^k}{\sum_{p=1}^{M} T^p} \right) \tag{3}$$

Where N^k is the number of vehicles composing movement K, N^p number of vehicles composing a movement p where p =[1..M], M is the number of all possible movements, T^k is the time elapsed since the movement K was last selected, α and β are weighting parameters allowing to favor one objective over the other. After setting a score for each movement, the master node combines all the simultaneous nonconflicting and slightly conflicting movements into phases and add up the corresponding scores to obtain a global score for each phase, the phase with the highest score is attributed a green light. Similar to [21] the green time of the selected phase is set to allow all the vehicles to pass the intersection and is equal to $T_G = T_s + T_h \cdot N_{max}$ where N_{max} is the largest number of vehicles found among the lanes composing the selected phase. As long as $T_G < T_G^{Max}$ T_G is extended by T_h for each vehicle detected by RN. Using SUMO (Simulation of Urban Mobility) software, the authors evaluated their adaptive algorithm on an intersection in the city of Amiens (France) and with the appropriate selection of T_G^{Max} and weighting parameters they obtained better AWT when compared to both the existing fixed time control of the Amiens' intersection and the algorithm proposed by Yousef et al. in [20]. The advantage of the architecture used in this work is that layer 3 and layer 4 roles can be performed by any Arrival Node (RN) allowing high fault tolerance and unlike the works in [19, 20], the TLC here is merely an actuator applying the light plan set by layer 4 sensor, however the authors didn't include the emergency vehicles and eventual accidents in their algorithms like in [20] and also the choice of parameters was based on experimentation only.

In [23] Faye et al. extended their work for multiple intersections in which they introduced two more new objectives in their score function to create green waves: 1) prevent overloading an intersection which already has enough vehicles and 2) take into account the number of vehicles coming from adjacent intersections, for these purpose the Departure Sensors (DS) were moved to the outgoing lanes instead of the incoming lanes in their previous work.

3.3 Fuzzy Logic

In [24] the authors use a WSN based on IEEE 802.15.4 communication protocol and four parallel fuzzy logic controllers to dynamically set the timing of the green light for four possible phases i.e. one controller for each phase. Similar to Faye et al. in [22], the authors use a hierarchical architecture where the sensor nodes are placed on the roadside of each lane and use magnetometers to collect traffic data.

The sensor nodes of each lane send the information to an aggregator node which assesses the queue length of the lane and transmit the information to the master node. Once the master node receives the number of vehicles on each lane from all the aggregator nodes, it sorts the different phases giving priority to the longest queue and apply the appropriate green light timing using the fuzzy logic controllers. The fuzzy controller is built following three steps: In the fuzzification process, the triangular input membership function characterizes the queue length of each lane as {Normal, Medium, Long} which corresponds to a number of vehicles per lane in the range of [16...80], the triangular output member function reflects the green light timing as {Min, medium, Max} corresponding to a range of [15s...90s]. In the second step, inference mechanism, an IF-Then rule is used to map the member functions. In the last step, defuzzification, the authors chose the Centroid Of Area (COA) method to obtain the crisp output representing the green light timing of the phase. Simulation was carried using MATLAB for the fuzzy logic controller and the TRUTIME toolbox to simulate IEEE 802.15.4 protocol. The authors compared their method with the fixed time control scheme and with three other fuzzy logic based methods from the literature. The multicontroller approach outperformed the previously mentioned methods in term of reducing the AWT especially in a high traffic situation. The authors suggested 240 sensor nodes per intersection, one for each vehicle detection, which is a huge number influencing the cost of the system. The intersection is assumed fixed meaning that the algorithm does not rely on a conflict matrix to dynamically compose the phases irrelative to the intersection configuration, also the control scheme adopted is cycle based and not phase based which makes the solution not fully adaptive.

3.4 Prioritizing Emergency Vehicles

Krishna et al. [25] developed a system to provide a free passage for emergency vehicles at intersections. Sonar sensors and a camera were used to detect the emergency vehicles and switch the traffic lights in favor of a green wave. A prototype of the system was built using two traffic junction nodes and a communication node. The junction node is built around Raspberry-Pi processor and uses a USB camera to confirm and identify a passing emergency vehicle after being detected by a sonar sensor. Upon confirmation of a passing emergency vehicle, the junction node sends a signal to the traffic controller to properly adjust the lighting and another signal to all junction nodes of adjacent intersections through communication nodes. The communication node is built around Arduino UNO and an RF module, it is used as a message repeater between two junction nodes. The communication between the intersection node and the junction node is based on the ZigBee protocol. Simulation on a four-intersection path shows that the proposed system reduces the journey time of the emergency vehicle by 3 min compared to conventional fixed time control. the algorithm proposed by the authors affects all adjacent intersections irrelative to where the emergency vehicle goes next and thus induce more traffic jam, also if two or more emergency vehicles are present on the different roads of an intersection, priority is given based on the arrival time to the intersection rather than the degree of emergency.

4 Discussion and Enhancement Factors

All the previously discussed works share the same objective of reducing the average waiting time AWT, so comparing them with each other based on AWT minimization would imply having access to their detailed algorithms and apply them to the same set of intersections which is out of the scope of this paper. Instead we will compare them based on cost, adaptivity to traffic variation, flexibility to cope with different intersection configuration (single or multiple intersections) and some other factors presented in Table 1 as pros and cons for each technique.

In the reviewed works, authors focused on how to reduce average waiting time AWT of vehicles and some prioritized emergency vehicles in their algorithms but they didn't take into consideration some other important factors like pedestrians crossing the roads as an AWT for pedestrians could be considered by introducing for example some weight sensors near traffic lights pols to estimate the number of pedestrians waiting for a red light to cross the road. Reducing congestion emissions is also an important factor that could be added in designing an adaptive system especially in cities where air pollution

Table 1. Comparative table.

Approach	References	Pros	Cons
Queuing Theory	Yousef et al. [19]	- Low Cost - Flexible (multiple intersections) - Applicable to different intersection configuration (Conflict Matrix)	- Not Fully Adaptive (Cycle Based) - Low Fault Tolerance - No emergency vehicle consideration - Famine situation is not considered
Scoring Function	Zhou et al. [20, 21]	- Flexible - Highly adaptive (phase based) - Special events consideration - Famine situation consideration	- Predetermined intersection Configuration and phases - Weighted Factors selected experimentally
	Faye et al. [22, 23]	- Flexible - Highly adaptive (phase based) - High Fault Tolerance - Low Cost	- No Special events consideration (accidents or emergency vehicles) - Weighted Factors selected experimentally
Fuzzy Logic	Collotta et al. [24]	- High Fault Tolerance and high performance (One controller for each phase and multiple sensors)	- High Cost - No Special events consideration - Single intersection only - Predetermined phases
Prioritizing Emergency Vehicles	Krishna et al. [25]	- Reduced Journey Time for Emergency Vehicles - Simple and Low Cost	- Increases Traffic Jams - For Multiple Emergency Vehicles, Priority is given based on arrival time

is a huge issue, remote sensing units could be placed on road sides to measure emissions from vehicles. Making use of studies like the work in [8], discussed in Sect. 2, for vehicle classification could be of great importance because treating a bus and a car alike is not practical especially if we seek reducing the average waiting time, as we all know a bus in peak hours is generally full of passengers going to work or school or returning from them, so adding priority to lanes with a bus or more could considerably reduce AWT per person rather than AWT per vehicle. Finally, when using a score function like the work in [20, 22], Sect. 3, the weighting parameters could be optimized using mathematical techniques instead of experimentation, this would lead to a better average waiting time AWT.

5 Conclusion

Managing intersections effectively using adaptive traffic light control systems as opposed to fixed time or preprogrammed control helps tremendously in alleviating congestion impacts in cities. In this paper we reviewed relevant works of designing an adaptive traffic control systems using WSN as a solution to the existing expensive techniques, we focused on the different sensing technologies employed to gather traffic information like AMRs and also on the algorithms and architectures used to increase throughput and reduce the average waiting time AWT including queuing theory, weighted score functions, fuzzy logic controllers. Furthermore, additional aspects for enhancing the performance of the reviewed works were discussed. As future work, we will try to design our own sensing node for traffic detection and also implement on it our own algorithm relying on the experience learned from the present state of the art.

References

1. Rodrigue, J.P.: The Geography of Transport Systems, 5th edn. Routledge, New York (2020)
2. Schrank, D., Eisele, B., Lomax, T.: 2019 urban mobility report. Texas A&M Transportation Institute, Texas, TX, USA (2019)
3. Inrix: Embouteillages : Une Facture Cumulee De Plus De 350 Milliards D'euros Pour La France Sur Les 16 Prochaines Annees, Inrix. https://inrix.com/press-releases/embouteillages-une-facture-cumulee-de-plus-de-350-milliards-deuros-pour-la-france-sur-les-16-prochaines-annees/. Accessed 19 May 2020
4. Remouche, K.: Embouteillages: ce que ça coûte: Toute l'actualité sur liberte-algerie.com, Embouteillages : Ce Que Ça Coûte. http://www.liberte.dz/actualite/embouteillages-ce-que-ca-coute-291453. Accessed 19 May 2020
5. HSPH: Emissions from traffic congestion may shorten lives, News. https://www.hsph.harvard.edu/news/hsph-in-the-news/air-pollution-traffic-levy-von-stackelberg/. Accessed 13 June 2020
6. Faye, S.: Contrôle et gestion du trafic routier urbain par un réseau de capteurs sans fil. Ph.D. dissertation, Paris Institute of Technology, Paris, France (2014)
7. Padmavathi, G., Shanmugapriya, D., Kalaivani, M.: A study on vehicle detection and tracking using wireless sensor networks. Wirel. Sens. Netw. **02**(02), 173–185 (2010)
8. Chen, X., Kong, X., Xu, M., Sandrasegaran, K., Zheng, J.: Road vehicle detection and classification using magnetic field measurement. IEEE Access **7**, 52622–52633 (2019)

9. Liu, M., Hua, W., Wei, Q.: Vehicle detection using three-axis AMR sensors deployed along travel lane markings. IET Intell. Transp. Syst. **11**(9), 581–587 (2017)
10. Santoso, B., Yang, B., Ong, C.L., Yuan, Z.: Traffic flow and vehicle speed measurements using anisotropic magnetoresistive (AMR) sensors. In: 2018 IEEE International Magnetics Conference (INTERMAG), Singapore, pp. 1–4 (2018)
11. Gajda, J., Stencel, M.: A highly selective vehicle classification utilizing dual-loop inductive detector. Metrol. Meas. Syst. **21**(3), 473–484 (2014)
12. Bhate, S.V., Kulkarni, P.V., Lagad, S.D., Shinde, M.D., Patil, S.: IoT based intelligent traffic signal system for emergency vehicles. In: 2018 Second International Conference on Inventive Communication and Computational Technologies (ICICCT), Coimbatore, pp. 788–793 (2018)
13. Datondji, S.R.E., Dupuis, Y., Subirats, P., Vasseur, P.: A survey of vision-based traffic monitoring of road intersections. IEEE Trans. Intell. Transp. Syst. **17**(10), 2681–2698 (2016)
14. Siemens: Deploying SCOOT in Seattle, Austin, TX, USA, Project White Paper (2017)
15. Siemens: Keeping Traffic Moving in Ann Arbor, Project White Paper, Austin, TX, USA (2016)
16. Zhao, Y., Tian, Z.: An overview of the usage of adaptive signal control system in the United States of America. Appl. Mech. Mater. **178–181**, 2591–2598 (2012)
17. NSW: SCATS and Intelligent Transport Systems. SCATS (2020). http://scats.nsw.gov.au/. Accessed 21 June 2020
18. Selinger, M., PTOE, Schmidt, L.: Adaptive traffic control systems in the United States. HDR Engineering (2009)
19. Yousef, K.M., Al-Karaki, J.N., Shatnawi, A.M.: Intelligent traffic light flow control system using wireless sensors networks. J. Inf. Sci. Eng. **26**, 753–768 (2010)
20. Zhou, B., Cao, J., Zeng, X., Wu, H.: Adaptive traffic light control in wireless sensor network-based intelligent transportation system. In: 2010 IEEE 72nd Vehicular Technology Conference - Fall, Ottawa, ON, Canada, pp. 1–5 (2010)
21. Zhou, B., Cao, J., Wu, H.: Adaptive traffic light control of multiple intersections in WSN-Based ITS. In: 2011 IEEE 73rd Vehicular Technology Conference (VTC Spring), Budapest, Hungary, pp. 1–5 (2011)
22. Faye, S., Chaudet, C., Demeure, I.: A distributed algorithm for adaptive traffic lights control. In: 2012 15th International IEEE Conference on Intelligent Transportation Systems, Anchorage, AK, USA, pp. 1572–1577 (2012)
23. Faye, S., Chaudet, C., Demeure, I.: A distributed algorithm for multiple intersections adaptive traffic lights control using a wireless sensor networks. In: Proceedings of the first workshop on Urban networking - UrbaNe '12, Nice, France, p. 13 (2012)
24. Collotta, M., Lo Bello, L., Pau, G.: A novel approach for dynamic traffic lights management based on wireless sensor networks and multiple fuzzy logic controllers. Expert Syst. Appl. **42**(13), 5403–5415 (2015)
25. Krishna, A.A., Kartha, B.A., Nair, V.S.: Dynamic traffic light system for unhindered passing of high priority vehicles: wireless implementation of dynamic traffic light systems using modular hardware. In: 2017 IEEE Global Humanitarian Technology Conference (GHTC), San Jose, CA, pp. 1–5 (2017)

Forwarding Strategies in NDN-Based IoT Networks: A Comprehensive Study

Adel Djama(✉), Badis Djamaa, and Mustapha Reda Senouci

Distributed and Complex Systems Lab., Ecole Militaire Polytechnique, Algiers, Algeria
{adel.djama,badis.djamaa,mustaphareda.senouci}@emp.mdn.dz

Abstract. Named Data Networking (NDN) is a new communication model that proposes to shift the manner to do networking by fetching data by names instead of host addresses. This data-driven architecture is considered promising for the Internet of Things (IoT) applications due to its inherent characteristics, such as naming, caching, and stateful forwarding, which give NDN the power to support natively, without adaptation mechanisms, the major requirements of the IoT environments. Nevertheless, particular management should be provided by the forwarding protocols when handling the limited resources of the IoT objects. This paper is devoted to investigating NDN-based forwarding strategies for the IoT, along with a comprehensive comparative study, followed by a simulation of some representative schemes. Insights on the main observations learned from the conducted evaluation are presented in the paper, highlighting the strengths, weaknesses, and suitability of every benchmarked solution in the context of NDN-IoT twinning.

Keywords: Internet of Things · Named data networking · Name-based forwarding · Wireless Ad hoc networks

1 Introduction

Interconnecting smart resource-constrained objects in the Internet of Things (IoT) is currently mostly supported by IP-based solutions, which rely on the adaptation of the original TCP/IP communication stack to fit IoT basic requirements. Nonetheless, these adaptation efforts have incurred management complexity and overload on the network resources [1].

Recent research have explored the aptitude of the Information-Centric Networking (ICN) paradigm in handling IoT requirements, which proposes to fetch data by names instead of host IP addresses. This new networking technology would provide natural support for the existing IoT applications where data is placed in the first plane.

Named Data Networking (NDN) [2] has been considered as the most prominent instantiation of the ICN, whose key features, namely naming, caching, packet level security, and stateful forwarding, make it very attractive for the IoT

B. Lejdel et al. (Eds.): AIAP 2021, LNNS 413, pp. 418–427, 2022.
https://doi.org/10.1007/978-3-030-96311-8_38

ecosystem. In such networks, sharing the same wireless communication medium in addition to the mobility of the nodes constitute a serious challenge to NDN-based forwarding mechanisms, which have to provide smart and lightweight techniques to handle efficiently the unreliable and ad hoc nature of these environments.

Indeed, NDN employs, in addition to the routing decision, a stateful forwarding plane where each node keeps track of received Interests to reply to them later once getting a positive or negative response. Moreover, thanks to its salient features, especially naming and caching, which are manageable directly on the network layer, multicast/anycast forwarding support is provided, hence offering a robust solution to cope with unpredictable topology changes of the IoT networks.

Nevertheless, through a unique external communication interface, imposed by such ad hoc networks, NDN forwarding operation faces a new kind of challenges, especially the broadcast storm phenomenon [3]. This latter problem is not adequately handled by the NDN original machinery, which might cause serious performance degradation to the network. To overcome this new challenge, the research community has devoted an effort axis focused on forwarding solutions, but which is still at a pre-maturing stage [4].

The rest of this paper is organized as follows. Named Data Networking and its peculiarities in IoT networks are introduced in Sect. 2, with a focus on the NDN forwarding machinery in the context of wireless ad hoc environments. This is followed, in Sect. 3, by a comprehensive state-of-the-art study on the existing NDN-based forwarding strategies devoted to wireless ad hoc networks in general and IoT in particular. In Sect. 4, a performance evaluation, via ndnSIM simulation platform, of some representative forwarding schemes is presented, along with an in-depth analysis and discussion. Finally, a conclusion summarizing the main obtained results and the lessons learned is given in Sect. 5.

2 Named Data Networking in the Internet of Things

NDN architecture proposes, through dissociating the content and its location, to retrieve data directly on the network layer, of the communication stack, by substituting source addresses by application data names. Consequently, this new paradigm leads to the shift of the communication model from location-centric to data-centric.

Two types of packets are used in NDN: Interest and Data. A consumer requests content by sending an Interest packet, which carries the targeted Data prefix name. A Data packet is returned by a producer, or any intermediate node having the requested Data in its cache, in response to that Interest, and follows the reverse path taken by the Interest to reach the consumer.

The structure of an NDN-IoT node is depicted in Fig. 1 which basically incorporates three data structures: the FIB (Forwarding Information Base), the PIT (Pending Interest Table), and the CS (Content Store). This latter is employed to store temporarily Data packets, thus allowing reducing request response time in the network. The PIT is used to save incoming interfaces of pending Interests to respond to them later, thus allowing achieving stateful forwarding feature. Whereas the FIB table contains Data prefix names with the corresponding output faces toward potential content providers. Besides, these three data structures are managed by a Forwarding Strategy engine, namely Ndn Forwarder Daemon (NFD) [5], to make the forwarding decisions about incoming Interest or Data packets.

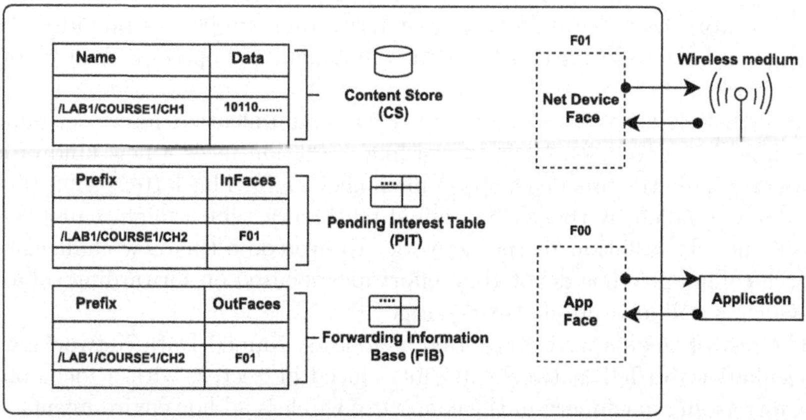

Fig. 1. NDN-IoT Node structure, adapted from [6].

In a NDN-based IoT environment, mostly characterized by wireless ad hoc deployment mode, the basic structure of the NDN node keeps the same form as the infrastructure architecture. Nevertheless, some specific characteristics, due to the wireless communication medium, affect the machinery of its forwarding daemon that faces a new range of challenges, such as broadcast storm problem, or intermittency caused by nodes' mobility in the network.

Furthermore, in such architecture, the NDN node is obliged to share one external ad hoc face with all its neighbors, through which all packet types are exchanged.

To cope with this new deal, NFD, in its latest releases, has been updated to support ad hoc communication mode, in addition to the wired mode, by adding specific exceptions in Data and Interest packets management events. We describe in the following some differences between the two communication modes.

For instance, forwarding received packets on the same incoming face has been enabled for both Data and Interest packets in ad hoc communication, since there is one external face. This situation is prohibited in point-to-point (wired) mode, where each communication link has its corresponding face. Besides, sending a Negative ACKnowledgment (NACK) packet is bypassed in ad hoc mode, so even if no valid next-hop exists in the routing table (FIB) of a node, no Nack packet is sent on the Interest incoming face. The reason for that is, as we saw earlier, the targeted content could be brought, to the requester, by another neighbor node having the same ad hoc face (outgoing face in the FIB table). Moreover, in ad hoc communication, when a forwarder node receives a new Interest packet, it keeps it in the PIT table and any subsequent Interest with a similar name is considered a loop (the packet is ignored). Whereas in point-to-point mode, duplicate incoming Interests from the same face is not considered a loop.

In summary, it can be observed that the new and flexible design of NDN architecture, thanks to its key features, especially naming, caching, and stateful forwarding, allows it to handle efficiently the challenging aspect of the IoT environment.

After highlighting NDN's main characteristics in the context of the IoT, specifically its forwarding engine, we will present in the next section the forwarding strategies that have been proposed in the literature.

3 Related Work

Many research studies have been devoted to handling forwarding issues in the context of NDN-based wireless ad hoc networks, including the IoT environments.

In [7], the authors present an Interest forwarding strategy based on randomized scheduling timers to reduce packet collisions, while exploiting the geolocation of the nodes to perform a distance-based data forwarding. In [8], a forwarding technique tailored for wireless ad hoc NDN networks is proposed, where Interest forwarding is based on the beacon messages that include the identifier of the sender and a bloom filter carrying the list of all its valid neighbors. Interest packet is further forwarded to nodes that are not included in the incoming bloom filter.

In [9], by drawing on Directed Diffusion protocol [10] principle, an enhancement of NDN forwarding scheme in WSNs is proposed. To do this, the original NDN Data packet has been modified to carry the ID of the sender, which is stored in a new data structure called Next Hop Table (NHT). This latter is exploited by nodes to manage incoming data retrieval queries. In a similar context, the same authors propose in [11] a content-centric architecture (E-CHANET) to handle the multihop wireless issue. In this solution, the node uses a new distance table in forwarding decisions, which stores the provider ID and the distance to the consumer. These two extra information are retrieved from the exchanged Interest and Data packets.

Moreover, a Neighborhood-Aware Interest Forwarding (NAIF) protocol designed for MANETs is proposed in [12], where the eligibility of a forwarder for a given prefix name, among its neighbors, is based on its content retrieval rate for that Interest and its distance to the consumer. Whereas in [13], a direction-selective forwarding strategy for content retrieval is proposed in a mobile cloud computing architecture, where the geographical coordinates of the neighbors are used by a forwarder node, in addition to new packet types (ACK and CMD), to select the relay nodes from the four quadrants of its transmission range.

On the other hand, authors in [14] designed a Reactive Optimistic Name-based Routing (RONR) mechanism, which minimizes the number of radio communications in IoT environments. For this purpose, only the first Interest query flooding is performed in the network, while the subsequent requests for the same prefix follow the footsteps, traced by the initial Data response, which are stored in the FIB table of the relay nodes.

Additionally, a Geographic Interest Forwarding (GIF) scheme for NDN-based IoT is proposed in [15]. In this solution, the nodes in the network discover their neighbors by means of HELLO messages, which include the ID of the node and its coordinates. Before sending Interest requests, a Producer Discovery phase is performed by the content producers to announce their availability to the potential consumers.

Besides, authors in [16] proposed a dual-mode Interest forwarding scheme (DMIF) for NDN-based WSNs, where both flooding and directive forwarding modes are used by the nodes in the network according to the FIB lookup for the incoming Interests. To manage Interest flooding, a TTL value is added to the Interest packet, which is dynamically tuned according to the network needs. A deferred timer is additionally used in Interest and Data forwarding phases, to counteract the broadcast storm problem in wireless communications.

In another research work [17], a hybrid forwarding strategy is proposed in wireless ICN, where an Ad hoc Dynamic Unicast (ADU) communication mechanism has been designed based on MAC notifications, allowing a dynamic alternation between unicast and broadcast modes. To achieve this, MAC addresses are disseminated in Data packets and stored in the FIB table, to serve as next-hops for subsequent queries.

While authors in [18] propose a reinforcement learning NDN-based forwarding solution for low-end IoT. The eligibility of a node to forward an Interest packet is based on a waiting time. This latter is calculated according to a cost field, carried in the Interest and Data packets, which jointly reflects the distance to the provider of the node and its eligibility to forward an Interest packet.

Lastly, authors in [19] introduce a Location-Based Deferred Broadcast (LBDB) scheme for ad hoc NDN networks, which relies on a transmission timer for the Interest forwarding phase. This timer is used to determine the priority of the forwarders and is based on the location information of the nodes and the data providers.

In sum, what we observed from our literature review, is that almost all the proposed forwarding solutions in ad hoc networks do not respect properly NDN native design, where additional fields are added to the original packets (Interest and Data). For instance, identifiers of nodes, and/or extra data structures are employed, to keep information about the network activity. Furthermore, deferred forwarding timers are often used in the proposals to avoid broadcast storm problems, alleviate the network, and improve its performances.

Based on that, and since NDN-based forwarding strategies for constrained IoT networks were not largely explored in the literature, we conduct, in the next section, an evaluation of three representative forwarding solutions, while considering the IEEE 802.15.4 as a wireless medium, by mean the official and the most used simulator in this area. The attempted goal is to assess different forwarding strategies and identify their weaknesses, strengths, and suitability in the context of a wireless-constrained IoT environment.

4 Performance Evaluation

We have learned from our previous study, that NDN-based forwarding solutions in IoT resort to the deferred broadcasting and/or modification of NDN primitives to overcome constraints imposed by wireless ad hoc communication links. Thus, we have chosen to conceive and examine the following forwarding protocols: (1) a Blind deferred Interest Forwarding (BF), inspired from [11], which uses collision avoidance timers for the Interest forwarding phase, based on random delays; and (2) a Geographic Interest Forwarding (GF), inspired from [15], which uses geographic coordinates of nodes to perform a greedy forwarding of the Interest packets to the Data providers; in addition to the Native Forwarding mechanism of NDN without modification, which we call (NF).

4.1 Simulation Platform and Parameters

To this end, ndnSIM [20], the official simulator of the NDN project, is chosen as an evaluation platform, which implements all the basic features of the NDN architecture and reproduces faithfully the holistic functioning of its forwarder engine NFD. Besides, we have selected the IEEE 802.15.4 communication standard, tailored for wireless low end and constrained IoT, as an underlay to the NDN layer. Concerning the network deployment, we have chosen a grid topology of 25 nodes including one consumer and one producer. The simulation time was set to 100 s, the Interest transmission rate was fixed at 10 p/s, and the transmission range of the nodes was varied from 10 m to 40 m.

Table 1 summarizes the simulation parameters.

Table 1. Simulation parameters.

Parameter	Value	Parameter	Value
NetDevice	LrWPAN (IEEE 802.15.4)	CS size	10
Area size ($m \times m$)	40×40	PIT size	10
Topology-Number of nodes	Grid-25	Interest packet size ($bytes$)	5
Simulation time (s)	100	Data packet size ($bytes$)	10
Interest Transmission Rate ($packet/s$)	10	Transmission Range (m)	10, 20, 30, 40

4.2 Performance Metrics

To pinpoint the behavior of the benchmarked solutions in the context of constrained wireless IoT network, we have considered the following performance metrics:

- *Sent packets:* The total number of Interest and Data packets that have been forwarded in the network;
- *Success rate:* The average success rate (satisfied Interests) of the consumers in the network;
- *Hop count:* The average hop count for all Data packets received by the consumers in the network; and
- *Retrieval time:* The average retrieval time of all Data packets received by the consumers in the network.

In the sequel, we will discuss the obtained results.

4.3 Obtained Results and Discussion

The performance metrics collected from the different simulations are depicted in Fig. 2.

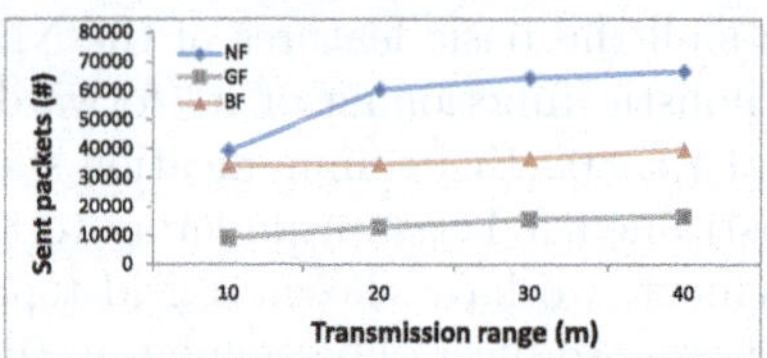

(a) Sent Interest packets

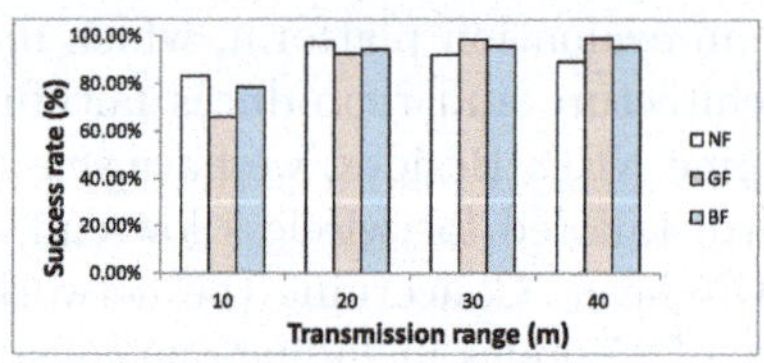

(b) Average success rate

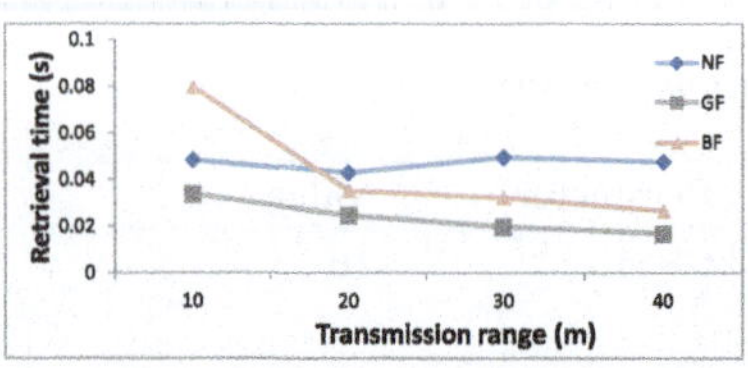

(c) Average retrieval time

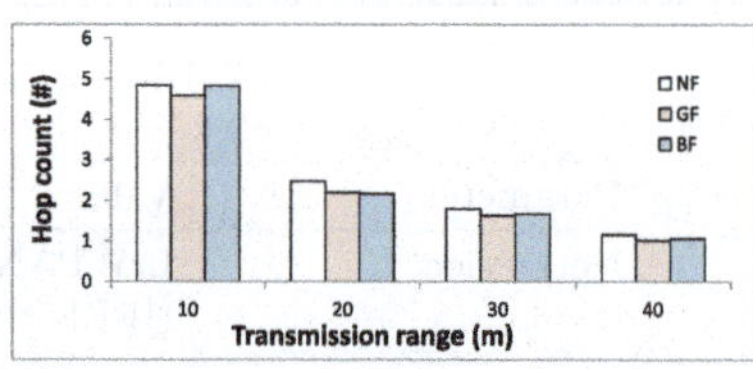

(d) Average hop count

Fig. 2. Impact of the transmission range on the benchmarked solutions.

The simulation results show that the number of sent packets (Interest and Data) in the network is huge in the case of NF, moderate in the case of BF, and very low in the case of GF, for all the transmission ranges (see Fig. 2a). The reason for this is that GF uses a single forwarding path in the content exploration phase, thanks to its knowledge of the producers' coordinates which allows reaching them optimally without flooding all the network.

Besides, the deferred Interest forwarding of BF permits to reduce the number of transmitted packets by enabling some nodes, having the lowest waiting time, to forward the Interest packets among all their neighbors. These latter, cancel forwarding operation once receiving the same Interest packet, within the waiting period, from the eligible neighbor. Lastly, for its part, without a specific mechanism to counteract the broadcast storm problem, NF floods the entire network by the Interest packets at every Interest transmission phase, which explains its worst performance regarding the number of transmitted packets in the network.

Furthermore, the success rate stats, as shown in Fig. 2b, reveal that for low transmission ranges values (up to 20 m), NF registers the best performances followed by BF and GF respectively; whereas, GF outperforms the two others for higher transmission range values (30 m and above). This can be explained by the unreliable wireless communication medium of the IoT, which causes packet loss. Indeed, the probability of success packet transmission is proportional to the transmission range; consequently, in low transmission ranges, one path forwarding scheme of GF will be penalized in terms of packet success rate compared to BF and NF, both of which use multipath forwarding. Nonetheless, for higher transmission range values, GF registered a better success rate than the two others, thanks to the geographic-based greedy forwarding technique which leads to less network overload and thus fewer packet collisions.

Regarding the retrieval time results, Fig. 2c shows that GF is better than BF and NF for all tested network densities. Indeed, the retrieval time metric is highly related to the induced overload in the network, i.e., the less network traffic, the best retrieval time is. On the one hand, GF exploits geographic-based forwarding to reach Data producers, allowing a reduction in Interest (re)transmissions and collisions, and thus a rapid content retrieval. On the other hand, NF's and BF's repeated Interest broadcast operations, whatever deferred or not, cause packet collisions and extension of their waiting time in the nodes' queues, especially with the concurrent access to the wireless communication medium of the IoT, which leads to important content retrieval delays.

Lastly, the greedy geographic forwarding mechanism of GF led to building optimized paths to the Data providers, hence, ensuring an average hop count to the producers almost equivalent to the two other strategies, which both, with no awareness of network topology, flood the entire network using multipath forwarding to retrieve the content (see Fig. 2d).

To sum up, the carried-out simulations show that the native forwarding machinery of NF falls in the broadcast storm problem, which was traduced by a huge number of transmitted and redundant packets in the network. Besides, the deferred forwarding technique of BF has allowed reducing flooding the entire network, where the traffic overload has been nearly halved compared to NF. Furthermore using the nodes' geo-coordinates has allowed GF to register better performances than NF and BF, especially in terms of traffic overload and success rate. Nevertheless, despite being closer to host-centric than the data-centric paradigm, this geographic knowledge requires additional modules (e.g., GPS) and extra data storing structures that could incur more complexity and overhead (heaviness) to the resource-constrained IoT nodes.

5 Conclusion

In this paper, we investigated NDN-based forwarding strategies in the IoT. After pinpointing the NDN core principle and its key strengths in handling IoT needs, we analyzed the state-of-the-art forwarding solutions that have been proposed in the literature in this area, along with an in-depth discussion focusing on the different used techniques allowing meeting IoT requirements related to wireless ad hoc environments. This was followed by a comparative study of representative NDN-based IoT forwarding schemes, where simulation outputs revealed that the native forwarding machinery of NDN falls in the broadcast storm problem, which can be reduced by the deferred forwarding technique. Also, the geographic forwarding solution registered almost the best network performances but which, however, could be less suitable to constrained-resources IoT deployments, due to its host-centric nature and the additional network topology knowledge over cost.

All in all, the conducted analysis in this paper shows that a well-designed NDN-based IoT forwarding protocol should be issued from a compromise between different techniques, such as the deferred forwarding and the topology network knowledge, in the aim to handle efficiently the challenging aspects of the IoT environments.

References

1. Djama, A., Djamaa, B., Senouci, M.R.: TCP/IP and ICN networking technologies for the internet of things: a comparative study. In: The 4th International Conference on Networking and Advanced Systems (ICNAS), Annaba, Algeria, 26–27 June 2019, pp. 1–6. IEEE (2019)
2. Jacobson, V., Smetters, D.K., Thornton, J.D., Plass, M.F., Briggs, N.H., Braynard, R.L.: Networking named content. In: Proceedings of the 5th International Conference on Emerging Networking Experiments and Technologies, pp. 1–12. ACM (2009)
3. Tseng, Y.-C., Ni, S.-Y., Chen, Y.-S., Sheu, J.-P.: The broadcast storm problem in a mobile ad hoc network. Wirel. Netw. **8**(2/3), 153–167 (2002)
4. Djama, A., Djamaa, B., Senouci, M.R.: Information-centric networking solutions for the internet of things: a systematic mapping review. Comput. Commun. **159**, 37–59 (2020)
5. NDN Forwarder Daemon. https://named-data.net/doc/NFD/current/. Accessed 21 Jan 2021
6. Zhang, L., et al.: Named data networking. ACM SIGCOMM Comp. Comm. Review **44**(3), 66–73 (2014)

7. Wang, L., Afanasyev, A., Kuntz, R., Vuyyuru, R., Wakikawa, R., Zhang, L.: Rapid traffic information dissemination using named data. In: Proceedings of the 1st ACM Workshop on Emerging Name-Oriented Mobile Networking Design - Architecture, Algorithms, and Applications, NoM '12, New York, NY, USA, pp. 7–12. Association for Computing Machinery (2012)
8. Angius, F., Gerla, M., Pau, G.: Bloogo: bloom filter based gossip algorithm for wireless NDN. In: Proceedings of the 1st ACM Workshop on Emerging Name-Oriented Mobile Networking Design - Architecture, Algorithms, and Applications, NoM '12, New York, NY, USA, pp. 25–30. Association for Computing Machinery (2012)
9. Amadeo, M., Campolo, C., Molinaro, A., Mitton, N.: Named data networking: a natural design for data collection in wireless sensor networks. In: 2013 IFIP Wireless Days (WD), pp. 1–6 (2013)
10. Intanagonwiwat, C., Govindan, R., Estrin, D., Heidemann, J., Silva, F.: Directed diffusion for wireless sensor networking. IEEE/ACM Trans. Netw. (ToN) **11**(1), 2–16 (2003)
11. Amadeo, M., Molinaro, A., Ruggeri, G.: E-CHANET: routing, forwarding and transport in information-centric multihop wireless networks. Comput. Commun. **36**(7), 792–803 (2013)
12. Yu, Y.T., Dilmaghani, R.B., Calo, S., Sanadidi, M.Y., Gerla, M.: Interest propagation in named data MANETs. In: 2013 International Conference on Computing, Networking and Communications (ICNC), pp. 1118–1122 (2013)
13. Lu, Y., Zhou, B., Tung, L.C., Gerla, M., Ramesh, A., Nagaraja, L.: Energy-efficient content retrieval in mobile cloud. In: Proceedings of the Second ACM SIGCOMM Workshop on Mobile Cloud Computing, MCC '13, New York, NY, USA, pp. 21–26. Association for Computing Machinery (2013)
14. Baccelli, E., Mehlis, C., Hahm, O., Schmidt, T.C., Wählisch, M.: Information centric networking in the IoT: experiments with NDN in the Wild. In: 1st ACM Conference on Information-Centric Networking (ICN-2014), Paris, France, September 2014. ACM (2014)
15. Aboud, A., Touati, H., Hnich, B.: Efficient forwarding strategy in a NDN-based internet of things. Clust. Comput. **22**(3), 805–818 (2019). https://doi.org/10.1007/s10586-018-2859-7
16. Gao, S., Zhang, H., Zhang, B.: Energy efficient interest forwarding in NDN-based wireless sensor networks. Mobile Information Systems 2016 (2016)
17. Amadeo, M., Campolo, C., Molinaro, A.: A novel hybrid forwarding strategy for content delivery in wireless information-centric networks. Comput. Commun. **109**, 104–116 (2017)
18. Abane, A., Daoui, M., Bouzefrane, S., Muhlethaler, P.: A lightweight forwarding strategy for named data networking in low-end IoT. J. Netw. Comput. Appl. **148**, 102445 (2019)
19. Kuai, M., Hong, X.: Location-based deferred broadcast for ad-hoc named data networking. Future Internet **11**(6), 139 (2019)
20. Named-data Project. https://named-data.net/. Accessed 21 Jan 2021

Dilated Convolutions Based 3D U-Net for Multi-modal Brain Image Segmentation

Ouissam Kemassi(✉), Oussama Maamri, Khadra Bouanane, and Ouissal Kriker

Computer Science and IT Department, Kasdi Merbah University, Ouargla, Algeria

Abstract. Several deep learning based medical image segmentation methods use U-Net architecture and its variants as a baseline model. This is because U-Net has been successfully applied to many other tasks. It was noticed that the U-Net-based models are unable to extract features for segmenting small masks or fine edges.

To overcome this issue, we propose a new 3D U-Net-based model, baptized Y- Net. In this model, we make use of dilated convolution which has shown its effectiveness in grasping different features at different scales. This allows us to capture more information from small anatomical parts.

Our model is assessed on MRbrains13 dataset for brain tissue segmentation task. Compared to the traditional UNet 3D, the obtained results show that the proposed model performs better in segmenting cerebrospinal fluid, white matter and gray matter tissues.

Keywords: Brain image segmentation · UNet 3D · MRI modalities · Dilated convolution

1 Introduction

In the field of medical image analysis, image segmentation is considered as the most challenging task that aims to identify the pixels of organs or lesions from background medical images produced from the process of medical imaging.

Image segmentation is frequently utilized in brain MRI analysis for measuring and visualizing brain structures, assessing brain changes, identifying diseased regions, surgical planning, and image-guided therapies.

Due to its importance, the automation of this task has been extensively studied over decades, where researchers aim to address this problem using new methods.

One of the most successful approaches that have emerged this last decade is deep learning-based models, which have performed a significant improvement in segmentation accuracy. These models are mainly designed using basic architectures like convolutional neural network (CNN), fully convolutional neural network (FCNN) [9] and U-net [14].

B. Lejdel et al. (Eds.): AIAP 2021, LNNS 413, pp. 428–436, 2022.
https://doi.org/10.1007/978-3-030-96311-8_39

In particular, it was recorded in many works that, because of the skip connections between the encoder and decoder part in its architecture, U-Net based models achieve high performance in segmenting medical images [1,17,18].

However, it was noticed that the U-Net-based models are unable to extract features for segmenting small masks or fine edges. Which makes the model unable to grasp small details and thus cannot capture accurately tiny anatomical structures [16].

To overcome this issue, we propose a new 3D UNet based model, baptized YNet.

In this model, we make use of dilated convolution which has shown its effectiveness in grasping features at different scales [6]. This allows to capture more information from small anatomical parts and thus, enhance the performance of the model.

In the remainder of this paper we first present in Sect. 2, some recent U-Net based models that were proposed to tackle brain image segmentation. We give a detailed description of the proposed model in Sect. 3 and in Sect. 4, experimental results are provided.

2 Related Work

Based on U-Net architecture, several previous works have proposed efficient models for brain image segmentation task.

Rehman et al. [13] proposed a 2D image segmentation method, called BU-Net, to contribute in brain tumor segmentation research, where residual extended skip (RES) and wide context (WC) are used along with the customized loss function in the baseline U-Net architecture. These modifications allow finding more differing features to get better segmentation performance.

For the task of Brain tumor segmentation, BU-Net was assessed on the high-grade glioma (HGG) datasets of the BraTS2017 Challenge as well as the test datasets of the BraTS 2017 and 2018 Challenge dataset.

Ozgun et al. [4], proposed a model for volumetric segmentation that learns from sparsely annotated volumetric images. This network extends the previous U-Net architecture from Ronneberger et al. [14] by replacing all 2D operations with their 3D counterparts.

To tackle the poor performance of the U-net architecture when segmenting small structures, Valanarasu et al. [16] proposed Kiunet, an overcompleted convolutional network, which achieves an improved performance comparing to all the recent methods with an additional benefit of fast convergence.

KiU-Net and KiU-Net3D were applied on five different datasets covering various image modalities. Brain MRI images from BraTS2020 Challenge were used to assess KiU-Net3D.

Chen et al. [2] proposed a novel separable 3D U-Net architecture using separable 3D convolutions. This method achieved a mean enhancing tumor, whole tumor, and tumor core Dice scores in Preliminary results on BraTS 2018 validation set.

This novel separable 3D U-Net architecture succeeded to overcome the issue of 2D convolutions, which cannot make a full use of the spatial information of volumetric medical images, by using 3D convolutions in such a way to prevent the high expensive computational cost and memory demand.

Peng et al. [11] proposed a Multi-Scale 3D U-Net architecture, which uses several U-Net blocks to capture long distance spatial information at different resolutions in order to extract and utilize sufficient features. The model was evaluated on the BraTs2015 dataset for the Brain tumor segmentation task.

Feng and Meyer [5] proposed to use a 3D U-Net structure on extracted 2D/3D patches to predict the class labels for all pixels in the patch, in order to improve the performance of U-Net for brain tumor images. Authors made use of BraRs2017 datasets to assess their model.

On the other hand, we can notice that many additional operations have been applied in order to enhance the performance of the proposed models. Among these operations, dilated convolution has shown its ability and potential in improving segmentation results. The advantage in using dilated convolution in image segmentation has been highlighted in [6], where it was shown that dilated convolution permits the increase of the receptive fields without losing resolution or coverage. This makes the model more efficient in capturing different features at different scales. Therefore, it can be very beneficial for segmenting small anatomical parts in biomedical domain [7,19].

In literature, dilated convolution has been used to improve the segmentation accuracy. For instance, Zhang et al. [19] proposed a new end-to-end network based on ResNet and U-Net, in which the usual pooling layer is replaced with convolutional layer which permits to reduce the information loss extent.

Lei et al. [7] proposed a dual aggregation network to adaptively aggregate different information from infant's brain MRI images. Based on 3D U-Net, authors used the dilated convolution pyramid down-sampling module to solve the problem of loss of spatial information on the down-sampling process. This model achieved the first place in Iseg-2019 challenge.

Li et al. [8] proposed a novel model called Multi modal aggregation Network (MMAN). This model is an innovative deep network that was designed to extract and aggregate multiscale features of brain tissues from multi-modality MR images in order to get a precise and perfect segmentation. This network is based on an inception model with a variable dilation.

In this paper, we propose a novel model, baptized Y-Net. The main purpose is to make use of dilated convolution benefits to enhance a U-Net architecture. the model is detailed in the following section.

3 Description of our model

Larger kernel receptive fields can increase network capacity to capture spatial context, which is profitable to reconstruct big and complex edge structures. However, using common convolutions require a large number of parameters to expand their receptive fields.

Dilated convolutions can then be used to increase the receptive field with a linearly increasing number of parameters [12]. They work by introducing "holes" in the kernel by inserting zeros into defined gaps to expand receptive field size.

Dilated convolutions can then view larger input image portions without requiring a pooling layer, resulting in no spatial dimension loss and reduced computational time [3].

The dilated convolution with the dilation rate r and a filter $w[s]$ with size s is formulated as [12]:

$$y[i] = \sum_{n=1}^{s} x[i + r.s]w[s].$$

where $x[i]$ denotes a 1D signal and $y[i]$ is the output of a dilated convolution. The standard convolution is then considered as a particular case of dilated convolution with a dilation rate $r = 1$.

Figure 1 illustrates the dilated convolution operation with rates $r = 1, r = 2$ and $r = 3$, respectively.

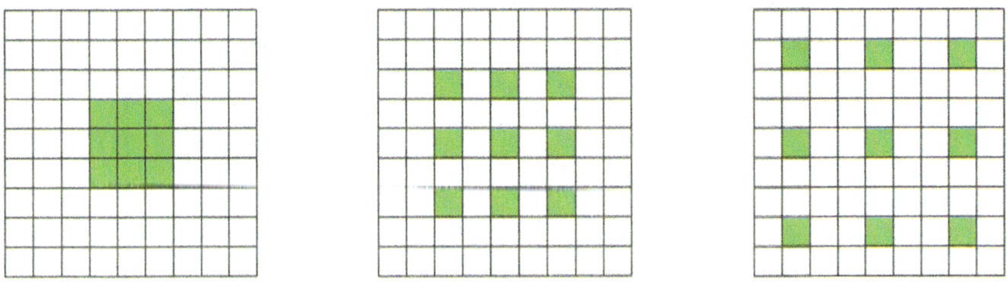

Fig. 1. Dilated convolution operation with a 3×3 kernel size. On the left, the dilation rate is $r = 1$, in the middle $r = 2$ and in the right $r = 3$.

In Our model, we make use of two convolution blocks with different convolution rates, as described below.

3.1 Network Architecture

Our proposed model contains three components: a regular encoder, a dilated encoder and a decoder.

- The regular encoder comprises four downsampling layers. Each layer uses two convolutions "conv3D" with a kernel size of $3 \times 3 \times 3$ voxels per block. The rectified linear units (ReLUs) is used as an activation function and a subsequent 3D-max pooling operation "MaxPool3D" is then performed.
- The dilated encoder includes four downsampling layers, where the usual convolution operation is replaced by dilated convolution with a predefined rate. This allows our model to enlarge the receptive field and thus capture fine details and accurate edges in the image.

In this encoder, each layer uses two convolutions "conv3D" with a kernel size of $3 \times 3 \times 3$ voxels per block and a dilation-rate size of $2 \times 2 \times 2$. A rectified linear units (ReLUs) is used as an activation function.
- The output features of the regular and the dilated encoders are aggregated via an additional block, then the result of this operation is inputted to the decoder component.
- The decoder comprises four upsampling convolution layers where each layer uses two convolutions with a kernel size of $3 \times 3 \times 3$ voxels per block, the rectified linear units (ReLUs) as activation functions.

The model architecture is illustrated in Fig. 2.

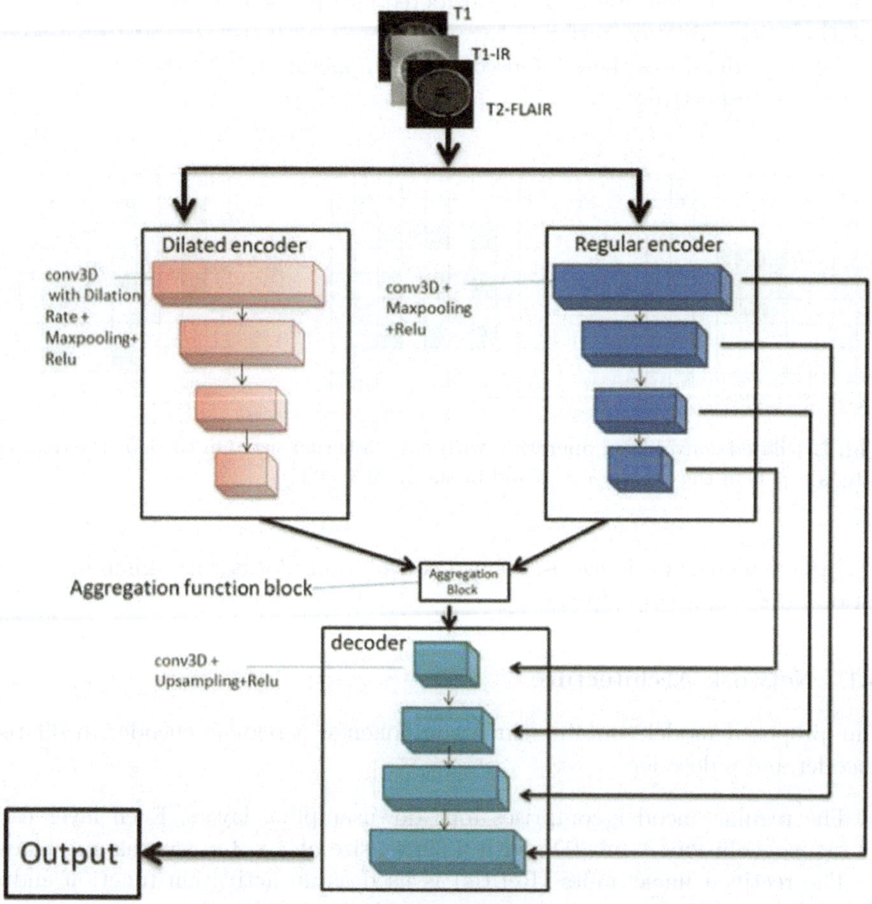

Fig. 2. Y-Net architecture

3.2 Training

The enhancement of training phase of our model is due to the aggregation of the different learned features that are provided by regular and dilated encoders.

This improvement in efficiency is due to dilated convolution that allows us to deal with object dependencies on different scales without reducing the image resolution.

The model is trained by minimizing the categorical cross entropy between prediction and label.

$$L = -\sum_{i \in \Omega} \sum_{c} y_{ic} \log p_{ic} + (1 - y_{ic}) \log(1 - p_{ic}) \quad (1)$$

where

- p_{ic} is the probability that the $i$the voxel belongs to the c-th class.
- y_{ic} is the corresponding ground truth.
- Ω denotes all pixels in predicted segmentation result p.

4 Experimental Results

In this experiment, we assess the performance of Y-Net compared to the U-Net 3D [4] for the task of brain tissue segmentation from MRbrainS13 dataset.

All the experiments were performed via Google Colab, since this service provides a powered GPU. Specifically, a processor Intel(R) Xeon(R) CPU 2.20 GHz and 13G RAM.

4.1 Data

In the MRBrainS Challenge, MRI data was recorded by multisequence (multi-modality) 3T MRI brain scans including T1-weighted scan, T1-weighted IR scan and T2-FLAIR scan [10].

The MRI data of 5 subjects (2 males and 3 females, varying degrees of atrophy and white matter lesions) were provided as training database. The MRI data of another 15 subjects were collected as the testing database of MRBrainS Challenge. The size of those modalities is $240 \times 240 \times 48$.

The segmented tissues are labeled as follows:

0: Background (everything outside the brain)
1: Cerebrospinal fluid (including ventricles)
2: Gray matter (cortical gray matter and basal ganglia)
3: White matter (including white matter lesions)

4.2 Data Augmentation

Since results of the testing dataset are not available, we make use of the training dataset to train and assess our model. For this purpose, we train our model and U-Net 3D on the data of 3 subjects then use the two other subjects' data for assessment.

We increase the number of training and testing data by extracting sub-volumes of size $64 \times 64 \times 16$ and feed them to the models by patches.

Y-Net and U-Net 3D are then trained for 40 epochs on patches of 5 samples each.

4.3 Evaluation Metrics

To measure the performance of the two models, we use the Dice Coefficient(DC), which is one of the most popular evaluation segmentation performance metrics. Dice Coefficient is a similarity measurement technique based on a statistical approach. It is defined by the overlap region between the predicted segmentation results and ground truth, its result is expressed as a percentage that varies from 0% (total mismatch) to 100% (perfect match) [15]. Formally, the DC formula is expressed as follows:

$$DC(G,S) = 2\frac{G \cap S}{G + S}.100\%.$$

4.4 Results

Results of the training of Y-Net and U-Net 3D for 40 epochs are summarized in Table 1.

As we can notice, Y-Net outperforms U-Net 3D in the segmentation of the 3 tissues, cerebrospinal fluid tissue (CSF), Gray Matter (GM) and White Matter (WM) tissues.

Table 1. Comparison of dice score for brain tissue segmentation (MRbrainS13 dataset).

	Cerebrospinal fluid (CSF)	Gray matter (GM)	White matter (WM)
Y-Net	64.92	71.02	80.00
U-Net 3D	63.19	71.15	79.20

5 Conclusion

In this paper, we presented a new U-Net based model for brain tissue segmentation. Y-NET aims to overcome the inability of U-Net models in extracting features for segmentation of small structures. To do so, we proposed to exploit the benefits of dilated convolution by integrating a dilated encoder to the usual architecture. Information from different convolution blocks is then aggregated and used to extract meaningful features at different scales.

Experimental results show the effectiveness of the proposed model for the segmentation of cerebrospinal fluid tissue, white matter and gray matter tissues, compared to U-NET 3D, on the MRbrains13 dataset.

However, In order to corroborate the model, further experiments must be conducted for brain tissue segmentation task and other tasks as well.

References

1. Chang, J., Zhang, X., Ye, M., Huang, D., Wang, P., Yao, C.: Brain tumor segmentation based on 3D UNET with multi-class focal loss. In: 2018 11th International Congress on Image and Signal Processing, BioMedical Engineering and Informatics (CISP-BMEI), pp. 1–5. IEEE (2018)
2. Chen, W., Liu, B., Peng, S., Sun, J., Qiao, X.: S3D-UNet: separable 3D U-Net for brain tumor segmentation. In: Crimi, A., Bakas, S., Kuijf, H., Keyvan, F., Reyes, M., van Walsum, T. (eds.) BrainLes 2018. LNCS, vol. 11384, pp. 358–368. Springer, Cham (2019). https://doi.org/10.1007/978-3-030-11726-9_32
3. Chim, S., Lee, J.G., Park, H.H.: Dilated skip convolution for facial landmark detection. Sensors **19**(24), 5350 (2019)
4. Çiçek, Ö., Abdulkadir, A., Lienkamp, S.S., Brox, T., Ronneberger, O.: 3D U-Net: learning dense volumetric segmentation from sparse annotation. In: Ourselin, S., Joskowicz, L., Sabuncu, M.R., Unal, G., Wells, W. (eds.) MICCAI 2016. LNCS, vol. 9901, pp. 424–432. Springer, Cham (2016). https://doi.org/10.1007/978-3-319-46723-8_49
5. Feng, X., Meyer, C.: Patch-based 3D U-NET for brain tumor segmentation. In: International Conference on Medical Image Computing and Computer-Assisted Intervention (MICCAI), pp. 67–72 (2017)
6. Hamaguchi, R., Fujita, A., Nemoto, K., Imaizumi, T., Hikosaka, S.: Effective use of dilated convolutions for segmenting small object instances in remote sensing imagery. In: 2018 IEEE Winter Conference on Applications of Computer Vision (WACV), pp. 1442–1450. IEEE (2018)
7. Lei, Z., Qi, L., Wei, Y., Zhou, Y.: Infant brain MRI segmentation with dilated convolution pyramid down sampling and self-attention. arXiv preprint arXiv:1912.12570 (2019)
8. Li, J., Yu, Z.L., Gu, Z., Liu, H., Li, Y.: MMAN: multi-modality aggregation network for brain segmentation from MR images. Neurocomputing **358**, 10–19 (2019)
9. Long, J., Shelhamer, E., Darrell, T.: Fully convolutional networks for semantic segmentation. In: Proceedings of the IEEE Conference on Computer Vision and Pattern Recognition, pp. 3431–3440 (2015)
10. Mendrik, A.M., et al.: MRBrainS challenge: online evaluation framework for brain image segmentation in 3T MRI scans. In: Computational Intelligence and Neuroscience 2015 (2015)
11. Peng, S., Chen, W., Sun, J., Liu, B.: Multi-scale 3D U-Nets: an approach to automatic segmentation of brain tumor. Int. J. Imaging Syst. Technol. **30**(1), 5–17 (2020)
12. Perone, C.S., Calabrese, E., Cohen-Adad, J.: Spinal cord gray matter segmentation using deep dilated convolutions. Sci. Rep. **8**(1), 1–13 (2018)

13. Rehman, M.U., Cho, S., Kim, J.H., Chong, K.T.: BU-Net: brain tumor segmentation using modified U-Net architecture. Electronics **9**(12), 2203 (2020)
14. Ronneberger, O., Fischer, P., Brox, T.: U-Net: convolutional networks for biomedical image segmentation. In: Navab, N., Hornegger, J., Wells, W.M., Frangi, A.F. (eds.) MICCAI 2015. LNCS, vol. 9351, pp. 234–241. Springer, Cham (2015). https://doi.org/10.1007/978-3-319-24574-4_28
15. Tiu, E.: Metrics to evaluate your semantic segmentation model. Towards datascience, recuperado de (2019). https://towardsdatascience.com/metrics-to-evaluateyour-semantic-segmentation-model-6bcb99639aa2#:~: text= Simply% 20put% 2C% 20the% 20Dice% 20Coefficient, of% 2 0pixels% 20in% 20both% 20images 8
16. Valanarasu, J.M.J., Sindagi, V.A., Hacihaliloglu, I., Patel, V.M.: KiU-Net: overcomplete convolutional architectures for biomedical image and volumetric segmentation. arXiv preprint arXiv:2010.01663 (2020)
17. Yang, B., Zhang, W.: FD-FCN: 3D fully dense and fully convolutional network for semantic segmentation of brain anatomy. arXiv preprint arXiv:1907.09194 (2019)
18. Zeng, G., Zheng, G.: Multi-stream 3D FCN with multi-scale deep supervision for multi-modality isointense infant brain MR image segmentation. In: 2018 IEEE 15th International Symposium on Biomedical Imaging (ISBI 2018), pp. 136–140. IEEE (2018)
19. Zhang, Q., Cui, Z., Niu, X., Geng, S., Qiao, Y.: Image segmentation with pyramid dilated convolution based on ResNet and U-Net. In: Liu, D., Xie, S., Li, Y., Zhao, D., El-Alfy, E.S. (eds.) ICONIP 2017. LNCS, vol. 10635, pp. 364–372. Springer, Heidelberg (2017). https://doi.org/10.1007/978-3-319-70096-0_38

Image Restoration Using Proximal-Splitting Methods

Nacira Diffellah[1(✉)], Rabah Hamdini[2], and Tewfik Bekkouche[3]

[1] Department of Electronics, Faculty of Technology, ETA Laboratory, University of Bordj Bou Arreridj, 34000 Bordj Bou Arreridj, Algeria
nacira.diffellah@univ-bba.dz

[2] Department of Automatics, Faculty of Technology, SET Laboratory, Univ Saad Dahlab Blida, 09000 Blida, Algeria

[3] Department of Electromechanics, Faculty of Technology, ETA Laboratory, University of Bordj Bou Arreridj, 34000 Bordj Bou Arreridj, Algeria

Abstract. In this paper, we focus on giving two fixed-point-like methods, using proximal operators, called forward-backward and Douglas-Rachford, for solving the restoration problem for grayscale images corrupted with Gaussian noise model. We discuss how to evaluate proximal operators and provide an example in reconstructed image. The main idea is to choose the classic variational model $TVL1$ for recovering a true image u from an observed image f contaminated with Gaussian noise. The objective function is a sum of two convex terms: the ℓ_1-norm data fidelity and the total variational regularization. The first term forces the final image to be not too far away from the initial image and the second term performs actually the noise reduction. Experimental results prove the efficiency of the proposed work by performing some test by changing the noise levels applied to different images. We notice that the Peak Signal-to-Noise Ratio ($PSNR$) is used to evaluate the quality of the restored images.

Keywords: Proximal operator · Fixed point · Splitting · Forward-backward algorithm · Douglas-Rachford algorithm · Image restoration · Total variation · ℓ_1-norm · PSNR

1 Introduction

Total variation [1] represents a powerful regularity measure in image restoration for recovering piecewise homogeneous areas with sharp edges [2,3]. Image restoration techniques are used to make the corrupted image as similar as that of the original image. Various methods available for image restoration such as inverse filter, Weiner filter, constrained least square filter, blind deconvolution method etc., and Many optimization approaches for regularization-based image inverse problems have been developed [4–6], our framework is to solve the restoration problem in the form of a convex optimization, we use in particular a proximal

B. Lejdel et al. (Eds.): AIAP 2021, LNNS 413, pp. 437–446, 2022.
https://doi.org/10.1007/978-3-030-96311-8_40

algorithm [7–9] that introduces the proximal operators of the objective terms. There are various restoration techniques for noise removal based on proximal filtering, forward-backward [10–12], and Douglas-Rachford algorithms [13,14] fall into the class of proximal splitting algorithms.

The remainder of the paper is organized as follows: first, in Sect. 2 definition and some properties of proximal operator, fixed points, Forward- backward splitting and Douglas Rachford splitting are provided. Then, in Sect. 3 we present the considered restoration problem using proximal splitting algorithms. Numerical results are reported in Sect. 4 where beneficial analysis are done. Some conclusions and future works are drawn in Sect. 5.

2 Proximal Splitting Methods

2.1 Proximal Operator

The proximal algorithm or proximal point method can be understood in many different ways, but we will start by thinking about as a regularizer that naturally damps its influence as the iterations proceed, and as a twist on gradient descent.

In proximal algorithms, the base operation is evaluating the proximal operator of a function, which involves solving a small convex optimization problem. Let $f : \mathbb{R}^n \to \mathbb{R} \cup \{+\infty\}$ be a closed proper convex function. The proximal operator $prox_{\gamma f}x : \mathbb{R}^n \to \mathbb{R}^n$ of f with parameter $\gamma > 0$ is defined by:

$$prox_{\gamma f}x = arg\min_{x_0} \left\{ f(x_0) + \frac{1}{2\gamma} \|x_0 - x\|_2^2 \right\} \tag{1}$$

where $\|.\|_2$ is the usual Euclidean norm.

The parameter γ controls the extent to which the proximal operator maps points towards the minimum of f, with larger values of γ associated with mapped points near the minimum, and smaller values giving a smaller movement towards the minimum.

The function minimized on the righthand side is strongly convex and not everywhere infinite, so it has a unique minimizer for every $x \in \mathbb{R}^n$.

$prox_f(x)$ is a point that compromises between minimizing f and being near to x. For this reason, $prox_f(x)$ is sometimes called a proximal point of x with respect to f. In $prox_{\gamma f}(x)$, the parameter γ can be interpreted as a relative weight or trade-off parameter between these terms.

To keep notation light, we write $\|.\|_2$ rather than $\|.\|$.

Throughout this paper, when we refer to the proximal operator of a function, the function will be assumed to be closed proper convex, and it may take on the extended value $+\infty$.

2.2 Fixed Points

The fixed points of the proximal operator of f are precisely the minimizers of f. In other words,

$$prox_{\gamma f}(x^*) = x^* \tag{2}$$

if and only if x^* minimizes f.

This implies a close connection between proximal operators and fixed point theory, and suggests that proximal algorithms can be interpreted as solving optimization problems by finding fixed points of appropriate operators.

2.3 Forward-Backward Splitting

Let $f_1 \in \mathbb{R}^n$, let $f_2 : \mathbb{R}^n \to \mathbb{R}$ be convex and differentiable with a β Lipschitz continuous gradient ∇f_2 , i.e., $\forall (x, y) \in \mathbb{R}^n \times \mathbb{R}^n$

$$\|\nabla f_2(x) - \nabla f_2(y)\| \leq \beta \|x - y\| \tag{3}$$

Where $\beta \in]0, +\infty[$. Suppose that $f_1(x) + f_2(x)$ as $\|x\| \to +\infty$. The problem is to

$$\underset{x\in\mathbb{R}^n}{\text{minimize}} \; f_1(x) + f_2(x) \tag{4}$$

Solution of Eq. (4) is characterized by the fixed point equation:

$$x = prox_\gamma f_1 (x - \gamma \nabla f_2(x)) \tag{5}$$

where $\gamma \in]0, +\infty[$. Equation 5 can be written as:

$$\begin{cases} x = prox_{\gamma f_1}(y) \\ y = x - \gamma \nabla f_2(x) \end{cases} \tag{6}$$

which motivates the following scheme (Algorithm 1)

Algorithm 1. Forward-Backward (Explicit-Implicit)

1: **Fix** $\varepsilon \in \left]0, \min\left\{1, \frac{1}{\beta}\right\}\right[$ **and** $x_0 \in H$
2: **For** $n = 0, 1, \ldots$ **:**

- $\gamma_n \in \left[\varepsilon, \frac{2}{\beta} - \varepsilon\right]$
- **Forward (implict) step :** $y_n = x_n - \gamma_n \nabla f_2(x_n)$
- $\tau_n \in [\varepsilon, 1]$
- **Backward (explict) step :** $x_{n+1} = x_n + \tau_n (prox_{\gamma_n f_1} y_n - x_n)$

2.4 Douglas Rachford Splitting

Let $f_1 \in \mathbb{R}^n$, $f_2 \in \mathbb{R}^n$ such that $\arg\min\; f_1 + f_2 \neq \varnothing$ and $f_1(x) + f_2(x)$ as $\|x\| \to +\infty$. The problem is to

$$\underset{x\in\mathbb{R}^n}{\text{minimize}} \; f_1(x) + f_2(x) \tag{7}$$

Solution of Eq. (7) is characterized by the two-level condition:

$$\begin{cases} x = prox_{\gamma f_2} y \\ prox_{\gamma f_2} y = prox_{\gamma f_1} (2prox_{\gamma f2} y - y) \end{cases} \tag{8}$$

which motivates the following scheme (Algorithm 2)

Algorithm 2. Douglas Rachford

1: **Fix** $\varepsilon \in]0,1[$, $\gamma > 0$ **et** $x_0 \in H$
2: **For** $n = 0, 1, \ldots$ **:**

- $y_n = prox_{\gamma f_2} x_n$
- $\tau_n \in [\varepsilon, 2 - \varepsilon]$
- $x_{n+1} = x_n + \tau_n \left(proy_{\gamma f_1} \left(2y_n - x_n \right) - y_n \right)$

3 Reconstructed Image Using Proximal Algorithm

3.1 Reconstructed Image Using Forward-Backward Splitting Algorithm

It is unconstrained problem with energy function split in two components:

$$\text{minimise } E = E_{reg}\left(u\right) + E_{data}\left(u\right) \tag{9}$$

The application of forward- backward algorithm to the problem (9) is a 4 steps process:

- Step 1: Create the energy that describe the quality image u

$$\min_u E\left(u\right) = \min_u \left\{ \frac{\lambda}{2} \left\| Ku - f \right\|^2 + \left\| \nabla u \right\| \right\} \tag{10}$$

with

$$E_{data}\left(u\right) = \frac{1}{2} \left\| K.u - f \right\|^2 \tag{11}$$

$$E_{reg}\left(u\right) = \left\| \nabla u \right\| \tag{12}$$

- Step 2: Define the gradient of $E_{data}\left(u\right)$:

$$\nabla E_{data}\left(u\right) = \lambda K^T \left(Ku - f \right) \tag{13}$$

- Step 3: Define the proximal operator of $E_{reg}\left(u\right)$

$$prox_{\gamma E_{reg}}\left(\nabla u\right) = \max\left(0, 1 - \frac{\gamma}{\left|\nabla u\right|}\right) \nabla u \tag{14}$$

- Step 4: Apply forward backward algorithm. Computes the following sequence of points:

$$u_{i,j}^{n+1} = u_{i,j}^{n} - \tau_n . prox_{\gamma E_{reg}} \left(u_{i,j}^{n} - \gamma \left(\nabla E_{data}\left(u\right) \right)_{i,j}^{n+1} \right) \tag{15}$$

$\tau_n > 0$ is step size.

3.2 Reconstructed Image Using Douglas-Rachford Splitting Algorithm

3.2.1 Problem with Constraints

In this section, the Douglas-Rachford splitting method is applied to solve the problem (16). We follow the steps below:

- Step 1: Create an image reconstruction problem with constraints that describe the quality image u

$$\begin{aligned} &\arg\min_{x} |\nabla u| \\ &such\ that \\ &\|Ku - f\| \leq \varepsilon \end{aligned} \tag{16}$$

with

$$E_{data}(u) = \|K.u - f\|^2 \tag{17}$$

$$E_{reg}(u) = \|\nabla u\| \tag{18}$$

where ϵ is an estimated upper bound on the noise level.

- Step 2: Define the proximal operator of $E_{reg}(u)$

$$prox_{\gamma E_{reg}}(\nabla u) = \max\left(0, 1 - \frac{\gamma}{|\nabla u|}\right) \nabla u \tag{19}$$

- Step 3: Define the proximal operator of $E_{data}(u)$
$E_{data}(u)$ is the indicator function of the set C define by $\|Ku - f\| \leq \varepsilon$. We define the *prox* of $E_{data}(u)$ as:

$$prox_{E_{data}}(z) = \arg\min_{u} \frac{1}{2}\|u - z\|^2 + i_C(u) \tag{20}$$

with $i_C(u)$ the indicator function of a set C gicen by the Eq. (21):

$$i_C(u) = \begin{cases} 0 & u \in C \\ +\infty & otherwise \end{cases} \tag{21}$$

This previous problem has an identical solution as:

$$\begin{aligned} &\arg\min_{x} \|u - z\|^2 \\ &such\ that \\ &\|Kz - f\| \leq \varepsilon \end{aligned} \tag{22}$$

It is simply a projection on the $B2 - ball$.

3.2.2 Problem Without Constraints

The application of Douglas- Rachford algorithm to the problem without constraints is a 4 steps process:

- Step 1: Create the energy that describe the quality image u

$$\min_u E(u) = \min_u \left\{ \frac{\lambda}{2} \|Ku - f\|^2 + \|\nabla u\| \right\} \tag{23}$$

with

$$E_{data}(u) = \frac{1}{2} \|K.u - f\|^2 \tag{24}$$

$$E_{reg}(u) = \|\nabla u\| \tag{25}$$

- Step 2: Define the proximal operator of $E_{data}(u)$

$$\Pr ox_{\gamma E_{data}} u = \frac{\lambda}{2} K^* (I + \gamma K^* K)^{-1} \tag{26}$$

- Step 3: Define the proximal operator of $E_{reg}(u)$

$$prox_{\gamma E_{reg}}(\nabla u) = \max\left(0, 1 - \frac{\gamma}{|\nabla u|}\right) \nabla u \tag{27}$$

- Step 4: Apply Douglas-Rachford algorithm: computes the following sequence of points, for any fixed $\gamma > 0$ and stepsize $\tau_n \in [\varepsilon, 2 - \varepsilon]$

$$y_n = prox_{\gamma E_{data}} u_n \tag{28}$$

$$u_{n+1} = u_n + \tau_n \left(proy_{\gamma E_{reg}} (2y_n - u_n) - y_n \right) \tag{29}$$

4 Numerical Results

We provide numerical results for five test images such as Cameraman, Lena, House, Boat, and Pepper (see Fig. 1), the pixel size of five test images is 256×256. (https://github.com/jianzhangcs/ISTA-Net/tree/master/Test_Image). A range of noise levels 10%, 20% and 30% are tested.

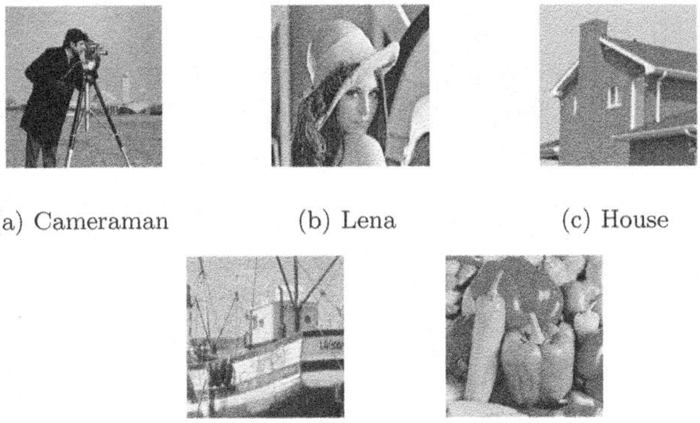

(a) Cameraman (b) Lena (c) House

(d) Boat (e) Pepper

Fig. 1. True images for Cameraman, Lena, House, Boat, and Pepper

The performance of all the filtering techniques are compared on the basis of the statistical parameter such as $PSNR$. The term Peak Signal-to-Noise Ratio ($PSNR$) is an expression for the ratio between the maximum possible value (power) of a signal and the power of distorting noise that affects the quality of its representation. The mathematical representation of the $PSNR$ is as follows:

$$PSNR = 20\log_{10}\left(\frac{256}{\sqrt{MSE}}\right) \tag{30}$$

where the MSE (Mean Squared Error) is:

$$MSE = \frac{1}{MN}\sum_{i=0}^{M-1}\sum_{j=0}^{N-1}\left|u\left(i,j\right) - \hat{u}\left(i,j\right)\right|^2 \tag{31}$$

u represents the matrix data of our original image, $\hat{u}$ represents the matrix data of our degraded image in question, M represents the numbers of rows of pixels of the images and i represents the index of that row, N represents the number of columns of pixels of the image and j represents the index of that column.

The peak signal-to-noise ratio (PSNR) between the restored image and the original image is selected as the performance index (Table 1).

Some visual results of the recovered images for the three algorithms are presented in Fig. 2. $DR2$ and FB not only removes lots of noise, but also preserves more details. We can see that $DR2$ and FB obtain a best compromise between noise-removal and detail-preservation compared with the $DR1$.

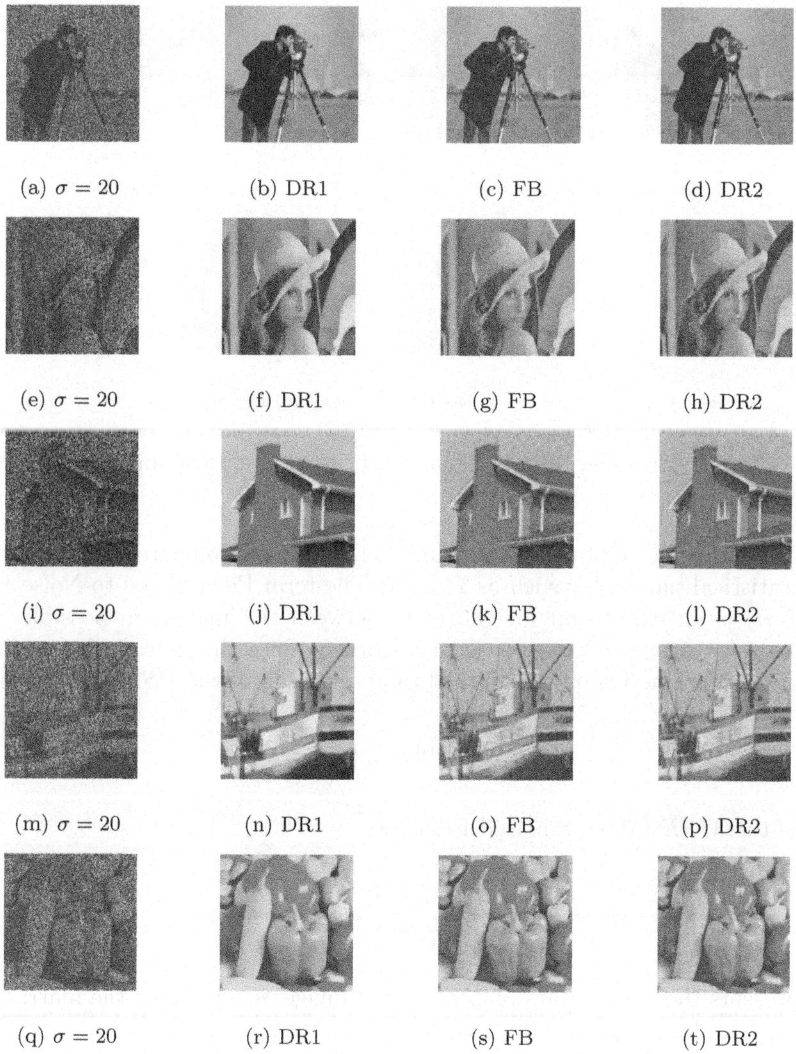

(a) $\sigma = 20$ (b) DR1 (c) FB (d) DR2

(e) $\sigma = 20$ (f) DR1 (g) FB (h) DR2

(i) $\sigma = 20$ (j) DR1 (k) FB (l) DR2

(m) $\sigma = 20$ (n) DR1 (o) FB (p) DR2

(q) $\sigma = 20$ (r) DR1 (s) FB (t) DR2

Fig. 2. Restored images with different proximal algorithms ($\sigma = 20$) for 'cameraman', 'lena', 'house', 'boat' and 'peppers'.

Table 1 lists the $PSNR$ values of restored images with random-valued additive gaussian noise level 10%, 20% and 30% for 'cameraman', 'lena', 'house', 'boat' and 'peppers' respectively. It presents the results of the three comparative denoising proximal algorithms FB, $DR1$ and $DR2$ for all test images.

Table 1. PSNR values of restored image for different percentage of additive noise

Image	Method	PSNR		
		10% Noise	20% Noise	30% Noise
Cameraman	$DR1$	26.4734	24.9357	24.2275
	FB	26.4920	25.7134	23.9520
	$DR2$	26.5372	25.7134	23.9520
Lena	$DR1$	26.8591	25.4343	23.5476
	FB	26.9945	26.1990	23.8520
	$DR2$	26.9116	26.1991	23.8520
House	$DR1$	30.5397	28.8260	26.6858
	FB	31.0293	29.0470	25.4867
	$DR2$	31.0293	29.0470	25.4867
Boat	$DR1$	27.3255	24.7160	23.4856
	FB	27.7091	25.8337	23.4413
	$DR2$	27.7285	25.8337	23.4413
Pepper	$DR1$	26.5146	25.2657	23.8551
	FB	26.8207	25.6298	23.4097
	$DR2$	26.8208	25.6298	23.4097

Its well-known that a higher $PSNR$ value provides a higher image quality. By comparing the $PSNR$ in Table 1, we can infer that $DR2$ and FB gives a better value of $PSNR$ than $DR1$.

5 Conclusion

Restoration techniques are used to improve the quality of corrupted image to ward its original form. In this paper proximal algorithms are addressed to image restoration. Three convex regularization approaches based on proximal algorithms: FB, $DR1$ and $DR2$ has been studied, developed, implemented then compared in order to restore data images degraded by additive noise. To improve the image quality, the performance of all filtering techniques are compared on the basis of the statistical parameters such as Peak Signal to Noise Ratio ($PSNR$) based on Mean Square Error (MSE).

Futures work include the investigation of this study to solve other problems using the proximal splitting algorithm like image denoising if K is identity, image deblurring if K is a blur, image inpainting if K is a diagonal matrix, where whose diagonal entries are either 0 or 1, keeping or killing the corresponding pixels or resolve compressive sensing problem when K is a set of random projections.

Acknowledgment. The authors would like to thank the organizers of the conference AIAP'2021 and the anonymous reviewers for their valuable comments and suggestions which greatly improved the quality of the paper. Authors would like to thank too the General Directorate for Scientific Research and Technological Development of the Algerian Republic in general and the ETA research laboratory of Bordj Bou Arreridj University in particular, for all material and financial support to accomplish this work.

References

1. Rudin, L.I., Osher, S., Fatemi, E.: Nonlinear total variation based noise removal algorithms. Physica D **60**(1–4), 259–268 (1992)
2. Weiss, P., Blanc-Féraud, L., Aubert, G.: Efficient schemes for total variation minimization under constraints in image processing. SIAM J. Sci. Comput. **31**(3), 2047 2080 (2009)
3. Hütter, J.-C., Rigollet, P.: Optimal rates for total variation denoising. In: Conference on Learning Theory, PMLR (2016)
4. Peyré, G., Bougleux, S., Cohen, L.: Non-local regularization of inverse problems. In: Forsyth, D., Torr, P., Zisserman, A. (eds.) ECCV 2008. LNCS, vol. 5304, pp. 57–68. Springer, Heidelberg (2008). https://doi.org/10.1007/978-3-540-88690-7_5
5. Goldstein, T., Osher, S.: The split Bregman method for L1-regularized problems. SIAM J. Imag. Sci. **2**(2), 323–343 (2009)
6. Afonso, M.V., Bioucas-Dias, J.M., Figueiredo, M.A.T.: Fast image recovery using variable splitting and constrained optimization. IEEE Trans. Image Process. **19**(9), 2345–2356 (2010)
7. Combettes, P.L., Wajs, V.R.: Signal recovery by proximal forward-backward splitting. Multiscale Model. Simul. **4**(4), 1168–1200 (2005)
8. O'Connor, D., Vandenberghe, L.: Primal-dual decomposition by operator splitting and applications to image deblurring. SIAM J. Imag. Sci. **7**(3), 1724–1754 (2014)
9. Condat, L., et al.: Proximal splitting algorithms: relax them all. arXiv preprint arXiv:1912.00137 (2019)
10. Daubechies, I., Defrise, M., De Mol, C.: An iterative thresholding algorithm for linear inverse problems with a sparsity constraint. Commun. Pure Appl. Math. J. Issued Courant Inst. Math. Sci. **57**(11), 1413–1457 (2004)
11. Figueiredo, M.A.T., Nowak, R.D.: An EM algorithm for wavelet-based image restoration. IEEE Trans. Image Process. **12**(8), 906–916 (2003)
12. Bect, J., Blanc-Féraud, L., Aubert, G., Chambolle, A.: A l^1-unified variational framework for image restoration. In: Pajdla, T., Matas, J. (eds.) ECCV 2004. LNCS, vol. 3024, pp. 1–13. Springer, Heidelberg (2004). https://doi.org/10.1007/978-3-540-24673-2_1
13. Combettes, P.L., Pesquet, J.-C.: A Douglas-Rachford splitting approach to nonsmooth convex variational signal recovery. IEEE J. Sel. Topics Sig. Process. **1**(4), 564–574 (2007)
14. Eckstein, J., Bertsekas, D.P.: On the Douglas-Rachford splitting method and the proximal point algorithm for maximal monotone operators. Math. Program. **55**(1), 293–318 (1992). https://doi.org/10.1007/BF01581204

Segmentation of the Breast Masses in Mammograms Using Active Contour for Medical Practice: AR Based Surgery

Mohamed Amine Guerroudji(✉), Kahina Amara, Djamel Aouam, Nadia Zenati, Oualid Djekoune, and Samir Benbelkacem

Centre for Advanced Technologies Development (CDTA), Baba Hassen, Algeria
mguerroudji@cdta.dz

Abstract. Images have been one of the most important ways humans have used to communicate and impart knowledge and information since the dawn of mankind, as an image can encompass a large amount of information concerning the quality of life linked to health and particularly in oncology precisely the breast cancer. New technologies such as Augmented Reality (AR) guidance allows a surgeon to see sub-surface structures, by overlaying pre-operative imaging data on a live laparoscopic video. The presence of masses in mammography is particularly interesting for the early detection of the breast cancer. In this article, we propose to use a mass detection system, based on two main axes: segmentation and pretreatment. The latter is based on the suppression of the noise by a Gaussian filter and mathematical morphology (white Top-Hat transform) in order to bring out all the spots (Clear Spots) possible to be pathologies. In the second axis, we are interested in the segmentation of pathologies in mammography images. This consists of segmenting the object of interest by active contour models (Chunming Li). Visually, the obtained results are very clear, and show the good performance of the new approach suggested in this work. This latter allows extracting successfully the masses starting from the mammography referents from the database Mini MIAS. The proposed breasts masses detection can, thus, provide an acceptable accuracy for an AR-based surgery or medicine courses with scene augmentation of videos, which provides a seamless use of augmented-reality for surgeons in visualizing cancer tumors.

Keywords: Augmented Reality · Mammography · Masses · Mathematical morphology · Transformation Top-Hat · Chunming Li · Active contour · Gaussian filter · Medical practice · Scene augmentation

1 Introduction

Recently, medical imaging (MI) is an extraordinaire Tool for clinical Diagnosis. It could be the best way to quick detect and localize complex human disease such as breast cancer. In fact, breast cancer is the first cause of death in women [1]. In Algeria, breast cancer registered 11,000 new cases per year, in the advanced stage of pathology [2].

B. Lejdel et al. (Eds.): AIAP 2021, LNNS 413, pp. 447–457, 2022.
https://doi.org/10.1007/978-3-030-96311-8_41

Efficient diagnostic is an important key to prevent breast cancer and the treatment could be more useful. One of the most important research in this area is about the detection and analysis of cancer. In this case, mammography remains an essential reference technique for breast exploration, the most effective in the field of surveillance and early detection of breast cancer [3–5], according to radiologist, mass and calcification are parts of quiet calcium, and there are also the celles is usuelle non-cancers but is an important indicator of breast cancer. If 80% of them are benign tumors, only 15% of breast tumors are malign, where the diagnostic was being by mammography [6]. In order to reduce the workload of radiologists, the design of a computer-assisted detection system (CAD) based on medical images processing algorithms is requested. It is based on a three-step workflow: detection, analysis and classification. Automated detection of breast diseases is always difficult, even for experienced radiologists. To date, some algorithms described in the literature have been applied to detect masses in the mammogram, whether benign or malign. Among the proposed methods, those that use the representation of mammography based on improved contrast by white Top-Hat transform of mathematical formation [7] and [8–14]. This paper attempts to introduce improvements based on the pathway and mathematical morphological factors applied to different gray levels images [6]. The latter provides time to extract the desired masses. In the second phase, we use a segmentation of the area of interest for the identification of masses, based on the active contour models (Chunming Li) [15]. The method suggested has been tested on several images from the Mini MIAS database of mammogram [16], the obtained results allowed higher and accurate segmentation rates. Fortunately, Augmented Reality (AR; i.e., synthetic vision) has been successfully implemented in some areas of surgery [17–19]. Augmented and Virtual Realities have invaded several Healthcare domains: Rehabilitation [20], emotion recognition [21, 22] and [23], educational technology by AR, for facilitating to explain complex medical situations for medical students [24], AR is an enhanced reality generated by a computer that is superimposed on a real-world environment. The purpose of AR in surgery is to combine preoperative data, such as MRI volumes, and fuse the data onto the intraoperative real-time environment [17]. For instance, in breast cancer surgery, AR can be used to generate a virtual scene that contains augmented videos of the patient's anatomy. These videos can act as a visual guide for surgeons in determining the tumor and lymph node location, which, in turn (Fig. 1), should improve the overall efficiency and safety of the procedure. Displaying AR content can be done in several ways, such as video-based, see-through based, or projection based.

The following of the paper is organized as follows: after giving a short review in breast pathology diagnostic in Sect. 1, we provide the proposed detection method. Tests and experiments of our approach is given in Sect. 3. Discussion and future work are presented in Sect. 4.

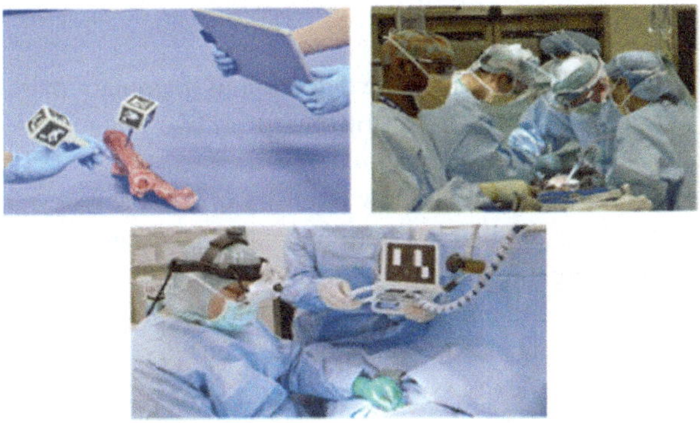

Fig. 1. Medical practice: AR-based surgery.

2 Proposed System

Figure 2 presents the proposed method for the detection of breast disease that helps improve and exposure masses in digital mammography. Here are the main steps conducted to achieve the proposed method. (1) Noise reduction by suppression of the noise by the Gaussian filter and the improvement of contrast between masses and the background-digitized mammography. (2) We will make an effort in the improvement of contrast by the operations of the mathematical morphology (by using a morphological transformation Top-Hat). In the third phase, the using a segmentation of the area of interest for the identification of mass. It exploits the result of description (which itself exploits the result of segmentation) to be able to decide the pathological nature of the mass, based on deformable models (active contour).

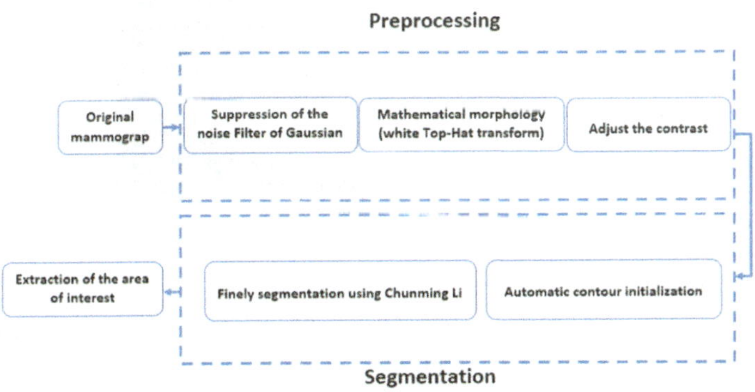

Fig. 2. Block diagram of the proposed algorithm.

2.1 Pretreatments Stage

Despite the efforts of the related work, the detection of mammary pathologies remains difficult. Thus, an additional research effort based on two main steps is examined to achieve the objective of the proposed method, as follows.

Step (1). Due to the small size and low intensity of the mass, designing a filter is a very difficult task, has been applied to filter the noisy images and preserving the details and contours while filtering the homogeneous areas. The main objective pursued by the Gaussian filter is to filter the image by the reduction of the noise in the presented areas while avoiding the smoothing of the contours. We consider the Gaussian distribution that is given by the following expression:

$$G(x, y) = \frac{1}{2\pi\sigma^2} e^{\frac{(x-\mu_1)^2 (y-\mu_2)^2}{2\sigma^2}} \tag{1}$$

with: μ: Average; and σ: Standard deviation.

It is clear that the Gaussian filter enhances the contrast of the contours with region smoothing and a partial reduction of the volume effect. Figure 3 shows of the typical histogram of a mammogram. Clearly, we distinguish three different classes of the pixels [23] and [25]:

- Class 1: the pixels of low of intensities.
- Class 2: the pixels having corresponding gray values within itself.
- Class 3: the pixels of high intensity, the large peak corresponding to the pectoral muscle, annotations and may be to breast lesions.

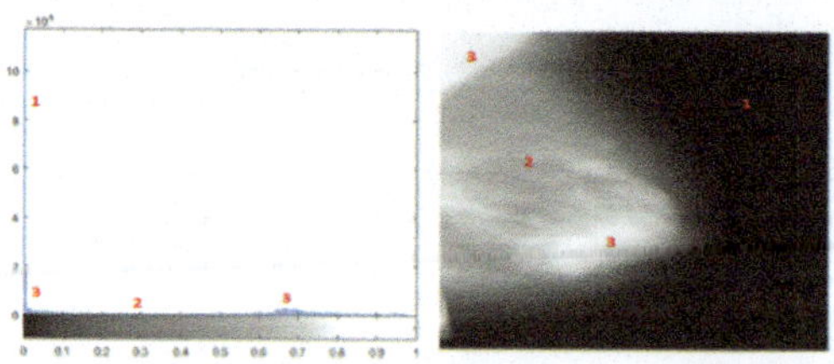

Fig. 3. The typical histogram of a mammogram.

Figure 4 shows the visual results when comparing the original mammography image before and after applying the Gaussian filter. It is clear, can easily be interpreted that the Gaussian filter accentuates the sharpness of the edges, the smooth areas, and even decreases the effect of partial volume.

Step (2). Mathematical morphology is a science of the form and structure. The basic principle of mathematical morphology is to compare the objects, which one wants to analyze with another object of reference, with size and form known, called structuring element. To some extent, each structuring element revealed the object in a new form. The fundamental operations of mathematical morphological are erosion and dilation [8]

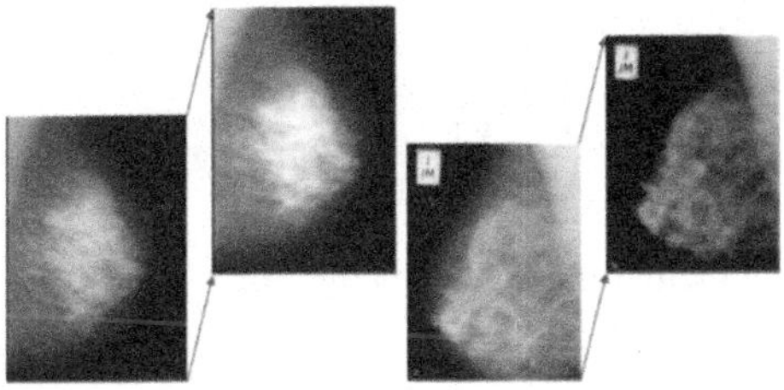

Fig. 4. Visual results comparison of the: (a) Original mammography image; (b) before and (c) after filtering.

and [25]. F is an image in levels of gray and E is a structuring element. Dilation and erosion of F(x) from y(x), noted F ⊕ E and F ⊖ E, are defined respectively as follows:

$$F \oplus E = \max_{u,v}(F(i-u, j-v) + E(u, v)) \tag{2}$$

$$F \ominus E = min_{u,v}(F(i+u, j+v) - E(u, v)) \tag{3}$$

With: F(x): Corresponding to the small clear zones of the image. x: x(i, j): Is a point of an image, and y: y(u, v): are the sizes of the structuring element E.

E(y): Structuring element.

2.2 Segmentation Method

In our work, we used the segmentation method based on the geometric deformable model (Li) [15] for the extraction of regions of interest. In order to eliminate non-suspect regions and for greater precision, the segmented image is presented to the expert in order to allow him to select from among the segmented regions, that which will be called the region of interest [15] and [26]. After selection of the region of interest by the expert, comes the role of the active contour detector. We used this algorithm in order to extract only the region of interest apart.

Chunming Li Model Algorithm. Below we present the most important equation used in this algorithm.

- Initialization: using the following equation

$$\text{if } \phi_0(x, y) = -\rho \qquad (x, y) \in \Omega_0 \tag{4}$$

$$\text{if } \phi_0(x, y) = 0 \qquad (x, y) \in \partial\Omega_0 \tag{5}$$

$$\text{if } \phi_0(x, y) = \rho \qquad \Omega - \Omega_0 \tag{6}$$

- Loop: for n = 1: Niter

Calculation of evolution function:

$$\frac{\partial \phi}{\partial t} = \mu\left[\operatorname{div}\left(\frac{\nabla\phi}{|\nabla\phi|}\right)\right] + \mathrm{ld}(\phi)\operatorname{div}\left(g\frac{\nabla\phi}{|\nabla\phi|}\right) + \mathrm{vgd}(\phi) \quad (7)$$

Evolve the model using the evolution equation:

$$\phi_{i;j}^{k+1} = \phi_{i;j}^{k} + \tau L\left(\phi_{i;j}^{k}\right) \quad (8)$$

Until convergence or until n = Niter.

- End of loop

Implementation Details of Algorithm. In this part, the proposed segmentation method has been applied to the mammographic images for the extraction of the area of interest. We first present the results obtained after the preprocessing and initialization of the contour towards the boundaries of the area of interest. Then we present the results obtained from the segmentation by the active contour approach (Chunming Li model). The contour is chosen automatically. The number of iterations for the detection of the active contour increases with the size of the area of interest [15]. This work was carried out within the IRVA team of the CDTA in Algiers, with a computer equipped with the Intel i3 processor (3.20 GHz) and 6 GB of RAM. And for programming, our application was coded by the Matlab programming language (R2020b). The diagram presented in Fig. 5 summarizes the several stages which we have proposed in an algorithm using Matlab functions for segmentation the masses is considered one of the most powerful programming languages in the medical field.

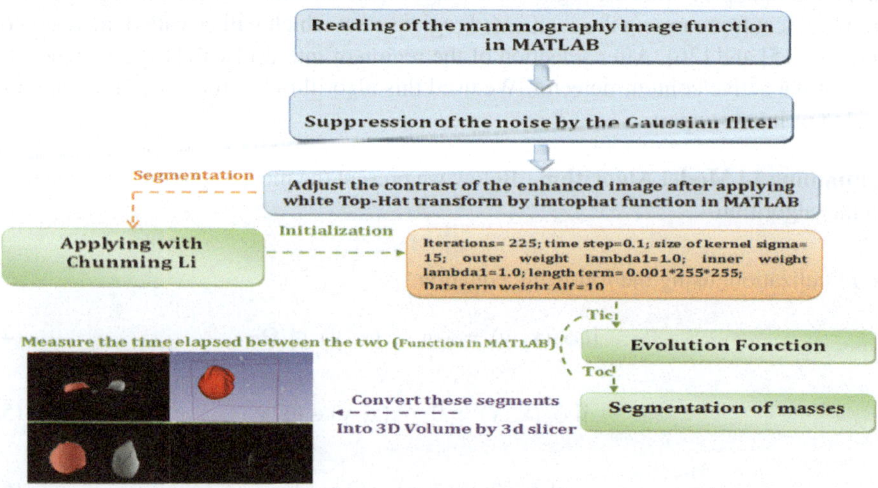

Fig. 5. Diagram of the proposed algorithme for the segmentation of masses.

3 Experimental Results and Discussion

In this work, the proposed method was tested using the mini-MIAS dataset from Mammography Image Analysis Society [16] to segment and illustrate primarily benign and malignant masses. A total of 66 images are examined with different mammary pathologies of the database mainly illustrate 34 benign masses and 32 malignant masses. The detection of masses is very complex due, on the one hand, of the diversity of their forms and, on the other hand, of the border badly definite between healthy tissue and the cancerous zone. For this reason, we propose an algorithm towards the improvement of the image mammography. To highlight masses; we selected several images of Mini MIAS database with dense fabric and the presence of masses. The results are illustrated by the figures below (see Fig. 6).

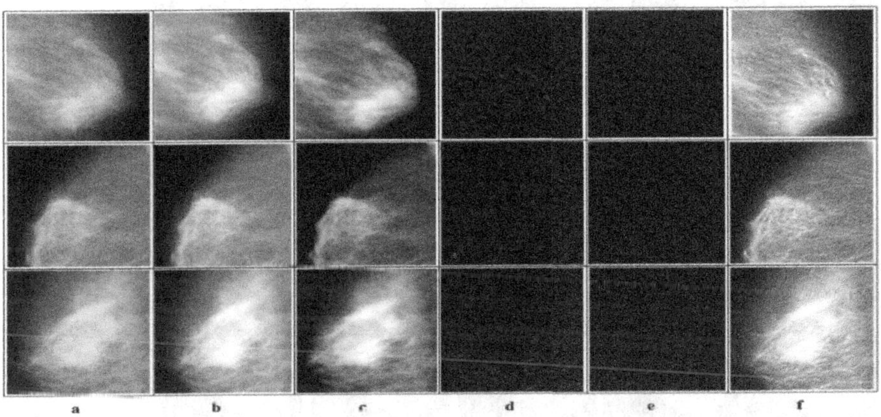

Fig. 6. Illustrate the visualization results of the pre process stage with: (a) Original image; (b) white Top-Hat transform; (c) Raising of contrast (two white Top-Hat transform + image original); (d) Complement of the raised image; (E) improvement of the intensity of the image; (f) Overlay of the previous process result with original image.

Noting that, this study confirms the performance of the image mammography improvement by the white Top-Hat transform for the detection of masses. In the second part of experimental results, the proposed segmentation method has been implemented in the context of identifying the main mammography tissues and extracting the mass area (see Fig. 7). This will subsequently allow the calculation the mass characteristics in order to classification that will be registered in the database.

For performance, comparison purposes our performance results with the performance results of mass detection studies in the literature Through Receiver operating characteristic (ROC), where availability analysis and graphs are commonly used in medical decision-making. The segmentation performance can be estimated by describing random errors (True positive; True negative; False positive; False negative), a measure of statistical changes (Eq. 9).

$$\text{Accuracy} = \frac{\text{TP} + \text{TN}}{\text{TP} + \text{TN} + \text{FP} + \text{FN}} \tag{9}$$

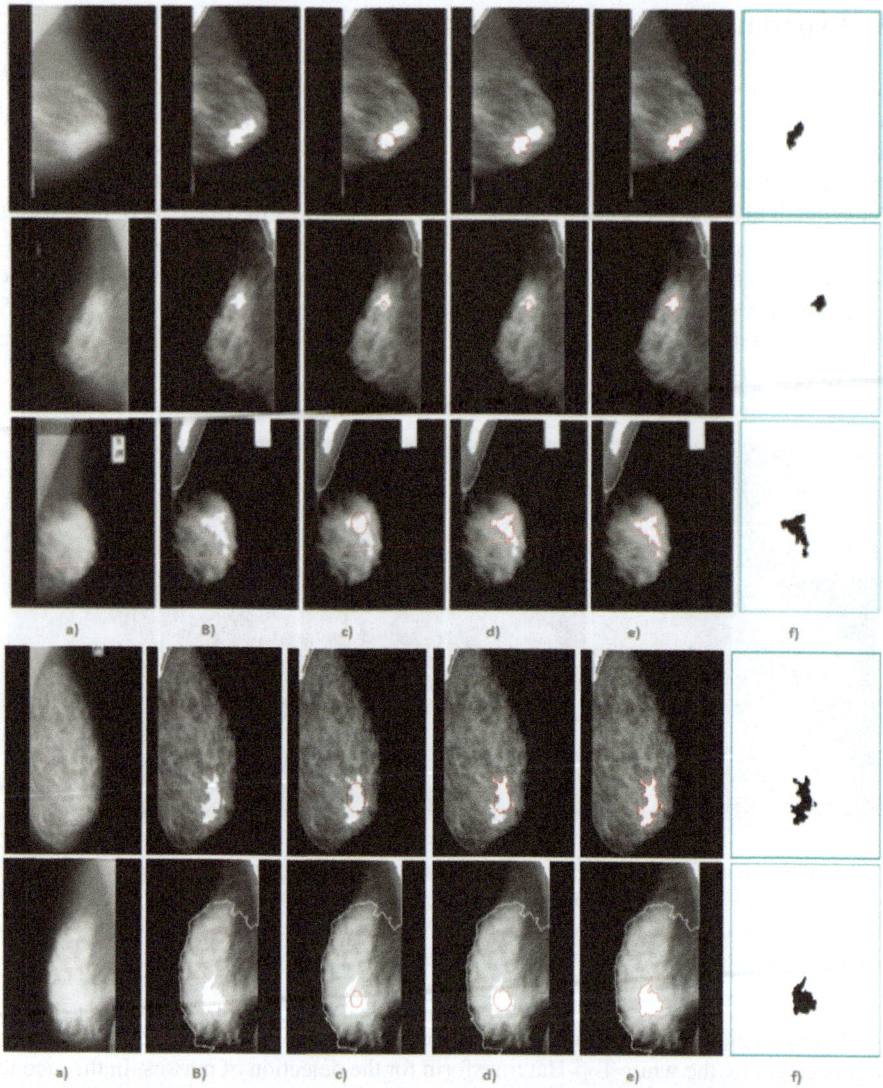

Fig. 7. The results of the Chunming Li model segmentation algorithm of the image. (a) Initial image (b) Mammography image after preprocessing; (c) Automatic contour initialization; (d) The evolution of the contour; (e) The final image with the Chunming Li model; (f) extraction of the separate area of interest.

In addition, the obtained results show that by applying the proposed enhancement and segmentation the contour of mass, with compared other algorithms results cited in the literature Table 1. The table summarizes the results of this work in terms of precision and shows the great performance of our approach.

Table 1. Comparison of results obtained through our approach with other ones in the literature.

Author	Segmentation method	Database	Accuracy
Rahimeh et al. [27]	**Region growing algorithm**	MIAS and DDSM	96.4%
Burcin et al. [28]	**Otsu algorithm**	MIAS	93.2%
Guerroudji et al.	**Active contours**	MIAS	97.09%

4 Conclusion

In computer vision, this vast field of research at the crossroads between mathematics, signal processing and artificial intelligence, the segmentation of images is a very delicate and by no means easy task. It requires precise knowledge of the images, their nature and the field of application. In this paper, we have developed a very efficient segmentation algorithm: Chunming Li, for the detection of breast pathologies. First, we applied a digital image processing technique in order to improve these images using mathematical morphology operations to improve contrast and background in the digital mammography image shot. Subsequently, we implemented a segmentation technique based on the active contours Chunming Li model for the detection of pathologies. The Chunming Li contour segmentation approach shows good results in locating the contours of the regions of interest if the initialization of these contours is not too far from the final contours. In future, in order to improve the accuracy as well as diagnostic. We will suggest an AR 3D visualization system and manipulation for Breast masses segmentation, which remains a very broad field of research.

References

1. Mossi, J.M., Albiol, A.: Improving detection of clustered microcalcifications using morphological connected operators. In: International Conference on Image Processing and Its Application on, pp. 498–501 (1999)
2. Huffpost, M.: http://www.huffpostmaghreb.com/2015/04/05/cancer-sein-algerie-n700-7174.html. Accessed 05 Apr 2015
3. Herman, C.Z.: The role of mammography in the diagnosis of breast cancer. In: Breast Cancer, Diagnosis and Treatment, pp. 152–172 (1987)
4. Guerroudji, M.A., Ameur, Z.: A new approach for the detection of mammary calcifications by using the white Top-Hat transform and thresholding of Otsu. Optik **127**(3), 1251–1259 (2016)
5. Guerroudji, M.A., Ameur, Z.: New approaches for contrast enhancement of calcifications in mammography using morphological enhancement. In: Proceedings of the International Conference on Intelligent Information Processing, Security and Advanced Communication, pp. 1–5 (2015)
6. Diaz-Huerta, C.C., Felipe- Riveron, E.M., Montao- Zetina, L.M.: Quantitative analysis of morphological techniques for automatic classification of micro-calcifications in digitized mammograms. Expert Syst. Appl. **41**(16), 7361–7369 (2014)
7. Nanayakkara, R.R., Yapa, Y.P.R.D., Hevawithana, P.B., Wijekoon, P.: Automatic breast boundary segmentation of mammograms. Int. J. Soft Comput. Eng. (IJSCE) **5**(1), 2231–2307 (2015)

8. Bai, X., Zhou, F.: Infrared small target enhancement and detection based on modified Top-Hat transformations. Comput. Electr. Eng. **36**, 1193–1201 (2010)
9. Bai, X., Zhou, F., Xue, B.: Toggle and top-hat based morphological contrast operators. Comput. Electr. Eng. **38**, 1196–1204 (2012)
10. Laine, S., Schuler, J., Fan, W.: Mammographic feature enhancement by multiscale analysis. Med. Imaging IEEE **13**, 725–740 (1994)
11. Veldkamp, W., Karssemeije, N.: Normalization of local contrast in mammogram. Med. Imaging IEEE **19**, 731–738 (2000)
12. Mcloughlin, K., Bones, P., Karssemeijer, N.: Noise equalization for detection of microcalcification clusters in direct digital mammogram images. Med. Imaging IEEE **23**, 313–320 (2004)
13. Duarte, M.A., Alvarenga, A.V., Azevedo, C.M., Calas, M.J.G., Infantosi, A.F.C., Pereira, W.C.A.: Segmenting mammographic micro calcifications using a semi-automatic procedure based on Otsus method morphological filters. Braz. J. Biomed. Eng. **29**, 377–388 (2013)
14. Diaz-Huerta, C.C., Felipe-Riverón, E.M., Montaño-Zetina, L.M.: Evaluation and selection of morphological procedures for automatic detection of micro-calcifications in mammography images. In: Alvarez, L., Mejail, M., Gomez, L., Jacobo, J. (eds.) CIARP 2012. LNCS, vol. 7441, pp. 575–582. Springer, Heidelberg (2012). https://doi.org/10.1007/978-3-642-33275-3_71
15. Li, C.: Level set evolution without re-initialization: a new variational formulation. In: Proceedings of the IEEE Computer Society Conference on Computer Vision and Pattern Recognition (CVPR05). University of Connecticut Storrs (2005)
16. Mini-MIAS: The mini-MIAS database of mammograms. http://peipa.essex.ac.uk/info/mias.htm. Accessed 2014
17. Tan, W., et al.: A novel enhanced intensity-based automatic registration: augmented reality for visualization and localization cancer tumors. Int. J. Med. Robot. **16**(2), 3167–4715 (2020)
18. Chauvet, P., et al.: Augmented reality in a tumor resection model. Surg. Endosc. **32**(3), 1192–1201 (2017). https://doi.org/10.1007/s00464-017-5791-7
19. Qian, S., Thomas, E., Doyle, R.S., Mona, A.-R.: Augmented reality based brain tumor 3D visualization. Procedia Comput. Sci. **113**, 400–407 (2017). ISSN 1877-0509
20. Masmoudi, M., Djekoune, O., Zenati, N., Benrachou, D.: Design and development of 3D environment and virtual reality interaction: application to functional rehabilitation. In: Proceedings of the International Conference on Embedded Systems in Telecommunications and Instrumentation, Annaba, Algeria, pp. 28–30 (2019)
21. Amara, K., Ramzan, N., Achour, N., Belhocine, M., Larbes, C., Zenati, N.: RGB-D and RGB data comparison for emotion recognition via facial expressions. In: IEEE/ACS 15th International Conference on Computer Systems and Applications (AICCSA) (2018)
22. Amara, K., Ramzan, N., Achour, N., Belhocine, M., Larbes, C., Zenati, N.: A new method for facial and corporal expression recognition. In: 2018 IEEE 16th International Conference on Dependable, Autonomic and Secure Computing, 16th International Conference on Pervasive Intelligence and Computing, 4th International Conference on Big Data Intelligence and Computing and Cyber Science and Technology Congress (DASC/PiCom/DataCom/CyberSciTech), pp. 446–450 (2018)
23. Amara, K., et al.: Towards emotion recognition in immersive virtual environments: a method for facial emotion recognition. In: CEUR ICCSA 2021, The 2nd International Conference on Complex Systems and their Applications, Workshop Proceedings (CEUR-WS.org, ISSN 1613-0073), Oum Elbougui, Algeria, vol. 2904, pp. 253–263 (2021)
24. Herron, J.: Augmented reality in medical education and training. J. Electron. Resour. Med. Libr. **13**(2), 51–55 (2016)

25. Hadjidj, I.: Analyse des Images Mammographiques pour Aide la Dtection du Cancer du Sein. Magister memory in biomedical electronics. Abou Bekr Belkaid University, Tlemcen Algeria (2011)
26. Liu, J., Liu, W.: Adaptive medical image segmentation algorithm combined with DRLSE model. Procedia Eng. **15**, 2634–2638 (2011)
27. Rahimeh, R., Mehdi, J., Shohreh, K., Keshavarzian, P.: Benign and malignant breast tumors classification based on region growing and CNN segmentation. Expert Syst. Appl. **42**(3), 990–1002 (2015)
28. Burcin, K., Vasif, V., Nabiyev, K.T.: A novel automatic suspicious mass regions identification using Havrda Charvat entropy and Otsu's N thresholding. Comput. Methods Programs Biomed. **114**(3), 349–360 (2014)

A Hybrid LBP-HOG Model and Naive Bayes Classifier for Knee Osteoarthritis Detection: Data from the Osteoarthritis Initiative

Khadidja Messaoudene[1](✉) and Khaled Harrar[2]

[1] LIMOSE Laboratory, University M'Hamed Bougara of Boumerdes, Boumerdes, Algeria
k.messaoudene@univ-boumerdes.dz

[2] LIST Laboratory, University M'Hamed Bougara of Boumerdes, Boumerdes, Algeria
khaled.harrar@univ-boumerdes.dz

Abstract. Knee OsteoArthritis (KOA) is a disease characterized by a degeneration of cartilage and the underlying bone. It does not evolve uniformly; it can stay silent for a long time and can quickly intensify for several months or weeks. For this reason, it is necessary to develop an automatic system for diagnosis and reduce the subjectivity in the detection of the disease. In this paper, we present a method for detecting knee osteoarthritis based on the combination of histograms of oriented gradient (HOG) and local binary pattern (LBP). Four classifiers including KNN, SVM, Adaboost, and Naïve Bayes were tested and compared for the prediction of the illness. A total of 620 X-Ray images were analyzed, composed of 310 images from healthy subjects (Grade 0), and 310 images from pathological patients (Grade 2). The results obtained reveal that Naïve Bayes achieved the highest performance in terms of accuracy (ACC = 91%) on the Osteoarthritis Initiative (OAI) dataset. The fusion of HOG and LBP features in KOA classification outperforms the use of either feature alone and the existing methods in the literature.

Keywords: Knee osteoarthritis · X-ray images · LBP · HOG · Naïve Bayes

1 Introduction

Knee osteoarthritis (KOA) is a chronic disease of the joint which progressively destroys the cartilage. It is often mistakenly thought to be associated with aging against which little can be done, whereas it is a real disease that causes disability in about 40% of adults over the age of 70 [1]. As for osteoporosis [2, 3], KOA is a highly prevalent health problem. KOA is typically diagnosed by radiography (X-ray imaging) as well as other imaging modalities like MRI and CT scan. Despite many limitations, conventional radiography (X-ray imaging) remains the first option and most widely utilized for OA because it is more inexpensive and accessible than other diagnostic modalities. The Kellgren and Lawrence (KL) scale is the most frequently used for defining the level of knee OA [4]. The grade in the KL classification system ranges from 0 to 4, according the intensity of OA. Figure 1 depicts the illness phases according to the KL categorization system.

B. Lejdel et al. (Eds.): AIAP 2021, LNNS 413, pp. 458–467, 2022.
https://doi.org/10.1007/978-3-030-96311-8_42

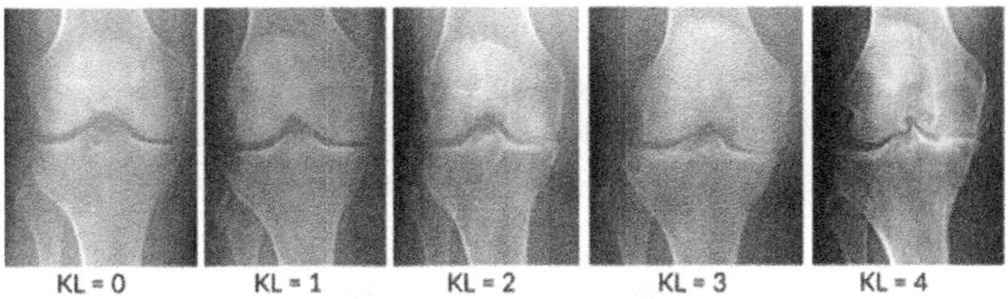

Fig. 1. Knee OA severity [5]

The treatment of knee OA depends on the quality of diagnosis that is why many researchers propose automatic systems aid diagnosis in rheumatology. In [6] the researchers proposed an approach for automatic localization of joint area in knee radiograph. They used the HOG and SVM classifier, the proposed methodology achieved an accuracy of 80%. Haftner et al. [7] describe a method of collecting additional information on the texture of the lateral and medial condyles of the distal femur. Shannon entropy and six other indicative features describing texture roughness and anisotropy were applied. Their framework selected an optimal combination of different texture parameters from six different regions for evaluation with various classifiers. They achieved an accuracy of 72%. Akter et al. [8] described an approach to extract texture features in radiographic images for osteoarthritis detection. The proposed method is based on Zernike orthogonal features and group method of data handling (GMDH) Neural Networks. This technique improved the detection accuracy by 82.8% for lateral images. In [9] the authors combined different texture descriptors (LBP and GLCM) with different classifiers (KNN, SVM, neural network) to determine the intact stage of knee osteoarthritis in radiographic images. The highest performance was obtained with a multilayer perceptron (MLP) classifier, with an overall accuracy of 90.2%.

In this paper, LBP and HOG methods are combined with the naive Bayes classifier on the OAI dataset to detect knee osteoarthritis in two stages of the disease: KL0 (normal case), KL2 (pathological case). First, LBP parameters are extracted from the images, then the HOG parameters are estimated, finally, several classifiers (Naive Bayes, SVM, Adaboost, and KNN) are carried out for the prediction of the disease. In the first stage, each model (LBP, or HOG) is tested and evaluated alone, then a combination of the two models is performed to improve the ability of the prediction. This is the first study to combine LBP and HOG for KOA detection.

This paper is structured as follows: Sect. 2 covers the material and approach and its extensions; Sect. 3 illustrates the results and discussion, and Sect. 4 summarizes the findings.

2 Material and Methods

2.1 Dataset

In this study, the data from the OAI was used. The OAI covers persons at risk of developing clinical tibiofemoral osteoarthritis. A total of 4,796 participants aged 45–79 years took part in the study between 2004 and 2006. The images were analyzed using the

Kellgren-Lawrence (KL) grading method [10]. The present study focuses on the early detection of knee OA. Therefore, only radiographs with a KL grade 0 (no OA) and a KL grade 2 (minimal OA) were considered. We used 620 radiographs of the knee in the lateral region. Figure 2 shows the ROI used in our study.

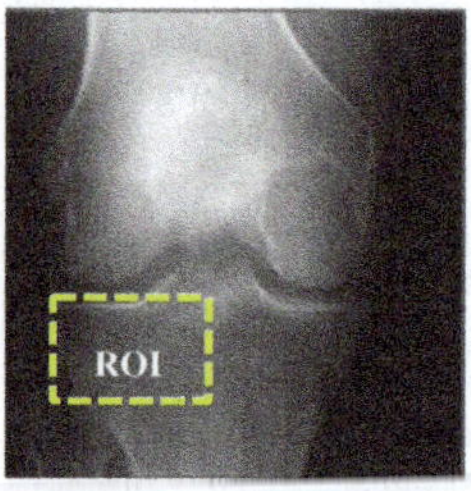

Fig. 2. Knee radiographic image.

2.2 Methods

The major goal of this study is to present a texture feature extraction technique that performs well in this situation. In our tests, we employed the LBP descriptor, the HOG, and a combination of them. Figure 3 depicts the design of our system. A brief overview of each phase of our method is provided below.

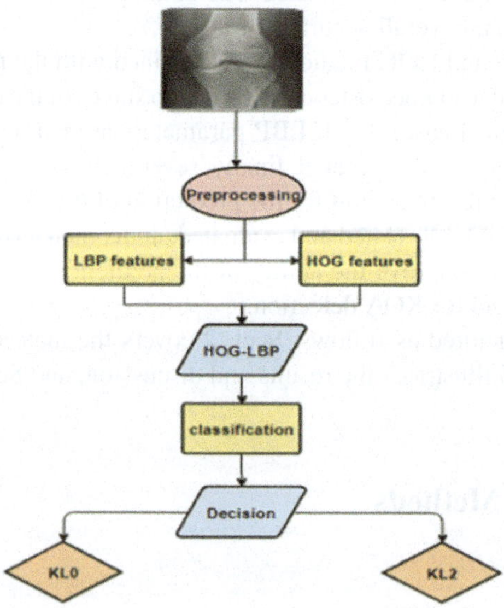

Fig. 3. Proposed classification system

Preprocessing

The anisotropic diffusion filter (ADF) has been effectively used in image processing to eliminate high frequencies while preserving the major existing objects without deleting substantial elements of the image content, often edges, lines, or other features crucial for image interpretation [11]. ADF is defined as:

$$\frac{\partial I}{\partial t} = div(c(x, y, t)\nabla I) = \nabla c.\nabla I + c(x, y, t)\Delta I \tag{1}$$

I is the input image, Δ represents the Laplacian, ∇ is the gradient, *c(x, y, t)* denotes the diffusion coefficient, div() is the divergence operators. Figure 4 shows the results of filtering.

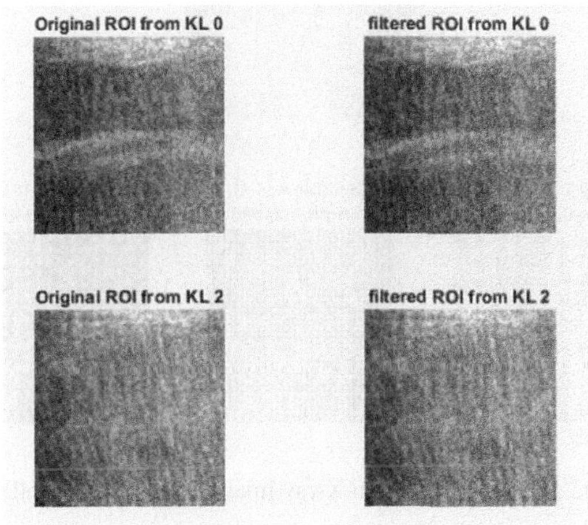

Fig. 4. Results of filtering

Histogram of Oriented Gradient (HOG)

The HOG method was suggested by Dalal and Trrigs in 2005 [12]. The original idea of this descriptor is that the local structure of the object is described by calculating the gradient distribution of the local intensities or the direction of the contours without knowing the localization of the gradient or the position of the contours in the image [13]. The HOG descriptors are the main features that encode object features into a sequence of specific numbers, which can be used to distinguish items from each other [14]. Gradients are the rate of modifications in local intensity at a specific pixel position. A gradient is a quantity of the vector that has both direction and magnitude. The pixel gradient magnitude $V(x, y)$ and direction $\alpha(x, y)$ are indicated in Eqs. (2) and (3) respectively:

$$V(x, y) = \sqrt{Vx(x, y)^2 + Vy(x, y)^2} \tag{2}$$

$$\alpha(x, y) = \arctan\left[Vx(x, y)/Vy(x, y)\right] \tag{3}$$

Figure 5 shows the HOG features extracted from an image using three different cell sizes. This figure shows the visualization of cell sizes [2 2], [4 4], and [8 8]. The size cell [2 2] contains more shape information than the size cell [8 8] in their visualization. In the latter case, the dimensionality of the feature vector using HOG increases compared to the former. A good choice is the cell size [8 8]. By using this size, the number of dimensions is limited, which speeds up the training process. It also contains enough information to visualize the shape of the mode image.

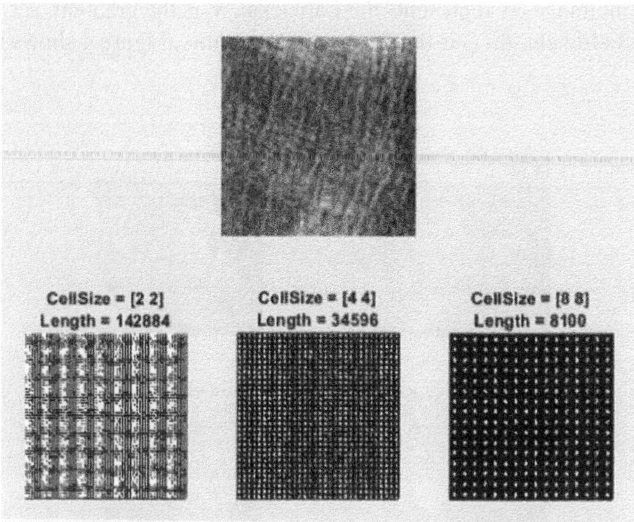

Fig. 5. HOG features of an X-ray image with different cell sizes

Local Binary Pattern (LBP)

Ojala et al. developed the LBP approach to measure texture patterns [15]. The LBP approach compares each neighboring pixel in neighborhood 3 × 3 against the center pixel to determine if it is 0 or 1. Each binary value is then multiplied by the corresponding weight. An LBP number for a unit of texture is obtained by adding up all the multiplications. LBP can generate up to 256 patterns.

$$LBP(x_c, y_c) = \sum S(i_a - i_c)2^n \tag{4}$$

Where i_a is the gray level of the pixel (x_c, y_c), i_c is the gray level of the circular neighborhood of the pixel (x_c, y_c), and S is the Heaviside function.

Classification

After the feature extraction step (HOG, LBP), we applied the Bayes model due to its speed and efficiency in the prediction of knee osteoarthritis. The naive Bayes system is a highly simplified Bayesian probabilities model. The naive Bayes classifier is considered one of the strongest independence assumptions [16]. This indicates that the probability of one characteristic has no influence on the results of the other. First, we tested several classifiers on the LBP parameters alone and then on the HOG. Then we combined the parameters (LBP-HOG) and tested the different classification models (Naive Bayes, SVM, Adaboost, and KNN).

Model Evaluation

Knowing a model's accuracy is necessary, but it is not sufficient to provide a full understanding of a model's level of efficiency. So, there are other measurement criteria that will help understand how performative the model is? The other metrics used in this study are: Precision, recall, ROC curve, MCC, etc.

Accuracy (ACC): a metric that allows a model to quantify the number of total accurate predictions.

$$ACC = \frac{TP + TN}{TP + TN + FP + FN} \tag{5}$$

Precision (Pr): as defined as the ratio of correct positive predictions to all positive predictions.

$$Pr = \frac{TP}{TP + FP} \tag{6}$$

Sensitivity (True Positive Rate (TPR)): is a measurement of the proportion of positives that are correctly identified.

$$TPR = \frac{TP}{TP + FN} \tag{7}$$

Specificity (True Negative Rate (TNR)): is the proportion of negatives that are correctly identified.

$$TNR = \frac{TN}{TN + FP} \tag{8}$$

FPrate (False Positive Rate (FPR)): is the percentage of negative values wrongly defined as positive in the data.

$$FPR = \frac{FP}{FP + TN} \tag{9}$$

F1-Score: is the weighted average between precision and sensitivity.

$$F1 - Score = \frac{2TP}{2TP + FP + FN} \tag{10}$$

Where TP is true positive, TN true negative, FP false positive, and FN false negative.

3 Results and Discussion

In the first test, we evaluated the performance of the LBP parameters, the HOG parameters, and the LBP-HOG system. To perform a comparison, we tested radiographic knee images taken from different subjects for the two stages of osteoarthritis (0 and 2). For each stage, 310 images were involved.

For each case (LBP, HOG, LBP-HOG), we tested four classifiers (Adaboost, Naive Bayes, SVM, and KNN). The results are presented in the following tables.

Table 1. Classification performance for LBP features

Classifier	TP	FP	FN	TN	Pr	FPR	TNR	TPR	F1-score	ACC
Naïve Bayes	**213**	**91**	**95**	**221**	**0.70**	**0.29**	**0.71**	**0.69**	**0.37**	**0.70**
SVM	195	115	121	189	0,63	0,38	0.62	0.62	0.43	0.62
Adaboost	155	145	153	167	0.52	0.46	0.54	0.50	0.48	0.52
KNN	150	158	165	147	0,49	0,52	0.48	0.48	0.50	0.48

Table 2. Classification performance for HOG features

Classifier	TP	FP	FN	TN	Pr	FPR	TNR	TPR	F1-score	ACC
Naive Bayes	216	93	105	206	0.70	0.31	0.69	0.67	0.37	0.68
SVM	**237**	**85**	**77**	**221**	**0.74**	**0.28**	**0.72**	**0.75**	**0.36**	**0.74**
Adaboost	195	142	118	165	0.58	0.46	0.54	0.62	0.50	0.58
KNN	158	148	162	152	0.52	0.49	0.51	0.49	0.49	0.50

Table 3. Classification performance for the combination of LBP-HOG features

Classifier	TP	FP	FN	TN	Pr	FPR	TNR	TPR	F1-score	ACC
Naive Bayes	**286**	**20**	**36**	**278**	**0.93**	**0.07**	**0.93**	**0.89**	**0.11**	**0.91**
SVM	257	36	48	279	0.88	0.11	0.89	0.84	0.18	0.86
Adaboost	192	114	134	180	0.63	0.39	0.61	0.59	0.42	0.60
KNN	248	62	69	241	0.80	0.20	0.80	0.78	0.29	0.79

Table 1 depicts a comparison of the four classifiers for the LBP parameters. As we can observe, Naive Bayes provided the best performance with a TPR of 0.69 and the lowest FPR (0.29). It outperformed the second-ranked method (SVM), by a significant margin. The worst performing method would be KNN with a low TNR (0.48), a high FPR (0.52), and a TPR of 0.48. The LBP model is shown to perform better with the Naive Bayes classifier.

The results of the classification using the HOG method are shown in Table 2. We can see that the combination of HOG parameters with the SVM model gave excellent results with an accuracy of 74% and a low FPR (0.28). The KNN classifier gave bad results in terms of FPR (0.49).

The combined performance of LBP-HOG with four classifiers is shown in Table 3. As can be seen, Naive Bayes provided the best performance with a TPR of 0.89 and the lowest FPR (0.07). It outperformed the second-highest ranked method (SVM), by a remarkable margin. On the other hand, Adaboost, with a TPR of 0.59 and an FPR of 0.39, is the worst-performing method in this case.

Regarding the F1-score, the same findings are noticed. The combination of LBP and HOG models provided the lowest rate (0.11), where LBP gave 0.37, and HOG achieved 0.36 of F1-score.

Through the results described in the previous Tables, it is clear that the combination of characteristics of the two models (LBP-HOG) achieved the best detection rate. The results obtained with the combination show a better performance than the systems based on each method alone.

Table 4 illustrates a comparison of our proposed method with the state of the art. Tiulpin et al. [6] used HOG and SVM to detect osteoarthritis and provided an accuracy of 80%. Haftner et al. [7] achieved a lower accuracy (72%) with entropy and LDA technique. Akter et al. [8] achieved an accuracy of 82.8% using Zernike and GMDH classifier. Peuna et al. [9] combined LBP and GLCM with MLP classifier and provided good results in terms of accuracy (90.2%). Examining Table 4, our proposed method achieved the highest accuracy with the combination of LBP and HOG descriptors and Naïve Bayes as classifier, where the rate achieved 91%.

Table 4. A comparison study with the state of the art

Author	Year	Method	Classifier	ACC (%)
Tiulpin et al. [6]	2017	HOG	SVM	80
Haftner et al. [7]	2017	Entropy	LDA	72
Akter et al. [8]	2019	Zernike	GMDH	82.8
Peuna et al. [9]	2021	LBP-GLCM	MLP	90.2
Proposed method	**2021**	**LBP-HOG**	**Naive Bayes**	**91**

4 Conclusion

This study offers an efficient and precise approach for the classification and identification of knee OA. The present work was carried out on a dataset composed of 620 radiographs of patients divided into 310 images of healthy subjects (Grade 0), and 310 images from patients suffering from KOA (Grade 2). Following the successful implementation of the proposed classification system using HOG and LBP methods with Naive Bayes classifier, we have demonstrated that the proposed system provided promising results in terms of classification of patients suffering from Knee OA with high accuracy (ACC = 91%).

We believe that our system can help and assists doctors in osteoarthritis diagnosis. In the future, we are planning to improve the feature extraction stage and the classification using other techniques. We are exploring other types of features to train classifiers and analyze the effects of other machine learning algorithms for the classification of knee OA images. Moreover, we are testing more images and we are working to assess other stages of OA (KL1, KL3, and KL4) to provide a reliable classification system.

References

1. Attur, M., Krasnokutsky-Samuels, S., Samuels, J., Abramson, S.B.: Prognostic biomarkers in osteoarthritis. Curr. Opin. Rheumatol. **25**, 136–144 (2013)
2. Harrar, K., Jennane, R.: Quantification of trabecular bone porosity on X-ray images. J. Ind. Intell. Inf. **3**(4), 280–285 (2015)
3. Harrar, K., Jennane, R.: Trabecular texture analysis using fractal metrics for bone fragility assessment. Int. J. Biomed. Biol. Eng. **9**, 683–688 (2015)
4. Kellgren, J., Lawrence: radiological assessment of osteo-arthrosis. Ann. Rheum. Dis. **16**(4), 494–502 (1957)
5. Bayramoglu, N., Nieminen, M.T., Saarakkala, S.: A lightweight CNN and Joint Shape-Joint Space (JS^2) descriptor for radiological osteoarthritis detection. In: Papież, B.W., Namburete, A.I.L., Yaqub, M., Noble, J.A. (eds.) MIUA 2020. CCIS, vol. 1248, pp. 331–345. Springer, Cham (2020). https://doi.org/10.1007/978-3-030-52791-4_26
6. Tiulpin, A., Thevenot, J., Rahtu, E., Saarakkala, S.: A novel method for automatic localization of joint area on knee plain radiographs. In: Sharma, P., Bianchi, F.M. (eds.) SCIA 2017. LNCS, vol. 10270, pp. 290–301. Springer, Cham (2017). https://doi.org/10.1007/978-3-319-59129-2_25
7. Haftner, T.S., Ljuhar, R., Dimai, H.P.: Combining radiographic texture parameters increases tibiofemoral osteoarthritis detection accuracy: data from the osteoarthritis initiative. Osteoarthr. Cartil. **25**, S261 (2017)
8. Akter, M., Jakaite, L.: Extraction of texture features from x-ray images: case of osteoarthritis detection. In: Yang, X.-S., Sherratt, S., Dey, N., Joshi, A. (eds.) Third International Congress on Information and Communication Technology. AISC, vol. 797, pp. 143–150. Springer, Singapore (2019). https://doi.org/10.1007/978-981-13-1165-9_13
9. Peuna, A., Thevenot, J., Saarakkala, S., Nieminen, M.T., Lammentausta, E.: Machine learning classification on texture analyzed T2 maps of osteoarthritic cartilage: oulu knee osteoarthritis study. Osteoarthr. Cartil. **29**(6), 859–869 (2021)
10. Eckstein, F., Wirth, W., Nevitt, M.: Recent advances in osteoarthritis imaging–the osteoarthritis initiative. Nat. Rev. Rheumatol. **8**(12), 622–630 (2012)
11. Perona, P., Malik, J.: Scale-space and edge detection using anisotropic diffusion. IEEE Trans. Pattern Anal. Mach. Intell. **12**(7), 629–639 (1990)
12. Dalal, N., Triggs, B.: Histograms of oriented gradients for human detection. In: Conference on Computer Vision and Pattern Recognition, vol. 1, pp. 886–893 (2005)
13. Pauly, L., Sankar, D.: Non-intrusive eye blink detection from low resolution images using HOG-SVM classifier. Int. J. Image Graph. Signal Process. **8**(10), 11 (2016)
14. Bhende, P., Cheeran, A.: A novel feature extraction scheme for medical X-ray images. Int. J. Eng. Res. Appl. **6**(2), 53–60 (2016)

15. Ojala, T., Pietikainen, M., Maenpaa, T.: Multiresolution gray-scale and rotation invariant texture classification with local binary patterns. IEEE Trans. Pattern Recogn. Mach. Intell. **24**(7), 971–987 (2002)
16. Al-Sharafat, W., Naoum, R.: Development of genetic-based machine learning for network intrusion detection. World Acad. Sci. Eng. Technol. **55**, 20–24 (2009)

RONI-Based Medical Image Watermarking Using DWT and LSB Algorithms

Aicha Benyoucef(✉) and M'Hamed Hamadouche

Faculty of Technology, M'Hamed Bougara University of Boumerdes, Boumerdes, Algeria
a.benyoucef@univ-boumerdes.dz

Abstract. In recent years, hiding information in medical images is the largest usage to secure this information or garnet the integrity of the owner, the embedding can distort the medical image and change the necessary health patient information. In this paper, we propose a robust method for medical image watermarking to secure patient data when it transmits in a non-secure channel. First of all, the original image is filtered by a sharpening filter for enhanced contrast then separated into two regions using snake segmentation. The embedding mark (Electronic Patient Record) is added to the frequency domain after applying Discrete Wavelet Transform (DWT) on the region of non-interest (RONI) using the last significa-tion bit (LSB). This region has a predominantly black background; the region of interest (ROI) has the necessary patient information. This method preserves a high-quality watermarked image, and it improves imperceptibility, security, and authentication. Our method is evaluated by Pick Signal to Noise Ratio (PSNR = 46.4039 for 512 * 512 bits image size), SNR, NC, and histogram analysis, and it's compared with existing schemes.

Keywords: Medical image · Region of non interest (RONI) · DWT · LSB algorithm · Digital watermarking · Authentication

1 Introduction

Modern health care's information infrastructure is based on digital information management. While recent developments in information technology provide new ways of accessing, managing, and moving medical information. Because of their ease of manipulation and replication, they also compromise their security [1]. Digital watermarking and steganography are the embedding of secret information methods (bits in general) into a host signal such as an image, video, audio, or database. These approaches are used to ensure the validity, integrity, or identity of the owners. Following that, the embedding mark is identified and extracted to disclose the owner/identity of the digital material [2].

Steganography of medical images represents a special case of steganography of images because of the sensitivity of the information on the patient's disease exists on it, to the well the sensitivity of the added patient recode. The medical image is divided into two regions: the first is the region of interest (ROI) which contains important information about the patient's disease, and helps doctors to interpret and diagnose it. The second

B. Lejdel et al. (Eds.): AIAP 2021, LNNS 413, pp. 468–478, 2022.
https://doi.org/10.1007/978-3-030-96311-8_43

region is the region of non-interest (RONI) which has a predominantly black background with some text content. So that if the RONI of the medical image is distorted or degraded by any modification, this deterioration has no influence on the patient's diagnosis and therapy [3].

In recent years many papers are published about the different steganography and watermarking algorithms of an image or their enhancement. The use of visible watermarking based on Region of Non Interest (RONI) for medical image authentication is proposed in [3]. In [5]; DWT-SVD with hamming code is used to build an improved non-blind, resilient, undetectable, and secure watermarking approach for concealing multi watermarks, the suggested method's confidentiality and compression performance are enhanced. Sivaganesan *et al.* [7] proposed innovative image watermark schemes for the protection of copyright protection and authentication, which are based on DWT, DCT, and LSB algorithms for embedding the watermark into the cover image. Khawatreh *et al.* [6] proposed a method in which they added a message to the color image by using the LSB2 algorithm and they encrypt it by blocking and reordering method to get a secure stego-image. For the medical image and patient's privacy, Ambika *et al.* [8] proposed a method based on effective selection of pixels for added an image continent the patient's information to the cover medical image in the frequency domain. This selection of pixels does by Elephant Herding-Monarch Butterfly (EH-MB) algorithm. Khalil [9] proposed a method based on the combination of steganography and cryptography of the medical image to analyze the degradation when the watermark was embedded in the frequency domain, and mention the relation between the PSNR value and the location of the secret message.

In this paper, we propose a new method based on separation of the image in two region using snake segmentation, and transform the RONI to the frequency domain by apply the second level of discrete wavelet transform (DWT). The obtain LL2 sub band are used such as a cover image to added the electronic patient record (EPR) into it using Last Signification Bit (LSB) algorithm. This method's aim is to achieve authentication, the security of the information hiding and imperceptibility without any distortion.

The remainder of this paper is structured as follows: Sect. 2 covers the same preliminaries. Section 3 discusses the suggested approach, displays the obtained results and the discussion. The final section of this paper contains the conclusion.

2 Preliminaries

2.1 Medical Imaging

Medical image is the distribution of numerous physical traits measured from the human body like organs and other tissue. It has been used to reveal qualities and image body extremities. Many techniques are used to produce images such as magnetic resonance, nuclear, ultrasound, tomography, and others, these techniques give a various image modality [21, 22].

2.2 Sharpening Filter

Sharpening is a critical pre-processing technique for bringing out edge details by increasing the contrast between dark and bright areas. It increases edge rigidity and accentuates the subtle characteristics that are already there [19]. Image blurring is a method of integrating or averaging image pixels in close proximity. This method is used to eliminate noise from images and smooth them down [20].

2.3 Snake Segmentation

Snakes or active contour models, meaning that they lock on to adjacent edges and properly localize them. To increase the capture zone encircling a feature, we utilize scale-space continuation. Snakes active contour provides a unified explanation of a variety of visual issues, including as edge recognition, line detection, and subjective contour detection, motion tracking, and so on [17]. A snake is an energy-minimization spline whose energy is proportional to its form and position within the image. Internal and external factors work together to regulate the snake's shape. External force directs the snake toward the image's features, while internal force acts as a smoothing restriction [16]. Medical imaging segmentation approaches can provide doctors and patients with an alternative computational tool to help diagnostic and health assessment progress, as well as propose the best therapy option [18] (Fig. 1).

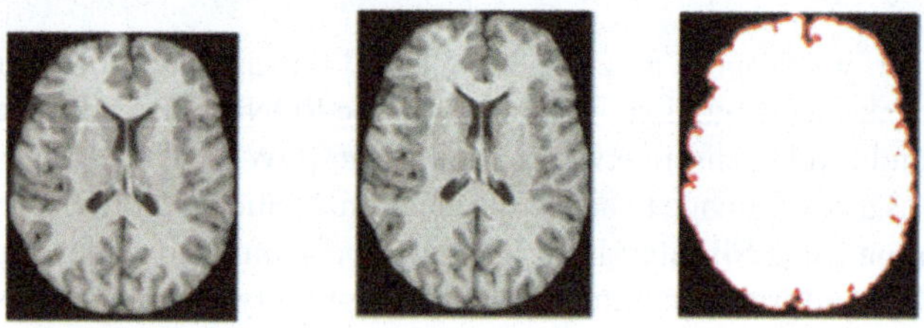

a b c

Fig. 1. Processing medical image; a: original image, b: filtered image, c: segmented image.

2.4 Discrete Wavelet Transform (DWT)

DWT is a mathematical technique for decomposing an image hierarchically. The image is decomposed into four sub-bands using DWT: LL1, LH1, HL1, and HH1 (approximation, horizontal, vertical and diagonal). Decompose approximation coefficient or one of the sub bands of level 1 into wavelet coefficients for multi-level. (Figure 3) The multi-resolution feature of the DWT assists in rapidly locating the regions where the watermark is embedding [10, 11].

In this paper, the wavelet type used is Haar wavelet, it proposed by Alfréd Haar in 1909, and it is the most basic form of wavelet, presented in discrete form [13, 14]. The Haar transform, like other wavelet transforms, decomposes an image into four sub-bands at the first level (Fig. 2.a), we used it in the LL1 sub-band for the second level of the Haar wavelet (Fig. 2.b). In (Fig. 2.c) the image represents the two-level together.

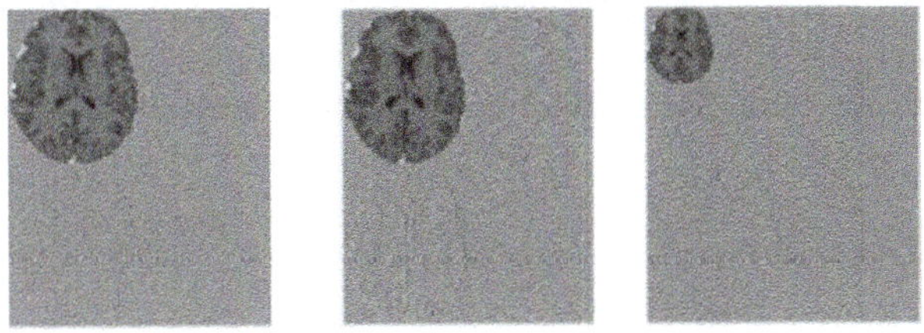

a b c

Fig. 2. Two level DWT decomposition of an image

2.5 Last Signification Bit (LSB)

One of the most effective methods in digital watermarking is LSB replacement which includes conventional LSB replacement. Figure 3 shows the principle of the algorithm LSB which was a simple and effective technique for hidden message bits in the cover image's least significant bits, in which the secret message bits exploited the last bit of the image [15].

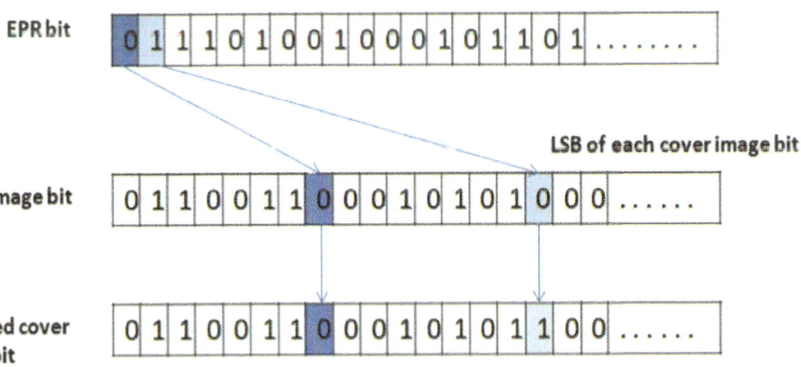

Fig. 3. The principle of the LSB algorithm

3 The Proposed Method

The proposed method is on the embedding of the EPR like a watermark in the frequency domain of the RONI. For the embedding algorithm section; the original medical image is processed and filtered to enhanced, before the separation of two regions. The application of DWT in level 2 can decompose the RONI into 4 sub-bands and use the LL2 sub-band to limit the location of the watermark. On the other hand, the EPR character is converted to the ASCII code than binary. The LL2 band is the band that contains the approximation coefficient, we will treat this part as an image and we will add the bits of the EPR bit by bit in it, it is an application of the LSB algorithm in the frequency domain. After the embedding of the watermark is done, apply the DWT inverse to get RONI watermarked then the medical image is watermarked by combining the two regions of the image (see bloc diagram on Fig. 4a). The steps of the watermark embedding are represented as follows.

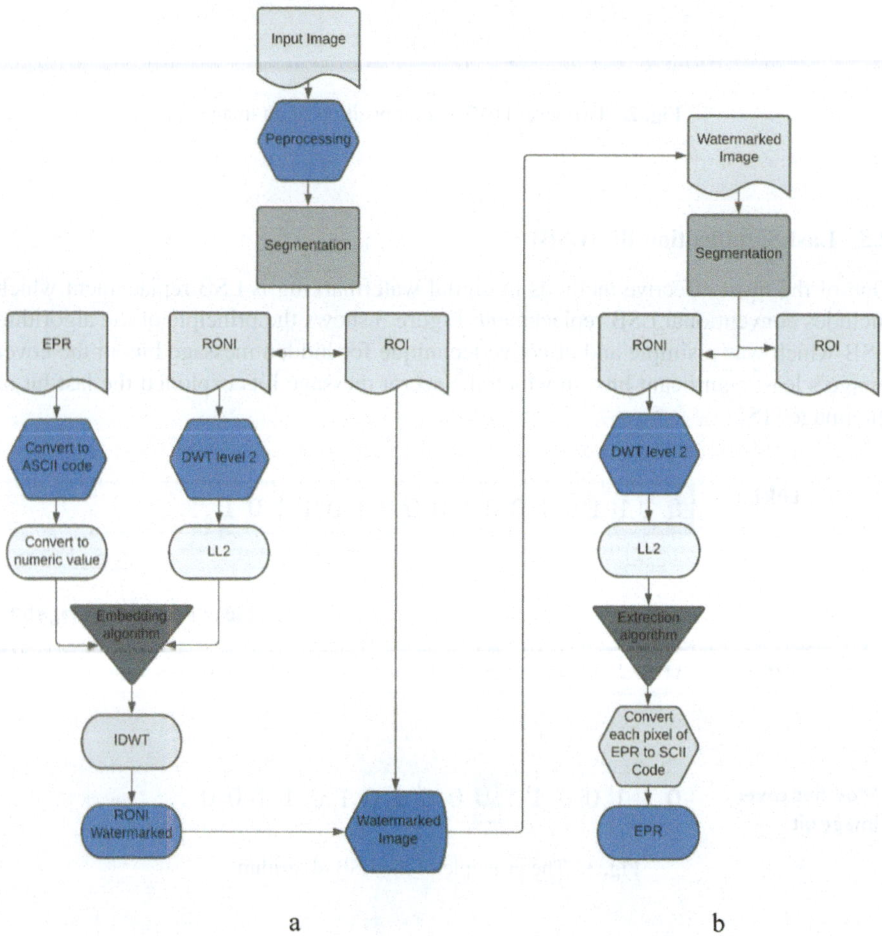

Fig. 4. Block diagram of the proposed method, embedding and extraction algorithm

Embedding algorithm

Step 1: read the medical image and apply the sharpen filter to enhance it
Step 2: apply the Snack segmentation on the image and separate the RON and RONI
Step 3:
- apply the second level of DWT on RONI for obtain LL2 sub bond
- read the EPR character and convert it to ASCII code then to binary

Step 4: calculate the LSB of each pixel of LL2 sub bond and replace it by each bits of EPR one by one
Step 5: apply the inverse of DWT on the RONI by replacing the LL2 by the LL2 obtained in step 4
Step 6: get the watermarked image by combining the two image areas ROI and RONI obtained in step 5

For the extraction algorithm section, get the watermarked image and separate the two regions, apply the DWT in the second level to obtain the LL2 sub-band which the watermark is excited and extract and convert it to the ASCII code than to character like it mention in (Fig. 3b). The steps of the watermark extraction are represented as follows:

Extraction algorithm

Step 1: read the watermarked image then separate the ROI and RONI
Step 2: apply the second level of DWT on RONI for obtain LL2 sub bond
Step 3: calculate the LSB of each pixel of LL2 sub bond
Step 4: retrieve bits and convert each 8 pixel into character and obtain the EPR

3.1 Measurements Metrics

For performance evaluation, the proposed approach is examined on grayscale medical images in different modalities like IRM (images 1, 2 and 4), PET (image 3), and OCT (image 5) images of various sizes, then we resize the small size image to be 512 * 512 pixels for more evaluation and comparison. The watermark tested has 12 characters, and the results are performed on the platform of Core i3, 1.70 GHz CPU, 4 GB RAM, MATLAB R2020a. For evaluating the watermarked image perceptibility, we use some quality metrics like:

- **Peak-Signal-to-Noise Ratio (PSNR):** The quality of the original and watermarked images is compared using this ration, which is measured in decibels. The PSNR and the mean-square error (MSE) are employed. The MSE is the difference in cumulative squared errors between the watermarked and original images, whereas the PSNR is the proportion of the original picture that is watermarked.

$$\boldsymbol{PSNR(I, Iw)} = 10 * \mathbf{log}_{10}\frac{(\boldsymbol{MAX_I^2})}{\boldsymbol{MSE}} \tag{1}$$

Where: $I(i,j)$, $Iw(i,j)$: Is represent the original image and the watermarked image respectively.

- **Signal to noise ratio (SNR):** In contrast to MSE, it calculate the similarity between the original image and the watermarked, it may also be computed by the following state, [4]

$$SNR = \frac{X^2(i, j)}{\sqrt{MSE}} \tag{2}$$

- **Normalized correlation (NC) analysis:** This metric indicate the resemblance factor between inserted and extracted watermark [12]. It is calculate by following equation:

$$NC\left(W, W'\right) = \frac{\sum_{i=1}^{m} \sum_{j=1}^{n} \left[W(i,j).W'(i,j)\right]}{\sum_{i=1}^{m} \sum_{j=1}^{n} [W(i,j)]^2} \tag{3}$$

Where: W, W′ are original and extracted watermark respectively.

When: NC = 1, it is the maximum attainable value which specifies that inserted and extracted watermark are impossible to differentiate. NC = 0, it is the minimum attainable value which specifies that original and extracted watermark are exclusively dissimilar.

Table 1. The PSNR and NC values of the proposed method

Images	Image size	PSNR	SNR	NC
Image 01	287 × 230	40.4146	36.1154	0.9986
	512 × 512	46.4039	42.0824	0.9997
Image 02	512 * 512	46.4039	37.0435	0.9999
Image 03	287 × 230	40.4146	34.0898	0.9992
	512 × 512	46.4039	40.0675	0.9998
Image 04	283 × 211	39.9792	33.1776	0.9996
	512 × 512	46.4039	39.4427	0.9999
Image 05	512 × 512	46.4039	35.3562	0.9999

3.2 Simulation Results

Table 1 shows the 3 parameters values PNSR, SNR, and NC for the 5 medical images with different modalities and sizes, the images (1, 3 and 4) are evaluated first in the average size, and his PSNR value is between 39.9792 and 40.4146, for better comparison, we resize this images to will be in 512 * 512 size like other, there PSNR value reach to 46.4039 with the same number of bit of EPR.

The human visual system (HVS) will struggle to detect changes between the original image and the watermarked one if the PSNR value is more than 30. So the imperceptibility of the watermark will better and make it more secure against attackers that are clearly verified in PSNR value of the proposed method.

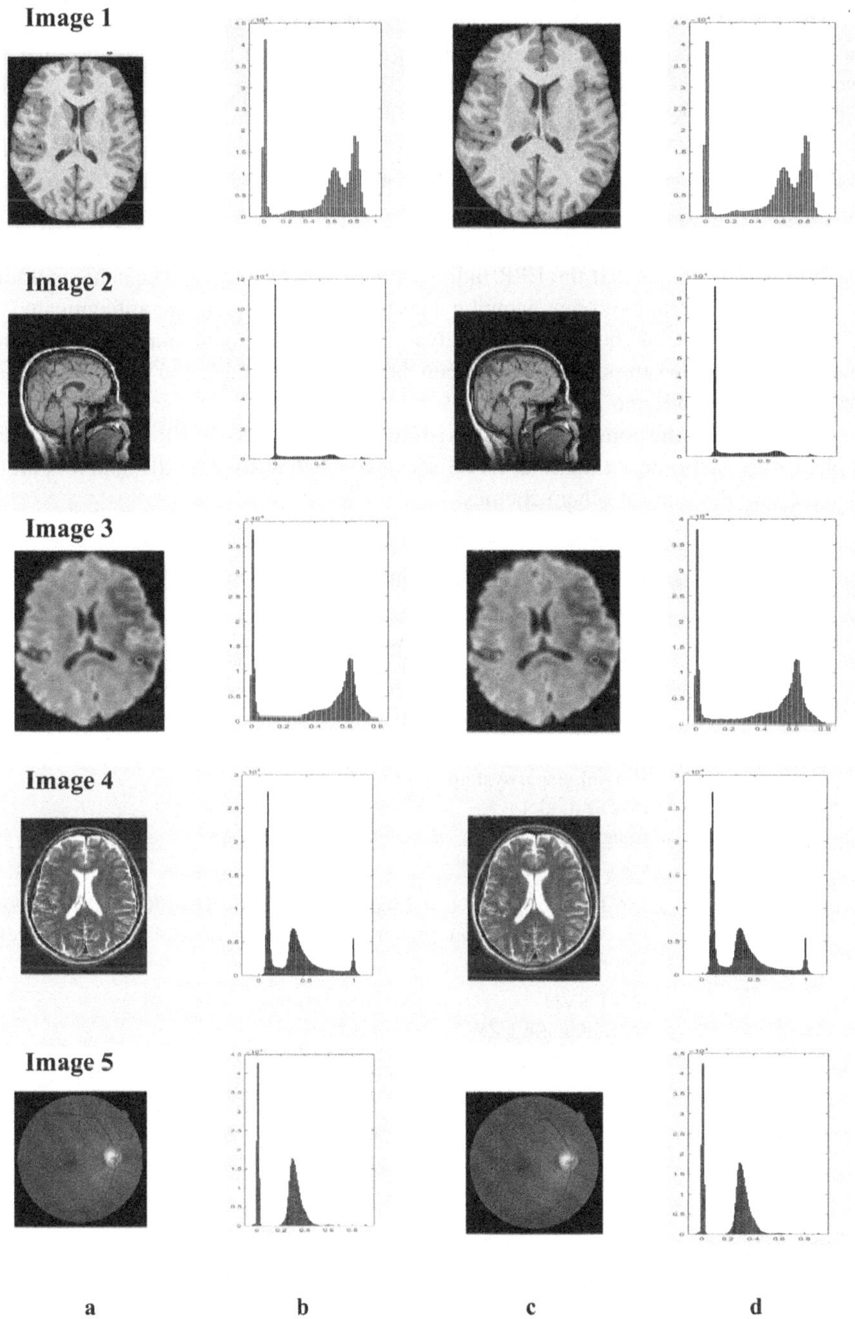

Fig. 5. Histogram analysis of original and watermarked image, a: Original medical image, b: Original medical image histogram, c: Watermarked medical image, d: Watermarked medical image histogram.

The SNR values are used in image watermarking to evaluate the performance of the method, and compare the quality images and distortion between the original and watermarked images. When the value of SNR is more than 32 dB it means that the watermarked image has an excellent quality. In the Table 1 the last value of SNR equal to 33.1776 dB and the more value is equal to 42.0824 dB, that's mean that the quality of medical image watermark excellent and the necessary information about the patient's illness is not changed.

The NC values of the proposed method are between 0.9986 and 0.9999, and they are reaching 1, this means that the EPR hiding in the medical image is the same extracted data from the watermarked one, which means that the hiding data are authenticated.

Figure 5 represents the histogram analysis of both the medical image and the watermarked one, it shows the similarity between them for each image, and that more explain the value of the SNR and PSNR.

Table 2 shows the comparison of 4 existing schemes [3, 06, 9, 10] when no attack is applied with the proposed method, its PSNS value reaches 46.4039 dB, and it is higher than the PSNR values of other schemes.

Table 2. Imperceptibility results comparison with literature methods

Paper reference	Techniue used	Application domain	PSNR
[9]	RC4 for encryption DFT	Medical image	44
[3]	Human visual system (HVS) model Image processing operations	Medical image	38.01
[8]	DWT EH-MB optimization algorithm	Medical and non-medical image	42.1776 (medical image) 38.01444 (Lena image)
[5]	DWT-SVD ECC, chaotic-LZW	Medical and Non medical image	44.1944
Our method (2021)	Snake segmentation DWT, LSB algorithm	Medical image	**46.4039**

4 Conclusion

The LSB algorithm is a technique used in the spatial domain for hiding data bits in the last signification bit of the cover image, to make the embedding data more robust, we apply it in the frequency domain. The approach is based on the Haar wavelet to transform the RONI to the frequency domain, and trait the LL2 band like the new cover image and hide the EPR in it by the LSB algorithm. For the watermarked original medical image we apply the IDWT and combine the ROI with RONI watermarked. The experimental results show that the proposed method has good values in SNR, PSNR, and NC. That represents the impeccability and security of the EPR embedding, excellent image quality without destroying the necessary health information which is excited in ROI. In the future work, we plan to embed more capacity data and apply different encryption systems to encrypt the image. And embedding more capacity data.

References

1. Assini, I., Badri, A., Safi, K.H., Sahel, A., Baghdad, A.: A robust hybrid watermarking technique for securing medical image. Int. J. Intell. Eng. Syst. **11**(3), 169–176 (2018)
2. Singh, P., Chadha, R.S.: A survey of digital watermarking techniques, applications and attacks. Int. J. Eng. Innov. Technol. **2**, 165–175 (2013)
3. Thanki, R., Borra, S., Dwivedi, V., Borisagar, K.: A RONI based visible watermarking approach for medical image authentication. J. Med. Syst. **41**(143), 1–11 (2017)
4. Prabu, S., Balamurugan, V., Vengatesan, K.: Design of cognitive image filters for suppression of noise level in medical images. Measurement **141**, 296–301 (2019)
5. Anand, A., Kumar Singh, S.: An improved DWT-SVD domain watermarking for medical information security. Comput. Commun. **152**, 72–80 (2020)
6. Khawatreh, S., Nader, J., Khrisat, M., Eltous, Y., Alqadi, Z.: Securing LSB2 message steganography. Int. J. Comput. Sci. Mob. Comput. **2**, 156–164 (2020)
7. Sivaganesan, S., Geetha, M., Gowthaman, T., Pradeepa, M.: Fingerprint based watermarking using DWT and LSB Algorithm. Int. J. Pharm. Res. Technol. **10**(2), 10–14 (2020)
8. Ambika, Biradar, R.L.: Secure medical image steganography through optimal pixel selection by EH-MB pipelined optimization technique. Health Technol. **10**, 231–247 (2020)
9. Khalil, M.I.: Medical image steganography: study of medical image quality degradation when embedding data in the frequency domain. Int. J. Comput. Netw. Inf. Secur. **9**(2), 22–28 (2017)
10. Kashyap, N.: Image watermarking using 3-level Discrete Wavelet Transform (DWT). Int. J. Mod. Educ. Comput. Sci. **3**, 50–56 (2012)
11. Anand, A., Singh, A.K.: An improved DWT-SVD domain watermarking for medical information security. Int. J. Comput. Telecommun. Ind. **152**, 72–80 (2020)
12. Singh, A., Dutta, M.K.: Lossless and robust digital watermarking scheme for retinal images. In: 4th International Conference on Computational Intelligence & Communication Technology (CICT), pp. 1–5. IEEE, India (2018)
13. Walker, J.S.: A Primer on WAVELETS and their Scientific Applications, 2nd edn. Taylor & Francis Group, USA (2008)
14. Mahmoud, M.I., Dessouky, M.I., Deyab, D., Elfouly, F.H.: Comparison between Haar and Daubechies wavelet transformations on FPGA technology. In: Proceedings of World Academy of Science, Engineering and Technology, vol. 20, pp. 1307–6884 (2007)
15. Majeed, M.A., Sulaiman, R.: An improved LSB image steganography technique using bit-inverse in 24 bit colour image. J. Theor. Appl. Inf. Technol. **80**(2), 342–384 (2015)
16. Zhou, W., Xie, Y.: Interactive medical image segmentation using snake and multiscale curve editing. Math. Methods Appl. Med. Imaging **2013**, 1–13 (2013)
17. Kass, M., Witkin, A., Terzopoulos, D.: Snakes: active contour models. Int. J. Comput. **1**, 321–331 (1988)
18. Rebouças Filho, P.P., Silva Barros, A.C., Almeida, J., Rodrigues, J.P.C., de Albuquerque, V.H.C.: A new effective and powerful medical image segmentation algorithm based on optimum path snakes. Appl. Soft Comput. J. **76**, 649–670 (2019)

19. Jeevakala, S., Therese, A.B.: Sharpening enhancement technique for MR images to enhance the segmentation. Biomed. Signal Process. Control **41**, 21–30 (2018)
20. Habeeb, N.J., Omran, S.H., Radih, D.A.: Contrast enhancement for visible-infrared image using image fusion and sharpen filters. In: International Conference on Advanced Science and Engineering, pp. 64–69. IEEE, Iraq (2018)
21. Toennies, K.D.: Guide to Medical Image Analysis, 2nd edn. Springer, London (2017)
22. Ahishakiye, A., Van Gijzen, M.B., Tumwiine, J., Wario, R., Obungoloch, J.: A survey on deep learning in medical image reconstruction. Intell. Med. **1**, 118–127 (2021)

Deep Learning for Seismic Data Semantic Segmentation

Mohammed Anouar Naoui[1(✉)], Nedioui Med Abdelhamid[1], Brahim Lejdel[1], Okba Kazar[2], Nacira Berrehouma[1], and Ridha Berrehouma[1]

[1] University Echahid Hamma Lakhdar, El-Oued, Algeria
[2] Université Mohamed Khider, Biskra, Algeria

Abstract. Drilling for oil and gas is an expensive and time-consuming process. Companies in the oil and gas industry invest millions of dollars in an effort to improve their understanding of subsurface components, and using traditional workflows for interpreting large volumes of seismic data is an important part of this effort. Manually defining links between geological characteristics and seismic patterns is required by geoscientists. As a result, geologists and oil and gas industry businesses resorted to a seismic survey, in which seismic waves provide a wealth of information about what is inside the earth without the need to dig. The main of this paper concerns the identification of salt layers of a seismic image by a computer which often coexist with gas and oil under the ground by proposing a deep Learning for seismic analysis. We propose U-net architecture to discover seismic data. Moreover, we study the data augmentation with U-net architecture. The result of data augmentation can perform 10% the U-net architecture model.

Keywords: Deep learning · Seismic · Salt identification · U-net architecture · Data augmentation

1 Introduction

The seismic survey is an important part of the entire process of petroleum exploration and production, and it can be done either onshore (land) or offshore (marine) by using explosive charges such as airguns for offshore exploration and dynamite or specialized trucks for onshore exploration, which contain a heavy plate vibrated on the ground surface to generate waves that bounce off underground rock formations (hydrophones and geophones) (Fig. 1).

During maritime seismic acquisition, the hydrophone is a device used to detect seismic energy in the form of pressure changes in water. The geophone is a surface seismic acquisition device. The geophone is a device that detects ground velocity produced by seismic waves and converts the motion into electrical impulses. It is employed in surface seismic acquisition, both onshore and offshore. The amount of time it takes for data to travel from the source to the receivers can reveal information about rock density and the presence of fluids or gases. This can aid in the formation of a subsurface image [1].

B. Lejdel et al. (Eds.): AIAP 2021, LNNS 413, pp. 479–485, 2022.
https://doi.org/10.1007/978-3-030-96311-8_44

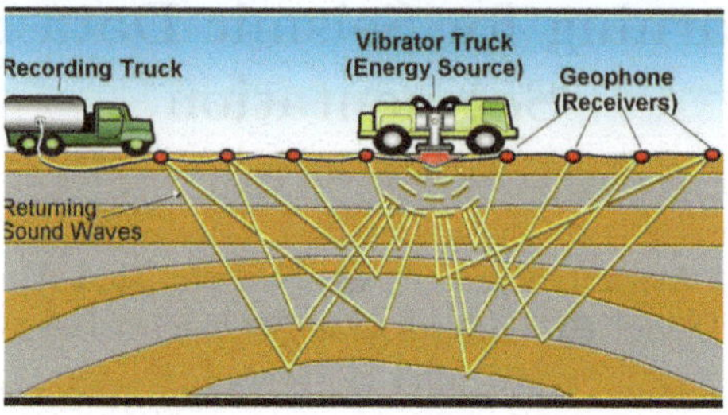

Fig. 1. Vibroseis research [14]

1.1 Salt Structures

Salt Structures are a common geological phenomenon that form as a result of the density difference between salt and the layer above it. Because salt is less dense than the layer above it, it pushes toward the top, forming a dome in the sedimentary layers above it. If the salt dome contains oil, it moves to the outer sides of the salt dome and is confined between the sedimentary layers, forming exploitable oil reservoirs. Because of the physical qualities of salt, identifying the salt structure is one of the challenges of seismic imaging, as the density of salt is typically 2.14 g/cc, which is lower than the density of most surrounding rocks (Fig. 2).

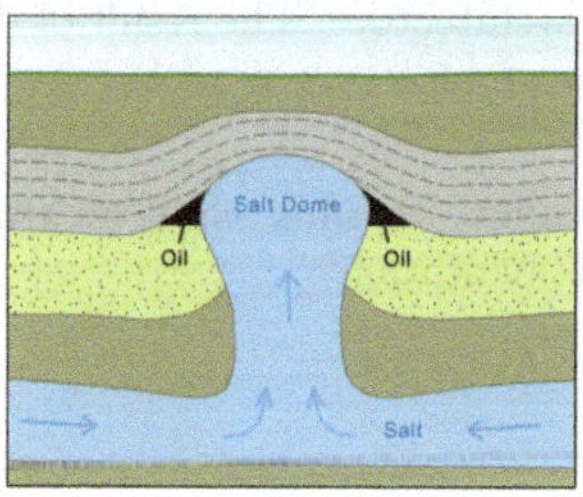

Fig. 2. Salt dome

Salt has a seismic velocity of 4.5 km/s [2], which is quicker than the rocks around it. At the salt-sediment interface, this difference causes a strong reflection. Salt is often an amorphous rock with little interior structure. This means that, unless there are sediments trapped inside the salt, there is usually little reflection. Seismic imaging can be hampered by salt's very high seismic velocity. The advantages of salt identification for oil and gas are [2]:

- Because salt is a good sealant, it is used to create the edges of many hydrocarbon traps. Without analyzing the salt contact, these traps cannot be properly mapped.
- Understanding salt geometry and evolution is therefore vital in forecasting reservoir placement in a salt basin. Salt structures can have a significant impact on sediment transport and, as a result, are fundamental influences on reservoir distribution.
- In many basins, salt's distinctive physical qualities make sub-salt and salt-flank imaging difficult. One method for overcoming these difficulties is pre-stack depth migration, which requires an accurate salt model. As a result, a substantial portion of current salt interpretation is focused on developing velocity models for pre-stack depth migration.

The paper is organized as follows, in Sect. 2 we present related works on seismic images. In the Sect. 3, we present our method of seismic image segmentation. In Sect. 4 we present evaluation and results. Finally, conclusion and future work.

2 Related Works

Olivier et al. [3] proposed a pre-processing step based on non-linear diffusion filtering, leading to a better detection of seismic faults. The non-linear diffusion approaches are based on the definition of a partial differential equation that allows to simplify the images without blurring relevant details or discontinuities.

Mikhail et al. [4] presents A Deep Learning Approach for Automatic Salt Deposit Segmentation, which shows how various unique deep learning approaches can be combined into a single neural network to achieve an excellent result. Authors proposed U-Net with ResNeXt-50 encoder pre-trained on ImageNet.

Milosavljević et al. [5] proposed a deep learning model. The architecture of the proposed network is inspired by the U-Net model in combination with ResNet and DenseNet architecture.

Yauhen et al. [6] propose a semi-supervised method for segmentation (delineation) of salt bodies in seismic images which utilizes unlabeled data for multi-round self-training. For the training model, authors used U-ResNet34 and U-ResNeXt50.

3 Proposition

Our proposed method for seismic data analysis is composed by the following steps:

- Data augmentation.
- U-net architecture for train and test data.

Data Augmentation. To artificially enhance the size of an actual dataset, data augmentation technics generate different versions of it. To deal with data scarcity and insufficient data diversity, computer vision and natural language processing (NLP) models employ data augmentation strategies [7–9]. A data-space solution to the problem of limited data is data augmentation. Data Augmentation is a term that refers to a range of strategies for increasing the size and quality of training datasets so that stronger Deep Learning models may be generated using them [7]. There are many strategies for image augmentation such as scaling (zoom in/zoom out), rotation, reflection, shear.

U-Net Architecture. Olaf Ronneberger et al. [10] created the U-net for Bio Medical Image Segmentation. There are two ways in the architecture. The contraction path (also known as the encoder) is the first path, and it is used to capture the image's context. The encoder is simply a convolutional and maximum pooling layer stack. The symmetric expanding path (also known as the decoder) is the second way, and it is employed to achieve exact localization via transposed convolutions. As a result, it is a fully convolutional network from beginning to end (Fig. 3).

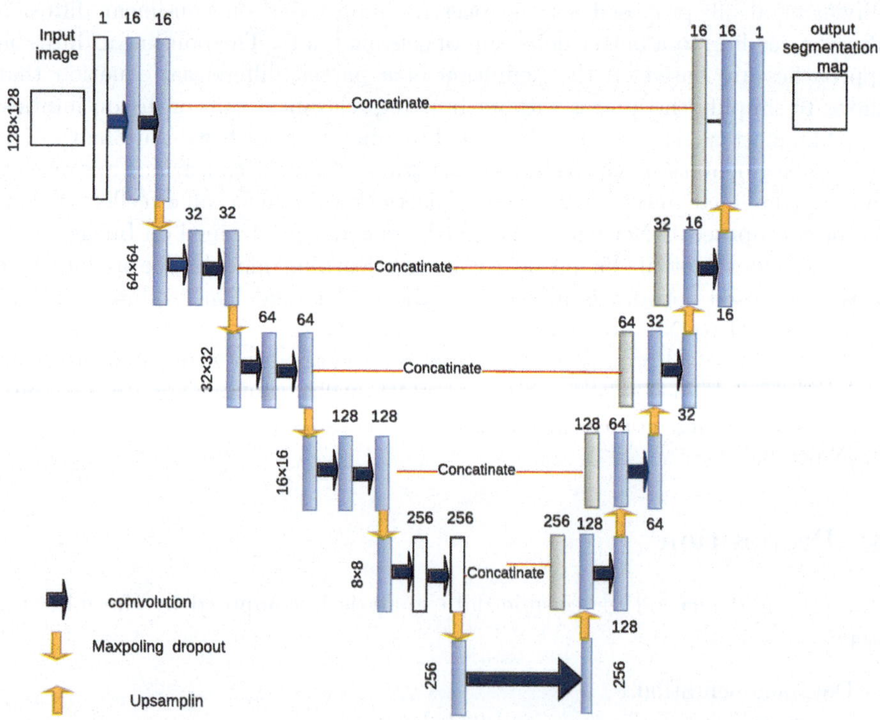

Fig. 3. U-net architecture

4 Evaluation and Results

For the evaluation of the proposed model, we used the Data [11] from TGS, a company of energy data and intelligence. The data consists of a collection of photos taken at various subsurface sites chosen at randomly. Images are 101 by 101 pixels in size, with each pixel labeled as salt or sediment. Each image includes the depth of the photographed place in addition to the seismic images. We are compared between U-net with augmented data and U-net without augmented data. The metric of the evaluation is mean average precision at different intersection over union IoU. The overlap between two borders is measured by IoU. It is used to determine the extent to which the anticipated border overlaps the ground truth (the real object boundary Fig. 4). In some datasets, IoU uses a predefined threshold to determine whether a prediction is true positive or a false positive [12].

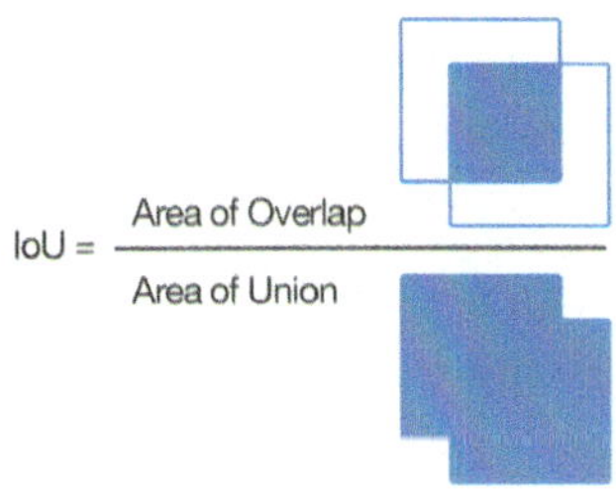

Fig. 4. Intersection over union [12]

For example The IoU is the area of overlap between the predicted and ground-truth bounding boxes B_p and B_gt divided by the area of union between them [12] (Table 1).

$$J(B_p, B_{gt}) = \frac{area(B_p \cap B_{gt})}{area(B_p \cup B_{gt})} \tag{1}$$

Table 1. Results of U-net

Model	IoU
U-net with augmented image	0.70
U-net without augmented image	0.62

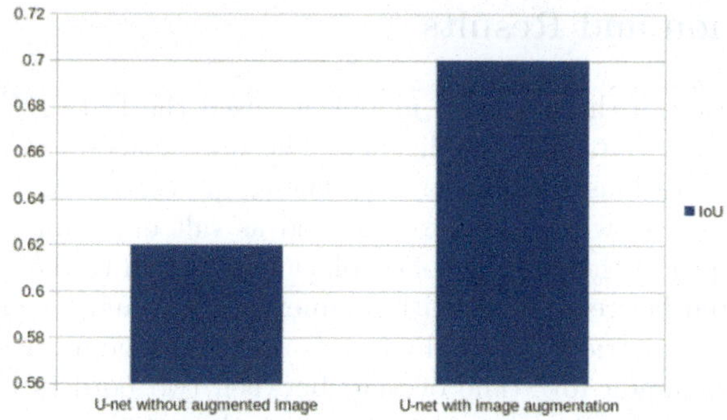

Fig. 5. Results of models

Result Discussion. We have compared between two method based U-net architecture, in the first we use data augmentation and in the second we used only U-net architecture without data augmentation. The IoU when used data augmentation is 0.70, and for second is 0.60. From the result, we summary the following (Fig. 5):

- U-net is an architecture can extract semantic segmentation.
- U-net needs images augmentation to perform the model.

5 Conclusion

Gas and oil discovery with artificial neuronal network become an interesting field to understand complex data and help expert decision. We are proposed U-net architecture for seismic data segmentation. Moreover, we studied the performance of data augmentation for U-net architecture. In the first use's case, we used U-net architecture with data augmentation, and in the second we used only U-net architecture for seismic semantic segmentation. The study illustrates the importance of U-net architecture with data augmentation. In the future work, we will propose other data augmentation technics.

References

1. Mondol, N.H.: Seismic exploration. In: Bjorlykke, K. (ed.) Petroleum Geoscience, pp. 375–402. Springer, Heidelberg (2010). https://doi.org/10.1007/978-3-642-02332-3_17
2. Jackson, M.P.A., Hudec, M.R.: Salt Tectonics: Principles and Practice, Tectonics, pp. 132–141. Cambridge University Press, Cambridge (2017)
3. Lavialle, O., Pop, S., Germain, Ch., et al.: Seismic fault preserving diffusion. Salt **61**, 132–141 (2007)
4. Karchevskiy, M., Insaf, A., Leonid, K.: Automatic salt deposits segmentation: a deep learning approach. arXiv preprint arXiv:1812.01429 (2018)

5. Milosavljević, A.: Identification of salt deposits on seismic images using deep learning method for semantic segmentation. ISPRS Int. J. Geo-Inf. 9–24 (2020). https://doi.org/10.3390/ijgi9010024
6. Babakhin, Y., Sanakoyeu, A., Kitamura, H.: Semi-supervised segmentation of salt bodies in seismic images using an ensemble of convolutional neural networks. In: Fink, G.A., Frintrop, S., Jiang, X. (eds.) DAGM GCPR 2019. LNCS, vol. 11824, pp. 218–231. Springer, Cham (2019). https://doi.org/10.1007/978-3-030-33676-9_15
7. Shorten, C., Khoshgoftaar, T.M.: A survey on image data augmentation for deep learning. J. Big Data **6**(1), 1–48 (2019)
8. Wong, S.C., et al.: Understanding data augmentation for classification: when to warp? In: 2016 International Conference on Digital Image Computing: Techniques and Applications (DICTA). IEEE (2016)
9. Mikołajczyk, A., Michał, G.: Data augmentation for improving deep learning in image classification problem. In: 2018 International Interdisciplinary PhD Workshop (IIPhDW). IEEE (2018)
10. Ronneberger, O., Fischer, P., Brox, T.: U-Net: convolutional networks for biomedical image segmentation. In: Navab, N., Hornegger, J., Wells, W.M., Frangi, A.F. (eds.) MICCAI 2015. LNCS, vol. 9351, pp. 234–241. Springer, Cham (2015). https://doi.org/10.1007/978-3-319-24574-4_28
11. TGS salt identification challenge segment salt deposits beneath the Earth's surface. https://www.kaggle.com/c/tgs-salt-identification-challenge/data. Accessed 4 Sept 2021
12. Padilla, R., Netto, S.L., da Silva, E.A.: A survey on performance metrics for object-detection algorithms. In: 2020 International Conference on Systems, Signals and Image Processing (IWSSIP). IEEE (2020)
13. Vibroseis Research. https://www.cem.utexas.edu/content/vibroseis-research. Accessed 4 Sept 2021
14. What is a Salt Dome. https://geology.com/stories/13/salt-domes/. Accessed 4 Sept 2021

Feature Fusion for Kinship Verification Based on Face Image Analysis

Fatima Zekrini(✉), Hassiba Nemmour, and Youcef Chibani

Laboratory d'Ingénierie des Systèmes Intelligents et Communicants (LISIC), Faculty of Electronic and Computer Sciences, University of Sciences and Technology Houari Boumediene (USTHB), Bab Ezzouar El-Alia BP. 32, 16111 Algiers, Algeria
{azekrini,hnemmour,ychibani}@usthb.dz

Abstract. This paper proposes the fusion of two new features for improving kinship verification based on face image analysis. Combined features are the Gradient Local Binary Patterns (GLBP), which associates gradient and textural information. The second descriptor is the Histogram Of Templates (HOT), which is a shape descriptor. These features are utilized with the support vector machines classifier to develop the kinship verification. Experiments are carried out on Cornell and Kinface W-II datasets. Results obtained highlight the effectiveness of the proposed system which provide competitive and sometimes better performance than the state of the art.

Keywords: Kinship verification · GLBP · HOT · SVM

1 Introduction

Automatic Kinship Verification from face images consists of determining whether a kin relation exists for a given pair of facial images. This task is useful in various applications such as finding missing children, WEB images annotation, and social media analysis. The underlying idea is that people from the same family share similar face features that cannot vary according to the age or the sex. Therefore, a Kin verification system is founded on comparing features of two image faces through simple dissimilarity metrics or by using dissimilarity learning techniques. Recall that in facial image analysis, we are usually able to extract multiple feature representations where various kinds of textural, gradient, and shape features are currently used with a notable success. So, compared to face recognition or verification, that are widely used in biometrics, the kin verification is considered as a new application, that derives from biometrical face analysis.

Recently, there has been a lot of efforts in developing methods of kinship verification systems. Mainly, proposed methods can be categorized into two classes that are feature-based methods and model-based methods [1, 2]. In the first approach, methods aim to extract discriminative information to preserve stable kin-related characteristics. Representative methods in this category include the Histogram Of Gradient (HOG) [1, 3], Salient Part [4], Self-Similarity [5], and Dynamic Spatio-Temporal Descriptor [6].

B. Lejdel et al. (Eds.): AIAP 2021, LNNS 413, pp. 486–494, 2022.
https://doi.org/10.1007/978-3-030-96311-8_45

In this respect, features are directly compared through distance measure to decide about the kin relation.

In the second approach, methods can be divided into two classes: Methods using metric learning and Methods using deep learning [7]. The aim of metric learning methods consists of extracting more pertinent kin decision that can reduce the distance between positive pairs (images representing a real kin relation), while enlarging the distance between negative pairs (images representing a fake kin relation). In this respect, several supervised classifiers were used in the state of the art such as the large margin nearest neighbor [8], information theoretic metric learning [9], metric embedding [10], pairwise constrained component analysis [11], and Support Vector Machines (SVM) [12]. Note that SVM stills one of the most effective classifiers and stills being the most commonly used.

Furthermore, with the huge performance of deep learning techniques, Convolutional Neural Networks (CNN) were used as a deep learning kin models [13, 14]. However, the CNN is commonly effective when handling the face images, while for kin verification, it should learn distance measures. For this reason, the verification scores derived from various CNN models are medium [15]. Therefore, the use of handcrafted features associated with machine learning techniques still remain an effective technique to develop kinship verification systems.

In this work, we propose the combination of two new features for improving the kinship verification. The first descriptor is the Gradient Local Binary Pattern (GLBP), which takes advantage from gradient and textural traits [16], and the Histogram Of Templates (HOT) [17], which is a shape descriptor. Both features were originally introduced for human detection, but they show satisfactory performance for other applications such as handwritten signature verification and document analysis [18, 19]. Presently, these descriptors are used to extract face features. The verification step is achieved by a SVM classifier that is trained to separate positive face image pairs from the negative ones. Experiments are conducted on two public datasets.

The rest of the paper is organized as follows: Sect. 2 details the proposed kinship verification system. Section 3 presents and discusses the experimental results. Section 4 concludes the paper.

2 Proposed Kinship Verification System

Commonly, a kinship verification system is composed of two main steps that are feature generation and distance metric learning (See Fig. 1). Given a set of training face images, we first extract features for each face image and consider couples of real and fake child-parent features, by using the difference between feature vectors. These features are then, trained by a classifier that decides whether there is a kinship relationship or not between the two face images. In this work, we propose to reinforce face features by combining two new descriptors. Precisely, we propose the Histogram Of Template (HOT), which is a shape descriptor, with the Gradient Local Binary Patterns (GLBP) that associates gradient and texture information. The distance metric learning is achieved by a SVM classifier.

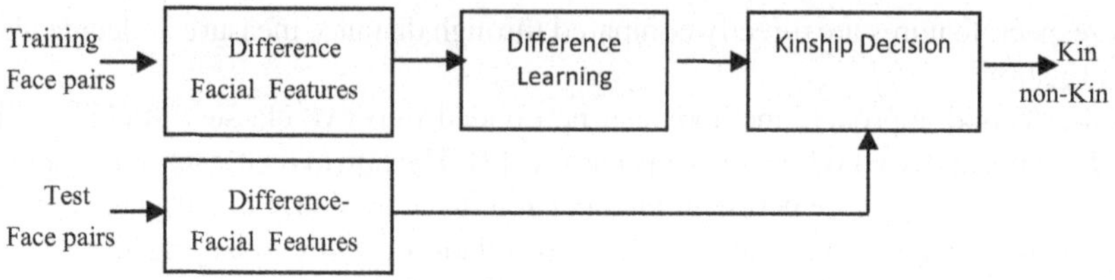

Fig. 1. Framework of kinship verification based on facial image analysis

2.1 Proposed Features

In order to get robust facial features, we propose to combine two descriptors that characterize different trait properties. Precisely, we use the Histogram Of Template (HOT) which is designed to highlight local shape features, while the second descriptor is the Gradient Local Binary Patterns (GLBP) that associates gradient and textural information. For each face image both features are independently computed and concatenated to form the face feature vector.

2.1.1 Gradient Local Binary Patterns

GLBP (Local gradient binary Patterns) was introduced for human detection [20]. Its basic principle is to calculated the gradient information at one to zero (or zero to one) transitions of the local binary patterns code. Recall that LBP aim to characterize the distribution of grey levels in the pixel surrounding by comparing the value of the grey level of a central pixel with the neighboring grey levels. For each pixel in the face image, GLBP is calculated according to the following steps:

1. Calculate LBP code
2. Calculation of width and angle such as:

   - The width value: corresponds to the number of "1" in the uniform LBP code and this number of "1" can vary from 1 to 7.
   - The angle value: corresponds to the freeman direction of the medium pixelin the one value area of the uniform LBP (See Fig. 2).

3. The width and angle values define the position in the GLBP matrix that is filled by the accumulation of gradient values calculated at one to zero (or zero to one) transition such as:

$$G = \sqrt{(I(X+1,Y) - I(X+1,Y-1))^2 + (I(X,Y+1) - I(X-1,Y+1))^2} \quad (1)$$

2.1.2 Histogram of Templates

HOT (Histogram Of Templates) was initially introduced to improve local shapes in human detection applications [21]. Afterwards it was successfully employed in various

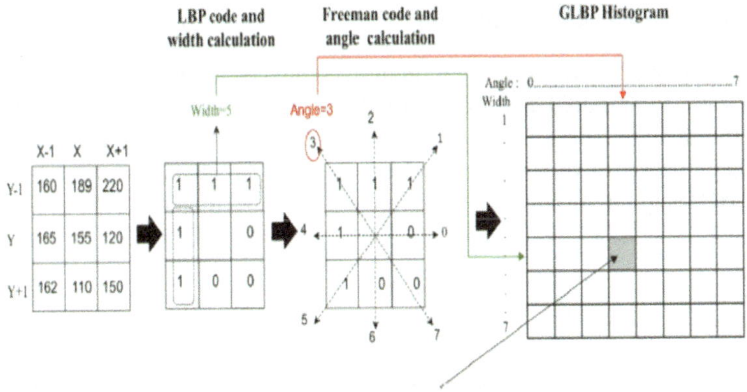

Fig. 2. GLBP calculation for a given pixel [20]

handwritten recognition tasks [22]. Roughly, HOT considers local shape orientations through relationships between pixels and their neighbors. This description is done using a set of 20 templates representing all possible orientations of a triplet of pixels (See Fig. 3).

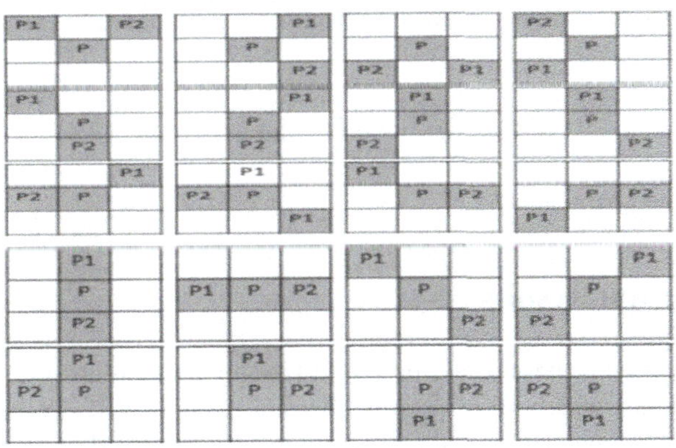

Fig. 3. Models used in the HOT calculation [21]

The generation of HOT characteristics consists in applying each template to all pixels of the face image. A pixel is said to fit a template if it verifies the following condition:

$$I(P) > I(P_1) \&\& I(P) > I(P_2) \quad (2)$$

I(P): Gradient intensity calculated by Sobel filtering.

Then, the histogram of templates contains the number of pixels that fit to each Template.

2.2 SVM-Based Kinship Verification

To develop the kinship verificatory, training face images are grouped into two classes. The first class is composed of truth child-parent image couples, while the second class contains the same number of false child-parent couples. Each couple is represented by the absolute difference vector calculated between the face features (that are generated by using HOT and GLBP), as described in the following equation:

$$Z_i = |A_i - B_i| \tag{3}$$

Where i = 1: N, and N is the size of feature vectors.

Then, a SVM classifier is trained on difference features. Recall that the training of SVM aims to find the optimal hyper-plane separating two classes. After the training, we get the following decision function:

$$f(A) = sign\left(\sum_{j=1}^{S_v} \alpha_j y_j K(A, B_j) + b\right) \tag{4}$$

y_j: Class label $\{+1, -1\}$.

S_v: Is the number of support vectors that represent training data for:

$0 \leq \alpha \leq C_j$: The bias b is a scalar while C is the cost parameter.

K is the SVM kernel. Presently, we employ the RBF (Radial Basis Function) kernel because of its proven performance in handwritten recognition [16]. This kernel is expressed as:

$$\left(K(A, B_i) = exp\left(-\gamma \|A - B_i\|^2\right)\right) \tag{5}$$

γ: user defined parameter.

3 Experimental Results

To evaluate the effectiveness of our proposed kinship verification system, we conduct experiments on two widely used datasets that are, KinFaceWII, Cornell KinFace[1,2]. Figure 1 presents some sample kin pairs from the KinFaceW-II and Cornell KinFace datasets. These datasets are composed of four kin relations classes that are: Father-Son (F-S), Father-Daughter (F-D), Mother-Son (M-S), Mother-Daughter (M-D). The KinFaceW-II contains 250 pairs for each category, while the Cornell set contains 150 pairs per category. In our experiments, 2/3 pairs for each category were used in the training stage, and the remaining pairs were used for performance assessment. Figure 4 presents some sample kin pairs from the two datasets.

The SVM-based kinship verification is developed by using the RBF kernel. User-defined parameters were experimentally tuning by considering cross-validation. Tables 1 and 2 summarize the verification accuracies obtained for the two datasets. Roughly, the verification accuracies are better on the Cornell dataset which contains a smaller

[1] https://www.ranksays.com/siteinfo/kinfacew.com.

[2] https://www.ranksays.com/siteinfo/Cornellkinface.com.

(a) (b)

Fig. 4. Some samples from the adopted datasets: (a) KinFace-WII [19], (b) Cornell dataset [6]

amount of data. For both sets, the most complicated task is the Father-Son verification, which is achieved by medium performance compared to the Father-Daughter kinship. Furthermore, the two proposed features provide approximately similar performance, where the difference in the average precision is 1% for the Kinface dataset and 1.42% for the Cornell corpus. Nevertheless, the proposed combination allows a significant improvement of the verification scores. Specifically, for the Kinface-WII dataset, the combination provides a gain of 5.16% in the average precision. For Cornell dataset set, for which individual features provide higher scores, the gain is about 2.51%. These outcomes highlight the complementarity between the two features, despite of having close individual precisions.

Table 1. Kinship results obtained for the Kinface-WII dataset (%)

Features	F-D	F-S	M-D	M-S	Average precision
GLBP	81.33	62.66	70	73.33	71.83
HOT	74.66	62	72.66	74	70.83
GLBP+HOT	84.66	70	75.33	78	76.99

Table 2. Kinship results obtained for the Cornell dataset (%)

Features	F-D	F-S	M-D	M-S	Average precisions
GLBP	87.5	80	88.88	90	86.56
HOT	87.5	80	94.44	90	87.98
GLBP+HOT	92.5	85	94.44	90	90.49

Furthermore, Table 3 reports some published results obtained on adopted datasets. As can be seen, our proposed system based on the GLBP and HOT fusion provides a competitive performance, since it outperforms several state-of-the-art systems. More precisely, the hierarchical representation learning proposed by Kohli et al. [34] achieved the best accuracy performance so far on several benchmark datasets. Nevertheless, our system achieves the second-best accuracy performance while being simpler to elaborate.

Table 3. Kinship results published in the state of the art (%)

Reference	Cornell	Kinface-WII
[24]	71.60	74.70
[25]	71.10	76.00
[26]	71.70	76.30
[27]	73.70	78.30
[28]	71.90	77.00
[29]	76.70	80.40
[30]	73.00	75.70
[31]	75.80	79.30
[32]	74.00	76.70
[33]	79.00	81.80
[34]	89.50	96.20
Our system	90.49	76.99

4 Conclusion

In this paper, proposed the combination of two descriptors to perform a robust kinship verification. The first descriptor is the Gradient Local Binary Patterns (GLBP) that associates the gradient and textural information, while the second descriptor is the Histogram Of Templates (HOT) which highlights local shapes. These features are concatenated to characterize the kinship relations in face images. The verification step is achieved by SVM classifier. Experiments conducted on two benchmark datasets, confirm the effectiveness of the proposed combination, which offers similar and sometimes higher performance than the state of the art. To improve again the verification scores, as a future work, we plan to associate other kinds of features such as CNN-based features, and use strength fusion rules such as the fuzzy integral combiners.

References

1. Fang, R., Tang, K.D., Snavely, N., Chen, T.: Towards computational models of kinship verification. In: Proceedings International Conference Image Processing, September 2010, pp. 1577–1580 (2010)
2. Lu, J., Zhou, X., Tan, Y.-P., Shang, Y., Zhou, J.: Neighborhood repulsed metric learning for kinship verification. IEEE Trans. Pattern Anal. Mach. Intell. **36**(2), 331–345 (2014)
3. Zhou, X., Lu, J., Hu, J., Shang, Y.: Gabor-based gradient orientation pyramid for kinship verification under uncontrolled environments. In: Proceedings ACM International Conference on Multimedia, pp. 725–728 (2012)
4. Guo, G., Wang, X.: Kinship measurement on salient facial features. IEEE Trans. Instrum. Meas. **61**(8), 2322–2325 (2012)

5. Kohli, N., Singh, R., Vatsa, M.: Self-similarity representation of weber faces for kinship classification. In: Proceedings IEEE International Conference Biometrics, Theory, Application System, September 2012, pp. 245–250 (2012)
6. Dibeklioglu, H., Salah, A. A., Gevers, T.: Like father, like son: facial expression dynamics for kinship verification. In: Proceedings IEEE International Conference on Computer Vision, pp. 1497–1504 (2013)
7. Weinberger, K.Q., Blitzer, J., Saul, L.K.: Distance metric learning for large margin nearest neighbor classification. In: Proceedings Advances in Neural Information Processing System, 2005, pp. 1473–1480 (2007)
8. Davis, J.V., Kulis, B., Jain, P., Sra, S., Dhillon, I.S.: Information theoretic metric learning. In: Proceedings ICML, pp. 209–216 (2007)
9. Köstinger, M., Hirzer, M., Wohlhart, P., Roth, P.M., Bischof, H.: Large scale metric learning from equivalence constraints. In: Proceedings IEEE Conference on Computer Vision and Pattern Recognition, pp. 2288–2295 (2012)
10. Mignon, A., Jurie, F.: PCCA: a new approach for distance learning from sparse pairwise constraints. In: Proceedings IEEE Conference on Computer Vision and Pattern Recognition, pp. 2666–2672 (2012)
11. Yeung, D.-Y., Chang, H.: A kernel approach for semisupervised metric learning. IEEE Trans. Neural Netw. **18**(1), 141–149 (2007)
12. Chapelle, O., Haffner, P., Vapnik, V.N.: Support vector machines for histogram-based image classification. IEEE Trans. Neural Netw. **10**, 1055–1064 (1999)
13. Ji, S., Xu, W., Yang, M., Yu, K.: 3D convolutional neural networks for human action recognition. IEEE Trans. Pattern Anal. Mach. Intell. **35**(1), 221–231 (2013)
14. Wayman, J.L.: Fundamentals of biometric authentication technologies. Int. J. Image Graph. **1**(1), 93–113 (2001)
15. Rachmadi, R.F., Purnama, I.K.E., Nugroho, S.M.S., Suprapto, Y.K.: Image-based kinship verification using fusion convolutional neural network. In: IEEE 11th International Workshop on Computational Intelligence and Applications, 9–10 November (2019)
16. Bouadjenek, N., Nemmour, H., Chibani, Y.: Robust soft-biometrics prediction from off-line handwriting analysis. J. Appl. Soft Comput. **46**, 980–990 (2016)
17. Serdouk, Y., Nemmour, H., Chibani, Y.: Handwritten signature verification using the quad-tree histogram of templates and a support vector based artificial immune classification. Image Vis. Comput. J. **66**, 26–35 (2017)
18. Serdouk, Y., Nemmour, H., Chibani, Y.: New gradient features for off-line handwritten signature verification. In: International Symposium on Innovations in Intelligent SysTems and Applications (INISTA), Madrid, 2–4 September (2015)
19. Bouibed, M.L., Nemmour, H., Chibani, Y.: New gradient descriptor for keyword spotting in handwritten documents. In: 3rd International Conference on Advanced Technologies for Signal and Image Processing – ATSIP 2017, 22–24 Fez–May (2017)
20. Jiang, N., Xu, J., Yu, W., Goto, S.: Gradient local binary patterns for human detection. In: IEEE International Symposium on Circuits and Systems (ISCAS), Beijing, pp. 978–981 (2013)
21. Tang, S., Goto, S.: Histogram of template for human detection. In: International Conference on Acoustics, Speech and Signal Processing, pp. 2186–2189 (2010)
22. Bouibed, M.L., Nemmour, H., Chibani, Y.: Writer retrieval using histogram of templates features and SVM. In: International Conference on Electrical Engineering and Control Applications, Constantine, 21–23 November, pp. 537–544 (2019)
23. Bertolini, D., Oliveira, L.S., Justino, E., Sabourin, R.: Texture-based descriptors for writer identification and verification. Expert Syst. Appl. **40**, 2069–2080 (2013)
24. Lu, J., Zhou, X., Tan, Y.-P., Shang, Y., Zhou, J.: Neighborhood repulsed metric learning for kinship verification. IEEE Trans. Pattern Anal. Mach. Intell. **36**, 331–345 (2014)

25. Weinberger, K.Q., Blitzer, J., Saul, L.K.: Distance metric learning for large margin nearest neighbor classification. In: Advances in Neural Information Processing Systems, pp. 1473–1480 (2005)
26. Davis, J.V., Kulis, B., Jain, P., Sra, S., Dhillon, I.S.: Information-theoretic metric learning. In: Proceedings of the 24th ACM International Conference on Machine Learning, pp. 209–216 (2007)
27. Yan, H., Lu, J., Deng, W., Zhou, X.: Discriminative multimetric learning for kinship verification. IEEE Trans. Inf. Forensics Secur. **9**, 1169–1178 (2014)
28. Yan, H., Lu, J., Zhou, X.: Prototype-based discriminative feature learning for kinship verification. IEEE Trans. Cybern. **45**, 2535–2545 (2015)
29. Lu, J., Hu, J., Tan, Y.-P.: Discriminative deep metric learning for face and kinship verification. IEEE Trans. Image Process. **26**, 4269–4282 (2017)
30. Zhou, X., Shang, Y., Yan, H., Guo, G.: Ensemble similarity learning for kinship verification from facial images in the wild. Inf. Fusion **32**, 40–48 (2016)
31. Xie, P.: Learning compact and effective distance metrics with diversity regularization. In: Appice, A., Rodrigues, P.P., Santos Costa, V., Soares, C., Gama, J., Jorge, A. (eds.) ECML PKDD 2015. LNCS (LNAI), vol. 9284, pp. 610–624. Springer, Cham (2015). https://doi.org/10.1007/978-3-319-23528-8_38
32. Mignon, A., Jurie, F.: CMML: a new metric learning approach for cross modal matching. In: Asian Conference on Computer Vision, South Korea, pp. 1–14 (2012)
33. Liong, V.E., Lu, J., Tan, Y.P., Zhou, J.: Deep coupled metric learning for cross-modal matching. IEEE Trans. Multimed. **19**, 1234–1244 (2016)
34. Kohli, N., Vatsa, M., Singh, R., Noore, A., Majumdar, A.: Hierarchical representation learning for kinship verification. IEEE Trans. Image Process. **26**, 289–302 (2017)

Image Processing: Image Compression Using Compressed Sensing, Discrete Cosine Transform and Wavelet Transform

A. Bekki[1(✉)] and A. Korti[2]

[1] Biomedical Engineering, Faculty of Technology Tlemcen, University of Abou Bekr Belkaid, 13000 Tlemcen, Algeria
amane.bekki@univ-tlemcen.dz

[2] Electronic Engineering, Faculty of Technology Tlemcen, University of Abou Bekr Belkaid, 13000 Tlemcen, Algeria
amel.korti@univ-tlemcen.dz

Abstract. Recently, the image quality and the speed of acquisition are very studied, particularly in the medical field. We try to find in our paper the most efficient compression method, which allows having good image quality with a short compression period. Reducing the compression time amounts to reducing the acquisition time.

In our article, we proposed three compression methods applied to the medical image: the discrete cosine transform (DCT) method, the wavelet transformation (DWT) and the compressed sensing method (CS). We studied also the acquisition time. In our results, we found that the DWT method gives better image quality compared to other methods. We found also that the CS method is faster than other methods.

Keywords: Compression image · Discrete Cosine (DCT) · Wavelet Transform (WT) · Compressed sensing (CS) · Medical image

1 Introduction

Image quality is very important in the medical field. it gives more information to the physical, anatomical or functional data of a human. Good medical image quality helps radiologists and doctors to give the correct diagnosis.

In addition, the acquisition time is very important in medical imaging.For faster decision about doctors and for the comfort of the patient, it is necessary to reduce the acquisition time.

Image compression is one of the most widely used in image processing techniques today. Its role is to reduce the size of the image in order to reduce the space. It facilitates processing with a reduced number of data in a short time. Different compression methods used such as the discrete cosine Transform (DCT) (Nasir Ahmed 1972), Wavelet transform (DWT) (Alfred Haar 1909) [2].

JPEG2000 (the Joint photographic Experts Group working group. (1997–2000), etc.)

B. Lejdel et al. (Eds.): AIAP 2021, LNNS 413, pp. 495–503, 2022.
https://doi.org/10.1007/978-3-030-96311-8_46

Unfortunately, the compressed image presents a loss of information, which degrades the image quality.

In this article, we improved the image quality by using less data. We used compression methods to choose the most relevant data. We chose three compression methods: DCT, DWT and CS.

In the following of this article, we presented compression methods used here and the different results obtained from the compression methods with discussion. We ended this article with a conclusion.

2 Methods

In our work, we used three compression methods: DCT, DCT and CS.

2.1 Discrete Cosine Transform (DCT)

The Discrete Cosine Transform (DCT) is a type of fast computing Fourier transform, which maps real signals to corresponding values in frequency domain. The DCT method works on the real part of the complex signal because most of the real-world signals are real signals with no complex components. We will discuss the implementation of DCT algorithm on medical Image Data.

The Fig. 1 represent the diagram that explains the steps in reconstructed compressed image using DCT. The information in the frequency domain DCT (i, j) is obtained from the discrete data of the image img (x, y) where the X and Y axes are the horizontal and vertical dimensions of the image [1, 3, 4]. In this article, we applied the DCT line by line on our image. The compressed image img ′(x, y) is obtained from the frequency data DCT (i, j) by applying the inverse transform DCT (i, j)$^{-1}$.

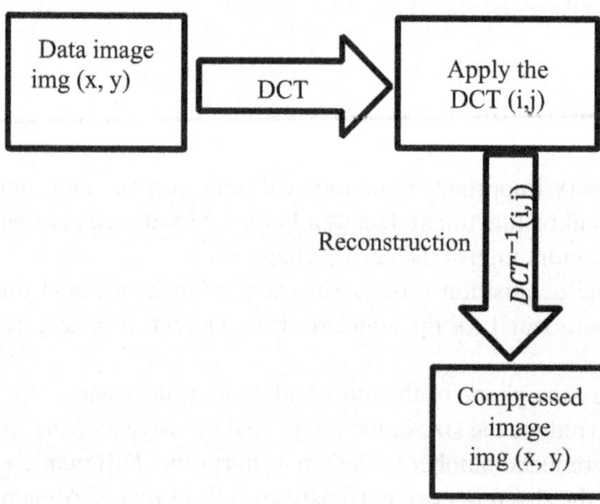

Fig. 1. Diagram for compressed image reconstruction using DCT.

2.2 Discrete Wavelet Transform (DWT)

The Discrete Wavelet Transform (DWT) is a representation of signals or images in a time-frequency form. The Fig. 2 represents the diagram, which explains the reconstruction of the compressed image using wavelets [6]. The time-frequency data obtained from the original image and the compressed image reconstructed from the time-frequency data.

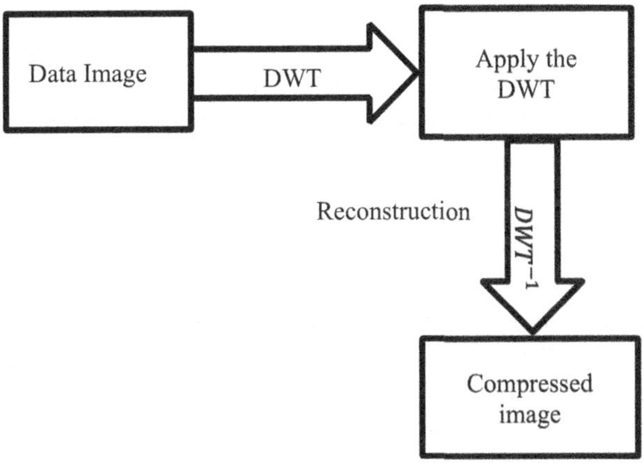

Fig. 2. Diagram for compressed image reconstruction using DWT.

2.3 Compressed Sensing (CS)

Compressed sensing (CS) is a newer compression method. It based on three essential points:

- Sparsity: the desired signal must have a sparse representation in a known transformation domain.
- Incoherency: The subsampled space should generate aliasing artifacts similar to noise in the compression transform domain.
- Nonlinear reconstruction: A nonlinear reconstruction is necessary to exploit the Sparsity while maintaining the consistency of the data acquires [7, 8, 9].

To meet the need for the Sparsity, we applied a mask on our data [5]. The Fig. 3 represent the diagram, which explains the steps of reconstructed compressed image using CS.

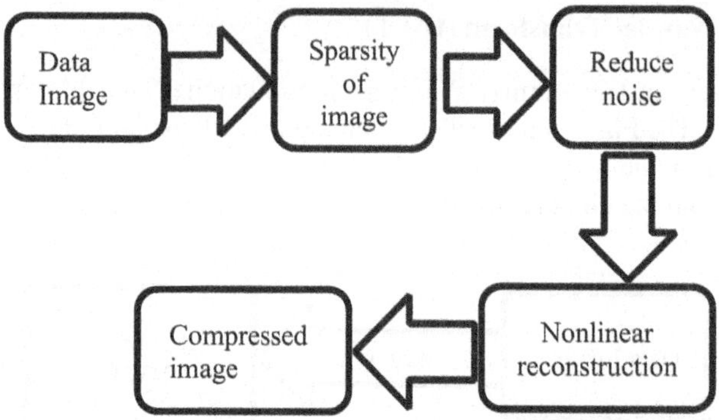

Fig. 3. Diagram for compressed image reconstruction using CS

3 Results and Discussions

In our applications, we used the Matlab 2014a language on a dual graphics card I5 PC. In order to study the performance of the compression methods, we chose a phantom image of size 512 * 512. The Fig. 4 (a) shows the phantom image used. To improve our results, we used the real data image [6]. The Fig. 4 (b) shows the real image used in this work. It is a T1-weighted image acquired from 1.5 T MRI machine (GE, Waukesha, WI) using an 8-channel head coil 3D and spoiled gradient echo sequence (SPGR). This image is acquired using the following parameters: TE = 8 ms, TR = 17.6 ms, an angle of 20°, and a field of view (FOV) of 20 cm × 20 cm × 20 cm with a size of 200 × 200 × 200 for a isotropic resolution 1 mm^3.

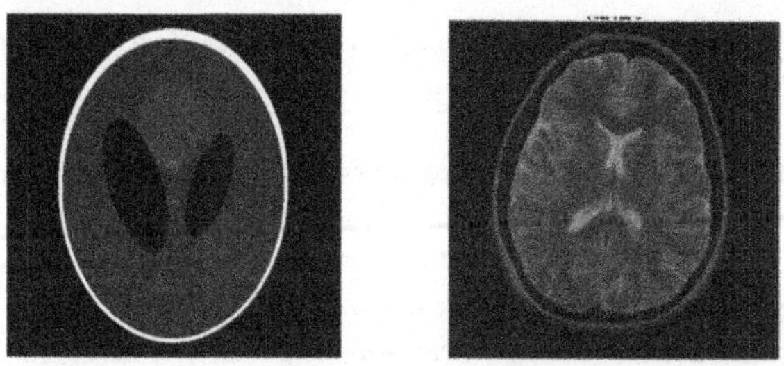

(a) (b)

Fig. 4. (a) Unreal data image, (b) real data image.

We applied the three compression methods: DCT, DWT and CS on images data used. We studied the results of the three compression methods by evaluating performance parameters such as PSNR and RLNE. We compared the three compression methods by studying the image quality and the compression time. We have chosen in the two compression methods: DCT and DWT different percentages in order to evaluate different thresholds. The goal is to choose the most correct threshold.

Algorithm 1 explains the different steps followed in each method:

Algorithm 1

- Load image data (phantom or MRI).
- Apply the DCT (or DWT) on the data.
- Take the absolute value of the new data.
- Select data on a table in descending order.
- Choose different percentage of the most relevant data (1%, 5%, 10%, 20%, 30% and 50%).
- Evaluate the threshold by eliminating unnecessary data.
- Apply the DCT^{-1} (or DWT^{-1}) on the data obtained after the threshold.

In the compressed sensing method, we used a mask to ensure the sparsity of the image. This mask chooses a large number of points at the center of the frequency data. These points correspond to the most relevant points. Algorithm 2 explains the different steps followed.

Algorithm 2

- Load image data (phantom or MRI)
- Apply a mask on the frequency data
- Nonlinear reconstruction to the new data.

In this part, we applied the DCT and DWT compression methods on the phantom and real images. We used different percentages: 1%, 5%, 10%, 20%, 30%, and 50% for both DCT and DWT methods. Figures 5 and 6 represent respectively reconstructed compression phantom and real images with DCT and DWT methods.

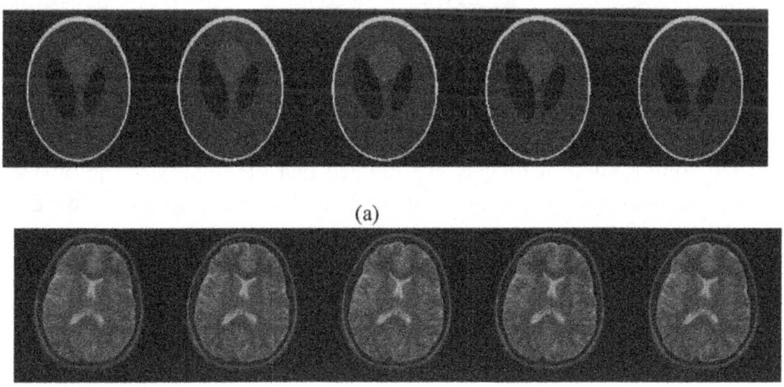

(a)

(b)

Fig. 5. Images compressed by the DCT method using different percentages 1%, 5%, 10%, 20%, 30%, and 50% (a) phantom (b) real.

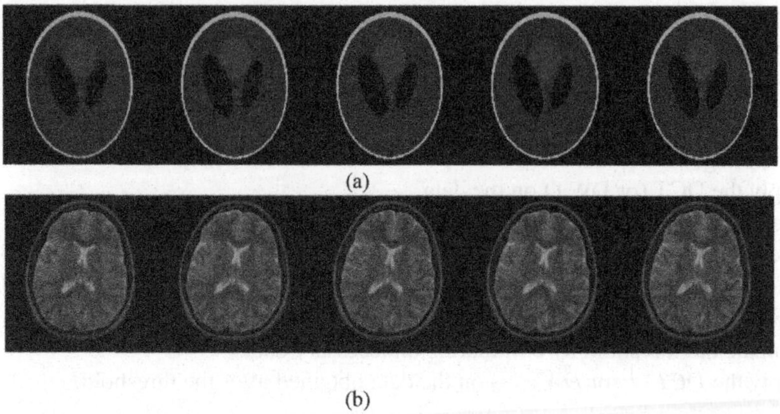

Fig. 6. Images compressed by the DWT method using different percentages 1%, 5%, 10%, 20%, 30%, and 50%. (a) phantom (b) real.

In this part, we applied Compressed Sensing compression method on both phantom and real images, we used a mask showed by the Fig. 7 (a). Applying the CS method on both phantom and real images, Figs. 7 (b) and (c) represent respectively phantom and real reconstructed images compressed with the CS method.

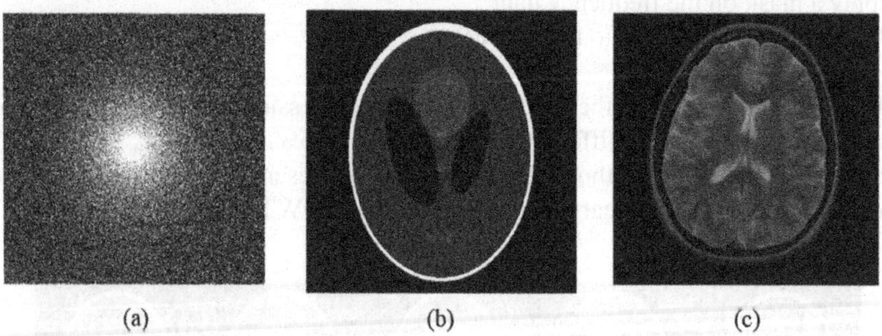

Fig. 7. CS method using (a) mask, reconstructed compressed image (b) phantom (c) real.

After the compression of images, we noticed that the quality of the phantom and real images reconstructed by the three methods is identical. Therefore, we choose in following applications the real image. Moving from one method to another, we noticed that the DCT and DWT methods give good quality images compared to the CS method.

Quantitatively, we studied the quality of the real images by evaluating the two parameters: the PSNR and the RLNE. The Table 1 compares the different results obtained by the three compression methods DCT, DWT and CS. The Table 2 shows the time required to compressed image for each method.

Table 1. Evaluation parameters obtained from compressed images by three compression methods: DWT and DCT using different percentages (1%, 5%, 10%, 30% and 50%) and CS.

		PSNR	RLNE
DCT	1%	27.6387	0.1837
	5%	33.4647	0.0939
	10%	35.2194	0.0767
	20%	37.7316	0.0575
	30%	40.0271	0.0441
	50%	44.8865	0.0252
DWT	1%	31.0985	0.1233
	5%	34.0302	0.0880
	10%	35.5564	0.0738
	20%	37.9702	0.0559
	30%	40.2243	0.0431
	50%	45.0828	0.0247
CS		30.9564	0.1254

Table 2. Compression times for the three methods

Method	DCT	DWT	CS
Compression time (s)	9.869361	5.707984	1.004419

From results obtained by the Table 1, we noticed that the DWT method improved compressed image quality with a high PSNR and a reduced RLNE compared to other compression methods. Qualitatively and quantitatively, the two images compressed by the two methods DWT and DCT give approximately same results. We notice that 10% of information is sufficient to obtain a good compressed image quality witch is approximately the same of the original image. The Fig. 8 shows correctly these images. Qualitatively, the DWT method improves the image quality much more with noise suppression.

In the CS method, the image quality degrades with noise occurring in the background of the image. It presents a reduced PSNR and high RLNE compared to parameters of the other methods.

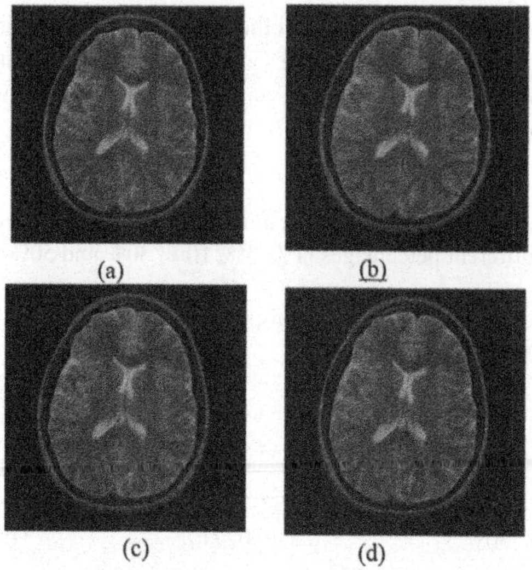

Fig. 8. (a) Original real data image, compressed image with 10% of data using (b) DCT and (c) DWT, (d) compressed image using CS.

We also noticed that the compression time by the CS method is reduced (approximately 1 s) compared to the other methods (approximately 5 s for WT and 9 s for DCT).

Reducing the compression time is very important in medical imaging; it will serve to reduce the acquisition time. Therefore, we noticed that the CS method is very important in medical applications.

4 Conclusion

Image compression is a very large field that uses several methods. The difficulty is to choose the most efficient method. In this paper, we noticed that the DCT and DWT compression methods with the right choice of thresholding give better results. We also noticed that the CS method presents a degraded image quality but it allows a very fast compression time compared to other methods.

These results have brought us to think of improving the CS method using different types of mask. We are also thinking of associating this method with other methods or with methods used in this article: CS-DWT or CS-DCT.

Acknowledgments. PSNR: Peak signal-to-noise ratio.

RLNE: relative error between image original and image compression.

DCT: discrete cosine transforms.

DWT: Discrete wavelet transform.

CS: compressed sensing.

References

1. Rao, K., Yip, P.: Discrete Cosine Transform, Algorithms, Advantages. Applications. Académique Press, London (1990)
2. Liu, T.H., Zhaiv, L., Gao, Y., Li, W., Zhou, J.: Image compression based on biorthogonaln wavelet transform. In: IEE Proceedings of ISCIT2005 (2005)
3. Mallat, S., Hwang, W.L.: Singularity detection and processing with wavelets. IEEE Trans. Inf. Theor. Singularity Detect. Process. Wavelets, **38**(2) (1992)
4. Telagarapu, P., Naveen, V.J., Prasanthi, A.L., et al.: Image compression using DCT and wavelet transformation. Int. J. Sig. Process. Image Process. Pattern Recogn. **4**(3) (2011)
5. Goyal, V.K., Fletcher, A.K., Rangan, S.: Compressive sampling and lossy compression. IEEE Sig. Process. Mag. **25**(2), 48–96 (2008)
6. Korti, A., Bessaid, A.: Wavelet regularization in parallel imaging. In: International Conference on Advanced Technologies for Signal and Image Processing (ATSIP), Fez, Morocco, 22–24 May 2017. IEEE (2017). ISBN 978-1-5386-0551-6
7. Donoho, D.L., Compressed sensing. IEEE Trans. Inf. Theor. **52**(4), 12891306 (2006)
8. Lustig, M., Donoho, D., Pauly, J.M.: Sparse MRI: the application of compressed sensing for rapid MR imaging. Magn. Reson. Med. **58**, 1182–1195 (2007)
9. Lustig, M., Donoho, D.L., Santos, J.M., et al.: Compressed sensing MRI. IEEE Sig. Process. Mag. **25**(4), 72–82 (2008)

An External Archive Guided NSGA-II Algorithm for Multi-depot Green Vehicle Routing Problem

Meriem Hemici[1], Djaafar Zouache[2(✉)], Brahmi Boualem[3], and Kaouther Hemici[4]

[1] Department of Mathematics, University of Mohamed El Bachir El Ibrahimi, Bordj Bou Arreridj, Algeria
[2] Department of Computer Science, University of Mohamed El Bachir El Ibrahimi, Bordj Bou Arreridj, Algeria
[3] Department of Operation research, University of Mohamed El Bachir El Ibrahimi, Bordj Bou Arreridj, Algeria
[4] Department of Economic Sciences, Commerciales and Management Sciences, University of El-Chahid Hama Lakhdar, El-oued, Algeria

Abstract. The paper proposes a new elitist non-dominated sorting genetic Algorithm 2, called External Archive Guided Elitist Non-dominated Sorting Genetic Algorithm 2 (EAG-NSGA-II) to solve the multi-depot green vehicle routing problem (MDGVRP). In the proposed algorithm, the elitist non-dominated sorting genetic Algorithm 2 (NSGA-II) is incorporated with adaptive local search to improve the solution's quality and greatly accelerate the convergence. To ensure a good balance between convergence and diversity, we employ the external archive based on adaptive epsilon dominance. The experiments are conducted over famous 11 Cordeau's data sets and compared against three well-known algorithms, including Strength Pareto evolutionary algorithm (SPEA2), Elitist non-dominated sorting genetic algorithm (NSGA-II), and Multi-objective evolutionary algorithm based on decomposition (MOEA/D). The results indicate that the proposed algorithm is highly competitive and outperforms the selected state-of-the-art multiobjective optimization algorithms.

Keywords: Multi-depot green vehicle routing problem · Multiobjective optimization · Evolutionary algorithms · Epsilon dominance · Crowding distance · Genetic algorithm

1 Introduction

Green Supply Chain Management (SCM) aims to integrate environmental thinking into supply chain management. Over the years, Green supply chain management (GrSCM) has gained increasing interest among researchers and practitioners of operations and supply chain management.

B. Lejdel et al. (Eds.): AIAP 2021, LNNS 413, pp. 504–513, 2022.
https://doi.org/10.1007/978-3-030-96311-8_47

The green vehicle routing problem (GVRP) is one of the most important problems in green supply chain management, this problem generates a set of routes with a set of the customer which consumes a determined demand corresponds to the quantity of product to deliver. A fleet of vehicles identical capacity vehicles is available at the depot to satisfy the demands of customers. A vehicle starts from a depot, serves customers one-by-one, and, ends its trip in the same depot. Instead of the single-depot problem, the multi-depot problem is proposed in the multi-depot green vehicle routing problem (MDGVRP), it is more complicated than other variants of the vehicle routing problems.

In the literature, there are many papers are addressed the green multi-depot vehicle routing problem. Erdoğan and Miller-Hooks [2] used a mixed-integer-linear programming (MILP) formulation and two heuristics. The first heuristic is a modified Clarke and Wright savings algorithm (MCWS) and the second heuristic is a density-based clustering algorithm (DBCA). Zhang et al. [5] proposed a Two-stage Ant Colony System (TSACS) that uses two distinct types of ants for two different purposes. The first type of ant is used to assign customers to depots, while the second type of ant is used to find the routes.

In this paper, we propose a new variant of elitist non-dominated sorting genetic Algorithm 2, called External Archive Guided Elitist Non-dominated Sorting Genetic Algorithm 2 (EAG-NSGA-II), to solve MDGVRP with two main contributions. The first one is the inclusion of an adaptive local search to greatly accelerate the convergence. The second is the employ of the external archive based on adaptive epsilon dominance to ensure a good balance between convergence and diversity. Epsilon dominance impacts the size of the archive; as ϵ increases, the size of the archive decreases.

The experimental results show that the proposed algorithm has a better performance compared to the selected state-of-the-art multiobjective optimization algorithms on famous 11 Cordeau's data sets.

The rest of the paper is structured as follows. Section 2 presents the multi-objective formulation of the considered MDGVRP. In Sect. 3, EAG-NSGA-II is proposed and illustrated. Computational Results in Sect. 4. Finally, we conclude this work and address some open problems in Sect. 5.

2 Multi-depot Green Vehicle Routing Problem

The MDGVRP can be defined as follows. Let $G = (V, A)$ be a directed graph, with $V = N \bigcup D$ is the vertex set, and A is the arc set. Vertex set $N = \{1, 2, .., n\}$ represents the customers to be served, and set $D = \{1, 2, ..., m\}$ represents the depots. Each Vertex $i \in V$ is geographically located at coordinates (x, y). The arc set A denotes all possible connections between the vertices V. Each customer $i \in N$ has a demand of goods $q_i > 0$. The travel costs or distance is associated with each arc $(i, j) \in A$ from $i \in N$ to $j \in N$,i.e. $d_{ij} = \sqrt{(x_i - x_j)^2 + (y_i - y_j)^2}$. Also, set M consists of homogeneous vehicles with capacity limit Q. To serve all customers in N, the MDGVRP constraints should be satisfied as follows:

(1) each vehicle starts and ends the route at the same depot.
(2) every customer vertex is visited on exactly one vehicle.
(3) the total load of vehicle k does not exceed Q.

Notations:
N: The number of customers.
M: The number of depots.
K: The number of vehicles.
Q: The capacity of vehicle.
d_{ij}: The traveling distance between customers i and j.
CCF: Carbon emission conversion factor.

Decision Variable:

$$x_{ij}^{k} = \begin{cases} 1 \; If \; the \; vehicle \text{ h } of \; depot \text{ d } visits \; customer \text{ j } from \; customer \text{ i} \\ 0 \; Otherwise \end{cases}$$

The mathematical model for the GMDVRP is given as follows:

$$min \; F_1 = \sum_{k=1}^{M} \sum_{i=1}^{N} \sum_{j=1}^{N} d_{ij} x_{ij}^{dk} \times CCF \tag{1}$$

$$min \; F_2 = \sum_{k=1}^{M} \sum_{i=1}^{N} x_{0i}^{dk} \tag{2}$$

$$\sum_{k=1}^{M} \sum_{i=1}^{N} x_{ij}^{dk} = 1 \quad \forall i \in \{1, ..., N\} \tag{3}$$

$$\sum_{i=1}^{N} x_{0i}^{dk} = 1 \quad \forall k \in \{1, ..., M\} \tag{4}$$

$$\sum_{i=1}^{N} x_{ih}^{dk} - \sum_{j=1}^{N} x_{hj}^{dk} = 0 \quad \forall h \in \{1, ..., N\}, \forall k \in \{1, ..., M\} \tag{5}$$

$$\sum_{i=1}^{N} x_{iN+1}^{dk} = 1 \quad \forall k \in \{1, ..., M\} \tag{6}$$

$$\sum_{i=1}^{N} q_i \sum_{j=1}^{N} x_{ij}^{dk} \preceq Q \quad \forall k \in \{1, ..., M\} \tag{7}$$

In this formulation, the objective function $(1, 2)$ refers to minimizing the total carbon emissions produced by all the vehicles and the number of vehicles respectively. The Eqs. (3) state that each customer must be visited once by one vehicle, Eqs. (4, 5 and 6) ensure that each vehicle leaves and returns to the depot. The Eqs. (7) state that the vehicle capacity should not be exceededs.

3 The Proposed Algorithm (EAG-NSGA-II) for MDGVRP

In this section, we present a new variant of elitist non-dominated sorting genetic Algorithm 2, called External Archive Guided Elitist Non-dominated Sorting Genetic Algorithm 2 (EAG-NSGA-II) for solving the MDGVRP. Firstly, an initial population P_0 of $\mid P_0 \mid$ feasible solutions is generated randomly. Then, two objectives functions are evaluated for each solution in the initial population. The first objective represents the total travel distances as given in Eq. (1) and the second objective is the number of vehicles as given in Eq. (2), where they must be minimized simultaneously, subject to constraints mentioned above (see Sect. 2). Our approach employs another external population called archive containing the nondominated solutions of the initial population of solutions. The solutions are then subjected to an iterative evolutionary process until the termination condition is met. The evolutionary process is performed by genetic operators, which are the tournament selection operator and the crossover operator called Best Cost Route Crossover (BCRC) to generate offspring solutions Q. The offspring solutions $Q^{'}$ are improved by adaptive local search. The offspring solutions $Q^{'}$ are used to update the external archive. Finally, a combination of the population and offspring population is performed to obtain a population $R = P \bigcup Q$ that is sorted according to the principle of dominance and classified into different fronts (F_i, i = 1, 2, ..., etc.) where all non-dominated solutions of the population are assigned the first front F_1, then they are removed from the population. All non-dominated solutions of the population are assigned the second front F_2, then they are removed from the population. And so on. This process is iterated until the entire population is sorted. The reader is referred to [1] for additional details about NSGA-II. The current population of size N is filled with the two steps is:

Step 1: Create a current population $P_{current} = \oslash$. Set $i = 1$. While $\mid P_{current} \mid + \mid F_i \mid < N$, do $P_{current} = P_{current} \bigcup F_i$ and $i = i + 1$.
Step 2: Sort the front F_i according to the crowding distances and include the $(N - \mid P_{current} \mid)$ solutions with the largest distance values in the population $P_{current}$.

The proposed algorithm EAG-NSGA-II for the MDGVRP is illustrated in Algorithm 1.

Algorithm 1. Pseudocode of EAG-NSGA-II algorithm for MDGVRP

Step I: Initialization
1: Generate an initial population P_0 of $| P_0 |$ feasible solution.
2: Evaluate the initial population P_0 with the two objective functions.
3: Initialize the external archive A by the non-dominated solutions set of the initial population.
Step II: New solution generation
1: **Selection:** Select parents Q from the population P by using tournament selection.
2: **Reproduction:** Apply crossover operators on Q to generate Q^*.
3: **Local search:** Apply adaptive local search to improve Q^*.
4: **Replacement:**

- $R = Combination(P, Q^*)$.
- Sort the non-dominated solution of $R = \{F_1, F_2, F_3, ...\}$.
- Create a current population $P_{current} = \oslash$, and i=1.

while $| P_{current} | + | F_i | < N$

- $P_{current} = P_{current} \bigcup F_i$.
- i=i+1.

endwhile
- Sort the front F_i according to the crowding distances.
- $P_{current} = P_{current} \bigcup F_i(1 : N - | P_{current} |)$.

5: **Update:**
Update of external archive by Q^*.
Step III: Stopping phase
1: If termination condition are satisfied, then output the external archive.

3.1 Solution Representation

The solution is represented by three basic steps (Scheduling, Grouping, Routing), like that seen in Fig. 1.

3.2 Crossover Operator

The crossover is an important operation in a genetic algorithm. Moreover, the crossover operators are responsible for producing the offspring for the selected parents with probability p_c so as to maintain the desired characteristics of the selected parents. The crossover (Best Cost Route Crossover (BCRC) [8] is used to minimizing the total carbon emissions produced by all the vehicles and the number of vehicles simultaneously by respecting the feasibility constraints. In practice, the Best Cost Route Crossover (BCRC) can be summarized as follows:

Step 1: Choose two parents P_1, P_2 from the tournament selection.
Step 2: Randomly select a depot from each parent and select a route from each parent under that depot in a random manner.
Step 3: Remove all customers belonging to route 1 from parent 1 and remove all customers belonging to route 2 from parent 2.
Step 4: Reinsert the removed customers from each parent to form the offspring. For every customer belonging to route 1 (or route 2).

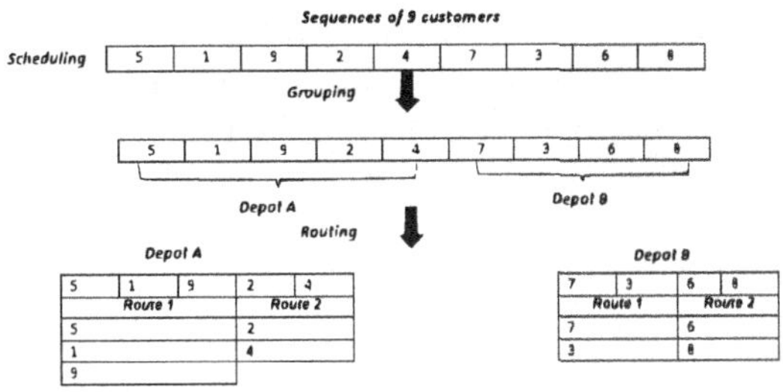

Fig. 1. Example of solution presentation.

- Compute the cost of insertion of route 1 (or route 2) into each location of parent 2 (or parent 1) and store the costs in an ordered list.
- For each insertion location, check whether the insertion is feasible or not.
- If there are no possible insertion locations in the unremoved route, then a new route was created.

3.3 Adaptive Local Search

The role of adaptive local search (ALS) is vital in EAG-NSGA-II in order to greatly accelerate the convergence. Two types of neighborhood structures are used in local search to further improve offspring solutions, which are interdepot and intradepot neighborhoods. The roulette wheel selection is used to decide which neighborhood method is applied to improve the select solution.

Interdepot neighborhood methods: These are constructed based on several basic operators, including swap, shift, 2-opt∗, Or-opt, and 2-opt.

Intradepot neighborhood methods: These are constructed based on several basic operators, including a cross, large neighborhood search (LNS), inversion, and 2-opt∗.

3.3.1 Updating the Archive

In this work, we use an external archive based on the adaptive parameter of ϵ. Initially, the external archive contains all the nondominated solutions of the initial population, since EAG-NSGA-II is an iterative algorithm, this external archive should be updated in each iteration; so, we employ Epsilon dominance approach to update the external archive's solutions.

For any archive solution A and offspring solution c, we associate an identification vector ($box = (box_1, box_2, ..., box_M)^T$, where M is the total number of objectives) as follows:

$$box_j(f) = \lfloor (f_j)/\log(1+\epsilon)\rfloor \tag{8}$$

where $\lfloor . \rfloor$ is the absolute value, f_j: the objective value jth of an archive solution and ϵ denotes the admissible error. The identification vector divides the whole objective space into hyper-box. An offspring solution c is compared with all the archive solutions A according to Epsilon dominance concept to decide if this solution is accepted into the archive as shown in the Algorithm 2. More precisely, the EAG-NSGA-II algorithm compares offspring solution with all the archive solutions according to three conditions [9]:

1. If the identification vector of the offspring solution dominates the identification vector of any archive solution, the offspring solution is accepted and the archive solution is deleted.
2. On the other hand, if the identification vector of the offspring solution is dominated by the identification vector of any archive solution, then it means that the offspring solution is Epsilon dominated by this archive solution and so the offspring is rejected.
3. If neither of the above two cases occurs, then it means that the offspring is equivalent with the archive solutions in terms of Epsilon dominance relation. We separate this third case into two conditions:

   3-1 If the offspring solution and an archive solution a are shared the same identification vector, so we have to verify if the offspring solution dominates the archive solution, or the offspring solution is non-dominated to the archive solution but is closer to the identification vector than the archive solution, then the offspring solution is accepted.

   3-2 In the event of an offspring solution not sharing the same identification vector with any archive solution, the offspring solution is accepted.

This means that the archive maintains the diversity by allowing only one solution to be present in each hyper-box on the Pareto front.

4 Computational Results

The experimental setup is outlined in this section, the results are presented, and the discussion is provided. The proposed algorithm is implemented in MATLAB on an Intel (R) Core (TM) i3 - 6006UCPU2:0GHz PC with a Windows 10 to 64 - bit operating system. The results presented below are based on the following set of EAG-NSGA-II parameters:

- Size of the population: 50.
- Maximum number of generation: 500.
- Probability of crossover: 0.9.
- Value of ε: 0.001.
- Value of CCF: 0.2

Algorithm 2. Pseudocode of Update the external archive procedure

Require: A: The external archive; f: The offspring solution;
Calculate the identification vector for f and all solutions of the archive A;
$D := \{\forall a \in A \mid box_f < box_a\}$
  if $D \neq \oslash$ **then**
    $A := A \cup f \setminus D$
  else
    if $\exists a : (box_a = box_f \wedge f < a)$ **then**
      $A := A \cup f \setminus \{a\}$
    else
      if $\nexists a : (box_a = box_f \vee box_a < box_f)$ **then**
        $A := A \cup f$
      else
        $A := A$
      end if
    end if
  end if

Table 1 shows the computational results for instances used by Cordeau et al. [3]. In presenting the Pareto-optimal set-based results, we provided the best solution (considering both the minimization of the CO2 emission and the number of vehicles) per a given problem. In some instances (like problems 1 and 3, 18, and 21), we have two or more solutions in each instance because neither one dominates the other. The boldface values in columns labeled indicate the problem instances where our algorithm outperforms the best-known results in the literature by reducing the number of vehicles or it has same the number of vehicles.

In the Fig. 2, we can see clearly that the convergence of the EAG-NSGA-II algorithm is better than NSGA-II, SPEA-II, and MOEA/D algorithms.

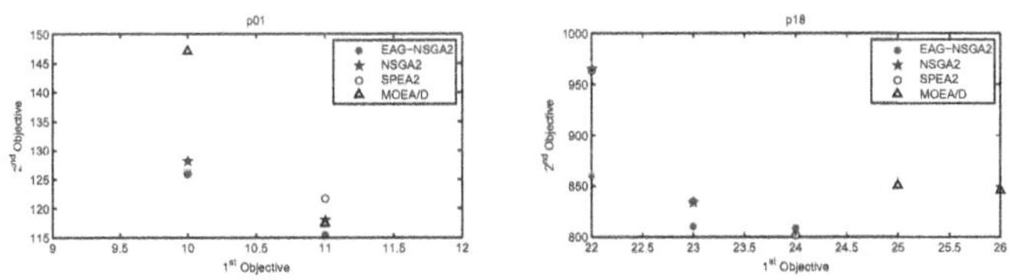

Fig. 2. Pareto set obtained by the multi-objective algorithms on some instances.

Table 1. Computational results for instances used by Cordeau et al. [3]

Instance	N	M	Objective 1, Total number of vehicles	Objective 2, CO_2 emission [kg]	Pareto's Front
p01	50	4	11 **10**	115.64 125.97	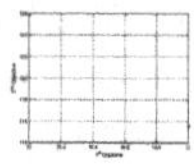
p02	50	4	**5**	96.46	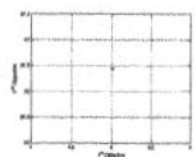
p03	75	5	11 **10**	130.12 135.45	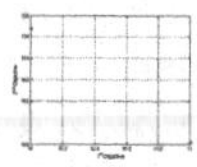
p04	100	2	**15**	208.22	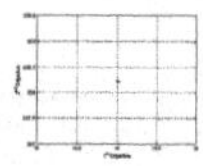
p05	100	2	**8**	157.60	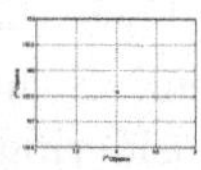
p06	100	3	**15**	183.56	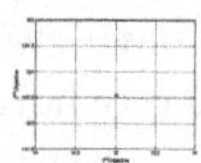
p07	100	4	16 **15**	187.51 193.72	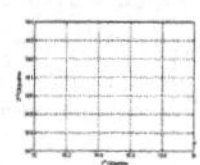
p12	80	2	**8**	264.15	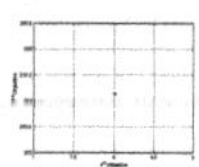
p15	160	4	**15**	519.11	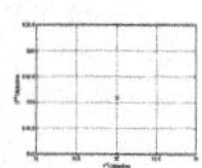
p18	240	6	24 23 **22**	810.11 811.66 861.52	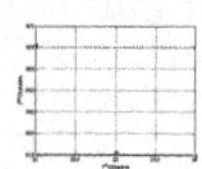
p21	360	9	38 37 36 35 34 **33**	1216.76 1219.86 1226.96 1252.55 1283.54 1486.46	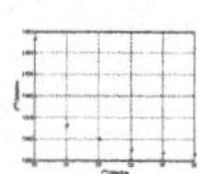

5 Conclusions

This paper proposed a new variant of elitist non-dominated sorting genetic algorithm II, called External Archive Guided Elitist Non-dominated Sorting Genetic Algorithm 2 (EAG-NSGA-II) for solving the MDGVRP. Adaptive local search was integrated to greatly accelerate the algorithm convergence. The external archive based on adaptive epsilon dominance was employ to ensure a good balance between convergence and diversity. EAG-NSGA-II was evaluated through using the famous 11 Cordeau's data sets. The experimental results proved that EAG-NSGA-II was effective to solve MDGVRP. EAG-NSGA-II was significantly superior over NSGA-II, SPEA-II, MOEA/D in all instances.

For future work, EAG-NSGA-II can be tested to solve other variants of the vehicle routing problems such as green multi-depot VRPTW. EAG-NSGA-II can be further developed using a novel update strategy to solve similar combinatorial optimization problems in the field of supply chain management.

References

1. Deb, K., Pratap, A., Agarwal, S., Meyarivan, T.: A fast and elitist multiobjective genetic algorithm: NSGA-II. IEEE Trans. Evol. Comput. **6**, 182–197 (2002)
2. Erdoğan, S., Miller-Hooks, E.: A green vehicle routing problem. Trans. Res. Part E: Logistics Transp. Rev. **48**, 100–114 (2012)
3. Cordeau, J. F., Gendreau, M., Laporte, G.: A tabu search heuristic for periodic and multi-depot vehicle routing problems. Netw. Int. J. **30**(2), 105–119 (1997)
4. Renaud, J., Laporte, G., Boctor, F.F.: A tabu search heuristic for the multi-depot vehicle routing problem. Comput. Oper. Res. **23**, 229–235 (1996)
5. Zhang, W., Gajpal, Y., Appadoo, S., Wei, Q.: Multi-depot green vehicle routing problem to minimize carbon emissions. Sustain. Multi. Digit. Publishing Inst. (2020)
6. Zitzler, E., Thiele, L.: Multiobjective evolutionary algorithms: a comparative case study and the strength Pareto approach. IEEE Trans. Evol. Comput. **3**, 257–271 (1999)
7. Zitzler, E., Laumanns, M., Bleuler, S.: A tutorial on evolutionary multiobjective optimization. In: Gandibleux, X., Sevaux, M., Sörensen, K., T'kindt, V. (eds.) Metaheuristics for Multiobjective Optimisation. Lecture Notes in Economics and Mathematical Systems, Springer, Berlin, Heidelberg (2004). https://doi.org/10.1007/978-3-642-17144-4_1
8. Ombuki, B., Ross, B.J., Hanshar, F.: Multi-objective genetic algorithms for vehicle routing problem with time windows. Appl. Intell. **24**, 17–30 (2006). https://doi.org/10.1007/s10489-006-6926-z
9. Mishra, S.K., Ganapati, P., Meher, S., Majhi, R.: A fast multiobjective evolutionary algorithm for finding wellspread pareto-optimal solutions. In: KanGAL Report No. 2003002, Indian Institute of Technology Kanpur, Citeseer (2002)

New Approach for Multi-valued Mathematical Morphology Computation

Samir L'haddad(✉) and Akila Kemmouche

Sciences and Technology, Houari Boumediene University, Algiers, Algeria
{slhaddad,akemmouche}@usthb.dz

Abstract. Mathematical Morphology (MM) is a useful tool for spatial image processing. It is based on an infimum operator (min) and a supremum operator (max) applied in local neighborhoods to detect pixel extremes. The MM was initially defined for mono-band images in which each pixel image is a scalar value and it is easy to find pixels extremes by the infimum and the supremum operators. However, in the case of multi-band images, where each pixel image is represented by a vector, establishing an order between image pixels in local neighborhoods by the infimum and supremum operators is not trivial. Many works discussed the feasibility to extend the MM to multi-band images but they did not lead to any consensual definition of the multi-valued mathematical morphology. Nevertheless, these existing works agree that the definition of the MM for multi-band images is based on the notion of vector ordering. In this paper, we propose a multi-valued MM operators computing by introducing a new vector ordering algorithm that allows extending the scalar MM to multi-band images. The proposed multi-valued morphological operations were tested in the experimental phase for the morphological descriptors computation. The obtained results based on use of the proposed vector ordering algorithm for the multi-valued MM computing improve the classification rates.

Keywords: Multi-band images · Vector ordering algorithm · Multi-valued mathematical morphology · Multi-valued Morphological Profile

1 Introduction

Mathematical Morphology (MM) is a useful tool for image processing. It allows the geometrical description of the structures present in the scene. The MM technique was originally developed by Matheron and Serra [1] and was formulated in terms of infimum (finding the min, noted $\wedge$) and supremum (finding the max, noted $\vee$) operators. Even if the MM is well defined for mono-band images, where each pixel is represented by scalar value and it is easy to designate the infimum and supremum pixels for compared pixels in a local neighborhood, there is no consensual definition of the MM for multi-band images in which each pixel is represented by a vector of p component (p is the number of bands, and each pixel have scalar value for each image band). A basic and frequently used idea to extend the morphological operation to the multi-dimensional

B. Lejdel et al. (Eds.): AIAP 2021, LNNS 413, pp. 514–523, 2022.
https://doi.org/10.1007/978-3-030-96311-8_48

data is to decompose the initial multi-band image into grayscale images (i.e. mono-band images), on which identical scalar morphological treatments are applied independently. The final result of the MM image processing is obtained by grouping the different isolated results (i.e. single results are assembled into a unique data set). The application of this so-called marginal strategy sometimes results in the creation of new pixel vectors that are not present in the original multi-band image (then the applied morphological treatment is not vectors preserving). Furthermore, this way of adapting the scalar MM to the multi-band images loses the correlations between bands [2]. To avoid these drawbacks, the definition of MM for a multi-band image requires non-scalar morphological approaches that consider the multi-band image as a single data block processed simultaneously. For this purpose, it is necessary to have an appropriate vector ordering strategy allowing vector space manipulation and selecting vectors extremes by the infimum and the supremum operators. The study of the appropriate vector ordering scheme to define the multi-valued MM has been studied for many years by several works. Barnett [2] has classified the existing vector ordering strategies for the MM extension into four groups: the marginal ordering strategy (M-ordering strategy), the conditional ordering strategy (C-ordering strategy), the partial ordering strategy (P-ordering strategy), and the reduced ordering strategy (R-ordering strategy).

The marginal ordering strategy (M-ordering strategy), as mentioned previously, consists of processing each band image separately and independently from the others by the same scalar morphological transformations. Despite its easy implementation, the marginal ordering strategy loses the bands correlation (relation between bands) and can lead to the appearance of some pixel vectors in the output result that are not present in the original image. This second disadvantage can be illustrated by taking the example of 3-dimensional vectors $X = (3, 8, 5)$ and $Y = (4, 7, 2)$. The $\bigwedge (X, Y) = (3, 7, 2)$ and $\vee(X, Y) = (4, 8, 5)$ (where $\bigwedge$ indicates the min operator and $\vee$ indicates the max operator). In this illustrative example, the vectors (3, 7, 2) and (4, 8, 5) are two new vectors that do not belong in the initial vectors set. Thus, for this example, the used marginal ordering is not vectors preserving. The reducing dimensionality transformation (with Principal Component Analysis PCA [3–5] or any other dimensionality reduction method) of the original multi-band image is also applied before a marginal treatment to ensure that the bands of the image are decorrelated and avoid the first problem of losing the bands correlation.

The conditional ordering strategy (C-ordering strategy) gives priority to some particular bands image in the vector ordering process. It uses a prioritization function to calculate the priority of each band (i.e. computing band weight) in the original image. The C-ordering approach is recommended when considering some image bands' weights are greater than others [6]. Thus, with a conditional ordering strategy, two vectors are ordered according to their scalar values for the most prioritized band. In the case of equality, the band with the next priority is considered, and so on. Two vectors are identical in a conditional approach if they are equal component by component (i.e. they have the same scalar values for all image bands). The main limit of the conditional ordering is linked to the difficulty to find the coherent band prioritization function. The lexicographic ordering strategy (L-ordering strategy) is considered as the most known derivative version of the C-ordering strategy. Its principle is taken from the classification of words by alphabetical order, where the first words sorting is based on the first letter, the undecided cases of the previous words classification are resolved by the second letter and so on. For the adaptation of the L-ordering to the vector ordering problem, the first image band is used for the first vectors ordering, the next one is used to solve the unresolved ex-aequo problems of the previous band and so on (1). In the L-ordering strategy, the ordering succession can be reversed by starting with the last band image and gradually advance to the first band image each time there is an indecisive case (2).

$$\begin{pmatrix} v_1 \\ v_2 \\ \vdots \\ v_n \end{pmatrix} < \begin{pmatrix} \mu_1 \\ \mu_2 \\ \vdots \\ \mu_n \end{pmatrix} \Leftrightarrow \begin{array}{c} v_1 < \mu_1 \; or \\ v_1 = \mu_1 \;\; and \;\; v_2 < \mu_2 \; or \\ \ldots \\ v_1 = \mu_1, \ldots\ldots, \;\; and \; v_{n-1} = \mu_{n-1} \; and \; v_n < \mu_n \end{array} \tag{1}$$

$$\begin{pmatrix} v_1 \\ v_2 \\ \vdots \\ v_n \end{pmatrix} < \begin{pmatrix} \mu_1 \\ \mu_2 \\ \vdots \\ \mu_n \end{pmatrix} \Leftrightarrow \begin{array}{c} v_n < \mu_n \; or \\ v_n = \mu_n \;\; and \;\; v_{n-1} < \mu_{n-1} or \\ \ldots \\ v_n = \mu_n, v_{n-1} = \mu_{n-1} \ldots\ldots, \;\; and \; v_2 = \mu_2 \; and \; v_1 < \mu_1 \end{array} \tag{2}$$

The Lexicographic order strategy remains the most used vector ordering strategy in the multi-valued MM definition. This can be explained by the fact that the L-ordering is vector preserving and allows to obtain a total order relation between compared vectors (i.e. there is no incomparability situation between two vectors and the unique case of equality between two vectors is the equality component by component between the compared vectors). In practice, the L-ordering strategy frequently corresponds to the exclusive use of the first prioritized bands to make vectors ordering decision. The remaining bands rarely participate in the comparison process [7, 8]. Thus, the lexicographic order is well suited for situations where the first image bands are those including the most relevant information. This case cannot be naturally present in the multi-band images but can be obtained after concentrated image information in the first bands image by projection techniques (such as a principal component analysis PCA or any other projection techniques). The lexicographic ordering and its variants were reported in various works like [9–14].

The partial ordering strategy (P-ordering strategy) clusters vectors into groups of equivalence according to a given criterion. With the P-ordering strategy, it is possible to compare vectors from two different groups, but not within the same group. Because of the impossibility to make an ordering relation between vectors of the same groups, this vector ordering scheme cannot guarantee the uniqueness of the extremes vectors (i.e. do not guarantee unique infimum vector and/or unique supremum vector). Thus, with partial ordering approaches, there is no total relation between vectors (i.e. some vectors are not comparable). The approaches presented in [15–20] are examples of using a P-ordering strategy to extend the MM to the multi-band images.

The reduced ordering strategy (R-ordering strategy) reduces vectors to scalar values easily comparable. The passage of an N-dimensional space to one-dimensional space can be obtained by a projection system or by a distance measurement from predefined reference. Once each pixel vector of the multi-band image is replaced by its associated scalar value, the created grayscale image (i.e. mono-band image) can be directly processed by any mono-band morphological transformations. The R-ordering approach using projection techniques, such as the PCA, to transform the multi-band image in one-band image causes losing too much information unlike exploiting distance measurement. Plaza et al. [21] proposed a vectors ordering algorithm using cumulative distances of each pixel vector from all the others. The presented reduced ordering strategy avoids using any predefined reference with which the vectors are compared. The proposed cumulative distances use two new metrics (Spectral Angle Distance "SAD" and Spectral Information Divergence "SID"). Other distances measurements from the predefined reference are also used in the R-ordering strategy like Euclidean distance, the Mahalanobis distance [22, 23], or other distance measurements [24–26]:

$$v \leq v' \Leftrightarrow \text{Distance}(v, v_{ref}) \leq \text{Distance}\left(v', v_{ref}\right) \quad (3)$$

Sometimes, the reduced ordering strategy leads to indecision cases when two different vectors have the same scalar values. In this situation, using an additional condition in the R-ordering approach, such as lexicographic ordering, is necessary to obtain a total order relation (i.e. any vectors are comparable) and decide between vectors having the same scalar estimation. After briefly introducing the four families of vector ordering strategies, we propose, in this work, a new multi-valued MM computation algorithm that extends the morphological operations to the multi-band image by processing all image bands simultaneously. For this purpose, a vector ordering algorithm is introduced for the multi-valued MM computation. This proposed algorithm allows finding the supremum and the infimum of pixel vectors in a local neighborhood by applying the infimum and the supremum operators on pixels that do not have a unidimensional structure. The proposed vectors comparison algorithm is based on outranking relations and operates the vectors ordering process by considering all image bands without excluding any of them. The presented vector ordering algorithm preserves also the original vectors (i.e. is vectors preserving) and does not omit the correlation between bands in the ordering process.

The rest of this paper is organized as follows: the proposed vector ordering algorithm is presented in Sect. 2. Experimental results obtained on the ROSIS sensor data sets from urban areas are presented in Sect. 3. Finally, conclusions are drawn in Sect. 4.

2 Vectors Ordering Based on Outranking Relations

By considering *m* compared vectors with N-dimensional structure (i.e. pixel vectors which constitute the N-bands image), the first step of the proposed vector ordering algorithm is binary comparisons of each pixel vector from all the others included in the vectors comparison set.

From two taken N-dimensional vectors X $(x_1, x_2, \ldots, x_N)$ and Y $(y_1, y_2, \ldots, y_N)$ belonging in the compared vectors set, two numerical values "α" and "β" are computed for the (X, Y) vectors pair.

The first numerical value "α" expresses the number of the vector components X which are greater than their equivalent components in the vector Y:

$$\alpha = \sum_{i=1}^{N} \Delta(x_i - y_i) \quad (4)$$

Where $\Delta(A) = 1$ if A is greater than 0 and $\Delta(A) = 0$ otherwise.

The second numerical value "β" represents the number of the vector components Y which are greater than their correspondent components in vector X:

$$\beta = \sum_{i=1}^{N} \Delta(y_i - x_i) \tag{5}$$

Note that "$\alpha = m - \beta$" and "$\beta = m - \alpha$".

It is considered that the vector X and the vector Y are equal (component by component) if "$\alpha = \beta = 0$".

When the "α" value is greater than "β", the vector X outrank the vector Y. Otherwise, the vector X is outranked by the vector Y.

This adopted "binary outranking relation" can lead to indecision cases (i.e. incomparability situation between the two compared vectors) between the two compared vectors when "α" and "β" are equal but not null. To resolve the undecided cases an additional ordering must be used. In our approach, the adopted additional ordering strategy is conditional ordering which is recommended when it is possible to find band prioritization function and give priority value for each band of the N-dimensional image. The associated priority function with the C-ordering process is the Improved Score Function (noted IFS function) introduced in [27, 28]. This IFS function was employed in [28] to estimate the discrimination power of the generated morphological descriptors. In this paper, the score function IFS is considered as a tool to give vector components prioritization (i.e. attribute the priority value of each image band) for the conditional ordering scheme. The use of the IFS function as a prioritization function in the conditional vectors ordering algorithm constitutes another contribution of this work.

To estimate the priority value of each band image, the IFS value is computed for each component of the multi-band image by the following formula:

$$IFS_i = \frac{\sum_{j=1}^{l} \left[avg\left(x_i^j\right) - avg(x_i)\right]^2}{\sum_{j=1}^{l} \frac{1}{n_j - 1} \sum_{k=1}^{n_j} [x_{k,i}^j - avg\left(x_i^j\right)]^2} \tag{6}$$

Where:

- IFS_i is the priority value (i.e. the weight) of the ith band image;
- "l" is the obtained objects classes number after a classification of the original multi-band image;
- n_j is the number of pixels belonging to the jth class;
- $avg(x_i)$ is the radiometric average value of the pixels in the ith band image;
- $avg(x_i^j)$ is the radiometric average value of the pixels in the ith band image belonging to the jth class;
- $x_{k,i}^j$ is the kth pixel of the ith band image belonging to the jth class.

Bands with a high priority value (i.e. with high IFS value) are more prioritized bands (i.e. the highest weight bands) in the conditional ordering process. Thus, incomparable pixel vectors with the binary outranking relation are initially ordered according to the scalar value of their first prioritized component. Vectors with the same value for the first prioritized component are ordered according to the scalar value of the next prioritized component, and so on. Therefore, the conditional ordering completes the ordering structure when the binary outranking relation between two compared vectors leads to an indecision situation (i.e. incomparability situation).

The second step of the proposed vectors ranking algorithm is the synthesis of the binary outranking comparisons to give the final ordering of the compared vectors and designate the two pixel-vectors extremes (i.e. designate the infimum pixel vector and the supremum pixel vector).

Note that the most outranked vector (i.e. lower ranked vector) is the infimum vector, while the outranking one (i.e. the higher ranked vector) is a supremum.

The evaluation of the proposed vector ordering algorithm is performed by generating the multi-valued Morphological Profile (MP) computed on all bands of the original image simultaneously. The MP is an association of morphological transformations that allows extracting structures of various sizes and shapes present in the original image (more details for the MP are given in [4, 30]). The MP was originally defined for the mono-band images. In the experiment section, the spatial characterization by MP is operated by the multi-valued MP and is built on all image bands simultaneously. For this purpose, various vector ordering algorithms (two classical algorithms and our proposed algorithm) are used to detect and extract objects of interest by the multi-valued MP.

The obtained results of the vector ordering algorithms used in the multi-valued computation are illustrated and discussed in the next section. The performance measurement of the exploited vector ordering algorithms is determined by SVM classification accuracy.

3 Experimental Results

In this section, performance measurements like classification Overall Accuracy (OA) and Kappa rate are used with the support vector machines (SVM) classifier [31] to measure the impact of the generated morphological descriptors by multi-valued MP on the classification improvement. Note that the three considered scenarios of vector ordering are:

- The lexicographic vector ordering strategy with decreasing priorities of the image bands;
- The lexicographic ordering strategy with increasing priorities of the image bands;
- The proposed vector ordering algorithm based on outranking relations and additional conditional ordering.

The three used vector ordering algorithms for the multi-valued MP computation are applied on all image bands simultaneously.

The experimental analysis was carried out on a multi-band image of Pavia university (northern Italy) acquired by the ROSIS sensor. This ROSIS image is composed of 610*610 pixels with 1.3 m of spatial resolution and 103 image bands. The image was proposed with a ground truth image that differentiates nine object classes (asphalt, meadows, gravel, trees, painted metal sheets, bare sol, bitumen, bricks, and shadows). The original image and its associated ground truths are shown in Fig. 1.

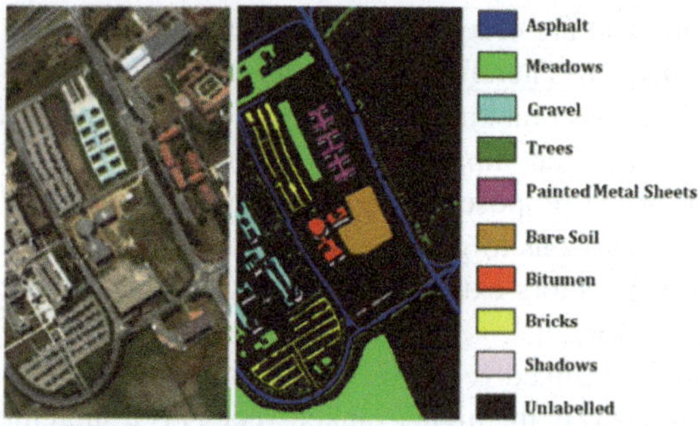

Fig. 1. Pavia University scene with the associated ground truth image.

The multi band image includes many highly correlated bands resulting in spectral redundancy. This spectral redundancy increases computational complexity and degrades classification accuracy [32]. For this reason, dimensionality reduction was achieved by the PCA projection technique and only the first decorrelated principal components of the original image are considered for the multi-valued MP computing.

The "spectral" classification (without considering spatial descriptors generated by multi-valued MP) of the reduced image gives 82.82% for the OA rate and 77.47% for the Kappa rate.

The use of the three vector ordering algorithms on the multi-valued MP computation produces different outcomes. A summary of results in terms of classification accuracies (OA and Kappa rate) is given in Table 1. The use of the vector ordering scheme improves the classification accuracies independently of the used vector ordering algorithm in the multi-valued MP computation.

As shown in Table 1, the classification accuracies obtained by the presented vector ordering algorithm is close to "the lexicographic ordering strategy with increasing priorities" with a small improvement in favor of the proposed vector ordering approach. The obtained results showed also a higher precision for "the lexicographic ordering strategy with decreasing priorities" in comparison to the other compared vector ordering approaches. This is probably due to the dimensionality reduction effect which concentrates the most information present in the image bands on the first bands that are the mainly considered bands in the lexicographic ordering.

Table 1. Classification accuracy using different vector ordering algorithms for the multi-valued MP computation.

Vector ordering algorithm	OA %	Kappa %
Lexicographic ordering with decreasing priorities	90.52	87.58
Lexicographic ordering with increasing priorities	85.35	81.05
Vector ordering algorithm based on outranking relations and conditional ordering	85.79	81.29

The resulting classification maps based on SVM classification of the multi-valued MP computed with the three different vector ordering algorithms is shown in Fig. 2.

(a) (b) (c) (d)

Fig. 2. Classification maps obtained with the SVM classifier. (a) Original multi-band image; (b) classification using the multi-valued MP computed by lexicographic ordering with decreasing priorities; (c) classification using the multi-valued MP computed by lexicographic ordering with increasing priorities; (d) classification using the multi-valued MP computed by the proposed outranking relations and conditional ordering.

4 Conclusion

Mathematical Morphology (MM) is an efficient tool for patterns and object recognition in digital image processing. The basic transformations of mathematical morphology are based on the search for the local minimum and local maximum in the predefined neighborhood. The MM was originally defined for mono-band images where each pixel is associated with a numerical value and the order relation between scalars pixel is natural. The application of the MM logic on multi-bands images is less trivial since there is no predefined order between vector values. In this paper, we have proposed a new vector ordering algorithm based on an idea of outranking relations between vectors. The presented vector ordering algorithm is completed by conditional ordering strategy to have a total order relation between compared vectors (i.e. all vectors are comparable). The experiments are interested in the characterization of the objects present in multi-band images by the multi-valued MP. Even if the classification results of the presented vector ordering algorithm get closer results to the widely used lexicographic approaches, it can be validated by the fact that the proposed algorithm is vector preserving, take into account the band correlation, and provides a total relation order between compared vectors. These proposed algorithms can be also used with other multi-valued morphological operators and is available for any type of multi-band images like color images.

References

1. Serra, J.: Image Analysis and Mathematical Morphology. Academic Press, London (1988)

2. Barnett, V.: The ordering of multivariate data. J. Roy. Stat. Soc. IRSS, Ser. A (Gen.), **139**(3), 318–355 (1976)
3. Li, J., Li, Y.: Multivariate mathematical morphology based on principal component analysis: initial results in building extraction. 20th ISPRS, **35**, 1168–1173 (2004)
4. Benediktsson, J.A., Palmason, J.A., Sveinsson, J.R.: Classification of hyperspectral data from urban areas based on extended morphological profiles. IEEE Trans. Geosci. Remote Sens. **43**(3), 480–491 (2005)
5. Fauvel, M., Chanussot, J., Benediktsson, J.A., Sveinsson, J.R.: Spectral and spatial classification of hyperspectral data using SVMs and morphological profiles. Int. Geosci. Remote Sens. Symp. **46**(11), 4834–4837 (2007)
6. Aptoula, E., Lefèvre, S.: A comparative study on multivariate mathematical morphology. Pattern Recogn. **40**(11), 2914–2929 (2007)
7. Hanbury, A., Serra, J.: Mathematical morphology in the CIELAB space. Image Anal. Stereol. **21**(3), 201–206 (2002)
8. Aptoula, E., Lefèvre, S.: On lexicographical ordering in multivariate mathematical morphology. Pattern Recogn. Lett. **29**(2), 109–118 (2008)
9. Angulo, J.: Unified morphological color processing framework in a Lum/Sat/Hue representation. In: Ronse, C., Najman, L., Decencière, E. (eds.) Mathematical Morphology: 40 Years On. Computational Imaging and Vision, vol. 30. Springer, Dordrecht (2005). https://doi.org/10.1007/1-4020-3443-1_35
10. Aptoula, E., Lefevre, S.: Pseudo multivariate morphological operators based on α-trimmed lexicographical extrema. In: 5th International Symposium on Image and Signal Processing and Analysis ISPA, pp. 367–372, Istanbul, Turkey (2007)
11. Aptoula, E., Lefèvre, S.: α-Trimmed lexicographical extrema for pseudo-morphological image analysis. J. Vis. Commun. Image Represent. **19**(3), 165–174 (2008)
12. Angulo, J.: Geometric algebra colour image representations and derived total orderings for morphological operators – Part I: Colour quaternions. J. Vis. Commun. Image Represent. JVCIR **21**(1), 33–48 (2010)
13. Gao, C.-J.Z.X.-H., Hu, X.-Y.: An adaptive lexicographical ordering of color mathematical morphology. J. Comput. (2013)
14. Lei, T., Fan, Y., Zhang, C., Wang, X.: Vector mathematical morphological operators based on fuzzy extremum estimation. In: 20th International Conference on Image Processing (ICIP), pp. 3031–3034. IEEE, Melbourne, Australia (2013)
15. Lezoray, O., Elmoataz, A., Meurie, C.: Mathematical morphology in any color space. In: 14th International Conference of Image Analysis and Processing - Workshops ICIAPW, pp. 183–187. Modena, Italy (2007)
16. Velasco-Forero, S., Angulo, J.: Supervised ordering in IRp: application to morphological processing of hyperspectral images. IEEE Sig. Process. Soc. (Trans. Image Process.), **20**(11), 3301–3308 (2011)
17. Velasco-Forero, S., Angulo, J.: Random Projection Depth for Multivariate Mathematical Morphology. IEEE J. Sel. Top. Sig. Process **6**(7), 753–763 (2012)
18. Aptoula, E., Courty, N., Lefevre, S.: An end-member based ordering relation for the morphological description of hyperspectral images. In: International Conference on Image Processing (ICIP), pp. 5097–5101. IEEE, Paris, France (2014)
19. Velasco-Forero, S., Angulo, J.: Vector ordering and multispectral morphological image processing. In: Celebi, M.E., Smolka, B. (eds.) Advances in Low-Level Color Image Processing. LNCVB, vol. 11, pp. 223–239. Springer, Dordrecht (2014). https://doi.org/10.1007/978-94-007-7584-8_7
20. Franchi, G., Angulo, J.: Ordering on the probability simplex of endmembers for hyperspectral morphological image processing. In: Benediktsson, J.A., Chanussot, J., Najman, L., Talbot,

H. (eds.) ISMM 2015. LNCS, vol. 9082, pp. 410–421. Springer, Cham (2015). https://doi.org/10.1007/978-3-319-18720-4_35
21. Plaza, A., Martinez, P., Perez, R., Plaza, J.: A new approach to mixed pixel classification of hyperspectral imagery based on extended morphological profiles. Pattern Recogn. **37**(6), 1097–1116 (2004)
22. Al-Otum, H.M.: Morphological operators for color image processing based on Mahalanobis distance measure. Opt. Eng. **42**(9), 2595–2606 (2003)
23. Garcia, A., Vachier, C., Vallée, J.-P.: Multivariate mathematical morphology and Bayesian classifier application to colour and medical images. In: Image Processing (Algorithms and Systems VI), vol. 6812 (1), p. 681203. SPIE, San Jose, CA, Astola (2008)
24. Angulo, J.: Morphological colour operators in totally ordered lattices based on distances: application to image filtering, enhancement, and analysis. Comput. Vis. Image Underst. **107**(1–2), 56–73 (2007)
25. Plaza, J., Plaza, A.J., Barra, C.: Multi-channel morphological profiles for classification of hyperspectral images using support vector machines. Sensors **9**(1), 196–218 (2009)
26. Sangalli, M., Valle, M.E.: Approaches to multivalued mathematical morphology based on uncertain reduced orderings. In: Burgeth, B., Kleefeld, A., Naegel, B., Passat, N., Perret, B. (eds.) ISMM 2019. LNCS, vol. 11564, pp. 228–240. Springer, Cham (2019). https://doi.org/10.1007/978-3-030-20867-7_18
27. Akay, M.F.: Support vector machines combined with feature selection for breast cancer diagnosis. Expert Syst. Appl. **36**(2), 3240–3247 (2009)
28. Jaganathan, P., Rajkumar, N., Kuppuchamy, R.: A Comparative Study of Improved F-Score with Support Vector Machine and RBF Network for Breast Cancer Classification. IJMLC **2**, 741–745 (2012)
29. Zemmoudj, S., Kemmouche, A., Chibani, Y.: Feature selection and classification for urban data using improved F-score with Support Vector Machine. In: 6th International Conference of Soft Computing and Pattern Recognition (SoCPaR), pp. 371–375. IEEE, Tunis, Tunisia (2014)
30. Pesaresi, M., Benediktsson, J.A.: A new approach for the morphological segmentation of high-resolution satellite imagery. IEEE Trans. Geosci. Remote Sens. **39**(2), 309–320 (2001)
31. Vapnik, V.: Statistical Learning Theory. Wiley, New York (1998)
32. Landgrebe, D.: Hyperspectral image data analysis. IEEE Sig. Process. Mag. **19**(1), 17–28 (2002)

Data-Intensive Scientific Workflow Scheduling Based on Genetic Algorithm in Cloud Computing

Siham Kouidri[1,2](✉) and Chaima Kouidri[1,2]

[1] Department of Computer Science, Faculty of Technology, University of Saida Dr Moulay Tahar, Saida, Algeria
ckouidri2014@gmail.com

[2] Department of Computer Science, Mustapha ISTAMBOULI University of Mascara, Mascara, Algeria
skouidri2000@gmail.com

Abstract. Cloud Computing is increasingly recognized as a new way to use on-demand, computing, storage and network services in a transparent and efficient way. Cloud Computing environment consists of large customers requesting for cloud resources. Nowadays, task scheduling problem and data placement are the current research topic in cloud computing. In this work, a new technique for task scheduling and data placement are proposed based on genetic algorithm to fulfill a final goal such as minimizing total workflow response time. The scheduling of scientific workflows is considered to be an NP-complete problem, i.e. a problem not solvable within polynomial time with current resources The performance of this proposed algorithm has been evaluated using CloudSim toolkit, Simulation results show the effectiveness of the proposed algorithm in comparison with well-known algorithms such as genetic algorithm with Random data placement.

Keywords: Cloud computing · Workflow scientific · Scheduling · Virtual machine · NP-Complet Problem · Data placement · Genetic algorithm

1 Introduction

Nowadays, many scientists and researchers are moving towards Cloud computing for achieving high performance. This paradigm brings a new operational model where resources are managed by specialized data centers and rented only under demand and for the period of time they need to be used, is becoming very attractive for companies and institutions.

Scientific workflows are a popular way of modeling applications to be executed in parallel or distributed systems like Clouds. Once the data-intensive scientific workflow is composed, one of the most challenging research topics is how to schedule the different tasks onto the available resources. Traditionally, in parallel and distributed systems, workflow scheduling has been targeted to optimize the performance, measured in terms of the makespan or time of completing

B. Lejdel et al. (Eds.): AIAP 2021, LNNS 413, pp. 524–533, 2022.
https://doi.org/10.1007/978-3-030-96311-8_49

all tasks. This problem has been shown to be NP-complete which is difficult to obtain exact optimal solution and is suitable for using intelligent optimization algorithms to approximate the optimal solution [1].

The advent of Cloud computing as a new model of service provisioning in distributed systems encourages researchers to investigate its benefits and drawbacks on executing scientific applications such as workflows. One of the most challenging problems in Clouds is workflow scheduling [2].

Scheduling is a process that maps and manages execution of inter-dependent tasks on distributed resources. Main motive of the scheduling is to allocate the suitable resources to the workflow tasks so that the execution of the workflow tasks completed within the deadline given by the customer. Suitable scheduling approach can have significant impact on the performance of the cloud computing.

The aim of this article is to present the scheduling and data placement problem as an optimization problem in the context of cloud-type platforms. We are going to propose a solution to minimize the runtime of the data-intensive scientific workflow. This solution can be in the form of an optimization algorithm, whether meta-heuristic (genetic algorithm), we propose to integrating task scheduling and data placement into one frame work for the sole goal of minimizing the total scientific workflow execution time of the cloud computing.

The rest of this paper is structured as follows: Sect. 2 explains some related works. Section 3 introduces our proposed approach. Section 4 evaluates the performance of simulation experiments using CloudSim. Conclusion and future works are presented in Sect. 4.

2 Related Work

Several works have been proposed to solve the scheduling problem in cloud computing.

Authors in [7] presents GHPSO algorithm to achieve the scheduling goals, this paper greatly improves the solution quality, so it can be used as an effective way to solve the cost minimization problem in cloud computing.

In [8] authors propose scheduling based on particle swarm optimization algorithm in cloud computing The FCFS algorithm is considered as an easy method in scheduling algorithms, where processes are ordered by arrival time and submitted to the virtual machine [9].

In this paper [10], authors propose a scheduling algorithm integrated with task grouping, priority-aware and SJF (shortest-job-first) to reduce the waiting time and makespan, as well as to maximize resource utilization. The authors of [11] propose the Max-min algorithm, the Max-min improvement is based on execution time instead of completion time as a basis for selection.

[13] propose the Round Robin (RR) algorithm focused on equity. RR uses the ring as a queue to store jobs. Each task in a queue has the same execution time and it will be executed in turn. If a job cannot be completed during its turn, it will be stored in the queue while waiting for the next turn. In addition to this, you need to know more about it.

A workflow tasks scheduling algorithm based on genetic algorithm in cloud computing is proposed in [14], In this algorithm, each task is assigned priority by an top-down leveling method for reducing the execution cost of workflow tasks scheduling, all workflow tasks are divided into the different levels, which can promote the parallel execution of workflow tasks. Authors in [15] present three scheduling algorithms are discussed such as LCFP, Min-Min, Max-Min and Genetic Algorithm. The idea of GA comes from natural selection which consists of population generation, selection, and mutation. This approach reduces the makespan of algorithm by using enhanced Max Min for initializing the population in GA.

[16] proposed a meta-heuristic based scheduling, which minimizes execution time and execution cost as well. An improved genetic algorithm is developed by merging two existing scheduling algorithms for scheduling tasks taking into consideration their computational complexity and computing capacity of processing elements.

In [17] A heuristic method to schedule bag-of-tasks (tasks with short execution time and no dependencies) in a cloud is presented. The number of virtual machines to execute all the tasks within the budget is minimum.

References	Algorithm used	Advantages	Disadvantages
[7]	particle swarm optimization	reduce the task average running time, and raises the rate availability of resources.	lay stress on resource load balance of dynamic task scheduling in cloud computing, and consider other resource in cloud computing, for example, how internet quality of service and data distribution influence task scheduling
[8]	Hybrid particle swarm optimization	minimize costs within a given execution time.	multi-target workflow scheduling in cloud computing
[9]	FCFS	minimum execution time or minimum and prioritized cost	allocating resources to customers will be one of the main concerns of resource providers
[10]	Priority-aware and SJF	reduce the waiting time and makespan, as well as to maximize resource utilization.	to gain more scientific and realistic results, he will collect real data sets and adopt the real cloud-based environment. It is also important to evaluate the proposed solution in a physical real run-time environment to observe its true efficacy for cloud-based systems.
[11]	Max-min	override the appropriate task allocation card on resources	reduces overall time and balances the load between resources
[13]	Round Robin	improve resource usage and response time	The drawback of RR is that the largest job takes enough time for completion.
[14]	Genetic Algorithm	minimize cost, minimize execution time.	They did not take into account data transfer for data-intensive applications
[15]	LCFP, Min-Min, Max-Min and Genetic Algorithm.	minimize execution time.	They did not take into account data transfer for data-intensive applications
[16]	Genetic Algorithm	minimize execution time.	They did not take into account data transfer for data-intensive applications

Fig. 1. A comparative study of tasks scheduling algorithm.

3 Problem Description

Scientific workflows are a set of computer tasks organized to perform a composite (complex) mission in different fields such as climate modeling, genome sequencing, seismic analysis and oil exploration. Scientific workflows include hundreds of interconnected computational tasks according to different dependency models. Workflow tasks typically require large input data files and/or perform an extraordinary number of instructions. These factors provide a high level of productivity for their activities. As a result, the optimal scheduling process becomes a complicated problem. In the distributed execution paradigm, the tasks of a scientific application are assigned to different data centers for execution. When a task requires data processing, displacement and availability becomes a challenge and therefore a very long response time. One of the most difficult issues in workflow planning is optimizing the movement of data and therefore the cost of running the workflow. Meta-heuristic scheduling schemes give the best result compared to heuristic algorithms. The genetic algorithm is one of the best meta-heuristic algorithms. A genetic algorithm (GA) is a research algorithm based on the principle of evolution and natural genetics. It combines the exploitation of past results with the exploration of new areas of the research space [18].

4 Description of Proposed Work

Scientific applications are generally represented by workflows. Data-intensive workflows describe a number of tasks and the data dependencies between the tasks of an application Fig. 1. It is advantageous to use the cloud to run complex science applications because of its large data and virtual machine resource pools. One of the most difficult problems in workflows scheduling are optimizing the cost of running the workflow and the cost of loading data. Figure 2 gives the steps of GA. A genetic algorithm (GA) is an algorithm based on the principle of evolution and natural genetics.

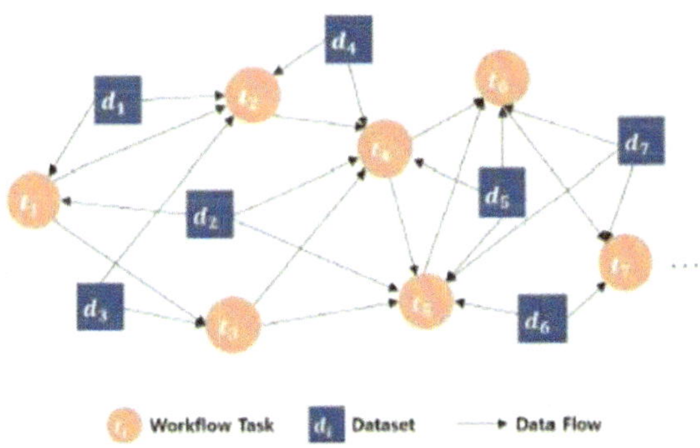

Fig. 2. Example of scientific workflow.

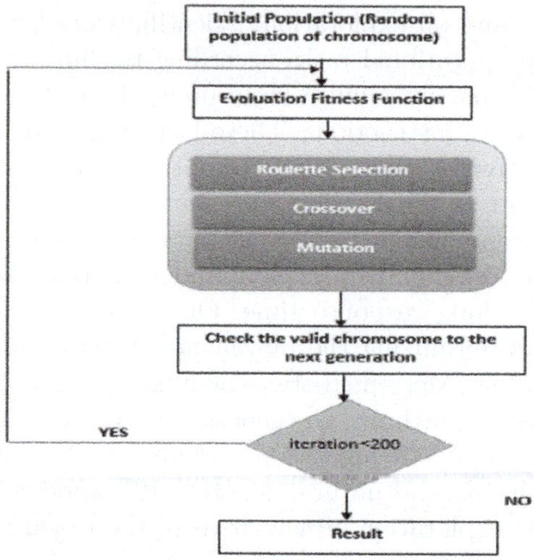

Fig. 3. Steps in genetic algorithm.

4.1 Chromosome Representation

The chromosome represents a complete sequencing of the workflow where each gene represents either a task and the virtual machine required to perform it or the location of a piece of data on it (VM). Each chromosome is a chain of genes encoding a specific solution. Figure 3 describes the modeling of a chromosome for a scientific application consisting of five tasks and six data that will be allocated on three virtual machines.

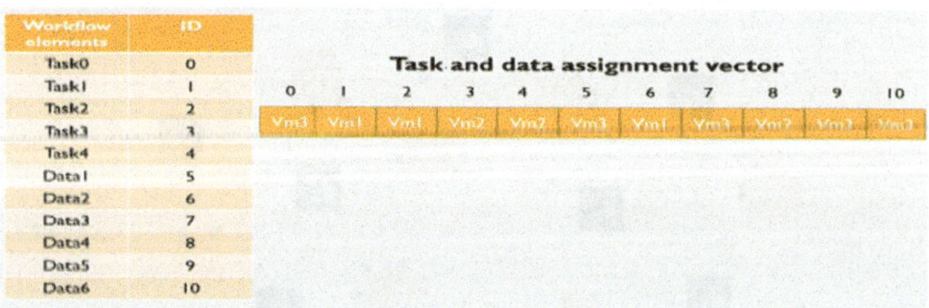

Fig. 4. Chromosome description.

4.2 Initial Population

We have merged Random solutions to generate the initial population. The success of each contribution to research is based on the problem modelling approach and on how to adapt the meta heuristic to the scheduling problem. which encode scheduling solutions to an optimization problem, evolves toward better solutions.

4.3 Fitness Function

The objective of the fitness function is to evaluate each chromosome in the population. In case of minimization problem, the best fit chromosome will have the lowest numeric value for the objective function.

We provide details on the objective function in the following:

$$MinFit = \sum (T\_exec_i + T\_dataAccess_i + T\_release_i)$$

Fit: the fitness function it represent the completion time of scientific workflow.

- T_exec_i : estimated execution time is measured based on the CPU processing capacity of the target virtual machine and the size of a task. It represents the processing time of a task on the VMj.

$$T\_exec_i = \frac{Length\_T_i}{VM\_capacity_j * PE\_Number\_VM_j}$$

Where:
$Length\_T_i$: is expressed as the number of instructions to be executed.
$VM\_capacity_j$: is the speed of a processor, is expressed in MIPS (Million Instructions Per Second);
$PE\_Number\_VM_j$: represents the number of processors of virtual machine VMj.

- $T\_dataAccess_i$: represents the estimate data access time to the data i.e. the processing time of local and remote data based on the following two formulas.

$$T\_dataAccess_i = \sum_{k=1}^{n} \frac{LocalDataSize_k}{DiskTransfertCapacityVM_j} + TRemoteDataAccess_j$$

$$TRemoteDataAccess_i = \sum_{p=1}^{L} \frac{RemoteDatasetSize_p}{BPvm_j} + \frac{RemoteDataSize_p}{DiskTransfertCapacityVM_j}$$

The principle of this estimate is to measure the processing time for data stored locally, and the time of missing data movement for the task and their treatment on vmj resource.

- $T\_release_i$: We define the resource release time as the waiting time to release the CPU resources and the data set necessary for the execution of the task as shown in the following formula:

$$T\_release_i = T\_release\_Processor_V Mi + T\_release\_Dataset$$

4.4 Selection

In the selection phase the Roulette wheel method is used. The roulette wheel selection operator minimizes the fitness function. The objective of the selection operation is to make duplicate copies of the good solution and eliminate bad solutions in a population, while maintaining the population size.

4.5 Crossover

Selecting individuals from the parental generation, new individuals are obtained. There are so many crossover operators which can be used to get the better results. In our case we have chosen the one-point crossover.

4.6 Mutation

There are several mutation operators based on the permutation-based representation of the schedule like Move, Swap, Move & Swap and Rebalancing. We chose simple Swap. The need for mutation operation is to keep diversity in the population.

5 Results and Discussion

The proposed Genetic Algorithm for Data-Intensive Scientific Workflow GA-DISW is developed using Cloudsim tools. it is an open source, scalable and low simulation overhead simulator [19].

In order to study the behavior of our proposal and analyze its results obtained by simulation, we compare them with other strategies. Several series of simulations have been launched. Simulation parameters are presented in Table 1.

Table 1. GA parameters.

Parameter	Value
Number of Cloudlets	[200,1000]
Number of VM	[10,80]
Number of Iterations	200
Population Size	100
Crossover Type	One-Point Crossover
Crossover Probability	0.5
Mutation Probability	0.5
Mutation Type	Simple swap
Termination condition	Number of iteration

We run each workflow instance through 4 simulation strategies:

Space shared: In space shared mode allocation of resources are done till the completion of task (i.e. no preemption of resources) [20];

Time Shared: in time shared mode preemption is allowed for resources continuously till the task completion [20].

GA: This strategy based on GA to submit the cloudlet in VM [14]. This approach does not take into consideration the movement of data.

GA-DISW: This simulation shows the performance of our approach.

1. **Response Time:**

Table 2. Response time (ms) vs Number of cloudlet.

N° Cloudlet (Tasks)	Algorithms			
	Space Shared	Time Shared	GA	GA-DISW
200	30500	12305	10526	**10365**
400	40035	18000	17325	**15230**
800	64500	31008	29652	**25303**
1000	80085	40000	35023	**33540**

In this first series of experiments, we measured the response time. For this simulation, we executed the simulation with the four approaches space shared, time shared, GA and GADISW. Simulation results were performed with the parameters described in the Table 1. We notice through the Table 2 that using GA-DISW, the response time of cloudlets decreases considerably in relation to space shared, time shared and GA.

2. **Number of displacement:**
In this second experiment, we measured the number of displacements of each data for the same workflow. For this simulation, we executed the simulation with the four approaches (space shared, time shared, GA and GA-DISW). The following Table 3 shows the resulting. We note with different data values, using the investment strategy reduces the number of moving data between VM compared to GA with random assignment

Table 3. NData vs Number of displacement.

N°Data	**Algorithm**	
	GA-Random Data Placement	GA-DISW
D1	30	25
D2	52	15
D3	10	09
D5	25	07
D6	18	12

6 Conclusion and Future Work

In this work, we realize a process of tasks scheduling based on genetic algorithm for data-intensive scientific workflow. Our goal is to reduce data movement and therefore improved response time of workflow.

Our strategy is tested in the CloudSim simulator and compared with Space Shared, Time Shared politics and GA with random data placement. The strategies can provide a better response time and a minimizing data displacement.

In future, we will increase the capabilities of the proposed model by enabling replication of data sets for workflow scheduling.

References

1. Durillo, J.J., Fard, H.M., Prodan, R.: Institute of Computer Science, University of Innsbruck Innsbruck, Austria, Document Text MOHEFT: A Multi-Objective List-based Method for Workflow Scheduling (2012)
2. Abrishami, S., Naghibzadeha, M., Epema, D.H.J.: Deadline-constrained workflow scheduling algorithms for infrastructure as a service clouds. Future Gener. Comput. Syst. **29**(1), 158–169 (2012)
3. Singh, P., Dutta, M., Aggarwal, N.: A review of task scheduling based on metaheuristics approach in cloud computing. Knowl. Inf. Syst. **52**(1), 1–51 (2017)
4. Yu, J., Buyya, R.: A taxonomy of workflow management systems for grid computing. J. Grid Comput. **3**, 171–200 (2005). https://doi.org/10.1007/s10723-005-9010-8
5. Jacob, J.C., et al.: Montage: a grid portal and software toolkit for science-grade astronomical image mosaicking. IJCSE **4**(2), 73–87 (2009)
6. Makhlouf, S.A.: Gestion des ressources dans les systmes grande chelle Application aux environnements en Cloud. Thesis, juin 2019
7. Marphatia, A.: Optimization of FCFS based resource provisioning algorithm for cloud computing. IOSR J. Comput. Eng. **10**(5), 1–5 (2013)
8. Devipriya, S., Ramesh, C.: Improved max-min heuristic model for task scheduling in cloud. In: International Conference on Green Computing, Communication and Conservation of Energy (ICGCE), pp. 883–888 (2013)
9. Mohapatra, S., Mohanty, S., Rekha, K.S.: Analysis of different variants in round robin algorithms for load balancing in cloud computing. Int. J. Comput. Appl. (2013)
10. Awad, A.I., El Hefnawy, N.A., Abdelkader, H.M.: Enhanced particle swarm optimization for task scheduling in cloud computing environments. Procedia Comput. Sci. **65**, 920–929 (2015)
11. Al-Husainy, M.: Tasks scheduling in private cloud based on levels of users. Int. J. Open Inf. Technol. (2017)
12. Alworafi, M.A., Dhari, A., Al-Hashmi, A.A., Darem, A.B.: An improved SJF scheduling algorithm in cloud computing environment. In: 2016 International Conference on Electrical, Electronics, Communication, Computer and Optimization Techniques (ICEECCOT), pp. 208–212 (2016)
13. Agarwal, A., Jain, S.: Efficient optimal algorithm of task scheduling in cloud computing environment. Int. J. Comput. Trends Technol. (IJCTT), **9**(7) (2014)
14. Cui, Y., Xiaoqing, Z.: Workflow tasks scheduling optimization based on genetic algorithm in clouds. In: 2018 the 3rd IEEE International Conference on Cloud Computing and Big Data Analysis (2018)
15. Singh, S., Kalra, M.: Task scheduling optimization of independent tasks in cloud computing using enhanced genetic algorithm. Int. J. Appl. Innovation Eng. Manage. (IJAIEM) **3**(7), 286–291 (2014)
16. Kaur, S., Verma, A.: An Efficient approach to genetic algorithm for task scheduling in cloud computing environment. Int. J. Inf. Technol. Comput. Sci. (IJITCS) **4**(10), 4–79 (2012)
17. Kaur, S., Verma, A.: An efficient approach to genetic algorithm for task scheduling in cloud computing environment. Inf. Technol. Comput. Sci. **10**, 74–79 (2012)
18. Zomaya, A.Y., Ward, C., Macey, B.: Genetic scheduling for parallel processor systems: comparative studies and performance issues. Parallel Distrib. Syst. IEEE Trans. **10**(8), 795–812 (1999)

19. Calheiros, R., Ranjan, R., Beloglazov, A., De Rose, C., Buyya, R.: CloudSim: a toolkit for modeling and simulation of cloud computing environments and evaluation of resource provisioning algorithms. Soft. Pract. Experience J. **41**(1), 23–50 (2011)
20. Pratap, R., Zaidi, T.: Comparative study of task scheduling algorithms through cloudsim. In: 7th International Conference on Reliability, Infocom Technologies and Optimization (ICRITO) (Trends and Future Directions), August 29–31 (2018)

Multi-robot Visual Navigation Structure Based on Lukas-Kanade Algorithm

Abdelfattah Elasri, Lakhmissi Cherroun(✉), and Mohamed Nadour

Laboratory of Applied Automatic and Industrial Diagnostics (LAADI), University of Djelfa, Djelfa, Algeria
{a.elasri,l.cherroun,m.nadour}@univ-djelfa.dz

Abstract. This paper presents an efficient control structure of two mobile robots based-visual navigation methods in an indoor environment. The proposed navigators are based on decision systems employed the necessary values estimated by a Lukas-Kanade (LK) algorithm of optical flow (OF) approach. The robots control systems use the generated motion values in order to detect and estimate the positions of the nearest obstacles and objects around each mobile robot. The multi-robot system task is to navigate autonomously in their environment safely without collisions. Obstacles are identified and detected by the employed cameras of each robot based on video acquisition and image processing steps. The efficiency of the proposed approach is verified in simulation using Visual Reality Toolbox. Simulation results demonstrate that the visual based control system allows autonomous navigation without any collision with obstacles.

Keywords: Multi-Robot · Visual · Camera · Decision · Lukas-Kanade · VRML

1 Introduction

Robotics is an important multidisciplinary field of science, based on mechanical aspects, automatic approaches and going up to higher-level aspects such as acquisition and perception, modeling of indoor and outdoor environments and decision-making techniques [1,2]. The research in the field of automatic systems focuses on the design of intelligent control systems based on efficient control techniques allowing the robotic machines to move and navigate in their environment without assistance or human intervention to accomplish the desired tasks or to achieve a desired goal [3–5]. One of the most important challenges and aspects in robotic fields is the autonomous and intelligent mobile robot navigation in unknown and complex environments, especially, in the case of applications that need multi-robot systems [5,7,9,10,18]. The task of autonomous navigation of a mobile robot must be able to take decisions to carry out movements according to the information on the position and the environment that it bypasses by endowing it with a capacity of perception, decision [8,9]. Many sensors may be

B. Lejdel et al. (Eds.): AIAP 2021, LNNS 413, pp. 534–547, 2022.
https://doi.org/10.1007/978-3-030-96311-8_50

utilized in robot applications (localization, interception, impediment avoidance or navigation) [2,7]. Vision structures are very effective and appealing sensors utilized in robotics [3,5]. The digital camera mimics herbal beings' imaginative and prescient to outline visible homes of the environment (colorings and shapes) [6]. For their use, PC imaginative and prescient is a thrilling device in many applications such as: photo-registration and enhancement, matching for stereo imaginative and prescient, sample recognition, and movement estimation and analysis [5,6,13]. Behavior-based navigation methodology presents a made tool to subdivide the world navigation task into tiny sub-tasks [3,7,16]: obstacle avoidance, wall following, goal seeking, target pursuing, and so on. Different strategies and approaches have been proposed for the obstacle avoidance task. Visual-based navigation technique is a powerful approach used in robotic applications [5,6,17,18]. The vision system will increase the utility of mobile robots in numerous domains [8–10]. Optical flow approach is one of the most known compact in vision techniques and estimation used in many applications such as in vehicle navigation, video detection, image reconstruction, object tracking, obstacle avoidance and motion estimation [4,5,11–13]. One of the most used algorithms of optical flow approach are Horn-Schunck algorithm [16] and Lucas-Kanade (LK) algorithm [4,13] which have proposed efficient techniques to estimate the motion of interesting features by comparing two consecutive images. In papers [6] and [7], surveys of visual navigation techniques are bestowed for robotic applications in indoor and outside environments. For the problem of avoiding obstacles, the author of [14] proposed real-time robot navigation based on the task of tracking objects and avoiding obstacles. The robot uses a stereo vision system to locate the desired target, while using a laser viewfinder to avoid collisions. In [2], the author used monocular vision based on color segmentation and edge detection to chock the obstacle avoidance task of the robot in a dynamic and changing environment. In the paper [15], authors have been studying the problem of avoiding obstacles using monocular cameras based on improving optical flow methods. The use of optical flow methods in robotic applications is outlined in [7,12]. In paper [16], the authors present an application of Horn-Schunck algorithm associated with type-1 fuzzy logic controller for visual based robot navigation. Whereas, in [17], they have employed Takagi-Sugeno Fuzzy logic systems in order to execute obstacle avoidance task. Authors in [18] deals with a vision-based mapping using Information Theory for a cooperative multi-robot systems in a 3D space. The main objective of this work is the development of an obstacle avoidance control strategy for a multi-robot system based on vision systems. Lukas-Kanade algorithm of optical flow approach is used as an estimator for objects and obstacles detection. This paper is organized as follows. Section 2 discusses a brief on optical flow approach and Lukas-Kanade algorithm. Section 3 presents the model of the simulated mobile robot. The proposed control structure is discussed in Sect. 4. Whereas, Sect. 5 will show the simulation results. Finally, conclusions are drawn in Sect. 6.

2 Brief on Optical Flow Approach

2.1 Definition

The optical flow describes the direction and time rate of pixels in a time sequence of two consequent images. A two-dimensional velocity vector, carrying information on the direction and the velocity of motion, is assigned to each pixel in a given place of the picture [16]. For making computation simpler and quicker, we may transfer the real world three-dimensional (3D+time) objects in a (2D+time) case. Then we can describe the image by means of the 2-D dynamic brightness function of location and time I(x, y, t). Provided that in the neighborhood of a displaced pixel, change of brightness intensity does not happen along the motion field, we can use the following expression [4,12].

$$(x, y, t) = I(x + \delta x, y + \delta y, t + \delta t) \tag{1}$$

Optical flow estimation is computationally demanding. At present, there are several groups of methods for its calculation. All methods are based on the equation. 1, and consequently the presumption of conservation of brightness intensity. - The optical flow determination is solved by the calculation of partial derivatives of the image signal [11]. There are two most used methods [13,16], namely: Lucas-Kanade, and Horn-Schunck.

2.2 Lucas-Kanade Algorithm

The Lucas-Kanade algorithm [13] makes a "best estimate" of a neighborhood's displacement by examining changes in pixel intensity that can be explained from the known intensity gradients of the image in that neighborhood. For a single pixel, we have two unknowns (u and v) and an equation (that is, the system is under-determined) [13]. We need a neighborhood to get more equations. By doing so, the system is determined, and we must find the least square solution. The Lucas-Kanade algorithm is an ancient method of observing optical information at interesting points in an image (i.e. those with sufficient information on the intensity gradient). It works for moderate object speeds. In Lucas-Kanade algorithm, we have an equation with two unknowns:

$$I_x u + I_y v + I_t = 0 \tag{2}$$

In order to calculate (u, v) for a pixel, we can follow the following steps: -If we use a 5×5 window, this gives us 25 equations per pixel.

- If we use a 5×5 window, this gives us 25 equations per pixel.

$$\underbrace{\begin{bmatrix} I_x(P_1) I_y(P_1) \\ I_x(P_2) I_y(P_2) \\ \vdots\vdots \\ I_x(p_{25}) I_y(P_{25}) \end{bmatrix}}_{\substack{A \\ 25\times 2}} \underbrace{\begin{bmatrix} u \\ v \end{bmatrix}}_{\substack{d \\ 2\times 1}} = - \underbrace{\begin{bmatrix} I_t(P_1) \\ I_t(P_2) \\ \vdots \\ I_t(P_{25}) \end{bmatrix}}_{\substack{b \\ 25\times 1}} \tag{3}$$

Now we have more equations than unknown parameters:

$$\underset{25\times2}{A}\ \underset{2\times1}{d} = \underset{25\times1}{b} \rightarrow \min imize\,\|Ad - b\|^2] \tag{4}$$

We can solve the problem using the least squares - the minimum solution of the least squares is given by the solution illustrated as:

$$\underset{2\times2}{(A^T A)}\ \underset{2\times1}{d} = \underset{2\times1}{A^T b} \tag{5}$$

$$\underbrace{\begin{bmatrix} \sum I_x I_x & \sum I_x I_y \\ \sum I_x I_y & \sum I_y I_y \end{bmatrix}}_{A^T A} \begin{bmatrix} u \\ v \end{bmatrix} = - \underbrace{\begin{bmatrix} \sum I_x I_t \\ \sum I_y I_t \end{bmatrix}}_{A^T b} \tag{6}$$

The summations are on all pixels in the K × K window.

2.3 Focus of Expansion (FOE)

For the translational motion of the camera, image motion everywhere is directed away from a singular point corresponding to the projection of the translation vector on the image plane. This point, is called the Focus of Expansion (FOE), it is computed based on the principle that flow vectors are oriented in specific directions relative to the FOE. Additionally, the FOE represents the projection point in the image, allowing obtaining information about the depth of some pixels and the FOE. This information is called Time To Contact (TTC) [16]. In a situation where the camera is moving forward, the Focus of Expansion point is shown as in the Fig. 1(b) (red circle).

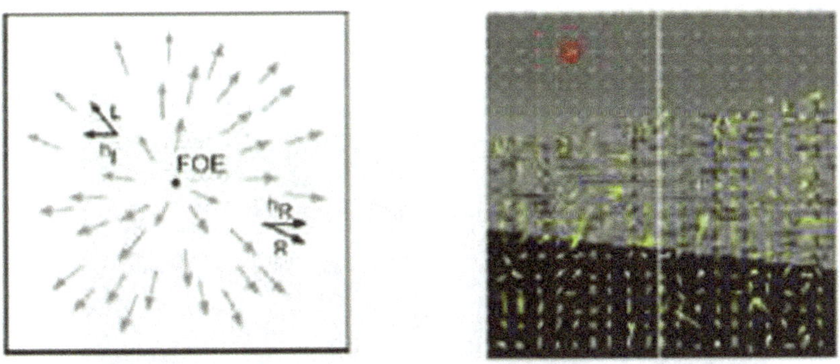

(a) **(b)**

Fig. 1. a. FOE calculation. b. Results of FOE.

2.4 Time to Contact (TTC)

The time-to-Contact (TTC) can be computed from the optical flow which is extracted from monocular image sequences acquired during motion. The image velocity can be described as a function of the camera parameters and split into two terms depending on the rotational (v_t) and translational components (v_r) of camera velocity (v) respectively. The rotational part of the flow field can be computed from proprioceptive data (e.g. the camera rotation) and the focal length. Once the global optic flow is computed, (v_t) is determined by subtracting (v_r) from (v). The TTC is computed as follows:

$$TTC = \frac{-z}{dz/dt} \tag{7}$$

Where Z is the distance camera-obstacle, and dZ/dt is the velocity of the robot camera.

3 The Mobile Robot

In this section, we present the simulated mobile robot in its environment using Virtual Reality Modeling Language (VRML). The used mobile robot is a cylindrical platform with two motorized wheels. In order to perceive its environment, the robot is endowed with a virtual camera, where the objective is to navigate autonomously. The simulated mobile robot is illustrated in Fig. 2(a). Using VRML library, we have created a virtual navigation environment that imitates the real space in 3D containing obstacles in the form of boxes, floor, walls and the goal. The designed 3D environment is depicted in Fig. 2(b). The robot motion is based on the nonholonomic kinematic model described as follows in the equation.

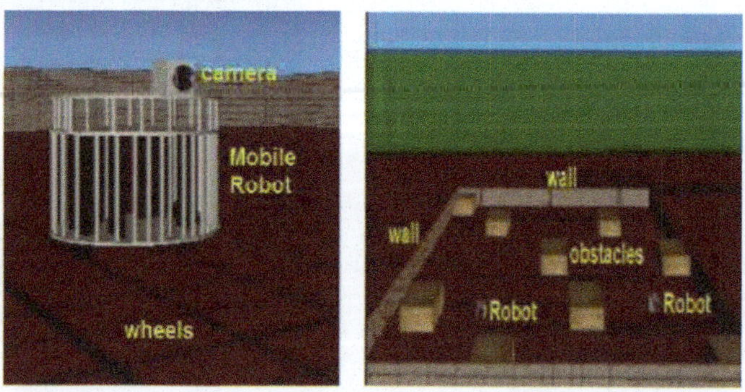

(a) (b)

Fig. 2. Structure of the simulated robot and its environment.

$$x_r = v * \cos(\theta_r) \tag{8}$$

$$y_r = v * \sin(\theta_r) \tag{9}$$

$$\theta_r = w \tag{10}$$

Where:

- (x_r) and (y_r) are the robot coordinates.
- (θ_r) is the heading angle.
- v is the linear velocity.
- w is the angular velocity calculated from the steering angle (α_r).

4 Control Structure

In this section, we present the proposed control structure for the multi-robot system in an indoor environment. Using the acquired images by the robot's cameras, the elaborated control system must infer the appropriate control action for the two mobile robots. We have simulated two wheeled mobile robots. In this application, we follow some assumptions such as:

- There isn't a final destination to reach ad a goal, but the guided mobile robots can navigate in their work-space by avoiding collisions.
- Robots are moving in the same indoor environment.
- The environment is considered static with unmoving obstacles and dynamic by considering the motion of the controlling mobile robots.
- We haven't considered pan-tilt motions of the used cameras.

The block diagram of this control strategy is presented in Fig. 3. It is composed of two control systems for each robot, a calculation block of Lukas-Kanade estimator, twos color cameras without considering pan-tilt motion. The environment and their components: robots, obstacles and walls are conceived using 3D form with VRML. At each time step for each motion action, robots must perceive their environment by capturing images of the surrounding space. Using Lukas-Kanade algorithms, right and left flow vectors are calculated to detect objects. So, the control systems can infer motion actions to guide the robot toward the next position by generating: the steering angle and the linear velocity. Acquired images are subdivided into parts in order to detect objects and to avoid them by executing the two possible actions; Turn right, and Turn Left.

5 Simulation Results

In this section, we will show the experimental results using 2D and 3D environments. Some tests are presented to verify the effectiveness of the proposed control strategy. As mentioned previously, the library VRML is used to conceive a virtual environment. The main task of the multi-.robot system is to navigate

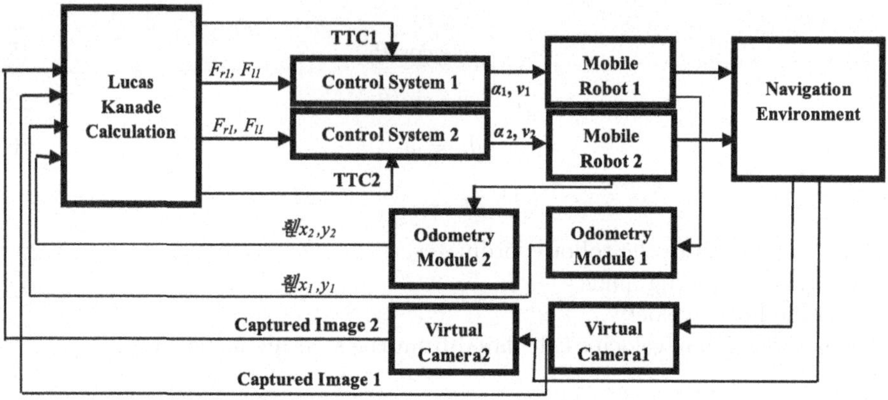

Fig. 3. Multi-robot visual navigator

autonomously and safely in its environment to accomplish a specified mission. We have considered firstly, the environment is free of obstacles enough the two controlled mobile robots as a first experiment. Secondly, we will simulate the control strategy in unknown environment with multiple surrounding obstacles (second experience). Before applying optical flow estimation on the captured frames, the image format is converted from RGB to gray as depicted in Fig. 4, because the intensity measurements acts well on grayscale frames. According to the employed methodology steps, Lukas-Kanade estimator has been used to calculate flow vectors and estimate the position of the obstacles. In the test experiments, we focus to the visual obstacle avoidance task by considering the following assumptions:

- The two used mobile robots have the same motion speed and the same characteristics;
- If the robot environment is free of obstacles, the mobile robot moves forward;
- Else, the robots turn left or turn right to avoid collisions.

To illustrate the robot's ability to detect and avoid obstacles, we have done the following experiments:

5.1 Experiment 1

For this experiment, we have simulated two mobile robots in an obstacle-free environment. The results of the simulation for this task are presented in Fig. 5 (a-b-c-d-e-f-g-h). We have illustrated at each sub-figure the position of two robots and their captured views (in top left and right respectively). This example shows the scenes of mobile robot navigation from a given position (x0, y0) in order to move freely in its environment without collision with obstacles. Each mobile robot considers the second one as a moving obstacle. So it must avoid the collision with it. This is illustrated in the depicted figures and frames. In Fig. 6, we show

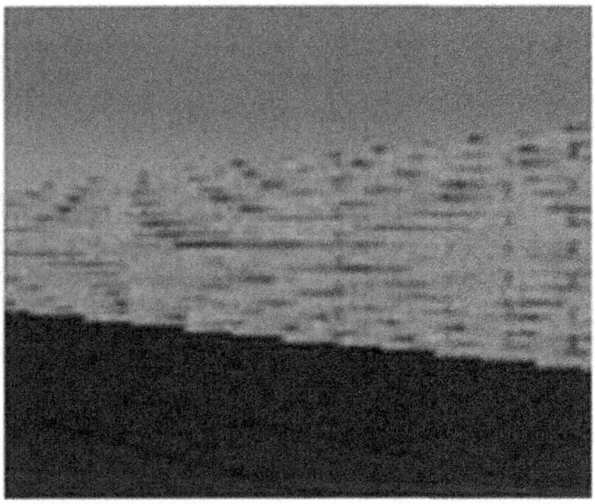

Fig. 4. Acquired gray image.

the executed paths by the two mobile robots in the 2D environment. The two robot positions are shown (The red path for the 1st robot and the black path for the 2nd robot). As can be seen, this navigation system is effective to accomplish this task with good performances. To present the moments of avoidance, the Time-To-Contact (TTC) is computed for each robot from the optical flow values using Lukas-Kanade algorithm as shown in Fig. 7 ((a) for the 1st robot, (b) for the 2nd robot). This variable gives information about the time and the number of avoidance actions.

5.2 Experiment 2

In this experiment, we will set the same conditions as experiment 1, but we have considered multiple objects and obstacles in the environment. The objective is to test the efficiency of the proposed control strategy for obstacle avoidance task in static and dynamic workspace. The results of the simulation of this task are presented Fig. 8 (a-b-c-d-e-f-g-h). Different frames for a number of positions of the two controlled mobile robots have been illustrated in a cluttered environment. Whereas, Fig. 9 shows the executed paths of the multi-robot system in 2D environment. The calculated Time-To-Contact during movement for each mobile robot is presented in Fig. 10. The control system is able to infer correct motion actions in order to guide the robot safely and autonomously. The obtained simulation results in these experiments show the effectiveness of the proposed control structure and acquisition step. The controlled mobile robots are able to navigate autonomously in the surrounded environment by the correct detection of all objects and obstacles.

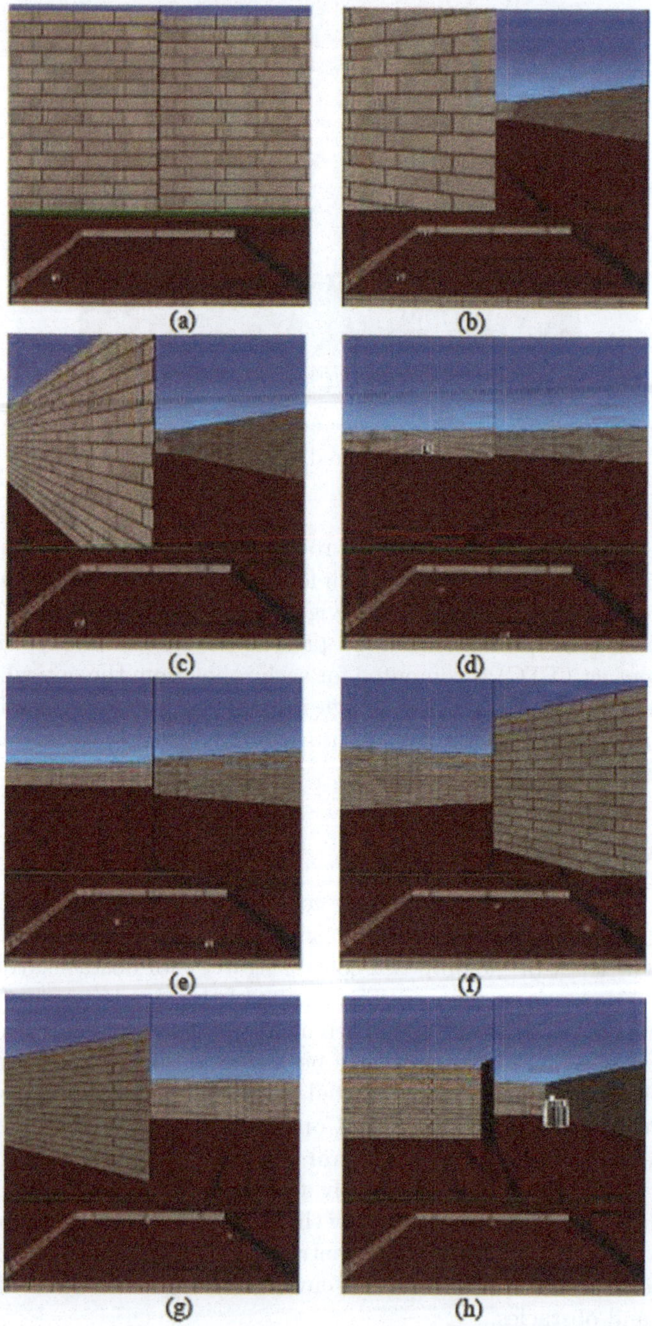

Fig. 5. Navigation frames of the 2 robots without obstacles (captured image of 1st robot in top left, captured image of 2nd robot in top right)

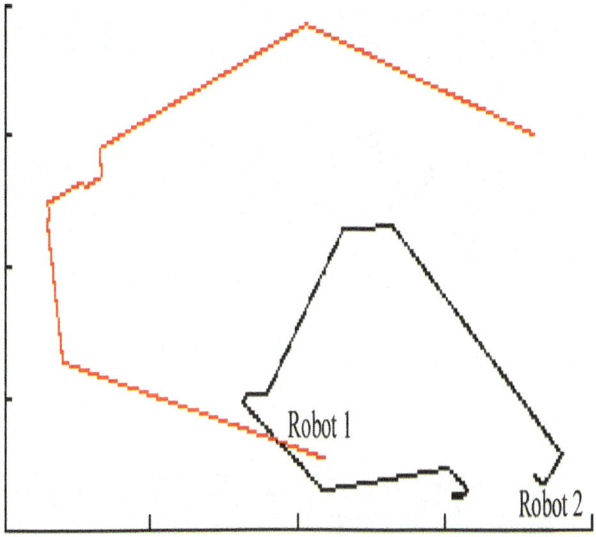

Fig. 6. Paths of the two robots in free environment

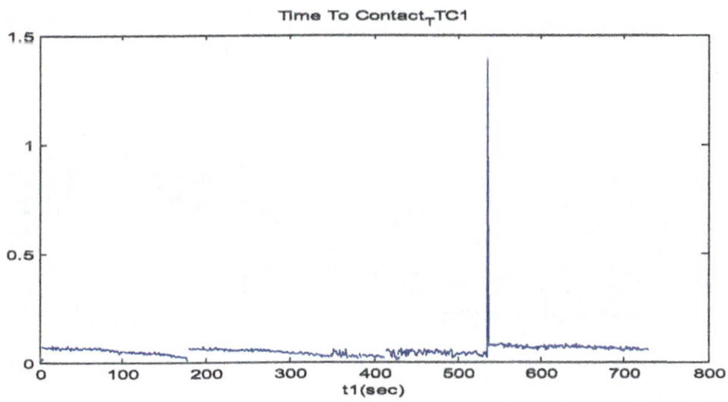

(a)

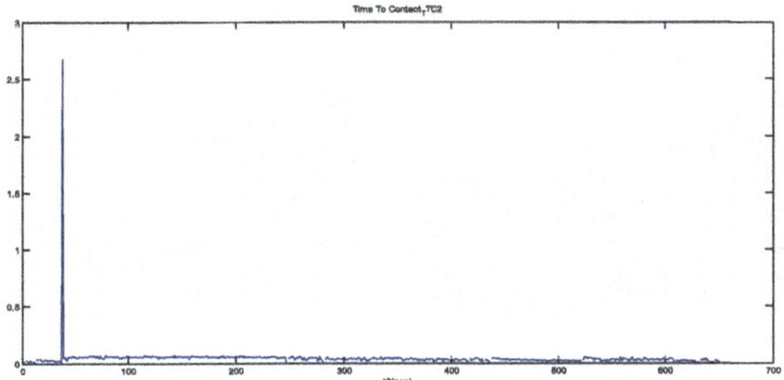

(b)

Fig. 7. Time to contact (a. for 1st robot and b. for 2nd robot).

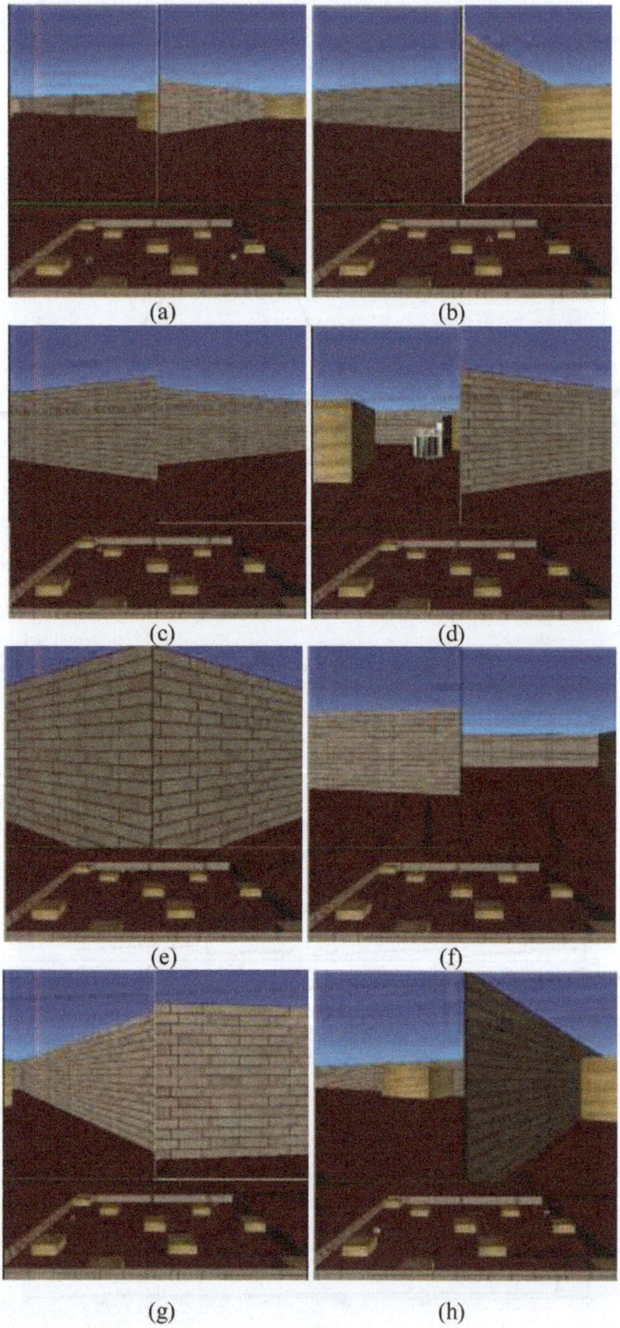

Fig. 8. Navigation frames of the 2 robots with obstacles (captured image of 1st robot in top left, captured image of 2nd robot in top right)

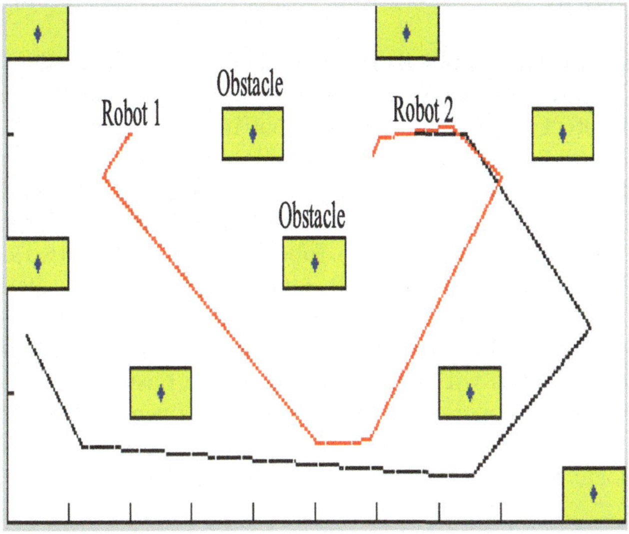

Fig. 9. Paths of the 2 robots in environment with obstacles.

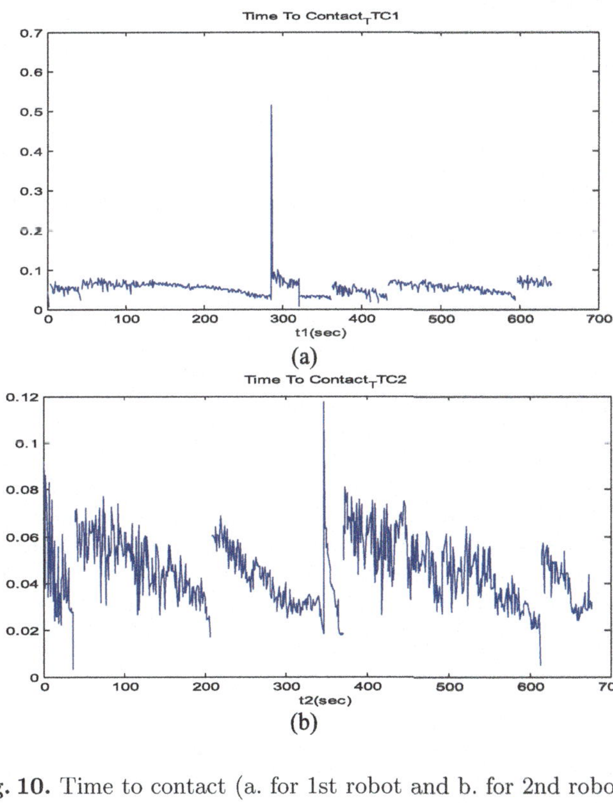

Fig. 10. Time to contact (a. for 1st robot and b. for 2nd robot)

6 Conclusion

In this paper, we have studied a multi-robot system controller in unknown environment based optical flow approach. Lukas-Kanade algorithm is used to estimate and detect objects and environment in order to elaborate an effective control action to move the two mobile robots. The proposed navigation structure is simulated in three dimensional environments using the VRML library. The acquired image in each time step of each mobile robot is divided into two parts right and left to guarantee the robot's motion in the two directions by executions: Turn Right and turn Left actions. The efficiency of the proposed approach is verified in simulation using Virtual Reality Toolbox. Simulation results demonstrate the efficiency of the elaborated visual-based control systems for autonomous motions without any collision with obstacles for the controlled mobile robots. In next work, the interest will be given to increasing numbers of robots in the navigation environment. Then, the use of a multi-agent system based fuzzy logic to coordinate actions between the controlled mobile robots.

References

1. Cuesta, F., Ollero, A.: Intelligent Mobile Robot Navigation. Springer-Verlag, Berlin Heidelberg (2005). https://doi.org/10.1007/b14079
2. Benn, W., Lauria, S.: Robot navigation control based on monocular images: an image processing algorithm for obstacle avoidance decisions. Math. Probl. Eng. 1–14 (2012)
3. Cherroun, L., Nadour, M., Boudiaf, M., Kouzou, A.: Comparison between type-1 and type-2 Takagi-Sugeno fuzzy logic controllers for robot design. Electrotehnică Electronică Automatică **66**(2), 94–103 (2018)
4. Aslani, S., Mahdavi-Nasab, H.: Optical flow based moving object detection and tracking for traffic surveillance. Int. J. Electr. Comput. Energ. Electron. Commun. Eng. **7**(9), 1252–1256 (2013)
5. Guzel, M.S., Bicker, R.: Vision based obstacle avoidance techniques. Recent Adv. Mob. Robot. (InTech), 83–108 (2011). https://doi.org/10.5772/25540
6. Desouza, G.N., Kak, A.C.: Vision for mobile robot navigation. A Survey. IEEE Trans. Pattern Anal. Mach. Intell. **24**(2), 237 (2002)
7. Font, F.B., Ortiz, A., Oliver, G.: Visual navigation for mobile robots: a survey. J. Intel Robot Syst. **53**, 263–296 (2008)
8. Gupta, M., Uggirala, B., Behera, L.: Visual navigation of a mobile robot in a cluttered environment. In: 17th World Congress of IFAC, Seoul, Korea (2008)
9. Tajti, F., et al.: Optical flow based odometry for mobile robots supported by multiple sensors and sensor fusion. Automatica **57**(1), 201–211 (2016)
10. Singh, P., Tiwari, R., Bhattacharya, M.: Navigation in multi robot system using cooperative learning: a survey. In: 2016 International Conference on Computational Techniques in Information and Communication Technologies (ICCTICT). IEEE (2016)
11. Corso, J.: Motion and Optical Flow. College of Engineering, in University of Michigan (2014)
12. Chao, H., Gu, Y., Napolitano, M.: A survey of optical flow techniques for robotics navigation applications. J. Intell. Robot. Syst. **73**, 361–372 (2014). https://doi.org/10.1007/s10846-013-9923-6

13. Lucas, B.D., Kanade, T.: An iterative image registration technique with an application to stereo vision (1981)
14. Tasalatsanis, A., Valavanis, K., Yalcin, Y.: Vision based target and collision avoidance for mobile robots. J. Intell. Robot. Syst. **48**(2), 285–304 (2007). https://doi.org/10.1007/s10846-006-9096-7
15. Wang, C., Liu, W., Meng, M.Q.H.: Obstacle avoidance for quadrotor using improved method based on optical flow. In: IEEE International Conference on Information an Automation, pp. 1674–1679, Lijiang, China (2015)
16. Nadour, Mohamed, Boumehraz, Mohamed, Cherroun, Lakhmissi, Puig, Vicenç: Mobile robot visual navigation based on fuzzy logic and optical flow approaches. Int. J. Syst. Assur. Eng. Manage. **10**, 1654–1667 (2019). https://doi.org/10.1007/s13198-019-00918-2
17. Nadour, M., Boumehraz, M., Cherroun, L., Puig, V.: Hybrid type-2 fuzzy logic obstacle avoidance system based on Horn-Schunck method. Electrotehnică, Electronică. Automatică (EEA) **67**(3), 45–51 (2019)
18. Rocha, R., Dias, J., Carvalho, A.: Cooperative multi-robot systems a study of vision-based 3-D mapping using information theory. In: Proceedings of the 2005 IEEE International Conference on Robotics and Automation, pp. 384–389 (2005). https://doi.org/10.1109/ROBOT.2005.1570149

Real-Time Speed Control of a Mobile Robot Using PID Controller

Sabrina MohandSaidi[1(✉)] and Rabah Mellah[2]

[1] Department of Electrical Engineering, L2CSP Laboratory, Mouloud Mammeri University, 15000 Tizi-Ouzou, Algeria
sabrina.mohandsaidi@ummto.dz

[2] Department of Automatic, L2CSP Laboratory, Mouloud Mammeri University, 15000 Tizi-Ouzou, Algeria

Abstract. This paper presents PID control of speed implemented by PC on a unicycle mobile robot. The designed control has been applied to a non-holonomic mobile robot. This type of robot is very popular because of simplicity of construction and interesting kinematic properties. The PID controller is designed to control the speed of two rear wheels. The accuracy of this control depends on PID parameters. So it is important that they are chosen with care. It makes the adjustment of the regulator parameters in real time. There is no such thing as perfect PID. It relies entirely on the requirements. The results obtained from the experiment show the efficiency of this control strategy, even in the case where we introduce a disturbance for the system such as overloading the robot.

Keywords: Real-time speed control · Non-holonomic mobile robot · PID controller

1 Introduction

The first attempts at mobile robots date back to the late 1960s. It is only during the 1990s that a significant amount of research effort has been devoted to the subject. Mobile robotics is clearly at the heart of technological innovation via companion robots, personal assistance robots or even robotic transport systems.

The key aspect of the mobile robot is its mobility; however its movement performance strongly affects the performance of the spots. Robots are designed to perform specific tasks in a dangerous and hostile environment. That is why it is important to move at an exact and well defined speed according to the task and the environment.

In recent years, researchers have shown increased interest in the field of mobile robot. Initially, the majority of research was focused on using kinematic models of mobile robots to develop and execute motion control. Much of the greater part of the literature on robotic has emphasized the importance for path following [1–3] and [4], obstacle avoidance in [5] and [6], speed control [7, 8], design and modeling [9–11] and [12], map building [13, 14] and [15]. A large part of the previous work applied in simulation. But the real system never responds in the same way and few studies are

B. Lejdel et al. (Eds.): AIAP 2021, LNNS 413, pp. 548–556, 2022.
https://doi.org/10.1007/978-3-030-96311-8_51

processed in real time [16] and [17]. The constraints of dc motors (current, voltage, and torque), the inertia of the robot, and the topography of the surroundings are not taken into consideration in some studies.

In the last decades the world of robotics has widely used the PID control approach [4, 7]. Because of its benefits, such as simplicity, robustness, and familiarity in the control community, this control strategy is still in use. Proportional, integral, and derivative are the three terms that make up a PID controller. The combined action of these three controllers yields a process control approach. PID controllers control process variables such as pressure, speed, temperature, and flow. As a result, a significant amount of time and effort has gone into determining the appropriate PID parameters for various process models. Several novel approaches for tuning PID controllers have been presented in the literature with the goal of improving on the Ziegler-Nichols (1942) method's performance.

Our work consists of the application of the PID control in real time for the speed of mobile robot. A simple model of mobile robot kinematics is utilized. This study proposes a novel PID tuning strategy considered as engineering process control method based on fundamental control tools. The non-holonomic mobile robot dr robot i90 has been used to experimentally test this controller

This paper begins by mobile robot description in Sect. 2. The third section presents the modeling of the mobile robot. Control system description is discussed in Sect. 4. Section 5 presents experimental approach and results. The final section gives a brief summary and discussion of the findings of this work and proposes future pursuit.

2 Mobile Robot Description

The dr Robot i90 is a complex tool for researchers with a totally wireless connection for building robotic applications including remote monitoring of varied surroundings, navigation, autonomous patrol, and additional use.

This robot is a lightweight device, weighing just 5 kg. Yet it can carry a payload of up to 15 kg. It measures 43 cm in width, 38 cm in length, and 30 cm in height. It has a top speed of 75 cm per second. This robot has two DC motors that allow it to move about in its surroundings, as well as integrated quadratic encoders on the driving wheels that offer a measure of incremental angles over a sample time [15, 18].

The PMS5005 robot card, designed to act as part of the WiRobot system, is the dr Robot i90's driving pilot element. It comes with built-in firmware for closed loop position, velocity sensor, and data collecting, and wired and wireless communication. WiRobot software development kit [19, 20] allows PC programs to connect with the PMS5005 firmware. Figure 1 depicts the robot's views.

Fig. 1. Front and side views of mobile robot.

3 Modeling of the Mobile Robot

3.1 Kinematic Model

The robot used is a wheeled platform, which can move thanks to two steering wheels located at the rear. The front wheel is a caster wheel. Its role is to maintain the platform in balance. The kinematic equations that express the motion of a robot are given in Eq. (1)

$$\begin{cases} \dot{x} = v\cos\theta \\ \dot{y} = v\sin\theta \\ \dot{\theta} = \omega \end{cases} \quad (1)$$

x and y are Cartesian coordinates of the center of the robot gear, θ is the orientation angle, ω and ν represent angular and linear velocities respectively [12].

3.2 Actuation Model

The actuation model consists of the representation of the robot's speed as a function of the driving wheel speeds and the geometric parameters of the robot [9].

$$v = \frac{v_r + v_l}{2} \quad (2)$$

$$\omega = \frac{v_r - v_l}{L} \quad (3)$$

L is the distance between steered wheels, v_r and v_l are velocities of right and left wheels representing inputs of the kinematic model [9]. The motion of a differential driving robot is characterized by two non-holonomic constraints, obtained by two main hypotheses:

Hypothesis I: No lateral slip. It simply implies that the robot is unable to move laterally in its local coordinate system, which is mathematically translated by the equation:

$$\dot{y}_{robot} = 0 \quad (4)$$

Hypothesis II: The pure rolling restriction denotes the fact that each wheel keeps contact with the ground at one point. There is no slippage of the wheel in its longitudinal or its orthogonal axis.

4 Control System Description

PID correction is a closed-loop control that is widely used in industry and academic research because it is simple to apply and provides good results. The proportional, integral, and derived actions comprise the PID control law. The objective of feedback control is to reduce the error signal; the difference between the measured value and the reference value. The proportional action is to generate an action that varies proportionally to the error signal [21]. The advantage of the integral controller is that it eliminates the regulation error which persisted with a proportional regulator alone [21]. The derived action makes it possible to anticipate the future variations of the measurement signal by applying a proportional action to its rate of variation. The derived action has a predictive effect [21]. Figure 2 shows the control scheme.

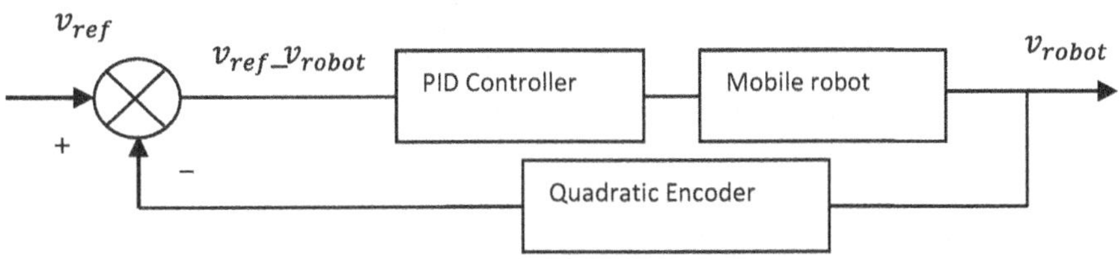

Fig. 2. PID Control scheme

The output of the PID controller is given by Eq. (5).

$$Output = K_p e(t) + K_i \int_0^t e(\tau)d\tau + K_d \frac{de(t)}{dt} \tag{5}$$

K_p, K_d *and* K_i are the proportional, derivative and integral parameters of the PID controller and e is the error and it is the difference between the reference speed and the measured speed [22].

A PID control consists of memorizing the error, the sum of the errors and the difference between the current error and the previous error. The PID regulation consists in the choice of the regulator parameters in such a way as to reduce the error to zero and to keep the system fast and stable. For the choice of the coefficients of the regulator we cannot apply Nichols-Ziegler because for this approach the system must be already regulated in closed loop and the fact of bringing the system to an oscillatory state risks destroying our robot. For this we proceeded to follow the flowchart given in Fig. 3 to design our PID regulator.

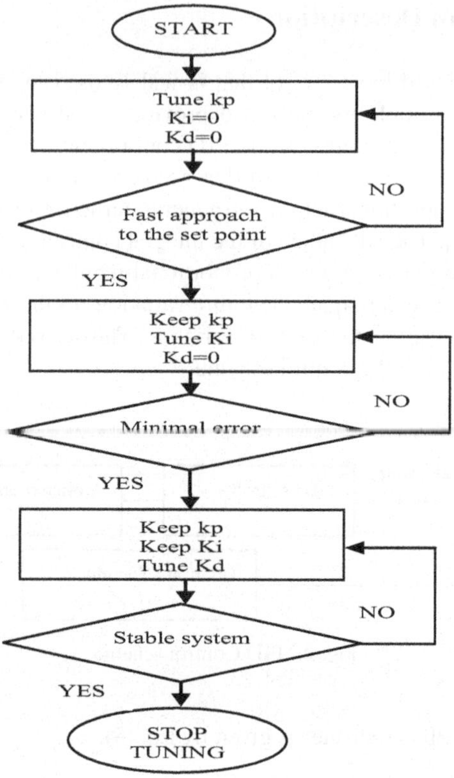

Fig. 3. PID controller design process

5 Experimental Results

This application is implemented using Matlab, and tested on real robot system in indoor environment using dr Robot i90. Matlab allows building these interfaces thanks to GUIDE (Graphical User Interface Development Environment). This tool is able to build high level applications. A graphical interface makes it possible to control an application interactively with the mouse rather than by launching the commands with the keyboard. It also makes it possible to click on images, graphs or objects to modify the value of a variable, to release functions or simply to make the information appear. The User Interface created for this work is shown in Fig. 4.

In the first time we tested the robot with all coefficients equal to zero and for two values of speed reference, 100 pulses per second and 500 pulses per second. We noticed that the speed for both cases is far from the desired speed with a very unstable system. Results are shown in Fig. 5 for $v_{ref} = 100$ pulses/s and for $v_{ref} = 500$ pulses/s in Fig. 6.

For the choice of our parameters we proceeded as follows. We have increased the value of K_p keeping the value of K_i and K_d equal to zero. We noticed that the speed is far from the reference but it becomes more stable for a value of $K_p = 5$. Afterwards

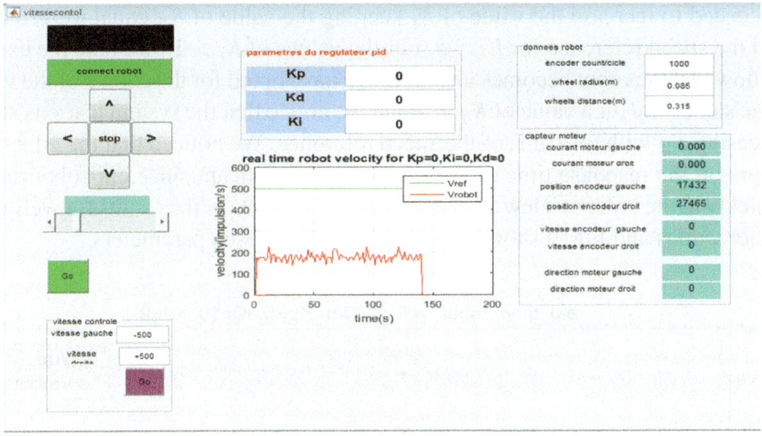

Fig. 4. Control user interface

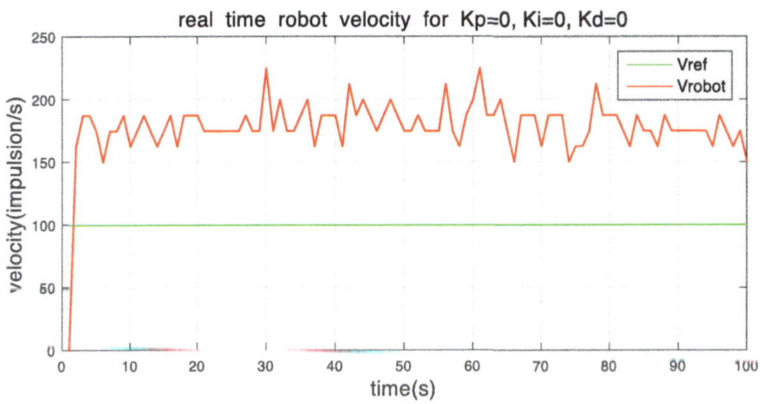

Fig. 5. Speed control with $K_p = 0$, $K_d = 0$ *and* $K_i = 0$ for $v_{ref} = 100$ pulses/s

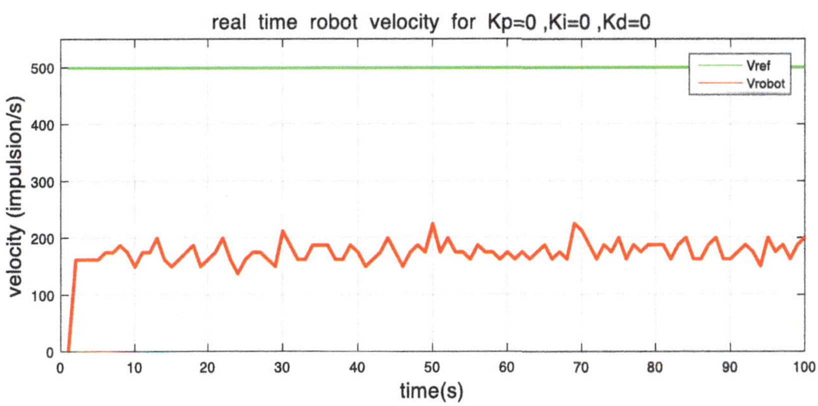

Fig. 6. Speed control with $K_p = 0$, $K_d = 0$ *and* $K_i = 0$ for $v_{ref} = 500$ pulses/s

we proceeded to increase the value of K_i keeping the value of K_d equal to zero and we reached the speed reference for $K_i = 5$. For the values of $K_i > 10$ we note the existence of overflow and the robot becomes unstable. We have opted for the choice of the value of K_d in the same way for a value of $K_d = 2$ and we notice that the system loses its stability by increasing K_d with overruns of the speed reference. We noticed that the effect of our regulators on the response time of the system is insignificant since our robot responds very quickly to the order of a few seconds. Figure 7 and 8 show the results for well-chosen parameter values and Fig. 9 shows results for poorly chosen parameters.

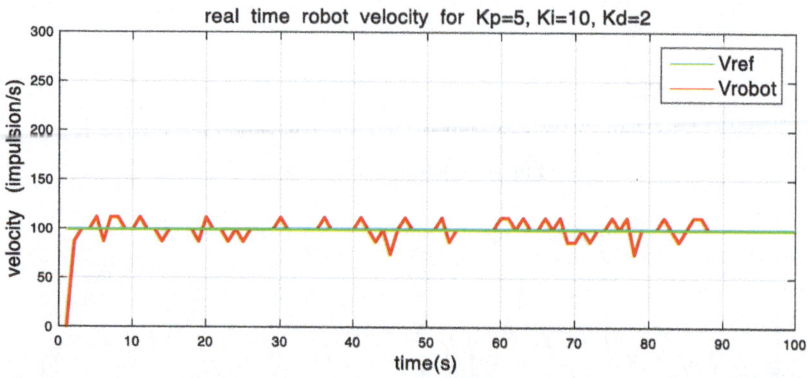

Fig. 7. Speed control with $K_p = 5, K_d = 2$ *and* $K_i = 10$ for $v_{ref} = 100$ pulses/s

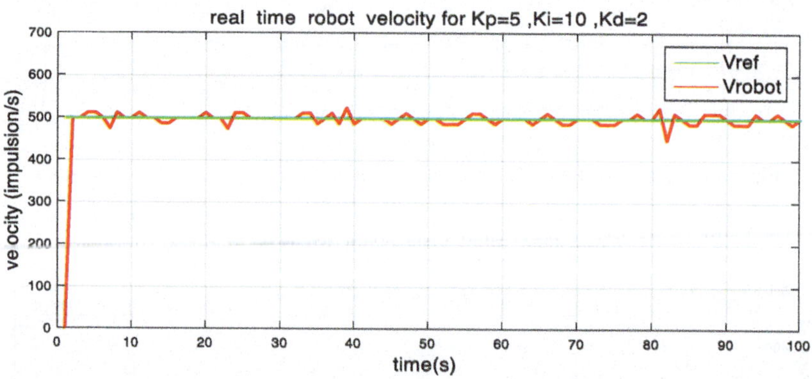

Fig. 8. Speed control with $K_p = 5, K_d = 2$ *and* $K_i = 10$ for $v_{ref} = 500$ pulses/s

In the last test we did not realize the test for $v_{ref} = 500$ pulses/s to avoid damage to our robot.

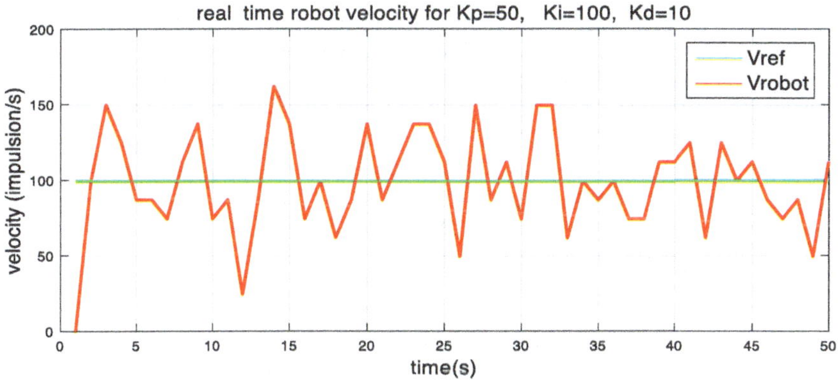

Fig. 9. Speed control with $K_p = 50$, $K_d = 10$ *and* $K_i = 100$ for $v_{ref} = 100$ pulses/s

6 Conclusion

In this paper we have proposed a controller that can be applied to a large class of systems. The application of this command on a non-holonomic mobile robot (real robot) made it possible to highlight the control with PID.

The PID controller mode has consequences if one mode dominates. Excessive proportional action causes faltering, excessive integral action causes overshoot, and excessive derivative action causes oscillations.

In this work we try to improve the speed performance of the robot i90 and we implement the regulator PID without calculation but by changing PID parameters while trying to keep our system stable and healthy. But the application of a fuzzy regulator for the choice of these parameters may be a better solution.

References

1. Matoui, F., Boussaid, B., Abdelkrim, M.N.: Distributed path planning of a multi-robot system based on the neighborhood artificial potential field approach. Simulation **95**(7), 637–657 (2019)
2. Kayacan, E., Chowdhary, G.: Tracking error learning control for precise mobile robot path tracking in outdoor environment. J. Intell. Rob. Syst. **95**(3), 975–986 (2019)
3. Chen, S., Xue, W., Lin, Z., Huang, Y.: On active disturbance rejection control for path following of automated guided vehicle with uncertain velocities. In: 2019 American Control Conference (ACC), pp. 2446–2451. IEEE, July 2019
4. Ng, K.H., Yeong, C.F., Su, E.L.M., Husain, A.R.: Implementation of cascade control for wheeled mobile robot straight path navigation. In: 2012 4th International Conference on Intelligent and Advanced Systems (ICIAS 2012), vol. 2, pp. 503–506. IEEE, June 2012
5. Cheribet, M., Laskri, M.T.: Évitement d'obstacles dynamiques par un robot mobile courrier du savoir, no. 14, Novembre 2012, Biskra, Algerie, pp. 119–126 (2010)
6. Zavlangas, P.G., Tzafestas, S.G., Althoefer, K.: Fuzzy obstacle avoidance and navigation for omnidirectional mobile robots. In: European Symposium on Intelligent Techniques, Aachen, Germany, pp. 375–382, September 2000

7. Sharma, S., Jain, S.: Speed control of mobile robotic system using PI, PID and pole placement controller. In: 2016 IEEE 1st International Conference on Power Electronics, Intelligent Control and Energy Systems (ICPEICES), pp. 1–5. IEEE, Newcastle upon Tyne, 2000, July 2016
8. Serrano, M., Godoy, S., Mut, V., Ortiz, O., Scaglia, G.: A nonlinear trajectory tracking controller for mobile robots with velocity limitation via parameters regulation. Robotica **34**(11), 2546–2565 (2016). https://doi.org/10.1017/S026357471500020X
9. Mahfouz, A.A., Aly, A.A., Salem, F.A.: Mechatronics design of a mobile robot system. Int. J. Intell. Syst. Appl. **5**(3), 23 (2013)
10. Scaglia, G., Montoya, L.Q. Mut, V., di Sciascio, F.: Numerical methods based controller design for mobile robots. Robotica **27**(2), 269–279 (2009)
11. Park, J.J., Lee, S., Kuipers, B.: Discrete-time dynamic modeling and calibration of differential-drive mobile robots with friction. In: 2017 IEEE International Conference on Robotics and Automation (ICRA), pp. 6510–6517. IEEE, May 2017
12. Aung, W.P.: Analysis on modeling and Simulink of DC motor and its driving system used for wheeled mobile robot. World Acad. Sci. Eng. Technol. **32**, 299–306 (2007)
13. Krivić, S., Mrzić, A., Osmić, N.: Building mobile robot and creating applications for 2D map building and trajectory control. In: 2011 Proceedings of the 34th International Convention MIPRO, pp. 1712–1717. IEEE, May 2011
14. Jia, S., Yang, H., Li, X., Fu, W.: LRF-based data processing algorithm for map building of mobile robot. In: The 2010 IEEE International Conference on Information and Automation, pp. 1924–1929. IEEE, June 2010
15. Mohand Saidi, S., Mellah, R.: Mobile robot environment map building, trajectory tracking and collision avoidance applications. In: 2019 International Conference on Advanced Electrical Engineering (ICAEE), pp. 1–5. IEEE, November 2019
16. Mendes Filho, J.M., Lucet, E., Filliat, D.: Real-time distributed receding horizon motion planning and control for mobile multi-robot dynamic systems. In: 2017 IEEE International Conference on Robotics and Automation (ICRA), pp. 657–663. IEEE, May 2017
17. Sanchez-Lopez, J.R., Marin-Hernandez, A., Palacios-Hernandez, E.R., Rios-Figueroa, H.V., Marin-Urias, L.F.: A real-time 3D pose based visual servoing implementation for an autonomous mobile robot manipulator. Proc. Technol. **7**, 416–423 (2013)
18. Dr Robot i90, (Wireless networked autonomous mobile robot with high resolution pan-Tilt-zoom camera) quick start guide, 2010–2013
19. WiRobot SDK application programming interface (API) reference manual, (for MS Windows), version: 1.3.0 (2010)
20. PMS5005 Sensing and Motion Controller, User Manual. Dr Robot, 25 Valley wood Dr. Unitn20, Markham, ON, L3R2, 5L9 Canada version: 1.0.5 (2006)
21. Flaus, J.M.: La régulation industrielle: régulateurs PID, prédictifs et flous. Hermes Science Publications (2000)
22. He, B., Adams, B.M.: Engineering Process Control. In: Balakrishnan, N., Colton, T., Everitt, B., Piegorsch, W., Ruggeri, F., Teugels, J.L. (eds.) Wiley StatsRef: Statistics Reference Online (2014)

A Novel Methodology for Geovisualizing Epidemiological Data

Fatiha Guerroudji Meddah(✉)

Département d'Informatique, Université des Sciences et de la Technologie d'Oran, Mohamed Boudiaf, USTO-MB, BP 1505, El M'naouer, 31000 Oran, Algérie

Abstract. Rapid advances in geographic information systems (GIS) and related technologies have created a potential for dynamic geovisualization methods to be integrated with GIS in support of a range of decision-making tasks. Cartographic visualization is then considered as an extension of spatial analysis, and GIS is configured as a spatial decision support system. This integration can add a great deal to epidemiologic research and is essential for health policy planning, decision making, and ongoing surveillance efforts. In this context, this paper presents a novel methodology to support interactive visual exploration and analysis of epidemiological data by coupling MapInfo GIS software with the Gastner area cartogram, a particular class of map type where some aspect of the geometry of the map is modified to accommodate the studied problem. The aim of the study is to help improve public health by identifying areas of exposure and risk on tuberculosis in the city of Oran in Algeria.

Keywords: GIS · Geovisualization · Cartograms · Tuberculosis · Epidemiology

1 Introduction

Tuberculosis (TB) remains among the 10 leading causes of death in the world and is a public health priority in Algeria, 23 000 cases in 2018 (www.aps.dz 2021). Today it is indisputable that tuberculosis is the subject of potential studies by the medical world and particularly by epidemiologists, whose primary objective is to find solutions through the analysis of statistical data. Identifying heterogeneity in the spatial distribution of (TB) cases and characterizing its drivers can help to inform targeted public health responses, making it an attractive approach.

However, common diseases such as tuberculosis are greatly impacted by geographical and environmental factors, we can contribute to improving public health with geovisualization solutions by identifying areas of exposure and risk, by providing relevant interpretable visual information essential for decision making.

To geovisualize (TB) epidemiological data, we propose in this paper a methodological approach integrating GIS and anamorphic maps: cartograms. The geographic information system (GIS) is an effective tool for the organization of diseases and health data. Cartograms are maps in which the real relationships of enumeration units are distorted based on a data attribute (Slocum et al. 2009) (Field 2017). Cartograms are of two

B. Lejdel et al. (Eds.): AIAP 2021, LNNS 413, pp. 557–564, 2022.
https://doi.org/10.1007/978-3-030-96311-8_52

types: area cartogram (Dorling 2011) and linear cartogram (Thomas 2018) depending upon the geographical feature being distorted.

Cartograms were used mainly for representing population density (Doll 2017) and electoral votes (Dominique 2005). They are employed to simultaneously convey two types of information: geographical and statistical.

However, in literature (Bhatt and others 13; Derryn et al. 2014; Nusrat 2016; Soetens 2017; Tran 2019), cartograms are also innovative mapping techniques that allow visualization of potentially complex health relationships but are underutilized in epidemiology. As shown in (Sui 2008), it is obvious that the use of cartograms in public health can affect our understanding of reality, both cognitively and analytically.

In this context, to facilitate public health intervention, to design new tuberculosis (TB) control strategies, and to identify when (TB) is transmitted in Oran, the main objective of this research, is to produce epidemiological cartograms in a form adapted to the perceived reality. In order to achieve this objective, the proposed approach was defined on a mathematical model based on Gastner Newman's algorithm and Bertin's graphic semiology.

The paper proceeds with four more sections. The following section describes a set of algorithms to construct a cartogram. The third section provides methodology for producing cartogram and discusses important design consideration. The results and discussion are recorded in the fourth and fifth sections respectively. Concluding remarks and future directions are offered in the final section.

2 Preliminaries

Several algorithms have been proposed in the literature to build cartograms:

Tobler's Algorithm (Tobler 1973, 1986, 2004), which was based on placing outlines on surface of variable or attribute being mapped, and repeatedly flatten the distribution based on regular sampling of the surface. But this method is not so successful as it produced poor shape of the county boundaries of US states after 99 iterations.

Dorling's Algorithm (Dorling 1993, 2011), replaces geometric features with circles. The size or radius of circle is proportional to the magnitude of the attribute being represented. This is for non-contiguous creating cartogram.

Dougenik's Algorithm (Dougenik 1895), was based on the gravitation concept. The polygons are shrunken and expanded at centroid to create cartogram. With iterations (approximately 8), the polygons converge much more rapidly. Other algorithms were based on some order in distorting the area of polygons: smaller area to larger area or smaller attribute value to larger attribute value. The disadvantage of these algorithms is that none of them preserved the shape of the geometric areas. Problems like overlapping regions, lack of readability, coordinate axes dependence were still persistent.

As a solution to these problems, Gastner and Newman (Gastner 2004) proposed a simple algorithm based on elementary physics. It is called diffusion algorithm and cartogram produced is called diffusion cartogram. In this algorithm, the original input map is projected onto a distorted grid, computed in such a way that the areas of the countries match the pre-defined values. A cartogram is created by allowing the flow of data from high-density area to low-density area until the density is equalized everywhere.

They express the problem as an iterative diffusion process, where quantities flow from one country to another until a balanced distribution is reached, i.e., the density is the same everywhere. This method allows for minimal cartographic error, while also keeping region shapes relatively recognizable. Over the last decade, this has become one of the most popular techniques to create cartograms. Its popularity is likely due to its shape recognizability, and the availability of the software to generate these cartograms (Nusrat 2016).

3 Proposed Methodology

In this work, we propose a new visualization methodology, by integrating MapInfo GIS and cartograms to assist the analysis and the exploration of tuberculosis data (Pulmonary data) set collected by local health services "DSP" (Direction de la Santé Publique) in Oran, the second largest city in Algeria, covering a total area of 2,121 km^2 with 26 municipality and a specific region named "Grande Sebkha".

"DSP" in Oran collates a large amount of data generated from each interaction a patient makes with the health service. This data set includes clinical information that can be used to analyze geographic trends of tuberculosis based on a history of four successive years (2014 to 2018).

The proposed method is a decision support tool to facilitate public health interventions and design new strategies to fight against tuberculosis (TB). The geographic information system (GIS) will make it possible on the one hand to organize the data of the (TB) to identify and control the evolution by taking appropriate measures with the discovery of spatial accumulation of the disease. On the other hand, cartograms are used to simultaneously convey the two types of resulting information: geographic and statistical. The visualization methodology is composed of successive steps as shown in the Fig. 1.

To build cartograms, we have needed not only to bring together the different types of geographic data in a database, but also convert them into a form suitable for thematic interpretation. After the pre-processing phase, the recovered data were structured in the form of geographic databases (spatial and non-spatial data) that we integrated into the MapInfo GIS.

The result is a map that contains a set of polygons where each polygon represents a municipality described by a set of data stored in the database. An example of this map for the cases observed in 2014 is shown in Fig. 2.

The data set stored in a MapInfo database were then converted to SHP file format to support the deformations necessary for the construction of the cartogram by the ScapeToad tool.

The deformation of the map is done in four steps. A computational grid is superimposed on the polygon layer. The value of the studied variable is calculated for each point of the Grid. This is a "rasterization" step. The grid is then distorted by the Gastner-Newman algorithm through a second, finer grid (grid of diffusion), each cell is enlarged or shrunk in such a way whether the density (value of the variable/area of the cell) is the same for all cells. In our case, three iterations were necessary to perform this deformation.

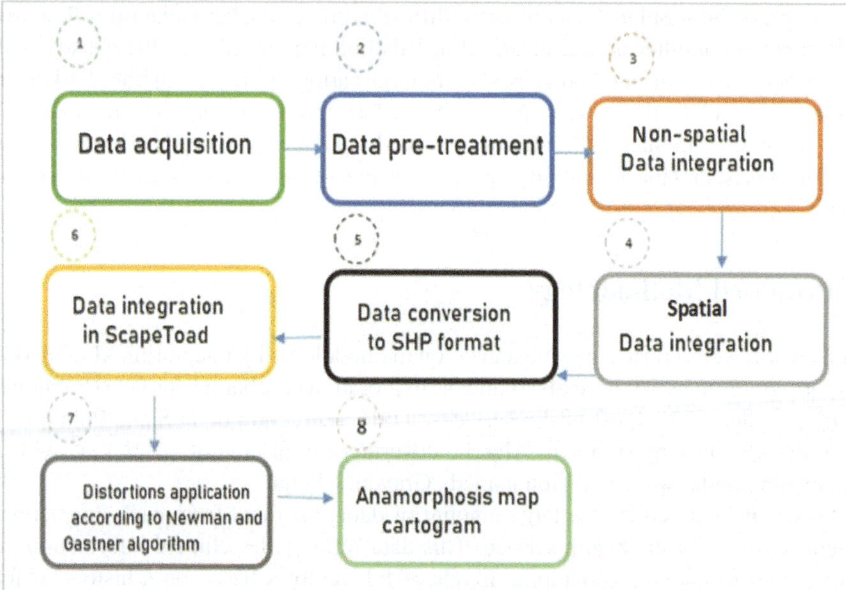

Fig. 1. Proposed methodology

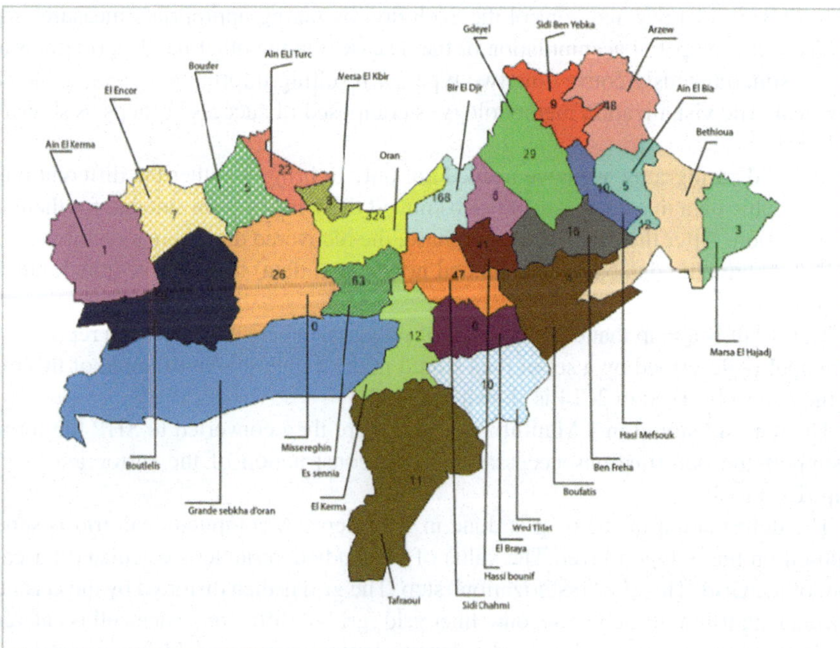

Fig. 2. Example of TB cases for year 2014

4 Results

The outcomes of the procedure outlined above are given at Fig. 3 which offers a series of cartograms of the number of cases of tuberculosis in the city of Oran during the period from 2014 to 2018. The size of each municipality is proportional to the number of cases reached.

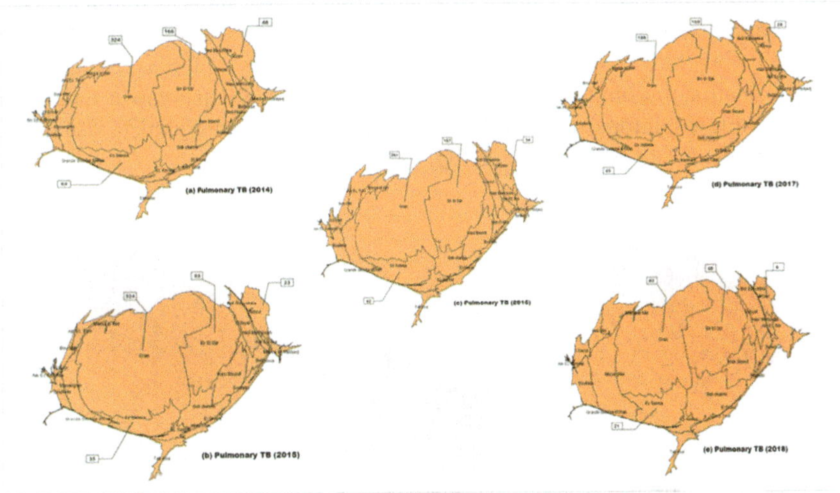

Fig. 3. Generated cartograms (a (2014), b (2015), c (2016), d (2017), e (2018))

In view of the cartogram of the year 2014, we note that the number of cases of people with pulmonary tuberculosis in the municipality of Oran is clearly higher than that of other municipalities with 324 cases. The municipality of Bir el Djir being classified second with 168 cases this which justifies the deformation in the cartogram.

Similarly, the cartogram of 2015 shows an improvement in the number of cases reached. Even if we note that the number of cases recorded in the municipality of Oran has remained stable, the improvement is clearly noted at the level of other municipalities, particularly in the municipality of Bir el Djir which has seen a reduction of more than almost 50% of cases reached, which involved a slight deformation.

As for the year 2016, the deformation of the Bir el Djir area resurfaced with an increase of 45% of cases, despite the stability of the cases observed in the municipality of Oran, the latter still remains in the lead with the most important deformation followed by the municipality of Bir el Djir.

In 2017, a slight decrease was observed in the two zones Oran and Bir el Djir, while an increase was reported in all other municipalities.

In 2018, the effort to fight this pathology was clearly observed in all the municipalities with significant reductions and very slight deformation.

5 Discussions

In his seminal work on the graphic semiotics, Jacques Bertin identified several preattentive visual dimensions across which sign vehicles differ, allowing for the theorization of syntactics for graphic sign systems (Bertin 1967, 1983). The original set of fundamental graphical elements, termed retinal or visual variables, included: location, size, grain, orientation, shape, color hue, and color value.

The absolute quantitative character of a statistical information is translated by the visual variable size. Cartograms use the visual variable size to signify the equalizing variable. In our work, to size we proposed the addition of color to reinforce the visualization of the information transmitted by the cartogram. The color as a qualitative variable will make it possible to reinforce the qualitative variable size to allow a complete visualization of the data. Results are as follows (see Fig. 4).

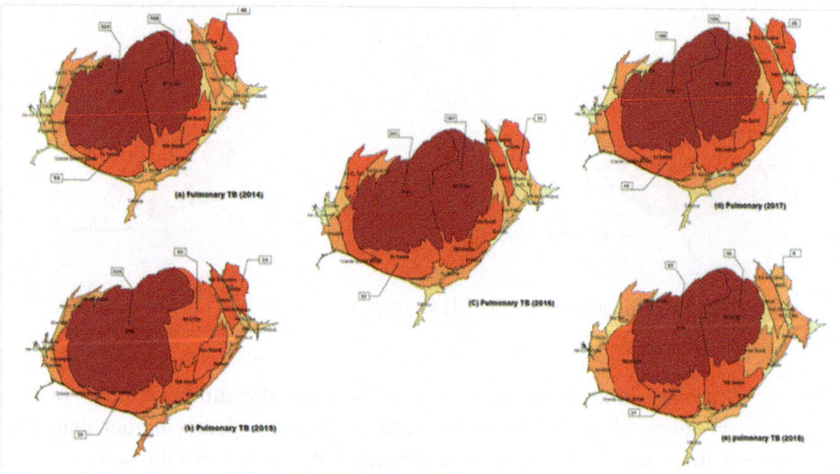

Fig. 4. New visualization of (TB) Data based on graphical semiology

These anamorphosis cartograms shown in Fig. 4 are more significant than the previous in terms of reading and interpretation. So, it is very easy to simultaneously make the link between the color and the size of the municipality generated in order to understand the change made which corresponds to the evolution of pulmonary tuberculosis.

6 Conclusion

Maps play an important role in geographic communication. They allow large amounts of data to be displayed in parallel and in a format understandable to humans. Cartograms are a special class of map type where some aspect of the map geometry is modified to accommodate the problem.

In this work, we proposed a new methodology to geovisualize epidemiological data based on MapInfo GIS software and cartograms. Cartograms are a powerful visual tool, both for communicating ideas and for facilitating data exploration. A visual assessment of the generated colored cartograms reveals several interesting features of the disease.

Thus, the proposed methodology would be a great aid to epidemiologic when the cartogram construction is integrated within GIS for a geovisualization process. It will make it possible to transform the geographic space into a functional space. Moreover, this method can be used to geovisualize the epidemiological data of any disease.

References

Bertin, J. (eds.): La Sémiologie graphique. Paris, Mouton (1967)

Bertin, J. (eds.): Semiology of Graphics. Gauthiers-Villars, Paris (1983)

Bhatt, S., et al.: The global distribution and burden of dengue. Nature **496**(7446), 504–507 (2013)

Çöltekin, A., Janetzko, H., Fabrikant, S.I.: Geovisualization. In: Wilson, J.P. (eds.) The Geographic Information Science & Technology Body of Knowledge (2018). https://doi.org/10.22224/Gistbok/2018.2.6

Derryn, A.L., Alan, J.P., Jake, T.C., Stuart, A.G., Edgar, S.S., Derek, B.: Using geographical information systems and cartograms as a health service quality improvement tool. Spat. Spatio-Temp. Epidemiol. 10, 67–74 (2014). https://doi.org/10.1016/j.sste.2014.05.004. ISSN 1877-5845

Döll, P.: Cartograms facilitate communication of climate change risks and responsibilities. Earth's Future **5**, 1182–1195 (2017). https://doi.org/10.1002/2017EF000677

Dominique, A.: L'intérêt De L'usage Des Cartogrammes: L'exemple De La Cartographie De L'élection Présidentielle Française De 2002. M@Ppemonde (2005)

Dorling, D.: From computer cartography to spatial visualization: a new cartogram algorithm. In: McMaster and Armstrong (eds.) Proceedings of the International Symposium on Computer-Assisted Cartography (Auto-Carto XI), ASPRS/ACSM, Bethesda, MD (1993)

Dorling, D.: Area Cartograms: Their Use and Creation, VL - 59 JO - Concepts and Techniques in Modern Geography (CATMOG) - 2011/04/24, SP - 252, EP - 260, SN - 9780470979587 (2011). https://doi.org/10.1002/9780470979587.Ch33

Dougenik, J.A., Chrisman, N.R., Niemeyer, D.R.: An algorithm to construct continuous area cartograms. Prof. Geogr. **37**(1), 75–81 (1985)

Field, K.: Cartograms. In: Wilson, J.P. (ed.) The Geographic Information Science & Technology Body of Knowledge (2017). https://doi.org/10.22224/Gistbok/2017.3.8

Gastner, M.T., Newman, M.E.J.: Diffusion-based method for producing density-equalizing maps. In: Proceedings of the National Academy of Sciences of The United States of America, vol. 101, no. 20, pp. 7499–7504 (2004). https://doi.org/10.1073/Pnas.0400280101

Kirby, R., Delmelle, E., Eberth, J.: Advances in spatial epidemiology and geographic information systems. Ann. Epidemiol. **27** (2016). https://doi.org/10.1016/J.Annepidem.2016.12.001

Laurini, R.: Geovisualization and chorems. In: Geographic Knowledge Infrastructure, pp. 223–246. Elsevier (2017). ISBN 9781785482434. https://doi.org/10.1016/B978-1-78548-243-4.50011-6

Nusrat, S., Alam, M.J., Kobourov, S.G.: Evaluating cartogram effectiveness. IEEE Trans. Vis. Comput. Graph. **24**, 1077–1090 (2018)

Nusrat, S., Kobourov, S.: The state of the art in cartograms. Comput. Graph. Forum **35**(3), 619–642 (2016). https://doi.org/10.1111/Cgf.12932

Röger, C., Krisp, J. M.: Using cartograms for visualizing extended floating car data (Xfcd). In: Proceedings of the ICA, vol. 2, 10 July 2019 (2019). https://doi.org/10.5194/Ica-Proc-2-107-2019

Sandul, Y., Vora, K., Carl, H., Ashish, U.: Geovisualization: a newer GIS technology for implementation research in health. J. Geogr. Inf. Syst. **7**(01), 20–28 (2015). https://doi.org/10.4236/Jgis.2015.71002

Selvin, S., Merrill, D., Schulman, J., Sacks, S., Bedell, L., Wong, L.: Transformations of maps to investigate clusters of disease. Soc. Sci. Med. **26**(2), 215–221 (1988)

Slocum, T.A., Mcmaster, R.B., Kessler, F.C. Howard, H.H.: Thematic Cartography and Geovisualization. Prentice Hall Series in Geographic Information Science. Pearson Prentice Hall (2009)

Soetens, L., Hahné, S., Wallinga, J.: Dot map cartograms for detection of infectious disease outbreaks: an application to Q fever, The Netherlands and Pertussis, Germany. Euro Surveillance: Bulletin Europeen Sur Les Maladies Transmissibles = Eur. Commun. Disease Bull. **22**(26), 30562 (2017). https://doi.org/10.2807/1560-7917.ES.2017.22.26.30562

Sui, D.Z., Holt, J.B.: Visualizing and analysing public-health data using value-by-area cartograms: toward a new synthetic framework. Cartographica: Int. J. Geogr. Inf. Geovis. (2008). https://doi.org/10.3138/Carto.43.1.003

Thomas, C.V.D., Dieter, L.: Realtime linear cartograms and metro maps. In: Proceedings of the 26th ACM SIGSPATIAL International Conference on Advances in Geographic Information Systems (SIGSPATIAL 2018), pp. 488–491. Association for Computing Machinery, New York (2018). https://doi.org/10.1145/3274895.3274959

Tingsheng, S., Duncan, I., Chang, Y.N., Gastner, M.: Motivating good practices for the creation of contiguous area cartograms. In: Bandrova, T., Konečný, M., Marinova, S. (eds.) Proceedings of the 8th International Conference on Cartographic GIS, vol. 1, pp. 589–598. Bulgarian Cartographic Association, Sofia (2020). https://tinyurl.com/icc8-2020-pdf. ISSN 1314-0604.

Tobler, W.R.: A continuous transformation useful for districting. Ann. NY Acad. Sci. **219**, 215–220 (1973)

Tobler, W.R.: Pseudo-cartograms. Cartogr. Geogr. Inf. Sci. 43–50 (1986)

Tobler, W.R.: Thirty-five years of computer cartograms. Ann. Assoc. Am. Geogr. **94**, 58–73 (2004)

Tran, N.K., Goldstein, N.D.: Jointly representing geographic exposure and outcome data using cartograms. Am. J. Epidemiol. **188**(9), 1751–1752 (2019). https://doi.org/10.1093/Aje/Kwz141

Ziqiang, L., Saman, A.: Diffusion-based cartogram on spheres. Cartogr. Geogr. Inf. Sci. **45**(5), 464–475 (2018). https://doi.org/10.1080/15230406.2017.1408033

www.aps.dz. Consulted 2021

MCBRA (Multi-agents Communities Based Routing Algorithm): A Routing Protocol Proposition for UAVs Network

Mohammed Chaker Boutalbi[1(✉)], Mohamed Amine Riahla[2], and Aimad Ahriche[1]

[1] Faculty of Hydrocarbons and Chemistry, Université M'hamed Bougara, Boumerdès, Algeria
mc.boutalbi@univ-boumerdes.dz

[2] Faculty of Technology, Université M'hamed Bougara, Boumerdès, Algeria
https://fhc.univ-boumerdes.dz
https://ft.univ-boumerdes.dz

Abstract. Group missions of autonomous aerial vehicles have attracted a big interest in the last years due to its vast coverage area and its capability of executing complex tasks in a short amount of time. To ensure a high QoS in this collaborative work, the designed routing protocol must overcome the constraints imposed by this type of wireless ad hoc network that are mainly driven by the rapid mobility of the UAVs. In this paper, we give a brief overview on flying ad hoc routing protocols, then, we theoretically present and argue about our bio-inspired routing protocol proposition that uses a smart multi-agent system communities followed with some adaptations of two existing routing protocols that we believe it can deliver a higher QoS

Keywords: FANET · UAVs network · Drone's fleet · Routing protocol · Multi-agent system

1 Introduction

Developing an adequate routing protocol becomes a challenge for researchers and developers in sync with the raised possibility of using swarm robotics of unmanned aerial vehicles (UAVs) in various application areas after being exclusive for military use like borders supervising [1], autonomous tracking [2], and surveillance [3]. Now, UAVs swarm is witnessing other applications such as the scientific field like environment sensation(wind estimation [4]), and in managing urban traffic where the UAVs act like relies nodes in VANETs (Vehicular Ad Hoc Networks) [5]. FANET (Flying Ad Hoc Network) comes as a particularity of MANET (Mobile Ad Hoc Network) with an extension of a high speed of the mobile nodes that can up to 6 times faster than the common speed of MANETs' nodes. This high dynamicity causes more difficulties concerning data

B. Lejdel et al. (Eds.): AIAP 2021, LNNS 413, pp. 565–573, 2022.
https://doi.org/10.1007/978-3-030-96311-8_53

routing process by affecting the link stability, the topology variation frequency, and causing high network fragmentation.

Many routing protocol solutions have been proposed in literature attempting to overcome the presented difficulties in FANET, but no one agreed that it fully does [6]. The conflicting constraints and the numerous scenarios in FANET applications forced the researchers to go toward tradeoff solutions looking to satisfy the question that says: which routing protocol that suits more application scenarios, and delivers an overall better quality of service. However, we can say that FANET constraints like the shared bandwidth, the limited energy resource, and the high dynamicity make this research field alive for a long time.

It becomes obvious that implementing traditional MANET routing techniques is not a sufficient solution. Therefore, including novel strategies is crucial in this case. That was the reason that has pushed more innovations and interests toward in this field of application. Briefly, this work presents an overview on routing in FANET in parallel with a theoretical UAVs routing protocol proposition. The paper is organized as follows: First, we discuss some of the proposed FANET routing protocols in the second Section. In the third Section, we present and argue bout our solution proposal. Finally, we conclude our paper in the last Section.

2 Related Works

In general, UAVs routing protocols have been classified by authors and reviewers into a set of protocols based on its used technique and tools. From the side of its appropriation for flying ad hoc network, in this section, we are going to present some of FANET routing protocols.

First, we are going to spot some topology-based routing protocols where each node has a routing table that contains paths that are based on the relaying nodes in the network. The metrics used here during the routing selection are the number of hops and the link state. This technique is perfect when dealing with static nodes, but using it alone in FANET where the topology changes rapidly costs a huge overhead and causes network congestion. For this reason, racing topology changes in FANET requires a set of adaptations that must be integrated. This category of protocols is divided into three main categories proactive, reactive, and hybrid routing protocols in which we are going to highlight some of these new propositions that are dedicated to FANET as follows.

A. I. Alshbatat and L. Dong proposed D-OLSR (Directional antenna OLSR) [7] as an extension of the OLSR [8]. D-OLSR uses Omni antennas for control packets exchange and directional antennas for data transfer to reduce signals interference among the nodes and to enhance the overall link state. The results showed that the network's overhead and latency were decreased. ML-OLSR (Mobile and Load-aware OLRS) [9] is proposed to improve the election of multi-points-relays. This protocol integrates the node's load and speed in the Hello message to avoid congested nodes in the decision making by adding a new metric named the stability degree and avoiding high relative speed nodes by adding a second metric named the reachability degree.

To be integrated in FANET, the authors in [10] proposed UE-DSR (UAV Energy Dynamic Source Routing). This reactive routing protocol includes ab energy balancing mechanism in the well-known routing algorithm the DSR [11] to prolongs the network lifetime. RE-DSR (restricted DSR) [12], this new DSR protocol can optimize the memory capacity of the node's and reduce the routing overhead by limiting the maximum hop count of the route request. For reducing packet collisions TS-AODV (time-slotted AODV) [13] uses a defined time to send data where one node can transmit its packets, the results show a high delivery ratio and low congestion average, but the risk of time synchronization deviation still be in real implementations.

ZRP (Zone Routing Protocol) [14] is a hybrid routing protocol. Unlike HWMP, ZRP doesn't switch from reactive to a proactive state. The developers used both routing approaches by dividing the coverage areas into small zones where the nodes proactively broadcast its routing table. When a node decides to reach an out-zone node, it uses the reactive approach and floods RREQ in the whole network to continue the communication with the same strategy. SHARP (Sharp Hybrid A Routing Protocol) [15] exploits the same method used in ZRP and improves the path discovery process when a node establishes a communication outside of the inter-zone. The word "sharp" is used to describe the intersection between the source node RREQ during its discovery process with the zone that contains the destination node. Instead of going inside the container zone, it is sufficient to reach one of the zone member nodes, then the last one gives the scheme toward the target directly and continues the routing process.

The following routing protocols are categorized as position-based routing protocols, in which this category is not entirely independent of the topology-based technique. The routing protocols here use the position information for supporting the discovery process of the destination node. In some propositions, it has been used for path maintenance as well. However, it is essential when dealing with high dynamic networks like FANET where the topology rapidly changes over time. This technique is not complicated when the positions of the nodes are known, but for autonomous moving objects, this case requires frequent position broadcasting. Examples of position-based protocols are given as follows:

The authors in [16] proposed MUDOR (Multipath Doppler Routing). Based on its name, the multipath refers to the application of the technique used in the famous routing protocol DSR that provides multipath routing choices. The Doppler word referes to the doppler effect of the electromagnetic waves fadding. This latter indicates the stability of the links and the lifetime of the relaying nodes in the transmission range and the routing table. GLSR (Geographic Load Sharing Routing) [17] is an improved version of the GPSR dedicated to FANET. GLSR exploits the geographic position of the moving nodes as well as it's speed to estimate the link stability. Also, it includes the node's buffer (queued packets) consumption and uses it as a metric when choosing the next-hop relay. GRG (Greedy Random Greedy) [18] switches to tow routing modes. The first one is the greedy mode in which GRG uses it during the regular situation when assuming that there are no disconnections between the nodes. If a disconnection appears, the last relay node tries to continue the process by randomly selecting another relay node using the RW algorithm. Reversely, RGR (Reactive Greedy Reactive) [19] uses the same strategy but differently. This efficient protocol is a combination of two routing modes. In the first phase, the AODV protocol is used to discover the destination node by flooding an RREQ in the network. Then, after having the destination response. In the case of a link failure, the second algorithm GGF (Geographic Greedy Forwarding) is activated to recover the broken links after having the address of the destination in the first phase (with the AODV), then it returns to the first mode.

3 Solution Proposition

Based on what we mentioned earlier and our work in [20] that illustrates how a FANET routing protocol proposition should cover, we recommend the use of a nature-inspired routing protocol to improve the QoS in this type of networks. The reasons behind this choice are driven from the difficulties imposed by the environment modeling where there is a lack of a mathematical formula that can describe the swarm behavior in a way that makes it impossible to cover all the scenarios that can happen during the mission, or the data transfer process. In drone's fleets applications: (i)The incoming events are not identified where, when, and how it can happen, (ii) the drone has a limited perception of its environment, (iii) it needs the collaboration of others drones to transfer its gathered DATA, (iv) it must react intelligently to the unexpected events during the mission execution. All these facts stand directly with the use of multi-agents system techniques. This promising technique has been invented to particularly deal with this sort of problems. According to the mentioned parameters in the latter section, Its strong points are its simplicity (few simple rules), scalability (maintain performance in both small and large agents number), flexibility (agents behave instinctively in a known manner), and its low cost.

The search for the optimum path in a continuous way in this type of network necessarily requires a complexity of calculation and a high cost of communication and energy. In these kinds of problems, even if the obtained path is a global optimum, this solution is subject to failure due to instability and high mobility. So the approximation or the convergence towards the global optimum (Multi-path routing) is favored in this case rather than risking and wasting time finding a non-guaranteed optimal solution. For these reasons, we decided that our proposition must be a multi-path routing protocol in which it is going to be a combination of three routing protocols: MMSR (Mobile Multi Agent System for Routing in Adhoc Network) [21], POSANT (Position Based Ant Colony Routing Algorithm) [22], and BIODRA(Bio-inspired On-Demand Routing Protocol) [23].

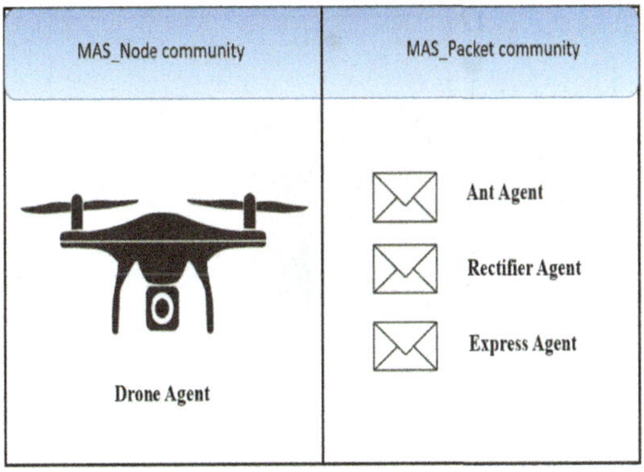

Fig. 1. MMSR agents communities.

MMSR is a multi-agent system routing protocol dedicated to MANET. It has two agents communities within, SMA-Node and SMA-Packet communities (see Fig. 1). The first one represents a physical agent group that is the moving nodes (robots), and the second community is a set of abstract agents in the form of packets where the latter has three agent families the Ant-agents, the Rectifier agents, and the Express agents. Each agent type has its own assigned functions (these functions are well detailed in [21]) in a way that each group is dedicated to accomplishing a specific purpose. Node agents and Ant agents are working collaboratively to build paths, Rectifier agents are made to update the routing table when a topology change is detected, and Express agents are issued to reduce the end-to-end delay between the sender and the receiver.

POSANT is also a multi-path routing protocol targeted for MANET which is based on the well-known ACO algorithm to establish its paths. Generally, the point that makes this algorithm different is its use of the position information to guide the ant agent to the destination node rapidly. The guidance process is based on creating a virtual zone within each node where the ant agent can put its pheromone trails on. The idea relays on the assumption that says that: The shortest path is more likely to form a position progression angel toward the destination node. Based on that and as Fig. 2 shows, we see three built zones; the belongingness of each node to a specific zone depends on is the relative angle between the claimant node and its destination node in which node H (neighbor node) belongs to zone1 if $\theta\text{H} <= \pi/4$, zone2 if $\pi/4 < \theta\text{H} <= 3\pi/4$ zone3 if $3\pi/4 < \theta\text{H} <= \pi$. After putting the pheromone on the corresponding zone, the quantity of the pheromone is multiplied by factors. Each factor ν_i is attached with its zone i in which $\nu_1 \geq \nu_2 \geq \nu_3$ as a method to enhance the probability of reaching shortest paths faster.

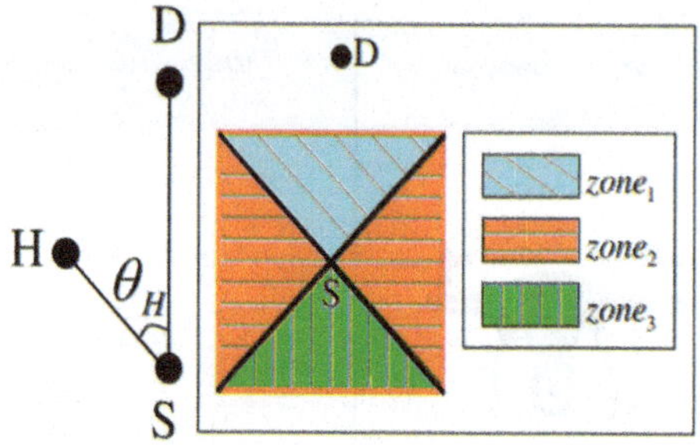

Fig. 2. POSANT virtual zones for pheromone trails.

Both propositions are trying to use the minimum control packets to avoid network overhead to provide a high QoS under the presence of MANET topology changes. Our proposed routing protocol is the integrity of these two algorithms with a new adaption to suit the requirement imposed in this high-speed network FANET. We maintain the same principal of MMSR, and we assign the virtual zones of the POSANT algorithm to the node agent where we keep the same fundamental number of the pheromone zones, and we divide zone number 2 into two zones with the same attributed coefficient ν_2. Finally, our main contribution is we add a function to the evaporation process that controls the pheromone quantity withing the Node agent in the ax of time by using the position and the direction information in order to reduce the frequency of releasing the Ant agent in the network where we can increase the amount of the pheromone and decrease it without the visiting of the ant agent to the node itself as Eq. (3). The former evaporation equation:

$$\psi_{\mathrm{it}} = (1 - \alpha) * \psi_{\mathrm{i(t-1)}} \tag{1}$$

Our adaptation:

$$\psi_{\mathrm{it}} = f(\overrightarrow{position_j}, \overrightarrow{velocity_j}) * \psi_{\mathrm{i(t-1)}} \tag{2}$$

$$\Delta \mathrm{q} = [\mathrm{C(i, j, d)^{*}E_j} + \mathrm{q}] * v_i \tag{3}$$

$f(\overrightarrow{position_j}, \overrightarrow{velocity_j})$ works for adjusting the pheromone evaporation in the axis of time by using the position and the velocity vectors of the corresponding node, and Eq. (3) express the integrity of POSANT zone coefficients to the quantity of the posed pheromone. The scenario comes as follows: Our proposed algorithm uses a small number of control packet in the proactive phase the same way as MMSR but in a lower amount. Here, we embrace the geographical information of the flying nodes within the virtual zones where each node releases just one ant agent (not 3 ant agents in POSANT case) from the zone that has been not received an ant-agent form it to increase the probability of visiting as possible as it can of nodes from different positions in the zone of interest. After this phase, we can say that the corresponding node has enough update information about its neighboring nodes. When information is about to be sent to a destination node, the same protocol as MMSR runs, also the pheromone quantity is multiplied by its attributed zone as mentioned in POSANT, as well as the node in the selected path start to align with each other using the algorithm that used in BIODRA (will be discussed in the next paragraph).

BIODRA is a reactive routing protocol proposed for FANET that uses the well-known AODV algorithm. The main idea of this protocol is when a connection starts between tow nodes the concerned nodes in the path and the neighborhood begin the align and get closer to each other as a flock using Craig Reynold flocking behaviors (see Fig. 3) to maintain enough time for passing the message and reduce packet loss. From our point of view, we see this routing protocol as a cross-layer mobility model more than a routing protocol because the main contribution can be applied in all routing algorithms. However, we also integrate this technique into our proposed routing protocol due to its efficiency in high dynamic networks like our case.

Fig. 3. Craig Reynold flocking boids.

4 Conclusion

In this paper, we briefly talked about FANET specifications and constraints. We reviewed a set of proposed solutions in the literature. We discussed some details mentioning a recent work of ours that refers to how a proposed routing protocol should be assessed in order to suit FANET applications. That was the structure that we used to introduce and argue about our routing protocol proposition that is based on a multi-agent system of multiple communities in which we believe that it is a promising solution that takes advantage of a large scale of used techniques and tools that can fit FANET requirements and balance its conflicting constraints.

References

1. Sun, Z., et al.: BorderSense: border patrol through advanced wireless sensor networks. Ad Hoc Netw. **9**(3), 468–477 (2011)
2. Pitre, R.R., Li, X.R., Delbalzo, R.: UAV route planning for joint search and track missions-an information-value approach. IEEE Trans. Aerosp. Electron. Syst. **48**(3), 2551–2565 (2012)

3. Qazi, S., Siddiqui, A.S., Wagan, A.I.: UAV based real time video surveillance over 4G LTE. In: Proceedings of the International Conference on Open Source Systems and Technologies (ICOSST), pp. 141–145, December 2015
4. Cho, A., Kim, J., Lee, S., Kee, C.: Wind estimation and airspeed calibration using a UAV with a single-antenna GPS receiver and pitot tube. IEEE Trans. Aerosp. Electron. Syst. **47**(1), 109–117 (2011)
5. Raza, A., et al.: An UAV-assisted VANET architecture for intelligent transportation system in smart cities. Int. J. Distrib. Sensor Netw. **17**(7), 1–17 (2021). 15501477211031750
6. Oubbati, O.S., et al.: Routing in flying ad hoc networks: survey, constraints, and future challenge perspectives. IEEE Access **7**, 81057–81105 (2019)
7. Alshbatat, A.I., Dong, L.: Cross layer design for mobile ad-hoc unmanned aerial vehicle communication networks. In: Proceedings of the International Conference on Sensor Networks Control (ICNSC), pp. 331–336 (2010)
8. Clausen, T., Jacquet, P.: Optimized link state routing protocol (OLSR). RFC, New York, NY, USA, Technical Report 3626 (2003)
9. Zheng, Y., Wang, Y., Li, Z., Dong, L., Jiang, Y., Zhang, H.: A mobility and load aware OLSR routing protocol for UAV mobile ad-hoc networks. In: Proceedings of the International Conference on Information, Communication and Computing Technology (ICT), pp. 1–7 (2014)
10. Li, J., Liu, X.C., et al.: A novel DSR-based protocol for small reconnaissance UAV Ad Hoc network. Appl. Mech. Mater. **568–570**(7), 1272–1277 (2014)
11. Johnson, D., Hu, Y., Maltz, D.: The Dynamic Source Routing Protocol (DSR) for Mobile Ad Hoc Networks for IPv4, Document RFC 4728 (2007)
12. Li, J., Zhang, X.L., et al.: A novel DSR-based protocol for signal intensive UAV network. Appl. Mech. Mater. **241–244**(12), 2284–2289 (2013)
13. Forsmann, J.H., Hiromoto, R.E., Svoboda, J.: A time-slotted on demand routing protocol for mobile ad hoc unmanned vehicle systems. In: Proceedings of the SPIE, vol. 6561, May 2007. Art. no. 65611P
14. Haas, Z.J., Pearlman, M.R.: ZRP: a hybrid framework for routing in ad hoc networks. In: Ad Hoc Networking, pp. 221–253. Addison Wesley, Boston (2001)
15. Ramasubramanian, V., Haas, Z.J., Sirer, E.G.: SHARP: a hybrid adaptive routing protocol for mobile ad hoc networks. In: Proceedings of the 4th ACM International Symposium on Mobile Ad Hoc Networking and Computing, pp. 303–314 (2003)
16. Sakhaee, E., Jamalipour, A., Kato, N.: Multipath Doppler routing with QoS support in pseudo-linear highly mobile ad hoc networks. In: 2006 IEEE International Conference on Communications, vol. 8. IEEE (2006)
17. Medina, D., Hoffmann, F., Rossetto, F., Rokitansky, C.-H.: A geographic routing strategy for north Atlantic in-flight Internet access via airborne mesh networking. IEEE/ACM Trans. Netw. **20**(4), 1231–1244 (2012)
18. Flury, R., Wattenhofer, R.: Randomized 3D geographic routing. In: Proceedings of the IEEE INFOCOM, pp. 834–842, April 2008
19. Shirani, R., St-Hilaire, M., Kunz, T., Zhou, Y., Li, J., Lamont, L.: On the delay of reactive-greedy-reactive routing in unmanned aeronautical ad-hoc networks. Proc. Comput. Sci. **10**, 535–542 (2012)
20. Chaker, B.M., Amine, R.M., Aimad, A.: A summary of the existing challenges in the design of a routing protocol in UAVs network. In: 2020 2nd International Workshop on Human-Centric Smart Environments for Health and Well-being (IHSH). IEEE (2021)
21. Riahla, M.A., et al.: A mobile multi agent system for routing in adhoc network. In: PECCS (2014)

22. Kamali, S., Opatrny, J.: POSANT: a position based ant colony routing algorithm for mobile ad-hoc networks. In: 2007 Third International Conference on Wireless and Mobile Communications (ICWMC 2007). IEEE (2007)
23. Bahloul, N.E.H., et al.: Bio-inspired on demand routing protocol for unmanned aerial vehicles. In: 2017 26th International Conference on Computer Communication and Networks (ICCCN). IEEE (2017)

A CNN Approach for the Identification of Dorsal Veins of the Hand

Abdelkarim Benaouda(✉), Aymen Haouari Mustapha, and Sarâh Benziane(✉)

Université des Sciences et de la Technologie d'Oran Mohamed-Boudiaf (USTO MB), El Mnaouar, BP 1505, Bir El Djir, 31000 Oran, Algeria

Abstract. In this paper, we proposed a dorsal hand vein recognition method based on Convolutional Neural Network (CNN). Firstly, implementation of raw images the region of interest (ROI) of dorsal hand vein images was extracted, and then contrast limited adaptive histogram equalization (CLAHE) and were used to preprocess the images. Next, the extraction of information using the Sato filter and the Otsu thresholding algorithm to create a new database containing only the processed images. Finally, CNN was applied for identification. The experimental results was has been optimized with Hyperparameter Optimization. The dorsal hand vein recognition rate reaches 99%.

Keywords: Dorsal hand vein recognition · Deep learning · Convolutional neural network · CLAHE · SATO · OTSU · Mask

1 Introduction

Biometric technology is one of the effective techniques for authenticating and identifying people. Biometrics is the computer science term for the field of mathematical analysis of unique human characteristics such as fingerprints, hand, palm and finger veins, eyes, voice, signature, gait and DNA. Biometric solutions have experienced accelerated growth in the global security market over the past few decades, primarily due to increasing public safety requirements against terrorist activities, sophisticated crimes and electronic fraud. Biometrics is the science of identifying a person based on their behavioral and physiological characteristics.

Biometric systems fall into two categories: physical and behavioral. Physical systems are related to body shape, such as fingerprints, facial recognition, DNA, vascular patterns, eye iris, vein pattern, etc. Behavioral biometrics are related to a person's behavior, such as voice, gait, signature, etc. We focus on the venous network of the back of the hand (i.e., dorsal hand) because it is clearly visible, easy to acquire, and efficient to process. Compared to other popular biometric features, such as face or fingerprints, the dorsal hand vein has several advantages.

It is also the best variant of biometric systems that require physical contact with the machine, as it extracts the vein pattern, the hand is not in contact with the device, the hand can easily stretch, and capturing the vein pattern can be done easily. Since the system is based on three features such as live body,

B. Lejdel et al. (Eds.): AIAP 2021, LNNS 413, pp. 574–587, 2022.
https://doi.org/10.1007/978-3-030-96311-8_54

internal veins and non-contact type, there is no possibility of tampering, and no misuse by criminals, so it can be used in places requiring high level of security to avoid crimes and frauds.

1.1 Related Works

The important characteristic of hand vein patterns is stability, which means that the structure of the hand and the configuration of the hand veins remain relatively stable throughout the life of the individual. For this reason, vein identification systems are considered a promising and reliable biometric. In this section, some of the vein identification systems are presented.

Huang et al. [1] proposed a method for identifying dorsal veins in the hand. A new process integrating together the holistic and local then hierarchically joint analysis with that of the surface modality, born by a reputable texture operator, that local binary patterns (LBP), binary coding (BC) and graph for decision generation by Factored Graph Matching(FGM). The results obtained are superior to those of the state of the art described so far in the work, which proves its effectiveness.

Traditional palm vein recognition algorithms use physical models, including minutiae, ridges, and texture, to extract features for matching. For example, the adaptive multispectral method [4], 3D ultrasound method [5], and adaptive contrast enhancement method [6] are applied to improve image quality. Ma et al. [7] proposed a palm vein recognition scheme based on an adaptive 2D Gabor filter to improve image quality. 2D Gabor filter to optimize parameter selection. Yazdani et al. [8] presented a new method based on wavelet coefficient estimation with an autoregressive model to extract the texture feature for verification. Some new methods have also been presented to overcome the drawbacks, including image rotation, shadows, darkness, and deformation [9,10]. However, as the database becomes larger, traditional palm vein recognition techniques are prone to have higher time complexity, which has a negative effect on practical applications.

Recently, deep learning, which is one of the most promising technologies, has disrupted traditional cognition and has also been introduced into the field of palm vein recog- nition. Fronitasari et al. [11] presented a palm vein extraction method that is a modified version of local binary pattern (LBP) and combined it with probabilistic neural network (PNN) for matching. In addition, supervised deep hashing technology has attracted more attention to large-scale image retrieval due to its higher accuracy, stability, and lower temporal complexity in recent years. Lu et al. [12] proposed a new deep hashing approach for evolutionary image retrieval by a deep neural network to exploit linear and nonlinear relationships. Liu et al. [13] proposed a supervised deep hashing (DSH) scheme for fast image retrieval combined with a CNN architecture. The superior performance of deep hashing approaches for image retrieval prompts researchers to expand the applications of deep hashing from image retrieval to biometrics.

Finger vein can be used in combination with other modalities to improve accuracy. Kang and Kang [18] presented a multimodal biometric recognition based on the fusion of finger vein and finger geometry. The recognition accuracy was significantly improved. Yang and Zhang [16] proposed a multimodal biometric approach based on the fusion of finger vein and fingerprint feature levels by CCA. The proposed SLPCCAM approach performed well for person recognition. Xi et al. [17] presented an effective personalized fusion approach combining finger outline and finger vein at the score level. The CCS-based weight fusion method exploits additional classification information and improves the recognition performance.

Recently, a number of multimodal works have attempted to combine ECG and other modalities to improve performance. ECG has been combined with electroencephalogram (EEG) for human identification [4]. Fatemian et al. [3] presented a decision-level fusion approach by combining ECG and phonocardiogram (PCG) to achieve improved identification performance. al [2] fused ECG with face to improve identification performance. A novel multimodal and behavioral biometric system that combines ECG and audio signal was proposed in [14]. The approach used popular statistical coefficients that were simple and computationally efficient. In [15], ECG signals were fused with a traditional fingerprint lividity detection approach to achieve better detection performance, and also with a fingerprint identification approach for higher recognition rate.

1.2 Dorsal Hand Vien Image Collection

The hardware configuration has a crucial role in the frame grabbing of veins. Two aspects can be underlined. The real camera used to take the stereotype has only one important parameter, the answer to the near infrared radiation [16]. The space resolution and the framework rate are of less importance since, for the acquisition of a model of vein, the image is necessary and the details are easily visible, even to a lower resolution. In [Cri] the design of the lighting system is one of the most important aspects of the process of frame grabbing. A good lighting system will provide precise data contrasts between the surrounding veins and fabrics while keeping Illuminations of the errors to a minimum. For a lighting system, they used IR LED because they offer a high contrast and easily available. But with a LED table formed using IR LED, they do not give a uniform illumination. Various arrangements were needed stamps LED to modify lighting. The LED is laid out like a simple or double table in 2D or rectangular table or concentric networks. Concentric arrangement table LED gives a better distribution of the light with only one or more concentric network of the LED and lens of the camera in the center of the image can acquire good contrasts (Fig. 1).

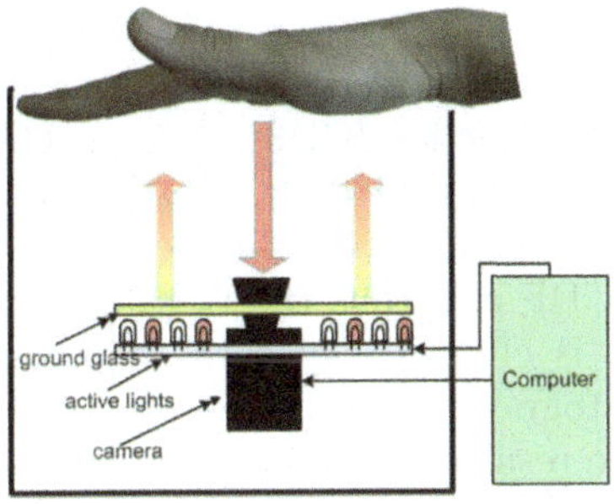

Fig. 1. Other system similar but which collects the veins of the fingers [HIT06].

1.3 Preprocessing and Extraction of the ROI

Preprocessing is the basis for feature extraction and matching. The quality of preprocessing has a significant impact on the recognition results. We mainly focus on the development of ROI extraction algorithms. This is indeed the main step of preprocessing, apart from other procedures such as image enhancement, image filtering, etc. The ROI is used to align different hand dorsal vein images and to segment the center for feature extraction. Most ROI extraction algorithms use the key points between the fingers to establish a coordinate system.

1.4 Aquiring the ROI Image

In vein images, the region of interest is only the region that contains the vein pattern information. We therefore extract the region of interest (ROI) from the image.

To speed up the processing time and to standardize the collection of vein images, Fig. 2 (b) was masked on the raw image taken and the boundaries of the hand surface were determined. Then, the outline of the image taken by the mask Fig. 2(c) is taken and the highest point and the point at the right end Fig. 2(d).

Then from these two points we take the x-coordinate of the highest point and the y-coordinate of the point on the right to have the center point Fig. 2 (e). And this point will be the center point of the square with 100 pixel to all directions (100 pixel to the top, left, right and bottom) Fig. 2(f) [personal contribution].

1.5 Image Processing

Ordinary AHE tends to overamplify the contrast in the near-constant regions of the image, because the histogram of these regions is very concentrated. As a result, AHE can cause noise amplification in the near-constant regions. Contrast-limited AHE (CLAHE) is a variant of adaptive histogram equalization in which the contrast amplification is limited, so as to reduce this noise amplification problem [19] (Fig. 3).

In CLAHE, the contrast amplification near a given pixel value is given by the slope of the transformation function. This is proportional to the slope of the cumulative distribution function (CDF) of the neighborhood and thus to the value of the histogram at that pixel value. CLAHE limits the amplification by cutting the histogram at a predefined value before calculating the CDF. This limits the slope of the CDF and thus the transformation function. The value at which the histogram is clipped, called the clipping limit, depends on the normalization of the histogram and thus the size of the neighborhood region. Common values limit the resulting amplification to between 3 and 4.

It is advantageous not to discard the part of the histogram that exceeds the clipping limit, but to redistribute it equally among all bins in the histogram [19].

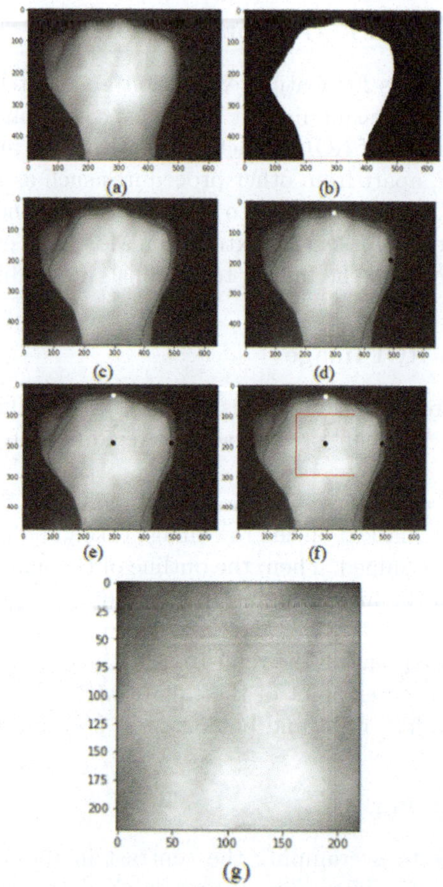

Fig. 2. (a) the original (raw) image (b) contour mask (c) the contour on the original image (d) extraction of the highest point and the point at the right end (e) the center point from the two points (f) the square obtained from the center point (g) the ROI image

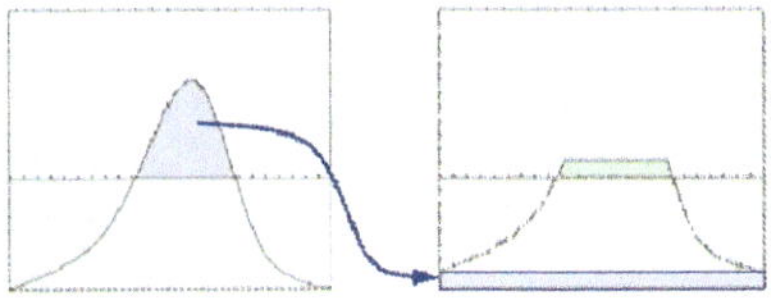

Fig. 3. Adaptive equalization of the contrast limited histogram

The redistribution will cause some bins to rise above the clipping limit (green shaded region in the figure), resulting in an effective clipping limit that is higher than the prescribed limit and whose exact value depends on the image. If this is not desirable, the redistribution procedure can be repeated recursively until the excess is negligible.

After acquiring the ROI image, we convert it from the BGR format to the LAB color format (which expresses color in three values: L: for perceptual brightness, and A and B: for the four unique colors of human vision: red, green, blue and yellow).

Then we separate the values and take only the L value. Then we create a CLAHE instance with the right parameters and apply it to the L value.

After that we take the equalized L-value and merge it with the A- and B-values and convert it back from LAB format to BGR format (Figs. 4, 5, 6 and 7).

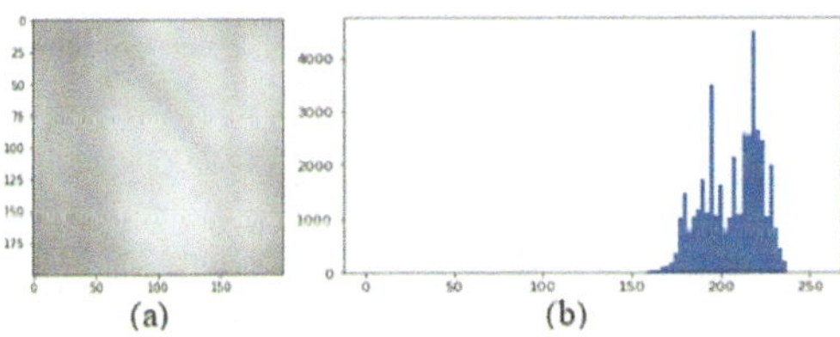

Fig. 4. a: ROI image b: image processed by simple histogram equalization c: image processed by CLAHE

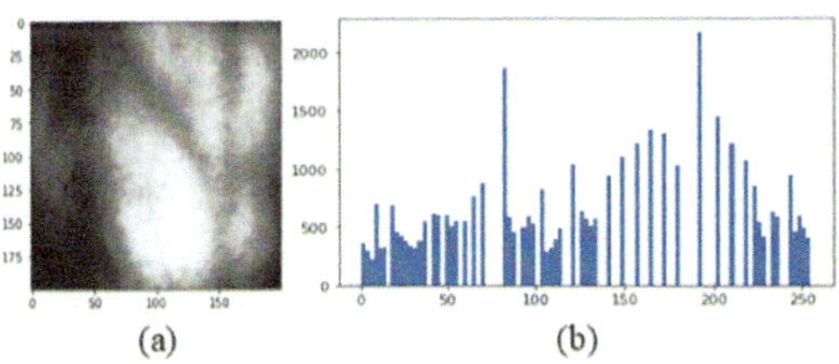

Fig. 5. a: ROI image b: image processed by simple histogram equalization c: image processed by CLAHE

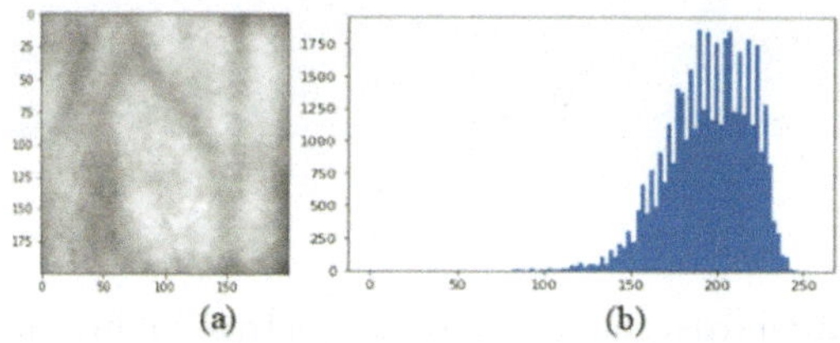

Fig. 6. a: ROI image b: image processed by simple histogram equalization c: image processed by CLAHE

1.6 Results

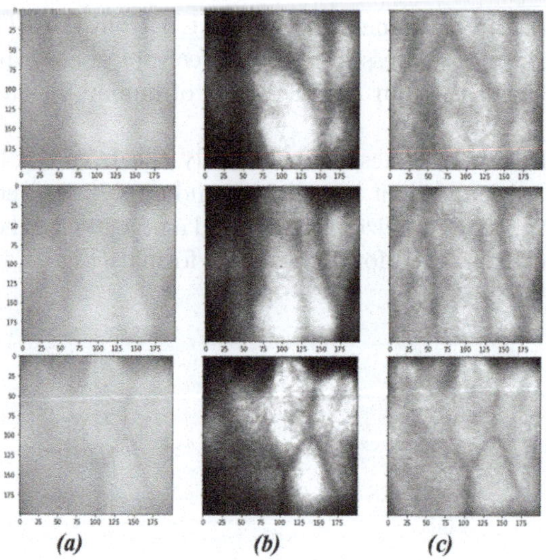

Fig. 7. a: ROI image b: image processed by simple histogram equalization c: image processed by CLAHE

2 Feature Extraction

2.1 SATO Filter

The second derivative has typically been used for line enhancement filtering. The Gaussian convolution is combined with the second derivative in order to tune the filter response to the specific widths of lines as well as to reduce the effect of noise. In the one-dimensional (1-D) case, the response of the line filter is given by:

$$R(x; \sigma_f) = -\frac{d^2}{dx^2} G(x; \sigma_f) * I \tag{1}$$

where * denotes the convolution, I(x) is an input profile function, and $G(x;\sigma_f)$ is the Gaussian function with the standard deviation σ_f defined as $(\frac{1}{\sqrt{2\pi\sigma_f}}\exp(-\frac{x^2}{2\sigma_f^2}))$. The sign of the Gaussian derivative has been inverted so that the responses have positive values for a bright line. We consider a profile having the Gaussian shape given by:

$$L(x;\sigma_x) = \exp(-\frac{x^2}{2\sigma_f^2}) \tag{2}$$

2.2 OTSU Thresholding Algorithm

Otsu's thresholding method corresponds to the linear discriminant criterion which considers the image to consist of only the object (foreground) and background, and the heterogeneity and diversity of the background is ignored [20]. Otsu set the threshold to try to minimize the overlap of class distributions [20]. Given this definition, Otsu's method segments the image into two light and dark regions T0 and T1, where region T0 is a set of intensity levels from 0 to t or, in ensemble notation, T0 = 0, 1, ..., t and region T1 = t, t + 1, ..., l − 1, l where t is the value of the threshold, l is the maximum gray level of the image (e.g. 256). T0 and T1 can be assigned to the object and the background or vice versa (the object is not necessarily always located in the bright region). The Otsu thresholding method scans all possible threshold values and calculates the minimum value for the pixel levels on either side of the threshold. The goal is to find the threshold value with the minimum entropy for the sum of the foreground and background. Otsu's method determines the threshold value based on the statistical information of the image where, for a chosen threshold value t, the variance of clusters T0 and T1 can be calculated. The optimal threshold value is calculated by minimizing the sum of the weighted cluster variances, where the weights are the probability of the respective clusters.

2.3 Experimental Results

We take the enhanced ROI image and convert it to gray scale to lighten it and reduce the amount of data. Then we apply the sato filter with the default settings as they are suitable for what we need. And after that we apply a median filter of kernel size 5 to smooth the edges of the image. Finally we use the Otsu algorithm to acquire the mask [personal contribution] (Fig. 8).

3 Classification

3.1 CNN Architecture

Set of parallel features maps are developed by sliding different kernels over the input imagesand stacked together in a layer which is called as Convolutional layer. Using smaller dimension as compares with original image helps the parameters sharing between the feature maps. In the case of overlapping of the kernel

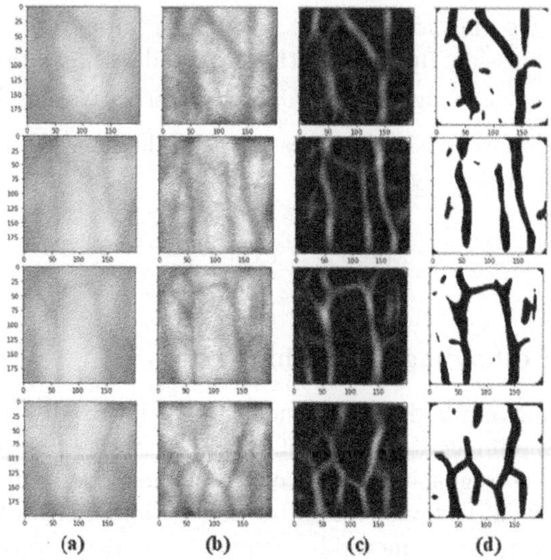

Fig. 8. a: ROI image b: image processed by simple histogram equalization c: image processed by CLAHE

with images, zero padding is used to adjust the dimension of the input images. Zero padding also introduce to control the dimensions of the convolutional layer. Activation function decides which neuron should be fired. The weighted sum of input values is passed through the activation layers. The neuron that receives the input with higher values has the higher probability of being fired. Different types of activation function are developed for the past few years that includes linear, Tanh, Sigmoid, ReLu and softmax activation functions. In practice, it is highly recommended that selection of activation function should be based deep learning framework and the field of application. Downsampling of the data is carried out in pooling layers. It reduces the data point and overfitting of the algorithm. Also, pooling layer reduces the noises in the data and smoothing the data. Usually pooling layer is implemented after the convolution process and non-linear transformations. Data points derived from the pooling layers are stretched into single column vectors and fed into the classical deep neural networks. The architecture of a typical ConvNet is given in the Fig. 9. The cost function, also known as loss function is used to measuring the performance of the architecture by utilizing the actual yi and predicted ŷi.

3.2 Experimental Results

The image pool is collected from a specified event as discussed in the database definition. The sampled images are divided into training and test data such that the training and test data contain 0.8% and 0.2% of the total amount of images. The training set is used for architecture training, and the test set is used for architecture validation. For computational simplicity, gray-scale images are used for network training. The input image windows are chosen from 64×64 pixel windows including the image information. However, the windows are carefully chosen to capture the rich amount of information about the data as well as to reduce the noise in the input and also to reduce the number of false positives. The image normalization procedure is performed in all training and test sets prior to array training. Batch gradient descent with the addition of the Adam optimizer is used to minimize the error.

The network is trained for 40 training epochs, and in each epoch the images in the training set are randomly shuffled. In practice, it is understood that the batch size 32 turns out to be effective. Therefore, the training batch size is fixed to 32. Along with the batch size hyperparameter, other hyperparameters such as activation layer types and cost function are chosen manually. The ReLU activation function is used in the intermediate layers and the softmax is introduced in the classification layer of the architecture. The categorical cross-entropy loss function is used to measure the error between the actual and predicted values. However, the hyperparameters such as the number of convolutional layers, dense layers, convolutional kernel size, dense layer size, dropout, weight regularization, learning rate are determined from the Bayesian optimization algorithm. The activation and pooling layers are implemented after each convolutional layer. The Bayesian optimization algorithm is effectively introduced to optimize a large number of hyperparameters of the search space.

For the effective implementation of the optimization algorithm, it is common to perform 10N trials where N is the number of hyperparameters. In this context, 1000 trials are performed for hyperparameter tuning. In the Bayesian optimization implementation, the number of convolutional layers, pooling layers and the number of dense layers vary from 1 to 6 with the interval of 1, the number of convolutional kernels varies from 32 to 526 with the interval of 32, the units of dense layers vary from 128 to 1004 with the interval of 128, and the continuous type hyperparameters such as dropout, regularization L1, L2 and learning rate vary from 0 to 0.2, 0 to 0.2, 0 to 0.2 and 0 to 0.2 respectively. No additional hyperparameters are considered and the default value of the Keras deep learning framework is used. From the range of values, the optimal values are determined with respect to the cost function. The resulting values are given in Table 2.

The finale ConvNet architecture Fig. 21 is developed with four convolutional layers followed by activation and pooling layers, four dense layers. The size of the architecture is determined by Bayesian optimization, resulting in a total of 11 layers excluding the output classification layers. The size of the convolutional kernels is 3×3 on each convolutional layer where in the pooling layers, the fixed pixel window size of 2×2 is selected. The input image window is selected as 64 $\times$ 64 and is fed into the architecture for training. The output of the architecture is developed to classify the 26 categorical images of the data sets. The developed architecture has about 2000 trainable parameters and took 2 min for training. The elementary features of the images are acquired in the middle layers, and the combined complex features are found in the final convolutional layer. Because the ReLU activation function removes unwanted features and noise, only natural image information is likely to be transferred through the layers. From layer 1 to 7, there is a series of successive convolutional layers and an activation layer followed by pooling layers. All convolutional layers are developed by employing 3×3 filters, known as convolutional kernels and connected to nonlinear ReLU activation functions. The output of the convolutional layers is used to form activation layers and is then subsampled using the max pooling operation in the pooling layers. The last four layers are fully connected dense layers that use the features extracted from the previous layers to classify the images. Each layer receives information from the previous layers (Table 1).

Table 1. Search space for hyperparameters.

Hyper-paramètres	Hyperparameter space	Type	Best fit
Input shape	[3,64,64]	Fixed	–
Number of Convolutional Layer	[1,2,3,4,5,6]	Discrete	[2]
Number of Convolutional kernel	[32,64,128,526]	Discrete	[32], [32], [64], [64]
Size of Convolutional kernel	(5,5)	Fixed	–
Activation unit in convolutional layer	ReLU	Fixed	–
Number of Pooling Layer	[1,2,3,4,5,6]	Discrete	[3]
Size of Pooling Layer	(2,2)	Fixed	–
Dropout	[0–0.2]	Continuous	[0]
Number of dense hidden layers	[1,2,3,4,5,6]	Discrete	[8]
Hidden layer size	[128,256,502,1004]	Discrete	[128]
Dropout	[0–0.2]	Continuous	[0.1]
Activation unit in dense layer	ReLU	Fixed	–
Weight regularization on dense layer, L1	[0–0.2]	Continuous	[0.2]
Weight regularization on dense layer, L2	[0–0.2]	Continuous	[0.2]
Learning rate	[0.01–0.00001]	Continuous	[0.00001]
Loss function	Cross-entrophy	Fixed	–
Activation unit in classification layer	Softmax	Fixed	–

3.3 Training of the Architecture

Convolutional layer 1 (Conv-1) is composed of 32 convolutional kernels of dimension $(32 \times 60 \times 60)$. Each unit of the convolutional kernels is connected to neighboring units of the input image (5×5), resulting in 747,674 trainable parameters.

A weighted sum of the feature map inputs with trainable bias addition is computed and passed through sigmoidal nonlinear activation functions. Then, subsampling of the convolved images is performed in pooling layer 1 (pooling1).

The output of the pooling layer is fed into the following convolutional layers. Convolutional layer 2 (Conv-2) contains 32 convolutional kernels and is formed with the dimension of 26×26. Each unit calculates the weighted sum and adds the resulting bias with 44 trainable parameters.

Then they are passed through nonlinear activation functions. The feature maps are subsampled from the total units in pooling layers 2.

Therefore, the feature extraction layers have two convolutional layers with two activation layers followed by two subsampling activation and pooling layers

Table 2. Search space for hyperparameters.

Couches Attributs	C-size	Act or P-size	Params	O/P shape
Conv2d	(5 × 5)	–	2432	(non, 60, 60, 32)
Activation	–	ReLU	0	(non, 60, 60, 32)
maxpooling2d	–	(2 × 2)	0	(non, 30, 30, 32)
Conv2d	(5 × 5)	-	51264	(non, 26, 26, 64)
Activation	–	ReLU	0	(non, 26, 26, 64)
maxpooling2d	–	(2 × 2)	0	(non, 13, 13, 64)
Flatten	–	–	0	(non, 10816)
Dense	–	–	692288	(non, 64)
Activation	–	ReLU	0	(13)
Dense	–	–	1690	(non, 26)
Activation	–	Softmax	0	(26)

where stable low-dimensional features are extracted and used for classification by the fully connected dense layer. These parameters are concatenated and vectorized into single-column vectors and fed into the classical multilayer perceptron, known as the fully connected dense layers.

From the Bayesian optimization, we find that two layers of dense layers are formed with 64 units in each layer.

The units in each layer are dropped with 0.2% of the total units and added with L2 and L1 regularization.

The fully connected dense layers act as a classifier, and the preceding convolutional layers act as feature extractors connected with vectorized parameters from the pooling layer-2. Similarly, the single unit in a dense layer is fully connected with all units in the previous dense layers. In these layers, the dot product is performed with the weight and input values and added with the bias values. The weighted sum of the input is passed through the nonlinear activation function (ReLU).

The final output of the dense layers is passed through the softmax activation where the categorical cross-entropy is implemented to measure the error between the actual and predicted values. As a result, the fully connected dense layers have 692288 trainable parameters. The network is trained using batch gradient descent, a modified version of the stochastic gradient descent algorithm. The Adam optimization algorithm is implemented to accelerate the convergence of the gradient. The performance of the network, such as loss and accuracy with up to 40 training epochs, is shown in Fig. 10.

The red curve represents the performance of the architecture on the training set, and the blue curve represents the performance of the architecture on the validation set. The early stopping algorithm is implemented to avoid overfitting.

The early stop tends to stop the training procedure if the performance of the test data is not improved after the fixed number of iterations. It avoids surfing

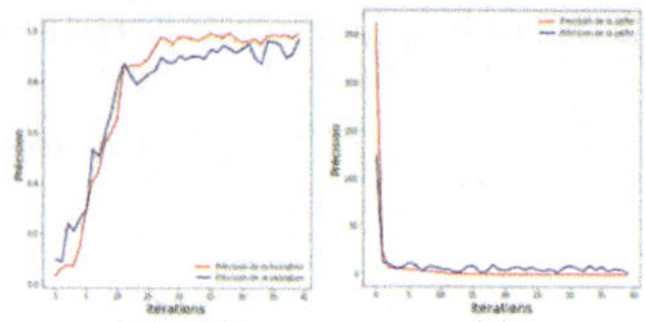

a:accuracy *b:loss*

Fig. 9. Learning curves a: learning and learning validation b: loss and loss validation

by attempting to automatically select the inflection point where performance on the test data set begins to decline while performance on the training data set continues to improve as the model begins to surfiter. Overall accuracies of 95% and 99.5% for the training and validation test images are achieved on the training architecture.

4 Conclusion

Biometrics is a growing technology. It is more and more used in applications related to security because of the advantages it offers in contrast to older methods

In this thesis, we have studied a biometric identity verification system based on the dorsal vein network of the hand. After a presentation of the general context of our study, the fundamental concepts of biometric systems and methodology are studied in the first chapter (state of the art). Then, we detailed the different components of the hand dorsal vein identification system proposed in Chap. 2 and 3. In Chap. 4 we tuned the neural network parameters in such a way that our system successfully identified the images in the database. In order to give more value to our work, we plan to test our system on large databases and by applying other feature extraction methods.

References

1. Huang, D., Zhu, X., Wang, Y., Zhang, D.: Neurocomputing. Dorsal hand vein recognition via hierarchical combination of texture and shape clues **214**, 815–828 (2016)
2. Zhu, X., Huang, D.: Hand dorsal vein recognition based on hierarchically structured texture and geometry features. In: Proceedings of the Chinese Conference on Biometric Recognition, pp. 157–164 (2012)
3. Wan, H., Chen, L., Song, H., Yang, J.: Dorsal hand vein recognition based on convolutional neural networks. In: Proceedings of the IEEE International Conference on Bioinformatics Biomedicine, pp. 1215–1221 (2017)
4. Jain, A.K., Ross, A., Prabhakar, S.: An introduction to biometric recognition. IEEE Trans. Circuits Syst. Video Technol. **14**(1), 4–20 (2004)
5. Liu, J., Xue, D.-Y., Cui, J.-J., Jia, X.: Palm-dorsa vein recognition based on kernel principal component analysis and locality preserving projection methods. J. Northeast. Univ. Nat. Sci. (China) **33**, 613–617 (2012)

6. Lajevardi, S.M., Arakala, A., Davis, S., Horadam, K.J.: Hand vein authentication using biometric graph matching. IET Biom. **3**, 302–313 (2014)
7. Chen, H., Lu, G., Wang, R.: A new palm vein matching method based on ICP algorithm. In: Proceedings of the 2nd International Conference on Interaction Sciences, Information Technology, Culture and Human, Seoul, pp. 1207–1211. ACM (2009)
8. Bhattacharyya, D., Das, P., Kim, T.H., Bandyopadhyay, S.K.: Vascular pattern analysis towards pervasive palm vein authentication. J. Univers. Comput. Sci. **15**, 1081–1089 (2009)
9. Xu, X., Yao, P.: Palm vein recognition algorithm based on HOG and improved SVM. Comput. Eng. Appl. (China) **52**, 175–214 (2016)
10. Elsayed, M.A., Hassaballah, M., Abdellatif, M.A.: Palm vein verification using Gabor filter. J. Sig. Inf. Process. **7**, 49–59 (2016)
11. Hartung, B., Rauschning, D., Schwender, H., Ritz-Timme, S.: A simple approach to use hand vein patterns as a tool for identification. Forensic Sci. Int. **307**, 110115 (2020)
12. Zhang, S.-X., Schmidt, H.-M.: Clinical anatomy of the subcutaneous veins in the dorsum of the hand. Ann. Anat. Anatomischer Anz. **175**(4), 381–384 (1993)
13. Ferrer, M.A., Morales, A., Ortega, L.: Infrared hand dorsum images for identification. Electron. Lett. **45**(6), 306–308 (2009)
14. Rahul, R.C., Cherian, M., Manu Mohan, C.M.: A novel MF-LDTP approach for contactless palm vein recognition. In: 2015 International Conference on Computing and Network Communications (CoCoNet), Trivandrum, India, pp. 793–798, December 2015
15. Mirmohamadsadeghi, L., Drygajlo, A.: Palm vein recognition with local texture patterns. IET Biom. **3**(4), 198–206 (2014)
16. Akbar, A.F., Wirayudha, T.A.B., Sulistiyo, M.D.: Palm vein biometric identification system using local derivative pattern. In: 2016 4th International Conference on Information and Communication Technology (ICoICT), Bandung, Indonesia, pp. 1–6, May 2016
17. Piciucco, E., Maiorana, E., Campisi, P.: Palm vein recognition using a high dynamic range approach. IET Biom. **7**(5), 439–446 (2018)
18. Tome, P., Marcel, S.: Palm vein database and experimental framework for reproducible research. In: 2015 International Conference of the Biometrics Special Interest Group (BIOSIG), Darmstadt, Germany, pp. 1–7, October 2015
19. Asmare, M.H., Asirvadam, V.S., Hani, A.F.M.: Image enhancement based on contourlet transform. Sig. Image Video Process. **9**, 1679–1690 (2014). https://doi.org/10.1007/s11760-014-0626-7
20. Otsu, N.: A threshold selection method from gray level histograms. IEEE Trans. Syst. Man. Cybern. **9**, 62–66 (1979)
21. Benziane, S.H., Benyettou, A.: Dorsal hand vein identification based on binary particle swarm optimization. J. Inf. Process. Syst. **13**(2), 268–284 (2017)
22. Hachemi-Benziane, S., Benyettou, A.: On the influence of anisotropic diffusion filter on dorsal hand authentication using eigenveins. Multidimension. Syst. Sig. Process. **29**(4), 1507–1528 (2017). https://doi.org/10.1007/s11045-017-0514-8
23. Benziane, S.: Uncontrained ear biometrics: survey research. Tianjin Daxue Xuebao Ziran Kexue yu Gongcheng Jishu Ban/ J. Tianjin Univ. Sci. Technol

A CBR Approach Based on Ontology to Supplier Selection

Mokhtaria Bekkaoui[1,2](✉), Mohamed Hedi Karray[3], and Sidi Mohammed Meliani[2]

[1] Higher School of Applied Sciences of Tlemcen, Tlemcen, Algeria
m_mekkaoui@mail.univ-tlemcen.dz

[2] Manufacturing Engineering Laboratory of Tlemcen (MELT), Tlemcen University, Tlemcen, Algeria

[3] Laboratory of Production Engineering (LGP), ENIT, University of Toulouse Tarbes, Toulouse, France

Abstract. Most previous research categorizes supply chain management (SCM) into the following three major parts: purchasing, manufacturing, and distribution. In the purchasing process, supplier selection represents a crucial step for enhancing firms' competitiveness. An increasing number of researches have been devoted to the development of different methodologies to cope with this problem. The majority of research explores the difference in the set of criteria for supplier selection. They used famous methods for selection such as AHP (Analytic Hierarchy Process); ANP (Analytic Network Process); DEA (Data Envelopment Analysis); TOPSIS (Technique for Order Preference by Similarity to Ideal Solution). Thus, this paper presents a CBR (Case Based Reasoning) approach in the context of supplier selection decision making.

Keywords: First supplier selection · Ontology · Experience feedback · CBR

1 Introduction

Problem of allocation or selection has been addressed in several research works. Among these, one can mention the use of fuzzy logic approach for modelling imprecise but known skills. Another approach is the one that is interested in scheduling preventive maintenance tasks on identical resources [1]. There is also another axis that focuses on the dynamic insertion of tasks in an ordinary scheduling for resources with one competence only [2]. However, in a production system, many studies are interested in the scheduling of human resource activities [3] where few approaches take into account the skill levels of resources [4]. One can still note the work of Gruat La Forme [5] who take into account the skill levels through a variable productivity rate in a multi-criterion problem.

Moreover, the best use of human resources is not limited only to the selection of the right knowledge in the right place at the right time, but also to find the experienced actor. This has the advantage of having a better control of activities. Hence, competence is not static anymore, but is rather dynamic over time, depending on the experience of the supplier,

B. Lejdel et al. (Eds.): AIAP 2021, LNNS 413, pp. 588–599, 2022.
https://doi.org/10.1007/978-3-030-96311-8_55

In the field of logistics, several research works, dealing with decision support systems for the problem of supplier selection, have been achieved. In this context, in order to meet the company's requirements, a selection model based on the exploitation of domain ontology to build a case base is proposed in this article. This case base generates new knowledge about competence according to the information gathered within the company. Such application supports the use of Experience-Feedback (EF) as a tool to aid the supplier' selection in order to have a better planning purchasing process.

2 State of the Art

Lima Junior [6] presented a comparative analysis of fuzzy TOPSIS methods (Fuzzy Technique for Order of Preference by Similarity to Ideal Solution) and Fuzzy AHP (Fuzzy Analytic Hierarchy Process) applied to the problem of Supplier Selection (SS). The comparison was based on several factors, such as the adequacy to changes of alternatives or criteria, agility in the decision process, computational complexity, adequacy to support group decision-making, number of alternative suppliers and criteria, and modeling uncertainty. In 2016, Bruno [7] proposed an integrated model, which combines two main approaches proposed in the literature, and which deals with the supplier selection (SS) problem, analytic hierarchy process (AHP) and the fuzzy set theory (FST). There is a deferent overview about criteria for supplier selection since the beginning of this area. A review is given by Weber et al. [8], covering 74 articles on selection of suppliers. The authors observe that aspects related to price, 'delivery', quality 'and production capacity' are the criteria most often considered in the literature.

However, the importance of criteria changes depending on the context industrial considered as presented in [9]. Where are the criteria classified as: logistics, technology, business and business cooperation. The aim was to create a model that distinguishes between qualitative and quantitative criteria.

An assignment is easy to achieve, but a good one is not always possible. Most of the above-mentioned studies are based on price and quality as a primary factors, in a problem of supplier assignment. Furthermore, as it is difficult to give precise numerical values to the concept of competence, almost all authors use fuzzy logic, initiated by Zadeh [10]. This is considered as the most appropriate theory in expressing the inherent imprecision. Unfortunately, to realize this theory, one needs to choose some rules of inference. This choice is neither exhaustive nor definitive, and provides a rough quantitative value that cannot be considered as accurate and reliable. Other studies are based on competence with a static level that is affected either directly or through fuzzy logic.

In summary, from the literature review presented above, it can be concluded that supplier selection criteria change over time, depending on the political, economic, social, and environmental characteristics of the business.

In light of the remarks of the above literature, one can easily notice that most systems focus on different criteria and lost the aspect of supplier competence. Thus, the objective of this article is to show the effectiveness and benefits of introducing the concept of dynamic competence for selecting supplier in a purchasing process. In other words, it is about assessing competencies in relation to past experience. This implies a mechanism that ensures the flexibility and adaptability of the process.

3 Proposed Supplier Selection Approach

Another vision is proposed in the present work to develop a decision support system. This approach focuses on the selection of supplier in a purchasing process, according to their level of competence. Competence has a dynamic character due to its evolution over time depending on the supplier's experience.

Knowledge Base and FOMES Ontology. Only a particular attention needs to be directed towards ontology. Ontology allows exchanging information and knowledge between actors in the same semantics [11, 12]. The use of ontology is due primarily to three factors: (i) the development of domain ontologies, (ii) the availability of software tools for building, editing and deployment of ontologies. (iii) the existence of a set of ontological model formalisms (OWL, PLIB, etc.) which have significantly contributed to the emergence of ontology-based application.

Several definitions of ontology are discussed in the literature. The best known is the one proposed by Gruber, who defined ontology as a 'formal and explicit specification of a shared conceptualization' [13]. In our work, we used FOMES (Feedback-CBR Ontology for Maintenance Expert Selection) [14]; it is an extended form of the ontology IMAMO (Industrial Maintenance Management Ontology) (Karray et al., 2012). FOMES is used to include all types of knowledge in the form of a set of concepts in the maintenance area as well as the dependencies existing between them. The Protégé 4.1 ontology editor was used to develop FOMES. Our model was extended to classes and datatypes in FOMES, so as to add some proprieties and change certain relations between the concepts. For example, the Data Type Level Commitment is an attribute that is added to the concept SUPPLIER. It shows the level of commitment of a supplier. It is a new criterion that expresses the dynamic expertise through experience. The concept Agenda was added, and connected with the SUPPLIER class, to help obtaining the schedule of every SUPPLIER. This is crucial for the best management of resource during a purchasing. Level data type is another attribute added to the class SKILL. This attribute allows determining the skill level of each actor.

Supplier Selection Process. The Supplier Selection operating mechanism is based on the dynamic assessment of the competence of a supplier, while paying more attention to the analysis of his past experiences. In such situation, the management of experiences focuses mainly on the practical knowledge resulting from each selection offer solving and on its modelization, i.e. specific knowledge linked to a purchasing activity, in a particular context. This implies an experience feedback process corresponding to a cycle of actions involving two main phases (Fig. 1) which are the capitalization process (acquisition of experience and memorization), and the operating process (reuse and adaptation) [15].

The capitalization phase aims to identify and extract knowledge. Therefore, this knowledge has got to be formalized and structured in the form of experience in order to make it easily accessible and reusable in solving new problem (in our study problem is mean new offer). The exploitation phase consists of finding the useful experience in the case base (experience base), adapt it to the problem to be solved and implement the solution from the experience base. Among the methods developed in the framework

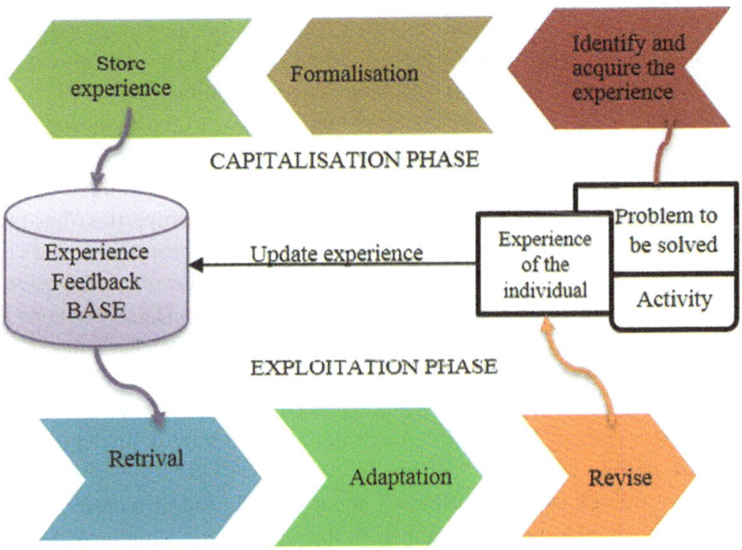

Fig. 1. Cycle of experience feedback

of engineering knowledge for solving problems based on experience, the method of Case-Based Reasoning (CBR), which is the most widely used in selection problems, was chosen for our work. In this method, elements of experience are expressed in the form of "cases".

Generally speaking, the CBR cycle may be defined by the following phases:

Retrieval phase: The most relevant and similar cases are recovered from the case base when a target problem (or a new case) appears.
Reuse/adaptation phase: It consists of building a solution to the problem of the target case. This solution is inspired by the most similar source case solution.
Revise phase: It tests the proposed solution in the real world or in a simulation and, if necessary, revises it in order to have a better solution.
Retain phase: It phase aims at storing the result obtained as a new case in the case base; this is done after validation of the solution.

Each phase of the CBR cycle is detailed in what follows.

Note: In order to use the CBR methodology, it is important to know that cases include problems and solutions. So, a case describing the problem formally is to be formulated first. The elaboration phase (Problem formulation) comes next as it is necessary to imply the construction of a new case. The case model used in our work is described in the next section.

Elaboration. The user (manager) must enter the new case using the same representation and level of detail of the cases stored in the case base.

A case base is a memory which contains a collection of cases used in the context of the CBR methodology which aims to perform a reasoning task. In general, a case in CBR consists of two essential parts, i.e. the problem and the solution. The case is represented as a set of descriptors. FOMES will be exploited in order to represent the case model previously investigated in the present work. First, a concept called Case was created; it puts together all the cases of the base. That concept is related to three concepts, i.e. "***Problem***", "***Solution***" and "***Evaluation***", using the object properties "has problem", "has solution" and "has evaluation", respectively. An Object property named "has GeographicalLocation" relates the concept "***Problem***" with "***GeographicalLocation***" which records the different zones and subzones of firm, with the two Datatypes zone and subzone. The different datatypes of the need, such as the quantity, price, urgency of the order, etc., are stored in the "***Need***" class. The equipment available is in the "***Equipment***" class, and each piece of equipment is part of a group of equipment (***Equipmentgroup*** concept), with the Object property "belongsTo". Each Equipment group is linked to the concept "***Domain***" by an object property "belongsTo".

Thus, the ontology stores information about the solution of a case in the concept "***solution***". When a new order is established, the search for a supplier is launched and it is stored in the class" ***Need***". This task is performed by supplier, who are found in the class supplier, which is defined by a set of Datatypes such as Supplier-ID, name, address, commitment level, etc. An Object property "has Skill" relates the skill in a specific domain to the concept "***Supplier***". To manage the availability of each Supplier, the object property "has Agenda" must be developed.

Finally, the ontology contains a concept called "***Evaluation***". This class contains information about the assessment of the case solved. Therefore, the concept Evaluation has a Datatype status (success/failure), coast (cost of invoice) and time (delivery time). The Object property "relatesTo" is associated with each Supplier evaluation.

In literature, there are two types of cases in case base; i.e. the source case and the target case. The source case is one in which the "problem" and "solution" parts are clearly indicated. Therefore, it is a case where one should be inspired to solve a new problem. However, the target case is a case that appears at the occurrence of a new problem, whose solution part is not indicated.

In our case base, a case describes a situation that is need to purchase an item. That case is determined by a list of descriptors, divided into two groups, i.e. one concerns the problem and one the solution.

The problem field consists of eight descriptors reflecting the description of a localized firm divided in two sub-parts; i.e. the Geographical Location that determines the zone and sub-zone of firm need (descriptors ds1 and ds2, respectively) and the need part, which consists of the Equipment marque, Equipment Type, estimated price, quantity, quality, the degree of urgency "it mean the limited time that the firm is needed the equipment" (descriptors ds3, ds4, ds5, ds6,ds7 and ds8, respectively). However, the solution field is composed of two descriptors that describe selected supplier (descriptor ds9). This formalism has been designed and developed in order to take into account all possible relationships between the constraint of firm and supplier.

Considering all these specificities, it becomes possible to schematically represent the structure of the case, as shown in Table 1.

Table 1. Case base

Case	Context								Solution
	Geographical		Need part						
	Zone	Sub zone	EQP Mar-	EQP type	Estimated price	Quantity	Quality	Emergency level	Supplier-ID
	ds1	ds2	ds3	ds4	ds5	ds6	ds7	ds8	ds9

In summary, in this section, the combination of the descriptors was defined so that any problem can be analyzed. The manager determines the first eight descriptors of the target case; they are also descriptors of the source case named (dt1 … dt8). From there, there will be an elaborated target case which allows us to go to retrieve the source cases most similar to that target case.

Elaboration. The purpose of the Retrieval step in case-based reasoning CBR is to recover one or more cases that can be reused in the context of a new problem. As it is generally unlikely that the case base contains a problem already processed, which exactly matches the new problem, the concept of similarity is therefore used. The similarity measure allows calculating the similarity between the descriptions of the two problems (source and target). In general, similarity can be of three types: (1) surface similarity, (2) derivative similarity and (3) structural similarity [16].

In the literature, many similarity measures, based on the taxonomic structure, have been illustrated. For example, the measure proposed by Wu and Palmer [17] and more recently that of [18–21] can be mentioned. It has been decided, in the present study, to use the similarity measure, proposed by Haouchine [22], and based on the information resulting from the FOMES ontology.

That measure is divided into two steps:

Retrieval Measure (RM)

To calculate the Retrieval measure (RM), it is possible to use the same principle of measure as that of Haouchine [22], by modifying the expression of that measure, which is the combination of various measures and results using formula (1):

$$R_M(S_1, T) = \frac{\sum_{i=1}^{P} sim(ds_i^{value}, dt_i^{value}) + \sum_{i=p+1}^{m} sim(ds_i^{value}, dt_i^{value})}{\sum_{i=1}^{j} sim_{presence}(ds_i^{value}, dt_i^{value})} \quad (1)$$

Where

S_1: First source case; **T**: Target case.
p: represents the number of location descriptors,
m: is the number of need-part descriptors (ds_3, ds_4, ds_8),

- ds_i^{value}, can have two possible values:

$$sim(ds_i^{value}, dt_i^{value}) = 1 \ \ when \ \ ds_i^{value} = dt_i^{value}$$
$$= 0 \ \ when \ \ ds_i^{value} \neq dt_i^{value}$$

- In case the descriptors are not all filled, a presence similarity measure is defined. It will take into account the presence of descriptors in the case.

$$sim_{presence}(ds_i^{value}, dt_i^{value}) = 1 \ if \ information \ is \ present \ both \ in \ S_1 \ and \ T$$
$$= 0 \ if \ \ information \ is \ not \ present \ in \ one \ descriptor$$

From the calculation of the overall similarity, a set of cases can be chosen which will take into account only the highest value.

Adaptation Measure (AM)

In order to select the most adaptable source case, among the retrieved cases, it was necessary to develop the "adaptation measure". This measure takes into account the urgency level by giving it a high priority. This weight is important in determining the urgent case. Indeed, a high weight is assigned to the top level of urgency (α). The presence of the descriptor value can also be taken into account, as this facilitates the adaptation.

$$A_M(S_1, T) = \frac{\sum_{i=p+1}^{j=m-1} sim(ds_i^{value}, dt_i^{value}) + sim(ds_m^{value}, dt_m^{value}) \times \alpha_m}{sim_{presence}(ds_i^{value}, dt_i^{value})} \quad (2)$$

Where

p represents the number of localization descriptor.
m: represents the number of Need-part descriptors.
$\boldsymbol{\alpha_m}$ is the associated weight according to the urgency level.

$$If \begin{cases} ds_m^{value} = dt_m^{value} = High \rightarrow \alpha_m = 2^2 \\ ds_m^{value} \neq dt_m^{value} = (High/Low \ \ or \ \ Low/High) \rightarrow \alpha_m = 2^1 \\ ds_m^{value} = dt_m^{value} = Low \rightarrow \alpha_m = 2^0 \end{cases}$$

Therefore, the source case with the largest value of the adaptation measure, among the retrieved source cases, will be the candidate selected for the next step.

Adaptation (Reuse). The second step in the CBR cycle is the reuse or the adaptation; this leads to propose a solution to a new problem from the solutions in the retrieved cases. In most situations, the authors simply use the substitution or transformation [16]. It is rarely possible to use a solution exactly as it is recorded. This happens if the new problem situation is not too different in essential aspects from the nearest neighbor selected from the case base. Then the recommendation is to adapt the recorded solution before reusing it to best suit the new problem. In some particular contexts, some approaches are proposed in the literature; they are based on the dependency relations (or correlative value) between the problem space and solution space of a given experiment [23]. Adaptation can be performed on different levels of granularity. In the present work, the solution part describes supplier, therefore the question is how can we assign a supplier in a new problem? Another question is who can be selected to the new problem? In this context, no problem if there is only one retrieved solution. In the present work, it is supposed that there are several retrieved cases. So, to generate a new solution, based on a previous solution, a specific approach is presented.

From all the recovered cases, the proposed adaptation is based on selection, described in the following:

Ranking with respect to case evaluation. This consists of selecting cases according to the evaluation criterion. Evaluation is a concept in FOMES ontology, which defines assessment of the case solved according to three criteria, which are time, cost, and an indicator that indicates whether the problem is resolved with success or no. Each criterion is represented by a DataType, i.e. delivery time, coast and state, respectively. The last criterion allows us to have a new list of cases (all cases that supplier has been delivered in the best condition). In order to achieve this goal, a description language is used to make rankings according to the queries.

Revise and Retain. During the revision phase of the CBR cycle, the solution proposed in the end of the adaptation phase is evaluated. This evaluation concerns the testing of the solution proposed in the real world. So, the revision phase consists of eventually continuing the development of the target solution, if necessary. Therefore, if supplier/new problem assignment is not satisfactory, then it is corrected. In the present situation, these corrections can be made by the manager who may give his own assessment relative to the selection provided via the CBR system. The case, reviewed and validated, can be applied; it becomes a new experience that must be capitalized in the case base.

4 Illustrative Example

In this section, our approach is applied to a real example. One starts by identifying the information resources and knowledge. These resources involve the history of documents about purchasing management, as well as the expertise of the domain. Our goal is to identify different equipment and constraint in the company. This statistical study allows collecting the information needed to enrich the FOMES ontology. Therefore, the attributes of the case are produced from these data. An attempt was made to reduce the number of descriptors in order to get a simpler table that meets our requirements.

Therefore, a case base was built, from which a sample of 31 cases was considered. For instance, case 1(or source 1) is interpreted with its problem and solution part see Table 2.

The principle of similarity calculation is applied to our case study, according to the approach proposed in Sect. 3.2.2.

Retrieval Measurement (RM). First, a target case is proposed to simulate the calculation of similarity (Table 2). For example, the calculation of local similarity for all descriptors, between the target and source1, is performed as follows:

Table 2. Example of source case and target

Case	Geographical location		Need-part						Solution
	ds1	ds2	ds3	ds4	ds5	ds6	ds7	ds8	ds9
Source S1	TLE	ZI	BMW	Inject	2M	900u	Medium	Hight	Supp_ad3
Taget T1	TLE	MG	W	Inject	1M	600u	Low	Hight	–

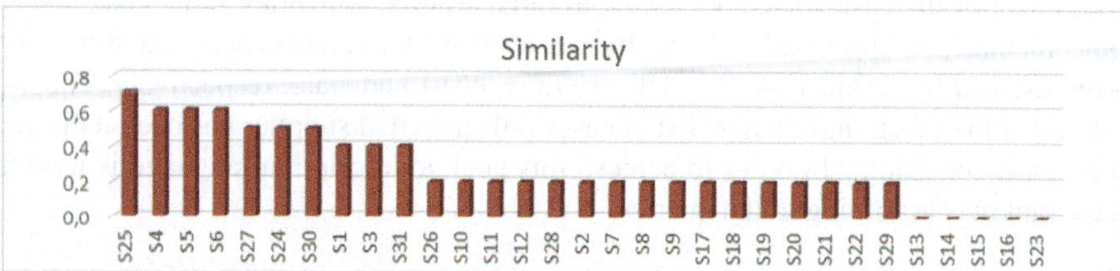

Fig. 2. Results of similarity calculations

$$R_m(\mathrm{S1},\mathrm{T}) = \frac{(1\times 1)+(0\times 1)+(0\times 1)+(1\times 1)+(0\times 1)+(0\times 1)+(0\times 1)+(1\times 1)}{8} = 0{,}4$$

All local similarities, between the source cases and Target, are calculated using the same formula (see summary of calculations in Fig. 2).

According to the results obtained, the retrieval source cases, which have the largest value of similarity measure (S25, S4, S5, S6, …, S31), are selected for the adaptation measure step. A total of 10 cases are selected.

Adaptation Measure (AM). Let's move to the second phase of similarity calculation. Remember that the objective is to choose, among the recovered cases, the closest one to our target case, which will be the candidate selected for the second stage of the CBR. Descriptor "ds8" is selected to be given a priority, by imposing a heavy weight. Because it is an urgent case, a quick decision selection in the solution space is needed.

Now the adaptation measure of the source cases most similar to target is calculated:

$$AM_{(S25,T)} = \frac{(0\times 1)+(1\times 1)+(0\times 1)+(0\times 1)+(0\times 1)+\left(1\times 2^2\right)}{6} = 0{,}8$$

Applying the calculations to our cases, the following results are obtained (Table 3):

One can note that the source cases S4, S27, S24, S30, S1, S3 and S31 have the different similarity measures; they are superior to 1. Moreover, cases S25, S5 and S6 have an adaptation measure less than 1. Therefore, the high degree of adaptation measure (AM) expresses the cases of perfect satisfaction; they will be selected for the adaptation phase.

Table 3. Adaptation measures

	S25	S4	S5	S6	S27	S24	S30	S1	S3	S31
AM	0,8	1,8	0,8	0,4	1,6	1,3	2	1,4	1,9	1,6

Adaptation. The purpose of the present study is to improve the selection process through the adaptation phase. It focuses firstly on a method based on ranking through SPARQL queries, secondly on OWL API, a high level Application Programming Interface (A.P.I.) for working with OWL ontologies.

Considering the results from the Retrieval step, the first step of ranking with respect to 'Evaluation' is applied. The suppliers must be recovered using the filter which is used in the SPARQL query to return only supplier who participated in the cases stored and evaluated successfully. With this query, all the suppliers are recovered and put in a list.

When this list is recovered, By using Monte Carlo simulation, the distribution of all possible outcomes of suppliers' ranking is generated by analysing a model several times, each time using input values selected by chance from the probability distributions of the factors (time and cost). Due to the different factors and because many values that each of these factors can take, there could be an infinite number of possible combinations that may affect supplier's selection.

The Monte Carlo simulation was run for 1000 iterations so as to generate the stochastic dataset for data mining. Each data record generated from Monte Carlo simulation, represents one evaluation of case solution, is classified into one of five groups: Supplier_1, Supplier 2, Supplier_9, Supplier_10 and Supplier_12 (suppliers' list obtained from the first step of ranking). The cost and time were generated based on a uniform distribution.

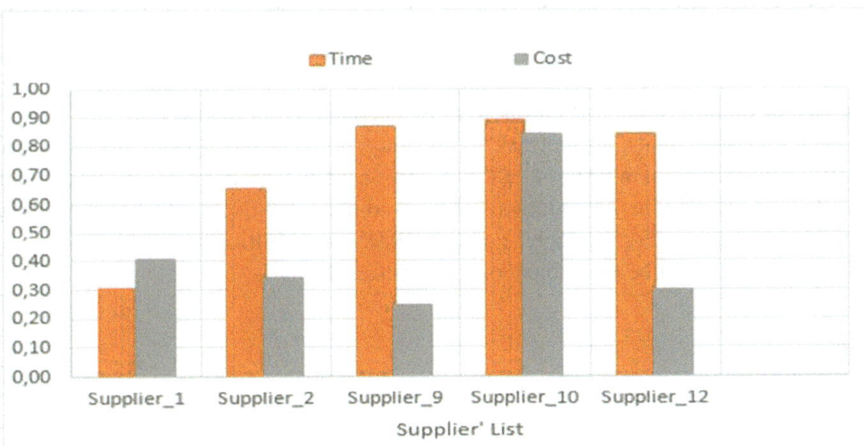

Fig. 3. Time and Cost calculations

According to the lot of-fold, the training dataset was randomly divided into five disjoint subsets (the five supplier). Figure 3 shows mean time and mean cost of each supplier. Based on the results obtained, when comparison is made between different supplier, it can be found that the first one in the list is supplier with the minimum value of time equal to (0,30). One notes that time and cost has higher values affected to Supplier_10, that mean he is the last one in supplier list, because he has the higher time and cost compared with others.

5 Conclusion

D The issue of selecting the actors involved in the conduct of industrial processes is highly justified by a research diversity which focuses on the combined integration of different criteria, in problems of allocation, planning and scheduling. This emancipation has led to a change in the formulations of these problems.

The research work, presented in this article, taken into account that different criteria are not always sufficient to meet the resolution requirements for purchasing problem with respect to time, cost and efficiency. As previously explained, other additional criteria are required to improve the selection process.

Furthermore, new criteria, such as the dynamic skill, should be integrated through experience. A methodology of experience feedback processes around the reasoning from cases should be considered. This methodology proposes an approach, developed in four phases, which allows capitalizing on experience.

Our proposal was structured to include three elements, which are the use of: (i) ontology FOMES as support for the domain, (ii) REX approach for the selection of supplier, and (iii) case-based reasoning (CBR) as a problem solving technique which is based on adapting the solution from past experiences. In this article, a formalization of the experience was developed. To recover the most appropriate information for the current case, a method of guided adaptation recovery was used and two similarity calculations were developed. To refine the selection of supplier, a ranking, based on the SPARQL query language, was developed in the third stage of the CBR cycle.

This work presented a sample application that allows integrating the proposed approach in situations that are similar to those encountered in industry.

In the end, some real contributions of this study should be mentioned. Since this is an experimental method, it must be validated through a large-scale application that could be validated for any domain. This method has been compared with other techniques related to problem selection. For example, Fuzzy TOPSIS and Multicriteria decision making (MCDM). Which will be the subject of future research.

References

1. Adzakpaa, K.P., Adjallaha, K.H., Leeb, J.: A new effective heuristic for the intelligent management of the preventive maintenance tasks of the distributed systems. Adv. Eng. Inform. **17**(3–4), 151–163 (2003)
2. Duffuaa, S.O., Al-Sultan, K.S.: A stochastic programming model for scheduling maintenance personnel. Appl. Math. Model. **23**(5), 385–397 (1999)

3. Tchommo, J.-L., Batiste, P., Soumis, F.: Etude bibliographique de l'ordonnancement simultané des moyens de production et des ressources humaines.In: Congrès International de Génie Industriel (2003)
4. Geneste, L., Grabot, B., Letouzey, A.: An assessment tool within the customer/sub-contractor negotiation context. Eur. J. Oper. Res. **147**(2) (2003)
5. Gruat-La-Forme, F.A., Botta-Genoulaz, V., Campagne, J.-P.: Problème d'ordonnancement avec prise en compte des compétences : résolution mono critère pour indicateurs de performance industriels et humains. J. Eur. des Systèmes Autom. **41**(5), 617–642 (2007)
6. Lima Junior, F.R., Osiro, L., Carpinetti, L.C.R.: A comparison between Fuzzy AHP and Fuzzy TOPSIS methods to supplier selection. Appl. Soft Comput. J. **21**, 194–209 (2014)
7. Bruno, G., Esposito, E., Genovese, A., Simpson, M.: Applying supplier selection methodologies in a multi-stakeholder environment: a case study and a critical assessment. Expert Syst. Appl. **43**, 271–285 (2016)
8. Weber, C.A., Current, J.R., Benton, W.C.: Vendor selection criteria and methods. Eur. J. Oper. Res. **50**(1), 2–18 (1991)
9. Çebi, F., Bayraktar, D.: An integrated approach for supplier selection. Logist. Inf. Manag. **16**(6), 395–400 (2003)
10. Zadeh, L.A.: Fuzzy sets. Inf. Control **8**(3), 338–353 (1965)
11. Karray, M.H., Chebel-Morello, B., Zerhouni, N.: A formal ontology for industrial maintenance. Appl. Ontol. **7**(3), 269–310 (2012)
12. Ruiz, P.P., Foguem, B.K., Grabot, B.: Generating knowledge in maintenance from experience feedback. Knowl.-Based Syst. **68**, 4–20 (2014)
13. Gruber, T.: A translation approach to portable ontology specifications. Knowl. Acquis. **2**(5), 199–220 (1993)
14. Haouchine, M.K.: Remémoration guidée par l'adaptation et maintenance des systèmes de diagnostic industriel par l'approche du raisonnement à partir de cas. L'UFR des Sciences et Techniques de l'Université de Franche-Comté (2009)
15. Armaghan, N., Renaud, J.: Experiences of feedback based on case-based reasoning and multicriteria aid approaches. In: 17th International Association for the Management of Technology (IAMOT) (2008)
16. Cheng, J.C.P., Ma, L.J.: A non-linear case-based reasoning approach for retrieval of similar cases and selection of target credits in LEED projects. Build. Environ. **93**(P2), 349–361 (2015)
17. Wu, Z., Palmer, M.: Semantic sand lexical selection. In: 32nd Annual Meeting of the Association for Computational Linguistics, New Mexico State University, Las Cruces, New Mexico, pp. 133–138 (1994)
18. Chebel-Morello, B., Haouchine, M.K., Zerhouni, N.: Reutilization of diagnostic cases by adaptation of knowledge models. Eng. Appl. Artif. Intell. **26**(10), 2559–2573 (2013)
19. Jabrouni, H., Kamsu-Foguem, B., Geneste, L., Vaysse, C.: Analysis reuse exploiting taxonomical information and belief assignment in industrial problem solving. Comput. Ind. **64**(8), 1035–1044 (2013)
20. Potes Ruiz, P.A., Kamsu-Foguem, B., Noyes, D.: Knowledge reuse integrating the collaboration from experts in industrial maintenance management. Knowl.-Based Syst. **50**, 171–186 (2013)
21. Akmal, S., Shih, L.H., Batres, R.: Ontology-based similarity for product information retrieval. Comput. Ind. **65**(1), 91–107 (2014)
22. Haouchine, M., Chebel-Morello, B., Zerhouni, N.: Adaptation-guided retrieval for a diagnostic and repair help system dedicated to a pallets 'ECCBR'. In: 9th European Conference on Case-Based Reasoning (2008)
23. Qi, J., Hu, J., Peng, Y.: Hybrid weighted mean for CBR adaptation in mechanical design by exploring effective, correlative and adaptative values. Comput. Ind. (2015)

Author Index

B. Lejdel et al. (Eds.): AIAP 2021, LNNS 413, pp. 601–602, 2022.
https://doi.org/10.1007/978-3-030-96311-8

The manufacturer's authorised representative in the EU is Springer Nature Customer Service Centre GmbH, Europaplatz 3, 69115 Heidelberg, Germany. If you have any concerns regarding our products, please contact ProductSafety@springernature.com

Printed and bound by CPI Group (UK) Ltd, Croydon, CR0 4YY
28/09/2026
02224057-0009